T0327401

Reliability and Risk

WILEY SERIES IN PROBABILITY AND STATISTICS

ESTABLISHED BY WALTER A. SHEWHART AND SAMUEL S. WILKS

Editors

David J. Balding, Peter Bloomfield, Noel A.C. Cressie, Nicholas I. Fisher, Iain M. Johnstone, J.B. Kadane, Geert Molenberghs, Louise M. Ryan, David W. Scott, Adrian F.M. Smith

Editors Emeriti

Vic Barnett, J. Stuart Hunter, David G. Kendall, Jozef L. Teugels

A complete list of the titles in this series appears at the end of this volume.

Reliability and Risk

A Bayesian Perspective

Nozer D. Singpurwalla
The George Washington University, Washington DC, USA

John Wiley & Sons, Ltd

Copyright © 2006 John Wiley & Sons Ltd, The Atrium, Southern Gate, Chichester,
West Sussex PO19 8SQ, England

Telephone (+44) 1243 779777

Email (for orders and customer service enquiries): cs-books@wiley.co.uk
Visit our Home Page on www.wiley.com

All Rights Reserved. No part of this publication may be reproduced, stored in a retrieval system or transmitted in any form or by any means, electronic, mechanical, photocopying, recording, scanning or otherwise, except under the terms of the Copyright, Designs and Patents Act 1988 or under the terms of a licence issued by the Copyright Licensing Agency Ltd, 90 Tottenham Court Road, London W1T 4LP, UK, without the permission in writing of the Publisher. Requests to the Publisher should be addressed to the Permissions Department, John Wiley & Sons Ltd, The Atrium, Southern Gate, Chichester, West Sussex PO19 8SQ, England, or emailed to permreq@wiley.co.uk, or faxed to (+44) 1243 770620.

Designations used by companies to distinguish their products are often claimed as trademarks. All brand names and product names used in this book are trade names, service marks, trademarks or registered trademarks of their respective owners. The Publisher is not associated with any product or vendor mentioned in this book.

This publication is designed to provide accurate and authoritative information in regard to the subject matter covered. It is sold on the understanding that the Publisher is not engaged in rendering professional services. If professional advice or other expert assistance is required, the services of a competent professional should be sought.

Other Wiley Editorial Offices

John Wiley & Sons Inc., 111 River Street, Hoboken, NJ 07030, USA

Jossey-Bass, 989 Market Street, San Francisco, CA 94103-1741, USA

Wiley-VCH Verlag GmbH, Boschstr. 12, D-69469 Weinheim, Germany

John Wiley & Sons Australia Ltd, 42 McDougall Street, Milton, Queensland 4064, Australia

John Wiley & Sons (Asia) Pte Ltd, 2 Clementi Loop #02-01, Jin Xing Distripark, Singapore 129809

John Wiley & Sons Canada Ltd, 22 Worcester Road, Etobicoke, Ontario, Canada M9W 1L1

Wiley also publishes its books in a variety of electronic formats. Some content that appears in print may not be available in electronic books.

British Library Cataloguing in Publication Data

A catalogue record for this book is available from the British Library

ISBN-13 978-0-470-85502-7 (HB)
ISBN-10 0-470-85502-9 (HB)

Typeset in 9.5/11.5pt Times by Integra Software Services Pvt. Ltd, Pondicherry, India

There are

NO FACTS

only

INTERPRETATIONS

Friedrich Nietzsche
German Philosopher
(1844–1900)

Contents

Preface

Over the past few years, there has been an increasing emphasis on what is commonly referred to as 'risk', and how best to manage it. The management of risk calls for its quantification, and this in turn entails the quantification of its two elements: the uncertainty of outcomes and the consequences of each outcome. The outcomes of interest here are adverse events such as the failure of an infrastructure element, e.g. a dam; or the failure of a complex system, e.g. a nuclear power plant; or the failure of a biological entity, e.g. a human being. 'Reliability' pertains to the quantification of the occurrence of adverse events in the context of engineering and physical systems. In the biological context, the quantification of adverse outcomes is done under the label of 'survival analysis'. The mathematical underpinnings of both reliability and survival analysis are the same; the methodologies could sometimes be very different. A quantification of the consequences of adverse events is done under the aegis of what is known as utility theory.

The literature on reliability and survival analysis is diverse, scattered and plentiful. It ranges over outlets in engineering, statistics (to include biostatistics), mathematics, philosophy, demography, law and public policy. The literature on utility theory is also plentiful, but it is concentrated in outlets that are of interest to decision theorists, economists and philosophers. However, despite what seems like a required connection, there appears to be a dearth of material describing a linkage between reliability (and survival analysis) and utility. One of the aims of this book is to help start the process of bridging this gap. This is done in two ways. The first is to develop material in reliability with the view that the ultimate goal of doing a reliability analysis is to appreciate the nature of the underlying risk and to propose strategies for managing it. The second is to introduce the notion of the 'utility of reliability', and to describe how this notion can be cast within a decision theoretic framework for managing risk. But in order to do the latter, we need to make a distinction between reliability as an objective *chance* or *propensity*, and *survivability* as the expected value of one's subjective probability about this propensity. In other words, from the point of view of this book

reliability is a chance* not *a probability,

and that probability, which describes one's uncertainty about reliability, is personal. It is this de Finettian-style perspective of an objective propensity and a subjective probability that distinguishes the material here from that of its several competitors. Hopefully, it changes the way in which one looks at the subject of reliability and survival analysis; a minor shift in paradigm, if you will.

With the above as a driving principle, the underlying methodology takes a Bayesian flavor, and the second aim of this book is to summarize this methodology in its broader context. Much of this Bayesian methodology is directed toward predicting lifetimes of surviving units. Here, the probabilistic notion of 'exchangeability' plays a key role. Consequently, a chapter has been devoted to this important topic, namely, Chapter 3.

The personalistic interpretation of probability is controversial, and unlike what is done by us, not all Bayesians subscribe to it. Thus it is deemed important to trace the historical evolution of personal probability, and to contrast it with the other interpretations. With that in mind, a brief history of probability has been included here. This historical material is embedded in a chapter that summarizes the quantification of uncertainty from a Bayesian perspective. Readers whose exposure to the essentials of Bayesian inference is limited may find the material of this chapter useful – so I hope. At the very least, Chapter 2 serves as the springboard in which the terminology and notation used throughout the book are established.

The heart of the book starts with Chapter 4, on stochastic models of failure. Included herein are mathematical concepts such as absolute continuity, singularity, the Lebesgue integral and the Lebesgue–Stieltjes integral, notions that may appear to be distracting. However, these are required for a better appreciation of some modern developments in reliability, in particular, the treatment of interdependence. They are included here because most analysts who do reliability, risk and survival analysis tend to be applied statisticians, engineers, and other physical and biological scientists whose exposure to concepts in mathematical analysis may be limited.

As a final point to this preface, I feel obliged to be upfront about some limitations of this book. For one, its current version does not have much by the way of examples, exercises or data. I hope this limitation will be addressed in a subsequent edition. For now, the focus has been on breadth of coverage and the spectrum of issues that the general topic of reliability and survivability can spawn. An example is the use of concepts and notions in reliability theory that are germane to econometrics and finance, in particular the assessment of income inequalities and financial risk. Chapter 11 is devoted to these topics; hopefully, it gives this book an unusual flavor among books in reliability and survival analysis. Another limitation of this book is that in some instances, I may have merely mentioned an issue or a topic without dwelling on its details in any substantive manner. This, I am aware, could frustate some readers. All the same, I have chosen to do so in order to keep the material to a reasonable size and at the same time maintain a focus on breadth and scope. As a compensatory measure, a large list of references has been provided. On the matter of maintaining breadth and scope, I may have also ventured into territory that some may consider unproven. My hope is that such excursions into the unknown may provide a platform to others for new and additional research. Finally, given the size of this book, and the amount of time it has taken to develop it, there are bound to be errors, typos, inconsistencies and mistakes. For this I apologize to the readers and ask for their tolerance and understanding. I would of course greatly value receiving comments pointing to any and all of the above, and advice upon how to improvise upon the current text.

The material of this book can be most profitably used by practioners and research workers in reliability and survivability as a source of information and open problems. It can also form the basis of a graduate-level course in reliability and risk analysis for engineers, statisticians and other mathematically orientated scientists, wherein the instructor supplements the material with examples, exercises and real problems.

The epitaphs in the chapter opening pages are not for real. All were given to me by a colleague, save for the one in Chapter 5 which is due to me. For this I ask forgiveness of Reverend Bayes.

Washington, DC
June 2006

Acknowledgements

Work on this book began in February 1994 at the Belagio Study and Conference Center of the Rockefeller Foundation in Belagio, Italy. The author thanks the Foundation for providing an environment conducive to jump-starting this project. Subsequent places wherein the author found habitats to work on this book have been The Santa Fe Institute, The Los Alamos National Laboratory, and the Departments of Statistics, University of Oxford, England, and the Université de Brentage-Sud, France. The author acknowledges with thanks the hospitality provided by these institutions, orchestrated by Sallie Keller-McNulty, Mary and Dan Lunn, and Mounir Mesbah. Of course, the author's home institution, The George Washington University, warrants a special acknowledgement for nurturing his interests, and for providing an atmosphere conducive to their development. The author's deep gratitude also goes to the sponsors of his research, The Office of Naval Research and the Army Research Office, particularly the latter for its continuous sponsorship over the past several years. Much of this work is embedded in the material presented here.

Since its beginnings in 1994, this book project has gone through several changes in title and publishing house. The project was initiated by John Kimmel, and the initial encouragement and guidance came from Sir David Cox of Oxford; for this I thank him. The book project underwent several publishing house changes until Adrian Smith navigated its safe landing in the hands of Sian Jones of John Wiley, UK; thanks to both. The persistent nags of Wiley's Kathryn Sharples forced the author to accelerate the writing in earnest and bring about the book's closure. Kathryn deserves an applause.

There are others who have directly or indirectly contributed to the completion of this project that the author acknowledges. The late Professor Louis Nanni got him interested in statistics, and Professors John Kao, Richard Barlow and (the late) Frank Proschan got him interested in reliability. Professors Denis Lindley and Jay Sethuraman contributed much to the author's appreciation of probability, which is really what the book is all about. Both occupy a special place in the author's heart and mind.

The nitty-gritty aspects of this book would not have been taken care of without the tremendous and dedicated help of Josh Landon. Later on, Josh was joined by Bijit Roy, and the two being masters in the art of manipulating equations and harnessing computers, provided invaluable support toward the book's completion; thanks to both.

Last but not the least, the author singles out two members of his family for their understanding and unconditional support. His sister Khorshed bore the brunt of his domestic responsibilities in India and freed him to pursue his professional career. His wife Norah did the same here in the US and spent, without anger or resentment, many an evening and weekend in isolation when the author sequestered himself in his basement cocoon. Thank you Norah; you can freely spend the royalties – if any!

Chapter 1

Introduction and Overview

G.H. Hardy: Here lies Hardy, with no apologies.

1.1 PREAMBLE: WHAT DO 'RELIABILITY', 'RISK' AND 'ROBUSTNESS' MEAN?

Words such as 'credibility', 'hazard', 'integrity', 'reliability', 'risk', 'robustness' and 'survivability' have now become very much a part of our daily vocabulary. For instance, the derogatory term **unreliable** is used to describe the undependable behavior of an individual or an item, whereas the cautionary term **risky** is used to warn of possible exposure to an adverse consequence. The term **survival** is generally used in biomedical contexts and is intended to convey the possibility of overcoming a life-threatening situation or a disease. **Robustness** encapsulates the feature of the persistence of some attribute in the presence of an insult, such as a shock, or an unexpectedly large change, such as a surge in electrical power, or an encounter with an unexpectedly large (or small) observation. Thus robustness imparts the attribute of reliability to a physical or a biological unit, and sometimes even to a mathematical or a statistical procedure. In what follows, we point out that all the above terms convey notions that are intertwined, and thus, in principle, they tend to be used interchangeably. Our choice of the words 'reliability' and 'risk' in the title of this book reflects their common usage.

It is often the case, even among engineers and scientists, that the above terminology is purely conversational, and is intended to convey an intuitive feel. When such is the case, there is little need to be specific. Often, however, with our increasing reliance on technology, and for decisions pertaining to the use of a technology, we are required to be precise. This has resulted in efforts to sharpen the notions of risk and reliability and to quantify them. Quantification is required for normative decision making, especially decisions pertaining to our safety and well-being; some examples are mentioned in section 1.4. When quantified measures of risk are coupled with normative decision making, it is called **risk management** (cf. The National Research Council's Report on "Improving the Continued Airworthiness of Civil Aircraft", 1998).

Historically, the need for quantifying risk pre-dates normative decision making. It goes back to the days of Huygens (1629–1695), who was motivated by problems of annuities. In the early 1930s, it was problems in commerce and insurance that sustained an interest in this topic. During

Reliability and Risk: A Bayesian Perspective N.D. Singpurwalla
© 2006 John Wiley & Sons, Ltd

the 1960s and the 1970s, quantified measures of reliability were needed to satisfy specifications for government acquisitions, mostly in aerospace and defense; and quantified measures of risk were needed for regulation, mostly drug approval, and for matters of public policy, such as reactor safety. During the 1980s, pressures of consumerism, competitiveness and litigation have forced manufacturers and service organizations to use quantified measures of reliability for specifying assurances and designing warranties. The coming era appears to be one of ensuring infrastructure integrity, infrastructure protection and with the advent of test-ban treaties, the stewardship of nuclear weapons stockpiles. Here again, quantified measures of reliability are poised to play a signal role.

The above developments have given birth to the general topic of **risk analysis**, which in the context of engineering applications takes the form of **reliability analysis**, and in the context of biomedicine, **survival analysis**. These two scenarios have provided most of the applications and case histories. The 1990s have also witnessed the use of risk management in business and finance – for acquisitions, bond pricing, mergers, and the trading of options – and in political science for matters of disarmament and national security. The applications mentioned above, be they for the design of earthquake-resistant structures, or for the approval of medical procedures, have one feature in common: they all pertain to situations of **uncertainty**, and it is this common theme of uncertainty that paves the way for their unified treatment. Uncertainty is about the occurrence of an undesirable event, such as the failure of an item, or an adverse reaction to a drug, or some other loss. Since a conversational use of the words 'reliability' and 'risk' conveys an expression of uncertainty, it is the quantification of uncertainty that is, *de facto*, the quantification of reliability and risk. To summarize, reliability and risk analysis pertains to quantified measures of uncertainty about certain adverse events.[1] However, since quantified measures of uncertainty are only an intermediate step in the process of normative decision making, one may take a broader view and claim that reliability and risk analysis is simply methods for decision making under uncertainty. This is the point of view taken in 'Risk: Analysis, Perception, and Management. Report of a *Royal Society Group*' (1992).

The quantification of uncertainty is an age-old problem dating back to the days of Gioralimo Kardano (1501–1575) (cf. Gnedenko, 1993), and decision making under uncertainty can trace its roots to von Neumann and Morgenstern's (1944) theory of games and economic behavior, if not to Daniel Bernoulli (1700–1782). It was Bernoulli who, in proposing a solution to the famous 'St. Petersburg paradox', introduced the idea of a utility, i.e. the consequence of each possible outcome in a situation of uncertainty. Thus, putting aside the matter of a focus on certain types of events, what is new and different about reliability and risk analysis, and why do we need another book devoted to this topic?

The answer to the first part of the above question is disappointing. It is that from a foundational point of view there is nothing special about problems in reliability and risk analysis that the existing paradigms used to quantify uncertainty cannot handle. The fundamental territory has been introduced, developed and explored by individuals bearing illustrious names like Bernoulli, De Moivre, de Finetti, Fermat, Huygens, Laplace, Pascal, and Poisson. I attempt to answer the second part of the question in the following sections, but I am unsure of success. This is because my main reason for writing this book is to articulate a way of conceptualizing the problems of reliability and risk analysis, and to use this conceptualization to develop a unified approach to quantify them. I warn the reader, however, that my point of view may not be acceptable to all, though my hope is that once the fundamentals driving this point of view are appreciated and understood – which I hope to do here – my position will be more palatable.

[1] Not to be considered as being synonymous with Heisenberg's 'uncertainty principle', which says that in the quantum mechanical framework the error (uncertainty) in the measurement of position multiplied by the error (uncertainty) in time measurement must exceed a certain constant called Planck's constant. This principle was enunciated by Werner Heisenberg in the mid-1920s.

1.2 OBJECTIVES AND PROSPECTIVE READERSHIP

The aim of this book is twofold. The first is to discuss a mathematical framework in which our uncertainty about certain adverse effects can be quantified, so that the notions of hazard, risk, reliability and survival can be discussed in a unified manner. The second aim is to describe several reliability and risk analysis techniques that have been developed under the framework alluded above. My intention is to focus strongly on the matter of how to think about reliability and risk, rather than to focus on particular methodologies. Since the quantification of uncertainty has been the subject of much debate, it is essential that the key arguments of this debate be reviewed, so that the point of view I adopt is put in its proper context. Thus we start off with an overview of the philosophical issues about the quantification of uncertainty, and decision making in the presence of uncertainty. The overview material should be familiar to most graduate students in probability and statistics; however, those who have not had exposure to Bayesian thinking may find it useful. The overview is followed by a description of the key ideas and methodologies for assessing reliability and risk. The latter material constitutes the bulk of my effort and should appeal to those with applied interests. However, the importance of the foundational material needs to be underscored; it sets the tone for the ensuing developments and provides a common ground for addressing the various applications.

Because of the current widespread interest in reliability, risk and uncertainty, the book should be of appeal to academics, students and practitioners in the mathematical, economic, environmental, biological and the engineering sciences. It has been developed while keeping in mind these multiple communities. The material here could possibly also be of some interest to quantitative philosophers and mathematically oriented specialists in the areas of medicine, finance, law, national security and public policy. However, by and large, the bulk of its readership would come from graduate students in engineering, systems analysis, operations research, biostatistics and statistics. With the above diversity of clientele, it is important to draw the reader's attention to one matter. Specifically, throughout this book, the uncertain event that I focus upon is the failure of items in biological and engineering systems, rather than, say, the occurrence of a financial or a strategic loss. This admission would appear to suggest that the material here may not be of relevance to risk analysts in business, finance and other such areas. This need not be true, because the manner in which I propose to quantify uncertainty is not restricted to a particular class of problems. The choice of applications and examples is determined by my experience, which necessarily is limited. All the same, I admit to the difficulty of using my approach for addressing risk problems that are basically communal or political. Matters pertaining to single issue campaigns involving extreme positions are not in the scope of our discussion.

1.3 RELIABILITY, RISK AND SURVIVAL: STATE-OF-THE-ART

Even though the first truly empirical mortality table was constructed as early as 1693 by Edmund Halley, formal material on survival analysis appeared in actuarial journals – mostly Scandinavian. The term 'hazard rate' seems to have originated there. Since the late 1960s the literature on reliability, risk and survival analysis has experienced an explosive growth. It has appeared in diverse and scattered sources, ranging from journals in philosophy, mathematics (predominantly statistics), biomedicine, engineering, law, finance, environment and public policy. In the last 30 years or so, there have been annual conferences and symposia devoted to the various aspects of these topics. More recently, journals that pertain exclusively to these subjects have also begun to appear. These conferences, journals and symposia have been sponsored by different professional groups, with different interests and different emphases.

The activities of the various interest groups have been unconnected because each emphasizes a particular type of an application, or a particular point of view or a particular style of analysis.

For example, risk analysis done by nuclear engineers and physicists is significantly different from survival analysis done by biostatisticians. The difference is due not just to the nature of their applications, but more so to their attitudes toward quantifying uncertainty. Physicists, being generally trained as objective scientists, have come to realize and to accept subjectivity in the sciences (cf. Schrodinger, 1947); consequently, their analyses of technological risks have incorporated subjective elements. By contrast, biostatisticians, being subjected to public scrutiny, have tended to be cautious and very factual with their analyses. Similarly, the type of reliability analysis done by, say, an electronics engineer differs from that done by an applied probabilist or a statistician. The former tends to emphasize the physics of failure and tends to downplay the mathematics of uncertainty; the latter two tend to do the opposite. Certainly there is a need for both, the physics of failure and the mathematics of uncertainty. A mathematical paradigm that can formally incorporate the physics of failure into the quantification of uncertainty would help integrate the activities of the two groups and produce results that would be more realistic. A goal of this book is to present a point of view that facilitates the above interplay.

A different type of situation seems to have arisen with software engineers working on reliability problems. They often do a credible job analyzing the causes of software failure, but then quantify their uncertainties using a myriad of analytical techniques, many of them ad hoc. This has caused much concern about the state-of-the-art of software risk assessment (cf. Statistical Software Engineering, 1996). Difficulties of this type have also arisen in other scenarios. A possible explanation is that most risk analysts tend to be subject matter specialists who concentrate on the science of the application and tend to accept analytical methodologies without an evaluation of their limitations and theoretical underpinnings.

A credible risk analysis requires an integration of the activities of the various groups, and one step toward achieving this goal would be to identify and agree upon a common theme under which the notions of risk, reliability and survival can be discussed and quantified, independent of the application. Our purpose here is to advocate such a theme and I endeavor to do so in a manner understandable by subject matter specialists who are mathematically mature; as a result, all risk analysts will have a common basis from which to start and a common goal to aim for.

1.4 RISK MANAGEMENT: A MOTIVATION FOR RISK ANALYSIS

We have seen that risk management is decision making under uncertainty using quantified measures of the latter. The purpose of this section is to elaborate on the above with a view toward providing a motivation for conducting risk analysis. A few scenarios are cited that are suitable candidates for risk management and, in one case, the steps by which one can proceed are outlined.

Most studies in risk management come into play because of the possible dangers faced whenever new technologies in engineering or medicine are proposed. New technologies advance the way we live, but occasionally at a price. In some cases, the price is unacceptable. The objective in risk management is to investigate the trade-off between the conveniences and the consequences, both tangible and intangible. The end result is often a binary decision to introduce, or not, a proposed technology. As an illustration, consider the following medical scenario: cholesterol-lowering drugs decrease the chances of heart attacks, but are expensive to administer, cause physical discomfort to the patient and often have side effects such as damage to the kidneys. If in an individual case, the possibility of the side effects materializing is small, if the patient is willing to bear the cost of the drug and is able to withstand its discomfort, then there is more to be gained by prescribing the drug, than by not. But to arrive at such a decision we need to quantify at least six uncertainties, four connected with the administration of the drug, and the remaining two if the drug is not administered (Figure 1.1). Risk analysis is the

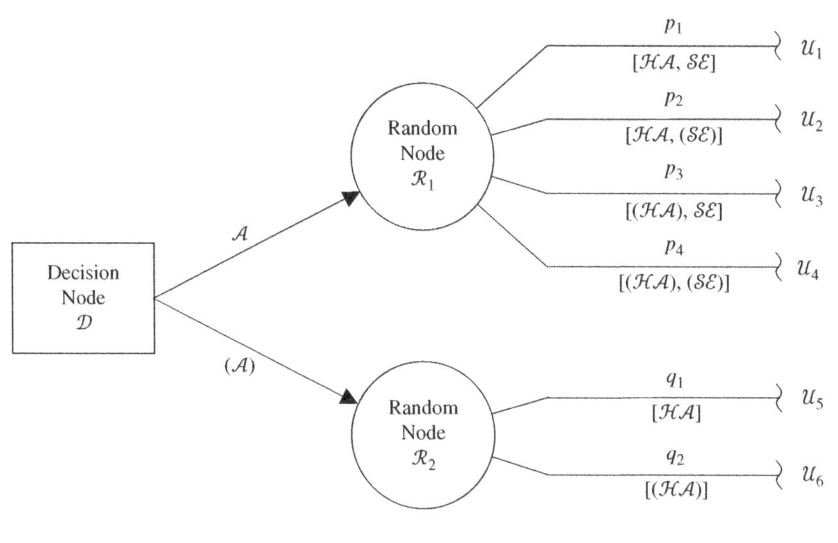

Key:

p_i and q_i: Probabilities;
$\mathcal{A}$: Administer Drug;
$\mathcal{HA}$: Patient Suffers a Heart Attack;
$\mathcal{SE}$: Patient Suffers Side Effects;
$\mathcal{U}_i$: Utilities;
$(\mathcal{A})$: Do not Administer Drug;
$(\mathcal{HA})$: Patient Avoids a Heart Attack;
$(\mathcal{SE})$: Patient Avoids Side Effects.

Figure 1.1 Decision tree for the cholesterol lowering drug problem.

process of quantifying such uncertainties. We also need to evaluate the patient's **utilities**, i.e. the consequences, usually expressed as costs, associated with each of the above six uncertainties. The utilities should include the cost of administering the drug, a cost figure associated with the discomfort of side effects, cost figures attached to suffering a heart attack, and the rewards of avoiding one. The assessment of utilities is a very crucial and, perhaps, the most difficult step in risk analysis. It is best performed by individuals knowledgeable in economics and the behavioral sciences. Determining an individual's utilities usually involves asking them to express preferences among different options. Clearly, assigning cost figures to the consequences of having a heart attack and to the merits of avoiding one is not a standard exercise, but one that nevertheless has to be addressed. The matter of utilities is discussed in some detail in section 2.8. Quantifying uncertainties is often a detailed task which, in our example, would start with the patient's medical history and would involve tracing the causes of a heart attack and the drug's side effects. A useful device that graphically portrays the causes and the sequences of events that lead to an undesirable event is a **fault tree**, also known as an **event tree**; see Barlow, Fussell, and Singpurwalla (1975) for examples on how to construct fault trees. The more detailed an accounting of the causes, the more credible is the quantification of uncertainty. Clearly, such a task would require the active participation of subject matter specialists, such as physicians and biochemists. The actual quantification should be done by someone knowledgeable in the calculus of uncertainties, which we hope that this book will help quantitative subject matter specialists to become. Thus, risk analysis is often a multi-disciplinary process involving participation by economists, mathematicians, social scientists, engineers and other subject matter specialists.

Figure 1.1 is a decision tree which shows the various steps that are involved in dealing with the cholesterol drug problem. The rectangle at the leftmost end of the tree denotes a decision node, which shows the two possible actions that a decision maker, usually the physician, can take: $\mathcal{A}$ – to administer the drug; or $(\mathcal{A})$ – not to administer it, the parentheses surrounding $\mathcal{A}$ denote its complement. The circles denote the random nodes, which show the possible (unpredictable) outcomes that can occur under each action taken by the decision maker. The notation $\mathcal{HA}$ denotes the event that the patient suffers a heart attack, whereas $(\mathcal{HA})$ denotes the event that the patient escapes it. Similarly, $\mathcal{SE}$ denotes the event that the patient experiences the drug's side effects, whereas $(\mathcal{SE})$ denotes the event that the patient does not experience side effects. Observe that a heart attack can occur whether or not the drug is administered, but that there can be no side effects when the drug is not administered. At the right-hand end of the tree, we indicate the patient's utilities that are encountered with each of the six terminal branches of the tree; these are denoted by the $\mathcal{U}_i$s. Important to Figure 1.1 are the numerical values that quantify the uncertainties associated with the events describing each of the six terminal branches; these have been denoted by the p_i's and the q_i's. A focus of this book is to describe procedures by which such numerical values can be assigned. Once this has been done, standard decision theory (cf. Lindley, 1985, pp. 139–159) prescribes a procedure by which the decision to administer the drug, or not, can be made. More details about this are given in section 2.9. The decision would depend on the assessed values of the patient's utilities and the numerical values of the assessed uncertainties. These of course would vary from patient to patient.

A similar type of analysis should be used for analyzing the risk of introducing the recently proposed 'fly-by-wire airplanes', in which the control of aircraft is under the direction of a computer. The advantage of such airplanes is less reliance on pilots who are prone to human error. Their main disadvantage is the possible presence of a fatal flaw in the software which directs the computer. What are the chances of having such a flaw and, should there be one, what are the chances of encountering it during flight? A numerical assessment of these chances, together with an assessment of utilities, would enable us to decide whether or not to commission the fly-by-wire airplanes. A less daunting example, also in connection with airplane travel, pertains to the current trend by aircraft manufacturers to equip large passenger jets with only two engines rather than the usual four, which is known as ETOPS (for extended twin engine operations). Whereas most people would prefer to fly in aircraft equipped with four engines, it is possible that having only two engines lowers the stresses on the rest of the airplane, making its overall reliability better than that of an aircraft with four engines. Decisions of this type should be supported by a formal exercise in risk management; see Appendix E of the National Research Council's 1998 Report mentioned earlier. Also see sections 10.5 and 10.6, wherein we describe decision making for allocating reliability in systems design and for system selection (procurement), respectively. In the context of reliability, decision trees also come into play when one considers life testing and the design of life testing experiments (sections 5.5 and 5.6).

Indeed, if one of the main purposes of doing reliability analysis is to facilitate a good engineering design, and because design involves a trade-off among alternatives, one may take the view that the purpose of a reliability study is to help make sound decisions in the face of uncertainty.

1.5 BOOKS ON RELIABILITY, RISK AND SURVIVAL ANALYSIS

In section 1.1, we raised the issue about the need for another book on reliability and survival analysis. I have given the reader some hints about my aims here but have said little about what is currently available and how it differs from what is planned here. In what follows, a broad-brushed perspective on published material on these topics is given.

The existing books and monographs on reliability, risk and survival analysis can be classified into the following three categories: 1) works that are heavily focused on the subject matter details of a particular application, say nuclear reactor safety; 2) works that develop models for quantifying uncertainties and which focus on the detailed mathematical structure of such models; and 3) works that emphasize statistical issues pertaining to the quantification of uncertainty and the treatment of data. The first category does not warrant concern vis-à-vis duplication because the material there does not advocate an overall theme for addressing a general class of problems, nor does it articulate any particular paradigm for thinking about problems in reliability and risk analysis. The treatment is generally on a case-by-case basis, and its greatest appeal is to practitioners whose interests are non-mathematical. In the second category, of which the two books by Barlow and Proschan (1965, 1981) are landmarks, the emphasis is on material that may be labeled 'academic'. The authors refer to their work as a 'mathematical theory of reliability', and correctly so. The main handicap of this second category is that the initial uncertainties are treated as being quantified (via probability), and the emphasis is on how these initial uncertainties propagate. That is, the initial probabilities (be they objective, subjective or logical) are assumed known. The attitude there is more in keeping with the Russian school of probability, wherein the source and nature of initial probabilities are not a matter of concern. Furthermore, the mathematical theory is not integrated into the broader framework of risk management. That is, its place in the context of decision making under uncertainty has not been sufficiently articulated. For us here, the real overlap – if any – is with respect to the third category, which also happens to be the biggest. Representatives of this group that have a focus toward engineering problems, data and applications are the books by Gnedenko, Belyaev, and Soloyev (1969) Mann, Schafer, and Singpurwalla (1974), Crowder *et al.* (1991) and Meeker and Escobar (1998); sandwiched between these are a myriad of others (cf. Singpurwalla 1993). I have mentioned these three books because the first is one of the oldest and the third, one of the latest. In the area of survival analysis, a classic source is the book by Kalbfleisch and Prentice (1980). Recent additions to this field include several books on counting process models, the one by Anderson *et al.* (1993) being encyclopedic. In all the above books, the statistical paradigm that is subscribed to is different from what we propose to do here. Thus, the possibilities of duplication appear to be minimum. As a final comment, in Barlow, Clarotti and Spizzichino (1993), the need to integrate reliability analysis into risk management has been recognized, but the material is more along the lines of a research monograph rather than an expository development.

1.6 OVERVIEW OF THE BOOK

Chapters 2 and 3 pertain to foundational issues; they present the underlying paradigm for our development. Chapter 2 starts off with a discussion of uncertainty and its quantification, leading up to decision making under uncertainty. In the interim we also present standard statistical notions such as 'inference', 'likelihood' and 'prediction'. Professionally trained statisticians, biostatisticians and applied probabilists may find little, if any, that is new to them here. By and large, the material of Chapter 2 is pitched toward engineers, operations research analysts and other mathematically oriented subject matter specialists. The material may also appeal to graduate students in the statistical sciences, and other statisticians who may not have had an exposure to subjective Bayesian thinking. Chapter 3 is specialized and pertains to the important notion of 'exchangeability', which plays a key role in selecting models for quantifying uncertainty. Most readers may want to skip this material on a first reading. Chapter 4 is basic and pertains to a discussion of standard notions in reliability and survival analysis. However, the perspective that we take in Chapter 4 is not traditional; the material given here should be viewed as being foundational to reliability and survival analysis. Chapter 5 presents a different perspective on the same topics that are covered in the books of category three mentioned before. Chapter 6 builds on the material of Chapter 5; it pertains to the propagation of uncertainty through a system of items. The material in the remaining chapters is mildly advanced and pertains to specialized topics, many of which may appeal to only certain segments of the readership. Thus, Chapter 7 focuses on dynamic environments and the use of stochastic process models, Chapter 8 on counting processes and event history data and Chapter 9 on non-parametric methods within the Bayesian paradigm. Chapter 10 pertains to the survivability of systems with interdependent failures and Chapter 11 describes the role of reliability and survival analysis in econometrics, asset pricing and mathematical finance. Our overall aim has been to give as broad a coverage as is possible, even if this means an occasional sacrifice of specifics. We compensate for this compromise of completeness by providing adequate references so that a reader can patch together a more complete picture.

Chapter 2

The Quantification of Uncertainty

Heisenberg: Here lies Werner Heisenberg – maybe.

2.1 UNCERTAIN QUANTITIES AND UNCERTAIN EVENTS: THEIR DEFINITION AND CODIFICATION

The previous chapter stated that an element that is common to all risk analyses is the presence of uncertain events. What do we mean by the term 'uncertain', and what is an **uncertain event**? We do have some intuitive notions about these but, to give them structure, we must start with the basics. We first distinguish between a scientist's appreciation of the world from that of, say, an artist. A scientist's appreciation of the world is in terms of things that can be measured. To understand and to manipulate things, a scientist must be able to measure them, or at least think of them as being measurable. Measuring means comparing to a standard, and some commonly used standards for measurement are those provided by the foot ruler, the scale, the thermometer and the watch. In effect, the act of measuring comes down to describing things by numbers; that is, quantifying them. We quantify because with quantification, we have at our disposal the full force of logic and the mathematical argument. We must of course recognize that not everything is amenable to measurement, and science excludes such entities from consideration. For example, works of art like literature, music and painting cannot be understood and manipulated by measurement, neither can be human feelings like anger and joy. This of course does not suggest that we cannot assess the utilities of things that cannot be quantified. Recall, from the example of Chapter 1, that risk analysis requires that we assess utilities for outcomes, such as the discomfort caused by a drug.

Given this background, we shall begin by noting that at any point in time, say $\tau \geq 0$, a scientist – henceforth us – contemplates a collection of things, called 'quantities', some whose numerical values are known to us and some whose values are not known, and may indeed never be known. The reference time τ plays an important role in our development, though by convention it is often taken to be zero. The collection of quantities known to us at τ will be denoted by $\mathcal{H}$, for history. Quantities whose values are unknown to us are called **random quantities** or **uncertain quantities** and, for convenience, are denoted by a capital letter, such as T. The numerical value that T can take will be denoted by the corresponding lower-case

Reliability and Risk: A Bayesian Perspective N.D. Singpurwalla
© 2006 John Wiley & Sons, Ltd

letter. When T becomes known (at some time later than τ) we shall replace it by its revealed value, say t. Our 'uncertainty' about T is our inability to assign a numerical value for t. The possible values that t can take depend on the situation being encapsulated. If t can take only discrete values, like say $t = 1, 2, 3, \ldots$, then T is said to be a **discrete** random quantity; if t is allowed to take any value in an interval, say $[0, \infty)$, then T is said to be a **continuous** random quantity. Thus, for example, T could denote the number of defectives in a particular batch, or the unknown time to failure of an item that was surviving at τ, or the unknown number of miles to the next failure of an automobile that was serviced at τ, or the remission time of a disease since a medical intervention at τ, or the amount of radioactive release, or the amount of loss incurred, etc. If the item fails after operating for 38 755 hours, that is at time $\tau + 38\,755$, then $t = 38\,755$ hours, and the continuous random quantity T has become known at $\tau + 38\,755$ hours. Similarly, if the automobile fails after accumulating 5000 miles since servicing, and if the time taken by the automobile to accumulate the 5000 miles is 5 months, then $t = 5000$ miles and the discrete random quantity T has become known at $\tau + 5$ months.

An important subclass of random quantities is that in which t is restricted to be binary, i.e. t can take only two values, say $t = 0$ and $t = 1$. Such random quantities are called **random events** or **uncertain events** and, for purposes of distinction, we denote them by the letter X, with X taking the value $x = 0$ or 1. It is often the case that random events are constructed from random quantities. For example, if the random quantity T denotes the time to failure of an item, and if we are interested in the proposition that T be at least t, then we may define a random event X, with $x = 1$, whenever $T \geq t$, and $x = 0$, when $T < t$. With this construction, our uncertainty about $(T \geq t)$ is synonymous with our uncertainty about $(X = 1)$. Because of this synonymity, we may also refer to $(T \geq t)$ as 'the event that T is at least t'. In general, random events are 'indicators' constructed for encapsulating situations that involve the truth or the falsity of a proposition. The general convention is that $x = 1(0)$ if the proposition is true (false). Examples of propositions that are common in risk analysis are: the patient will have a reaction to a drug; a business decision will lead to a loss; an earthquake will occur within the next five years; an automobile will provide service between 70 and 100 thousand miles, and so on.

We close this section on the codification of uncertain quantities and uncertain events by emphasizing two points. The first is that we have been talking about the known and unknown quantities for a 'particular' scientist, say us, and that we are referring to our state of knowledge at a 'particular' time τ. What is unknown to us at τ may very well be known to another scientist, and also what is unknown at τ may become known to us later. The important point is that what we are talking about is personal to us, with respect to both our state of knowledge and our time of reference. These matters play a key role when we quantify uncertainty, because it is our uncertainty that we will be quantifying, not someone else's. More about this essential point is said later, in section 2.3.2.

2.2 PROBABILITY: A SATISFACTORY WAY TO QUANTIFY UNCERTAINTY

Continuing with the setup of the preceding section, let us focus attention on some reference time τ at which there is an uncertain quantity T that we are interested in and have at our disposal history, or background knowledge, $\mathcal{H}$. Even though we are uncertain about T, i.e. are unable to assign a numerical value for t, we are not completely ignorant about T. At the very minimum, we know the range of values that t can take, and since it is reasonable to suppose that $\mathcal{H}$ gives us information about T, we may be able to guess which values of t are more likely to arise (when T will reveal itself) than the others. Our aim is to quantify this uncertainty or partial knowledge about T, in the light of $\mathcal{H}$, at time τ. By this, we mean that we want to associate a number with every value of t. What should we call this number and what should its properties be?

There have been several suggestions. However, Lindley (1982a) asserts that **probability** is the only satisfactory way to quantify our uncertainty. Furthermore, under some mild but reasonable assumptions, probability is inevitable (cf. Lindley, 1982b). To describe the properties of this number called 'probability', it is convenient to introduce some additional notation. When T is discrete, we write $P^{\tau}(T=t; \mathcal{H})$ to denote the number we associate with the event that when T reveals itself, its numerical value will be t, and that this assessment is made at time τ, in light of background knowledge $\mathcal{H}$. It is common to suppress τ by setting it equal to zero, so that $P^{\tau}(T=t; \mathcal{H})$ is simply $P(T=t; \mathcal{H})$. We refer to $P(T=t; \mathcal{H})$ as 'the probability that the random quantity T takes the value t' or equivalently, 'the probability of the event $T=t$'. Often $P(T=t; \mathcal{H})$ is abbreviated as $P_T(t; \mathcal{H})$ or simply $P(t; \mathcal{H})$. Analogously, for a random event X, $P_X(x; \mathcal{H})$ [or simply $P(x; \mathcal{H})$] denotes the probability that when X is revealed, it will take the value x. Irrespective of whether T is discrete or continuous, $P(T \leq t; \mathcal{H})$ denotes the probability that T is at most t; it is called the **distribution function** of T. It is common to abbreviate $P(T \leq t; \mathcal{H})$ as $F_T(t; \mathcal{H})$, or simply $F(t; \mathcal{H})$. If $f_T(t; \mathcal{H})$, the derivative of $F_T(t; \mathcal{H})$, exists for (almost) all values of t, and if the latter is given by an indefinite integral of the former, then $F_T(t; \mathcal{H})$ is said to be **absolutely continuous**, and $f_T(t; \mathcal{H})$ is called the **probability density function** (or simply, the **density**) generated by $F_T(t; \mathcal{H})$ at t (section 4.2). As was done with $F_T(t; \mathcal{H})$, the subscript T may be omitted – when there is no cause for confusion – and the probability density at t simply written as $f(t; \mathcal{H})$. In all of the above, we have adopted the convention that the semicolon separates the knowns from the unknowns; this convention is not standard. Finally, we wish to emphasize that probability makes sense only for events whose disposition is unknown to the assessor of probability at the time of making the probability assessment.

The setup described above generalizes when we have two (or more) uncertainties, say $(T_1 \leq t_1)$ and $(T_2 \leq t_2)$. Specifically, if $P(T_1 \leq t_1, T_2 \leq t_2; \mathcal{H}) = F(t_1, t_2; \mathcal{H})$ is absolutely continuous, then its derivative $f(t_1, t_2; \mathcal{H})$ exists for (almost) all values of t_1 and t_2, and $f(t_1, t_2; \mathcal{H})$ is known as the **joint probability density function** (or simply the **joint density**) generated by $F(t_1, t_2; \mathcal{H})$; more details are given in section 4.2. Note that $P(T_1 \leq t_1, T_2 \leq t_2; \mathcal{H})$ represents the event $P(T_1 \leq t_1 \text{ and } T_2 \leq t_2; \mathcal{H})$.

To describe how different uncertainties relate to each other, we restrict attention to the case of random events and extend the above setup by contemplating, at time τ, two (or more) random events, say X_1 and X_2, with X_i taking values x_i, $i = 1, 2$. Note that each x_i is either 0 or 1. We state below the rules (or the calculus) of probability. Often these rules are stated as axioms; however, as is pointed out in section 2.2.2, they can be motivated by both behavioristic as well as operational arguments.

2.2.1 The Rules of Probability

For coherence, a notion that will be explained in section 2.3.2, all probability specifications must obey the following rules (stated here for the case of random events):

Convexity:

$$0 \leq P(X = x; \mathcal{H}) \leq 1.$$

Addition: If X_1 and X_2 are mutually exclusive, that is both X_1 and X_2 cannot logically take the same value (be it 0 or 1), then

$$P(X_1 = x_1 \text{ } or \text{ } X_2 = x_2; \mathcal{H}) = P(X_1 = x_1; \mathcal{H}) + P(X_2 = x_2; \mathcal{H}).$$

Multiplication:

$$P(X_1 = x_1 \text{ and } X_2 = x_2; \mathcal{H}) = P(X_1 = x_1 \mid X_2 = x_2; \mathcal{H})\ P(X_2 = x_2; \mathcal{H}), \text{ or equivalently,}$$
$$= P(X_2 = x_2 \mid X_1 = x_1; \mathcal{H})\ P(X_1 = x_1; \mathcal{H}),$$

where $P(X_1 = x_1 \mid X_2 = x_2; \mathcal{H})$ is known as the **conditional probability** that $X_1 = x_1$ given that $X_2 = x_2$, and in light of $\mathcal{H}$. It is typical to refer to $P(X_1 = x_1 \mid X_2 = x_2; \mathcal{H})$ as the conditional probability of X_1 given X_2. The interpretation of conditional probabilities is subtle; it is in the subjunctive mood. Specifically $P(X_1 = x_1 \mid X_2 = x_2; \mathcal{H})$ denotes the probability of the event $(X_1 = x_1)$, in light of $\mathcal{H}$, if it were to be (i.e. supposing) that X_2, when it reveals itself, is x_2. The conditioning event $(X_2 = x_2)$ is not known to be true (or false) at time τ, when X_1 is assessed; all that is known at time τ is only $\mathcal{H}$. Consequently, the event $(X_2 = x_2)$ is to the left of the semicolon whereas $\mathcal{H}$ is to the right, and the vertical slash preceding the conditioning event signals its subjunctive feature. Finally, X_1 and X_2 are said to be **mutually independent** if $P(X_2 = x_2 \mid X_1 = x_1; \mathcal{H}) = P(X_2 = x_2; \mathcal{H})$ and also if $P(X_1 = x_1 \mid X_2 = x_2; \mathcal{H}) = P(X_1 = x_1; \mathcal{H})$.

The convexity rule states that all probabilities are numbers between 0 and 1. The addition and the multiplication rules specify how the various uncertainties combine or cohere. A collection of uncertainty statements that obey the above rules are, according to Lindley (1982a), **coherent**. The notion of conditional probability has been the subject of some discussion. Of particular concern are issues such as protocols for obtaining new information and the timing of events; the writings of Shafer (1982b, 1985) make interesting reading. By a repeated application of the addition rule, we can see that for any finite collection of mutually exclusive random events, say $X_1, \ldots, X_n$, $n < \infty$,

$$P(X_1 = x_1 \text{ or } X_2 = x_2 \text{ or } \ldots X_n = x_n; \mathcal{H}) = \sum_{i=1}^{n} P(X_i = x_i; \mathcal{H});$$

this is known as the **finite additivity** property of probability. Under **countable additivity** we allow n to be infinite, so that

$$P(X_1 = x_1 \text{ or } X_2 = x_2 \text{ or } \ldots X_n = x_n; \mathcal{H}) = \sum_{i=1}^{\infty} P(X_i = x_i; \mathcal{H}).$$

When countable additivity is allowed, and the multiplication rule is taken as a definition of conditional probability, the above three rules are generally known as the **Kolmogorov axioms**.

2.2.2 Justifying the Rules of Probability

A question arises as to why probabilities should combine in the manner shown above. For example, in Zadeh's (1979) theory, his 'possibilities' do not combine using the addition rule, and Shafer (1976), following Dempster (1968) and Smith (1961), extends the idea of using a single number for describing uncertainties to two numbers, called upper and lower probabilities. The same is also true of Jeffrey (cf. Diaconis and Zabell, 1982), who uses a rule of combination that is unlike those used in probability.

There are two lines of reasoning that justify the calculus of probability given before. The first is an 'axiomatic' one, and the second a pragmatic one. The axiomatic argument, first proposed by Ramsey in 1931, and further developed by Savage (1954) and by DeGroot (1970), is an argument in mathematics. It proceeds by searching for simple, self-obvious truths about uncertainty, taking these as axioms, and then developing a system of theorems that result in the above rules. The second, a more operational argument, is due to de Finetti (1974, ch. 4). It is called the **scoring rule argument**, and goes along the following lines:

A person contemplating an uncertain event X, say a proposition, assigns a number p to describe the person's uncertainty about the truth of the proposition. The person receives a penalty score of an amount $(p-1)^2$ if the proposition turns out to be true, and an amount p^2 if the proposition turns out to be false. The person should choose p in a manner that minimizes the penalty score. Thus if the person were to be certain that the proposition is true, then p will be specified as 1, whereas if the opposite were to be the case, then p would be zero. Now suppose that such a person contemplates several uncertain events $X_1, X_2, X_3, \ldots$, and assigns to each the corresponding numbers $p_1, p_2, p_3, \ldots$. Then, it is an argument due to de Finetti that if the scores for the different events are additive, then in order to avoid a sure loss, that is, a gamble in which the bettor loses irrespective of the outcome, the p_i's must obey the rules of probability given above; that is, the p_i's must indeed be probabilities. It is for this reason that probability and the calculus of probability are said to be inevitable for specifying uncertainty. The quadratic scoring rule mentioned above has been used to rate meteorological predictions and is known as the **Brier score**. Savage (1971) and Lindley (1982b) have generalized de Finetti's result to scores other than the Brier score. Many prefer the abstract axiomatic argument over the pragmatic scoring rule argument of de Finetti. One reason is that people do not like to be scored and may therefore not agree to specify numbers in the context of being scored. The second, and perhaps more important reason, is that the scoring rule argument can be used to extend the addition (and also the multiplication) rule to the case of only a finite number of events. Consequently, according to this argument the resulting probabilities are only finitely additive, not countably additive (or σ-additive), as in the Kolmogorov axiomatization. This means that countable additivity has to be introduced as an additional axiom, or by some extra axiom that is effectively equivalent. One such equivalent axiom is the axiom of conglomerability (cf. Lindley, 1997a). According to this axiom, if the events $\{X_i = x_i\}$, $i = 1, 2, \ldots$, are a countable partition of an event C, and if for some event A, $P(A|(X_i = x_i) \cap C) = a$, for all i, then $P(A \mid C) = a$. That is, a statement true for every member of a partition is true overall. As an aside, it is useful to note that not everyone is willing to accept the Kolmogorov axiomatization of probability. de Finetti for one rejected countable additivity (cf. Lane, 1987) and Hartigan (1983), among others, retains countable additivity but rejects the convexity rule, and develops a theory of improper probability. More on this is said in sections 2.5, 5.2.3 and 5.3.2.

2.3 OVERVIEW OF THE DIFFERENT INTERPRETATIONS OF PROBABILITY

In the previous section, we have outlined the premise that the probability of an uncertain event is a number between zero and one that behaves according to certain rules, and which is specified by a person with background knowledge $\mathcal{H}$ who is contemplating the event at some time τ. In Chapter 1, we stressed the fact that, in any problem involving decision under uncertainty, the numerical values assigned to the uncertain events determine the optimum decision to be taken. However, we have said little about how these numerical values are arrived upon, whether they are unique and how to make them operationally meaningful. For this, we need to review the several interpretations of probability and, in order to gain their broader appreciation, trace their historical development. Such material is relevant because the interpretation of probability that we adopt will, to a great extent, determine the methodologies that we will advocate, the problems we are able to address and the results we are able to obtain. Thus, whereas the interpretation of probability may not be of much concern to a mathematical probabilist, to a user of probability, like an engineer, a statistician or an operations research analyst, the interpretation is germane. The material of section 2.3.1, on the historical development of probability, can be skipped without a loss in continuity.

2.3.1 A Brief History of Probability

Probability can trace its roots to the times of Gioralimo Kardano (1501–1575), an ardent gambler who, in his treatise, 'Book on Dice Tossing', did consider the ratio of successes to the total number of trials but never fully explored the usefulness of this idea. So, who was it that introduced the idea of a random event, guessed that it was rational that probability be a number between zero and one and made precise the addition and the multiplication rules for calculating the probabilities of complex events? To answer these questions, we need to trace the development of probability, which in the beginning appears to have evolved along two paths: fair price and belief.

The word 'probability', derived from the Latin verb *probare*, means 'to prove'. Until about 1689, probability was not a number between zero and one; probability was an argument or a belief for which there are good proofs (cf. Shafer, 1996). It is widely believed that numerical probability was born in 1654 when Blaise Pascal (1623–1662), in correspondence [initiated by Chavalier de Méré (1607–1684), a philosopher and a man of letters] with Pierre Fermat (1601–1665), was the first to propose a correct solution to the problem of points, or fair prices. But Pascal and Fermat did not use the word 'probability' in their letters. They were not thinking about probability, which, as stated before, was then a qualitative idea used for evaluating an argument or evidence. They were thinking about problems of equity or fair prices. Shafer (1990), who provides an interesting perspective on the evolution of mathematical probability, and upon whose writings much of what we say here is based, describes the problem of points as follows:

$\mathcal{A}$ and $\mathcal{B}$ are playing a game contending on equal chances. They have both put five chips on the table, and have agreed that the winner gets the entire stake of ten chips. The game consists of several rounds of play, the winner of each round gains a point, and the first person to gain three points wins the game. At a certain time, when $\mathcal{A}$ needs two more points to win and $\mathcal{B}$ needs only one, $\mathcal{A}$ has to leave the game. $\mathcal{C}$ is prepared to take $\mathcal{A}$'s place in the game. What is the fair price that $\mathcal{C}$ should pay $\mathcal{A}$ for his position in the game? Pascal's argument was that $\mathcal{C}$ should pay $\mathcal{A}$ 2.50 units; this answer was different from the more traditional answer of 3.33 units. Probability theory got started from this kind of reasoning. Christian Huygens rediscovered Pascal and Fermat's reasoning (their correspondence was published in 1679), and wrote it up in 1657 in the widely circulated *De ratiociniis in ludo aleae*. Later on, Huygens and Montmort found fair prices for positions in more and more complicated games and also for practical applications such as annuities and insurance.

It would be nearly 60 years after the Pascal–Fermat correspondence that the idea of fair price would be tied up with that of probability (or belief) in James (or Jakob) Bernoulli's masterpiece *Ars Conjectandi*, published posthumously in 1713 by his nephew Nicholas Bernoulli.

Bernoulli's introduction of probability was motivated by his desire of applying the Pascal–Fermat–Huygens theory of fair price and expectations to problems other than games of chance where the qualitative idea of probability was traditionally used. He reasoned that, just as how the rounds won and lost in a game of chance entitle one to a fair price, the arguments that one finds for and against an opinion entitle one to a portion of certainty. Arguments in practical problems earn a certain portion of certainty, just as a position in a game of chance earns a certain portion of the stake. This portion is the opinion's probability. This analogy between the rounds in a game of chance and arguments about an opinion also appears to have been made by Pascal's friends in a widely used textbook called the *Port Royal Logic* (Arnauld and Nicole, 1662). According to Shafer (1986), the English cleric George Hooper used the word 'probability' to refer to a number between zero and one in a 1699 anonymous article in *The Philosophical Transactions of the Royal Society*. However, it was Bernoulli who made the case for the existence and meaningfulness of numerical probabilities, and went beyond the analogy between arguments and rounds to develop numerical rules for the combination of arguments, showed the inapplicability of the addition rule to non-disjoint events, gave his formula for the binomial distribution, and went on to prove his great theorem, the law of large numbers (section 3.1.4). This law was motivated by Bernoulli's recognition that, in practical problems, unlike games of chance, fair prices could not be deduced from assumptions about equal possibilities; probabilities in practical problems would have to be found (approximated) from observation. He proved that, if a large number of rounds are played, then the relative frequency with which an event occurs will approximate the relative ease with which it will happen, namely, its probability. Thus Bernoulli, to whom probability was a piece of the certainty (and not a relative frequency), bound up together fair price, belief and frequency. Bernoulli subscribed to Newton's notion of **metaphysical determinism**; i.e. an unconditional belief in the powers of omnipotence and formal logic as an instrument for cognition and description of the external world. To him, probability was not just chance; it was a state of our knowledge. **Chance** was viewed as the absence of a divine plan or design.

Bernoulli's idea of approximating probability by a frequency was taken up by De Moivre (1718) who made it the basis of his famous book *The Doctrine of Chances*. However, other elements of Bernoulli's strategy, like his rules for combining arguments that were based on the traditional view of probability theory in philosophy, theology and the law, were nowhere to be seen in De Moivre's book. Indeed, it was De Moivre who introduced the concept of independent random events and proposed both the summation and the multiplication rule of probability. However, to De Moivre, the rules for combining probabilities had to be derived from Huygen's rules, which could be justified directly using the idea of expectations under fairness. This is contrary to current thinking in which rules for expected values are derived from rules for probabilities. Despite the great influence that De Moivre's book had during the eighteenth century, Bernoulli's rules for combining probabilities survived that century in the writings of philosophers such as Lambert and Diderot (cf. Shafer, 1996). It was almost 75 years later, when Laplace (1795) (in his *Essai de Philosophique sur de Probabilite*) discussed Bayes' formula and the formula of total probability, that Bernoulli's rules for combining arguments disappeared.

After Bernoulli and De Moivre, the next important step was taken by Thomas Bayes (1702–1761), whose famous essay on inverse probability was published in 1764. It appears (cf. Gnedenko, 1993) that Bayes' main contribution was an elucidation of the multiplication rule, which allows one to calculate conditional probabilities from unconditional probabilities. This formula was crucial for Laplace's development of the law of total probability and Bayes' law – discussed later. Shafer (1982b) gives a fascinating account of Bayes' treatment of conditional probability, and Stigler (1982) interprets Bayes' essay from the angle of predictive inference.

After Bernoulli, Laplace (1749–1827) was the next great philosophical mind to come to grips with probability. Like Bernoulli, Laplace was a determinist who regarded numerical probability as a degree of certainty and to him the doctrine of chances was the universal tool for measuring partial knowledge. Assured by his success in giving a probabilistic justification to the method of least squares, Laplace associated mathematical probability with the very idea of rationality. Thus, to Laplace, probability was a 'rational belief' and the rules of probability were self-evident so that the additivity of probability no longer needed to be derived from fair price and expectation – it was axiomatic, a natural and obvious property of rational belief. Similarly, to him, rules for the values of expectation arose from viewing probability as rational belief. Thus Laplace's picture of probability differed from De Moivre's in several ways, the most salient differences being an expanded number of applications (which now included the combination of observations and of testimony) and the absence of the requirement that observations, or equally likely cases, were needed to get started (Shafer, 1996). Poisson (1781–1840) did much work on the technical and practical aspects of probability, his main contributions being a generalization of Bernoulli's theorem. Like Laplace, he too was a determinist and one who also greatly expanded the scope of applications of probability.

Laplace's definition of probability as a degree of belief was criticized in the 1840s by empiricist philosophers, like John Venn, and by mathematicians like Richard von Mises (1883–1953), who were equally unhappy with Bernoulli's notion of probability as ease of happening. Probability made sense to the empiricists only if it was empirically defined as a frequency. Thus, Bernoulli's theorem (which is (incorrectly) interpreted by some as probability equals frequency) was nonsense. To them, frequency itself was the starting point, and was the definition of probability. However, in doing so, the empiricists were willing to leave vast domains outside their range of application. If, in a problem such as the authorship of the gospels or station blackout in a nuclear reactor, one could not conceive relevant limiting frequencies, then it was not a problem for probability.

The modern era in probability began in the early 1930s after Kolmogorov axiomatized probability and freed it from the paradoxes and confusions of interpretation. He treated events as sets and probabilities as numbers assigned to these sets which obey certain rules (or axioms), the main one being additivity (section 2.2.1). The key to this adaptation was the philosophy of the German mathematician David Hilbert, who believed that mathematics was a formal exercise without an essential connection to reality. The perspective given by Kolmogorov is far from the setting of Pascal and Fermat, where repeated rounds of a game are played and, hence, prices and probabilities change. In Kolmogorov's scheme, neither price nor repetition is fundamental; they are both arbitrary elements added on top of the axioms which are foundational. All the same, Kolmogorov's intent was to view these axioms as providing a mathematical foundation for the frequency definition of probability.[1] Of course, their neutrality makes them susceptible to a degree-of-belief interpretation as well.

Our brief history of probability would end here, except for the resurgence of subjective ideas during the past 40 years. This happened in 1954 with Savage's publication of *Foundations of Statistics*. Motivated, among other things, by difficulties and limitations of the empiricist philosophy, the view that probability is a degree of belief was revived. The foundation for this subjectivist revival was provided in the 1920s and the 1930s by the philosopher Frank Ramsey and the mathematician-philosopher Bruno de Finetti, who gave the degree of belief an empiricist interpretation by insisting that people be willing to bet on their beliefs.

2.3.2 The Different Kinds of Probability

In tracing the historical development of probability we have seen that it has had various interpretations, starting from Bernoulli's notion of probability as a part of certainty, to De Moivre's equally likely cases, Laplace's degree of knowledge, the empiricists' relative frequency and finally to the subjectivists' degree of belief. This has prompted many, like Good (1950) and Savage (1954), to classify the different kinds of probability, and to discuss the nature of each. The purpose of this section is to examine this classification, so that the relevance of each type to problems in reliability, risk and survival analysis can be judged; see, for example, Bement, Booker, Keller-McNulty and Singpurwalla (2003). Following Good (1965, p. 6), we shall broadly classify probabilities as being either logical, physical, or psychological, with some having further sub-classifications. The former two are called objective probabilities, and the third is called epistemological probability.

Logical Probability

A logical probability (or what is also known as credibility) is a rational intensity of conviction, implicit in the given information such that if a person does not agree with it, the person is

---

[1] Though it was Kolmogorov (1963) who stated: 'The frequency concept, based on the notion of limiting frequency as the number of trials increases to infinity, does not contribute anything to substantiate the applicability of the results of probability theory to real practical problems where we have always to deal with a finite number of trials. The frequency concept applied to a large but finite number of trials does not admit a rigorous formal exposition within the framework of pure mathematics'.

wrong. This notion of probability has its roots in antiquity (cf. Savage, 1972); its more recent leading proponents have been Carnap (1950), Jeffreys (1961) and Keynes (1921), all of whom formulated theories in which credibilities were central. Savage (1972) calls this the 'necessary' or the 'symmetry' concept of probability and does not subscribe to its existence. According to this concept, there is one and only one opinion justified by any body of evidence, so that probability is an objective logical relationship between an event $\mathcal{A}$ and evidence $\mathcal{B}$. Savage and also Good claim that both Keynes and Carnap have, in the latter parts of their work, either renounced this notion or have tempered it. The logical probability concept has not been popular with statisticians, though Jeffreys has used it to address many practical problems. Other objections to this notion of probability have recently been raised by Shafer (1991). Cowell *et al.* (1999) claim that no satisfactory theory or method for the evaluation of logical probabilities has yet been devised.

Physical Probability

A physical probability (also called propensity, material probability, intrinsic probability or chance) is a probability that is an intrinsic property of the material world, just like density, mass or specific gravity, and it exists irrespective of minds and logic. Many people subscribe to the existence of such probabilities, especially those who, following Venn (1866) and von Mises (1939), interpret probability as a relative frequency. This happens to be the majority of statisticians; the influential Neyman–Pearson approach to statistical inference is based on a relative frequency interpretation of probability (Neyman and Pearson, 1967). The words 'frequentist' or 'frequentist statistics' are often used to describe such statisticians (and their procedures). Much of the current literature in reliability and survival analysis, including several government standards for acceptance sampling and drug approval, has been developed under the paradigm that probability is a relative frequency. Here, the probability of an event is the long-run frequency with which the event occurs in a certain experimental setup or a certain population. Specifically, suppose that a certain experiment is performed n times under 'almost identical' conditions, and suppose that an event of interest $\mathcal{A}$ occurs in k of the n trials of the experiment. Then the relative frequency of the event $\mathcal{A}$ is the ratio k/n. If as n increases, this ratio converges to a number, say p, then p is defined to be the probability of the event $\mathcal{A}$.

There have been several criticisms of this interpretation of probability. For one, the concept is applicable to only those situations for which we can conceive of a repeatable experiment. This excludes many ordinary isolated events, such as the release of radioactivity at a nuclear power plant, the guilt or innocence of an accused individual, or the risk of commissioning a newly designed aircraft or a newly developed medical procedure. There are many events in our daily lives whose probabilities we would like to know, but these probabilities would not be available in the frequency sense. Another objection to this notion of probability is that the conditions under which the repeatable experiments are to be performed are not clear. What does it mean to say that the experiments are to be performed under almost identical conditions? If they are performed under exactly identical conditions we will always get the same outcome, and the ratio k/n will equal to 1 or 0. How much deviation should we allow from the conduct of one experiment to the next? Finally, how large should n be allowed to get before the limit p is obtained, and how close to p should the ratio k/n get? Whereas the relative frequency view of probability makes Kolmogorov's axioms easy to justify, there are still some concerns about the adequacy of this point of view for interpreting conditional probability and independence (cf. Shafer, 1991). Kolmogorov simply defined conditional probability as the ratio of two unconditional probabilities, but offered no interpretation that could be used to assess conditional probabilities directly. A consequence is our inability to interpret the law of total probability, mentioned in section 2.4. These and other concerns, such as inadmissibility (cf. Basu, 1975; Cornfield, 1969; Efron, 1978), have recently caused many statisticians, and also others, such as economists, engineers, and operation research analysts, to rethink this empiricist view of probability, and to explore other alternatives. In this book, we will not subscribe to the relative frequency interpretation of probability (footnote 2); this is a feature that distinguishes this work in reliability from that of the others.

Psychological Probability

A psychological probability is a degree of belief or intensity of conviction that is used for betting, or making decisions, not necessarily with an attempt at being consistent with our other opinions. When a person uses a consistent set of psychological probabilities, then these probabilities are called **subjective probabilities**; see below. A consistent set of probabilities is one against which it is not possible for an opponent to make a selection of bets against which you are bound to loose, no matter what happens. Both de Finetti (1937) and Savage (1972) regard subjective probability as the only notion of probability that is defensible, and most Bayesian statisticians subscribe to this notion of probability. Savage (1972) calls subjective probability personal probability, and describes it as a certain kind of numerical measure of the opinions of somebody about something. The point of view of probability that we strive to adopt here is subjective or personal; thus it behooves us to elaborate on this notion and to discuss its pros and cons.

Making Personal (or Subjective) Probability Operational

The notion of a personal probability of an event was made operational by de Finetti (1937), who defined it as the price that you would pay in return for a unit of payment to you in case the event actually occurs. Thus, for example, if you declared that your personal probability for some event, say $\mathcal{A}$, is .75, then this means that you are willing to put \$0.75 on the table on $\mathcal{A}$, if the person you are betting with is willing to put \$0.25 on the table against $\mathcal{A}$. If the event $\mathcal{A}$ occurs, you win the other person's \$0.25; if $\mathcal{A}$ does not occur, you lose your \$0.75 to the other person. Put another way, when you declare that your personal probability for an event $\mathcal{A}$ is .75, then you are de facto paying \$0.75 for a ticket that returns \$1 if $\mathcal{A}$ happens. It is important to recognize that in taking the above bet, you are also willing to take the other side of the bet. That is, you are willing to paying \$0.25 for a ticket that returns \$1 if $\mathcal{A}$ does not happen. Therefore, according to de Finetti, the probability of a proposition is the price at which you are neutral between buying and selling a ticket that is worth \$1 if the proposition is true, and is worthless otherwise. Alternatively put, probability is a two-sided bet. Since there is no physically realizable way to exchange more than a finite number of dollars, de Finetti insisted that the number of transactions be limited to a finite number of sales and purchases. Finally, even though we used the US monetary unit of a dollar to illustrate the mechanics of betting, the basic requirement of de Finetti is that the betting be done in terms of any desirable monetary unit. A consequence is that probability is a unitless quantity. This feature is germane to the material of section 4.4.

Probabilities and odds on (odds against) are related. Specifically, when we say that the odds on (against) an event $\mathcal{A}$ are x to y, we are implying that we are willing to pay an amount $x(y)$ now, in exchange for an amount $x + y$ should $\mathcal{A}$ occur. The odds of x to y on are the same as the odds y to x against. Thus, probabilities and odds are related in the sense that the odds against an event $\mathcal{A}$ are $(1 - P(\mathcal{A}))/P(\mathcal{A})$ and $P(\mathcal{A}) = y/(x + y)$. Consequently, we can change from odds to probability and vice versa (Lindley, 1985).

Conditional probabilities can also be made operational via the above betting scheme. Specifically, suppose that our knowledge of an event $\mathcal{A}$ were to precede our knowledge of an event $\mathcal{B}$. Then the conditional probability $P(\mathcal{B} \mid \mathcal{A}; \mathcal{H})$ is the degree to which you currently believe in $\mathcal{B}$ if in addition to $\mathcal{H}$ you were also to learn that $\mathcal{A}$ has occurred. That is, it is the amount you are willing to pay now for a \$1 ticket on $\mathcal{B}$ right after $\mathcal{A}$ happens, but with the provision that all bets are off if $\mathcal{A}$ does not happen. Finally, the notion of independent events can be easily explained in the context of bets. To say that any two events $\mathcal{A}$ and $\mathcal{B}$ are independent means that a knowledge of the occurrence or the non-occurrence of $\mathcal{A}$ will not change our bets on $\mathcal{B}$, and vice versa. That is, the probability we would assign to the event $\mathcal{B}$ will be the same, irrespective of us being informed of whether $\mathcal{A}$ has occurred or not occurred.

Since there is no such thing as the right amount that one should put on the table, personal or subjective probabilities are not objective. However, employing personal probabilities requires one to be honest with oneself, in the sense that it takes a lot of self-discipline not to exaggerate the probabilities that one would have attached to any hypotheses before they were suggested. Also, when we theorize about personal probabilities, we theorize about opinions generated by a 'coherent person', that is a person who does not allow book to be made against them. By making a book against a person we mean that if we offered this person various contingencies we can, by some sleight of hand, sell a bill of goods such that the person will be paying us money no matter what happens. A coherent person is one who declares personal probabilities that are consistent, and personal probabilities will be consistent if they are coherent, as defined at the end of section 2.2.1.

There are other aspects of personal probability that are important to note. The first is that personal probabilities, all of them, are always relative to one's state of knowledge. That is, their specification at any time τ depends on the background information $\mathcal{H}$ that we have at τ. If $\mathcal{H}$ were to be different, our specified personal probability is also likely to be different. When $\mathcal{H}$ changes with the passage of time, either due to added knowledge or due to actual data, the personal probabilities may also change. Bayes' law, which will be discussed later, gives us a prescription of precisely how to change our personal probability in light of new information to include new data. Contrast this to probabilities based on relative frequencies; they, being independent of minds and logic, are always absolute. The second point to note is that the notion of personal probability encompasses the logical probability notion of symmetry (or equally likely cases) considered by De Moivre and others. The reason is that symmetry is a judgment, and is therefore personal or subjective. It is physically impossible to make coins and dice that are perfectly symmetrical, so the judgment of symmetry is one of practical convenience, and thus personal. Furthermore, to say that something is symmetrical or equally likely implies that it is equally probable, and to define probability in terms of its likeliness could be viewed as circular reasoning. Thirdly, the notion of personal probability does not exclude from consideration information pertaining to relative frequency; it simply incorporates frequencies into the background information $\mathcal{H}$. Our final point pertains to the issue of consistency and coherence into the specification of personal probabilities. This may be difficult to enforce in real life, and psychologists such as Tversky and Kahneman (1986) have given many examples showing that people do not conform to the rules of probability in their actual behavior. Consequently, some regard the theory of personal probability and the rules of probability (i.e. the Kolmogorov axioms) as being primarily a **normative theory** – that is, a theory which prescribes rules by which we ought to behave or strive to behave and not by which we actually behave. For a discourse on the elicitation of personal probabilities (Savage, 1971).

2.4 EXTENDING THE RULES OF PROBABILITY: LAW OF TOTAL PROBABILITY AND BAYES' LAW

Mathematical probability theory deals with developing the necessary techniques for calculating the probabilities of complicated events based on probabilities of simpler events. The theory per se does not concern itself with the interpretation of probability. Of course, when one uses probability for addressing a specific problem and for communicating the result to others, the interpretation of probability becomes crucial. The foundation for the theory is the three rules, or axioms, that were given in section 2.2.1. From these, several other rules can be derived, of which two stand out as being important. The first is **the law of total probability**, and the second is **Bayes' law**. For reasons that will be explained later, the law of total probability is also known as the law of the extension of conversation. As a historical note, it appears to be likely (Stigler, 1983) that Thomas Bayes (1702–1761) may not have been the one to derive the law that bears his name; it could have been a Cambridge mathematician by the name of Saunderson. However, Laplace (1774), unaware of Bayes' earlier work, is often credited with both Bayes' law (or what is also known as the law of inverse probability) and the law of total probability (Stigler, 1986). In this section, we motivate and develop these laws using the notation and terminology of sections 2.1 and 2.2, except that here X is generic and not restricted to denoting only random events.

With respect to the introductory material of section 2.1, we note that in general, X and $\mathcal{H}$ should embrace all that we don't know and do know, and so they could be multidimensional and continuous. However, in the interest of simplifying matters, we shall assume X to be discrete, so that we may use summation instead of the more general integration.

2.4.1 Marginalization

Suppose that an outcome $\boldsymbol{X}$ is a vector that can be partitioned as X_1 and X_2, and it is only an appreciation of X_1 (in light of $\mathcal{H}$) that is of concern to us. For example, suppose that $\boldsymbol{X}$ denotes the size of the next person that we encounter, with X_1 and X_2 denoting the person's height and weight respectively. Then, by the addition rule of probability, we can easily verify the marginal distribution of X_1 as

$$P(X_1; \mathcal{H}) = \sum_{X_2} P(X_1, X_2; \mathcal{H}),$$

where $\sum_{X_2}$ denotes the summation over all values, say x_2, that X_2 can take. We may now state

2.4.2 The Law of Total Probability

Applying the multiplication rule to the right-hand side of the previous expression, we see that

$$P(X_1; \mathcal{H}) = \sum_{X_2} P(X_1 \mid X_2; \mathcal{H})\ P(X_2; \mathcal{H}),$$

which is the law of total probability. This law is also known as the **law of the extension of conversation** because it suggests that when required to assess our uncertainty about X_1 in the light of $\mathcal{H}$ alone, we may find it easier to assess X_1 if, in addition to $\mathcal{H}$, X_2 were also known; that is, we have extended our discussion from $\mathcal{H}$ to both X_2 and $\mathcal{H}$. As an illustration of the use of this law, suppose that a scientist was asked to assess the temperature at a certain spot on another planet given all the relevant information $\mathcal{H}$ that we currently have. Now it is possible that the scientist may find it easier to assess the temperature if additional information X_2, say the thermometer readings on a space probe, were also available. However, since X_2 is not available, we weight the scientist's assessment of X_1 under both X_2 and $\mathcal{H}$, by the last term of the above law, and sum over all possible values that X_2 can take. The law of total probability plays a key role in our development of probability models; this will be seen later, in section 2.6.

2.4.3 Bayes' Law: The Incorporation of Evidence and the Likelihood

Experimentation, observation and data collection are some of the most important tools of science. Their purpose is to expand the background information because quantities that were previously unknown now become known. Let X_1 and X_2 be two unknown quantities, and suppose that the background information is $\mathcal{H}$. Let $P(X_1,\ X_2; \mathcal{H})$ describe our uncertainty about X_1 and X_2, and suppose now that the previously unknown quantity X_2 were to become known, but in actuality, it is not known. Then by an application of the multiplication rule, our appreciation of X_1, were we to know X_2 and also $\mathcal{H}$, would be given by

$$P(X_1 \mid X_2; \mathcal{H}) = \frac{P(X_1, X_2; \mathcal{H})}{P(X_2; \mathcal{H})}.$$

However, by the marginalization rule, $P(X_2; \mathcal{H}) = \sum_{X_1} P(X_1, X_2; \mathcal{H})$, and so

$$P(X_1 \mid X_2; \mathcal{H}) = \frac{P(X_1, X_2; \mathcal{H})}{\sum_{X_1} P(X_1, X_2; \mathcal{H})}.$$

The above relationship provides a prescription for passing from our original uncertainty about both X_1 and X_2 to only that of X_1, were X_2 to become known. It tells us how we should incorporate the observation X_2 into our appreciation of X_1. Applying the multiplication rule to the numerator and the denominator of the above expression, we get

$$P(X_1 \mid X_2; \mathcal{H}) = \frac{P(X_2 \mid X_1; \mathcal{H})\, P(X_1; \mathcal{H})}{\sum_{X_1} P(X_2 \mid X_1; \mathcal{H})\, P(X_1; \mathcal{H})} \tag{2.1}$$

which is known as Bayes' Rule, or the **law of inverse probability**; the latter name is a consequence of the fact that the positions of X_1 and X_2 get reversed as we go from the left-hand side of (2.1) to its right-hand side.

Bayes' Rule, as another rule of probability, is a straightforward mathematical result. In essence, the rule also follows as a theorem from the assumption of coherence and the subjective interpretation of conditional probability; however, Hacking (1975) disagrees and sees the law as an assumption. Since the denominator of (2.1) can be interpreted as a normalizing constant, i.e. a constant whose role is to ensure that $P(X_1 \mid X_2; \mathcal{H})$ is between zero and one, we may rewrite (2.1) as

$$P(X_1 \mid X_2; \mathcal{H}) \propto P(X_2 \mid X_1; \mathcal{H})\, P(X_1; \mathcal{H}), \tag{2.1a}$$

where '$\propto$' indicates proportionality.

A use of Bayes' Rule as a vehicle of statistical inference and data analysis (see, for example, Gelman *et al.*, 1995) has raised issues that are the subject of much debate. To indicate the nature of some aspects of this debate, suppose that X_2 has in actuality revealed itself as say x_2. Then what can we say about X_1 in the light of x_2 and $\mathcal{H}$? Alternatively put, how should we update our quantification of uncertainty about X_1 from $P(X_1; \mathcal{H})$ to $P(X_1; x_2, \mathcal{H})$?

One possibility is to simply re-assess X_1 in light of both x_2 and $\mathcal{H}$. That is, to make x_2 a part of $\mathcal{H}$. Let us denote this assessment as $P^*(X_1, \mathcal{H})$. The second possibility is based on the notion that prior to observing x_2 we had said – using Bayes' Rule – that were X_2 to reveal itself as x_2, we would use (2.1) to describe our uncertainty about X_1. We are now obliged to do what we have said we will do, and so now

$$P(X_1 = x_1; x_2, \mathcal{H}) \propto P(X_2 = x_2 \mid X_1 = x_1; \mathcal{H})\, P(X_1 = x_1; \mathcal{H}). \tag{2.2}$$

When $P^*(X_1, \mathcal{H}) = P(X_1 = x_1; x_2, \mathcal{H})$, we say that the **principle of conditionalization** has been invoked (Howson and Urbach, 1989, p. 68). Thus implicit in the use of Bayes' Law, (2.1) is an adherence to the principle of conditionalization. However, it is not mandatory that the principle of conditionalization be upheld. There is precedence from quantum mechanics for violating this principle. Specifically, in the famous 'double slit experiment'[2] it has been experimentally shown that the probability of some event, say B, when an event A always occurs is not equal to the conditional probability of B given A found from an experiment in which A occurs in some replications and the complement of A occurs in other replications.

Equation (2.2) relates the two uncertainties about X_1, $P(X_1 = x_1; \mathcal{H})$ which is prior to observing x_2 and $P(X_1 = x_1; x_2, \mathcal{H})$ which is posterior to observing x_2, via the term $P(X_2 = x_2 \mid X_1 = x_1; \mathcal{H})$. Since X_2 has been observed as x_2, and since probability makes sense only for unknown entities, $P(X_2 = x_2 \mid X_1 = x_1; \mathcal{H})$ should not be interpreted as a probability; consequently, it need not obey the rules of probability given in section 2.2.1. It is for this reason that $P(X_2 = x_2 \mid X_1 = x_1; \mathcal{H})$ is called the likelihood of $X_1 = x$, which for a fixed (observed) value of $X_2 = x_2$. The likelihood is to be interpreted as a weight that is assigned to $P(X_1 = x_1; \mathcal{H})$, our uncertainty about X_1 (prior to observing x_2), for describing our uncertainty about X_1 subsequent to observing x_2. The likelihood shows us the role of the data (new information) in relating our prior and posterior uncertainties about X_1. Since X_1 is unknown, and X_2 is observed as x_2, $P(X_2 = x_2 \mid X_1 = x_1; \mathcal{H})$ can be viewed as a function of x_1 for a fixed x_2 (and $\mathcal{H}$). This function is known as the **likelihood function** of X_1 for a fixed x_2, and can be interpreted as a weight function. It is convenient to denote the likelihood function $P(X_2 = x_2 \mid X_1 = x_1; \mathcal{H})$ by $\mathcal{L}(x_1; x_2, \mathcal{H})$. The likelihood, originally noted by Gauss and rediscovered by Fisher, can be interpreted as a scale of comparative support lent by the known x_2 to the various possible values of the unknown X_1 (Singpurwalla, 2002). Finally note that, in the above, it is only x_2 that matters and not the other possible values that X_2 could have taken; these are irrelevant. For this reason, the likelihood is said to be sufficient. The above feature constitutes the essence of what is known as the likelihood principle, more about which is in Berger and Wolpert (1988).

2.5 THE BAYESIAN PARADIGM: A PRESCRIPTION FOR RELIABILITY, RISK AND SURVIVAL ANALYSIS

The Bayesian paradigm for statistical inference and noncompetitive decision making is straightforward to enunciate. Essentially, it is a probabilistic view of the world which says that all uncertainty should only be described by probability and its calculus, and that probability is personal or subjective. Statisticians who subscribe to this paradigm are called Bayesians, and Bayesian statistics (or procedures) are statistical techniques developed in adherence to the Bayesian paradigm.

The formal use of Bayes' law is simply one aspect of the Bayesian paradigm. A mere use of this law or a use of background information does not necessarily imply an adherence to the Bayesian paradigm. A deeper characterization of the Bayesian paradigm is the exploitation and systematic use of the concept of subjective or personal probability. This provides a framework for statistics and decision making that is complete, logical and unified. Adherence to this paradigm eliminates from consideration other ways of describing uncertainty, such as by confidence limits, maximum likelihood estimation, significance levels and hypothesis testing with Type I and Type II errors, not to mention alternate ways of describing and combining uncertainties such as by Zadeh's (1979) possibility theory, by *Jeffrey's Rule of Combination* (R. Jeffrey, 1965) and by *Dempster's Rule of Combination* (Dempster, 1968). Some of the above procedures, in considering data that would have been observed but was not, do not adhere to the likelihood principle.

---

[2] I wish to thank Professor J.K. Ghosh for drawing my attention to the double split experiment.

Why is the above paradigm relevant for problems of reliability, risk and survival analysis? From a philosophical point of view, the answer is a natural one: that the Bayesian paradigm is founded on the logical framework of the calculus of probability. From a pragmatic point of view, one may say that, in risk analysis, we are often dealing with one-of-a-kind situations so the notion of relative frequency is not always relevant (Bement, Booker and Singpurwalla, 2002; and Singpurwalla, 2002). Another argument is that, many times, there are no direct previous data to go by so all assessments of uncertainty can only be based on background information alone; the Bayesian paradigm allows for this. Finally, risk, reliability and survival analyses are most credibly performed when subject matter experts are required to play a key role; the Bayesian paradigm enables the formal incorporation of the expertise of the experts into the analysis, via a consideration of prior probabilities.

The point of view that we take here is Bayesian. However, we want to qualify what we are attempting to do with some caution. Even among those who view probability as being subjective, there have been concerns about the appropriateness of Bayes' law as a mechanism for 'changing one's mind'. Many of these concerns have to do with the adequacy of conditioning as an exclusive model for belief revision. Difficulties in updating when new information arises as a result of introspection, unanticipated knowledge or probable knowledge have been cited. Diaconis and Zabel (1982), Lane (1987) and Shafer (1981), among others, have written on such topics and related issues. Some of these concerns go to the heart of the subject by questioning the nature of the Kolmogorov axioms and the rules of probability (section 2.2.1). Under debate is also the issue of whether probability needs to be countably additive or only finitely additive. A consequence is compatibility with improper probabilities, unless the axiom of conglomerability gets introduced (section 2.2.1). In de Finetti's theory of personal probability, all that is needed is finite additivity. Frequentists do not always describe uncertainty by probability, but subscribe to the Kolmogorov axiomatization of uncertainty. The above difficulties, plus the normative nature of Bayesian inference, have provided many statisticians a reason for not adopting the Bayesian paradigm; see, for example, Efron (1986). The concerns of many of these authors are worthy of consideration.

The material that follows is foundational. It provides a deeper appreciation of some of the most commonly used techniques in reliability and survival analysis. It should not be viewed as being detracting.

2.6 PROBABILITY MODELS, PARAMETERS, INFERENCE AND PREDICTION

Much of the statistical analyses that one encounters, save that which is called 'exploratory data analysis' and 'computational statistics', pertain to inference about 'parameters'; these are generally denoted by Greek symbols. This is generally true of both Bayesian and frequentist statistics, and more so with the latter. By statistical inference, we mean statements of uncertainty; these may entail estimation and the testing of hypotheses. Parameters appear in 'probability models' which have played a key role in both reliability and survival analysis. Examples of some well-known probability models for continuous random quantities and their parameters are: the normal distribution with location μ and scale σ; the uniform distribution on (α, β); the exponential distribution with mean θ; the Weibull distribution with scale η and shape β, and so on. In this section, we address the important question, where do probability models come from? With the computer revolution, exploratory data analysis and computational statistics may be paradigms of the future. However, as of now, they are a form of art, intuition and innovation. They are therefore not a part of our discussion.

2.6.1 The Genesis of Probability Models and Their Parameters

We start by considering an uncertain quantity X with background information $\mathcal{H}$; our aim is to specify $P(X; \mathcal{H})$. In general, the dimensions of X and $\mathcal{H}$ could be very large, particularly of $\mathcal{H}$, since it includes all that we know. We therefore seek ways of simplification. For this, we may appeal to the law of the extension of conversation and write that, for some unknown quantity θ which we for now assume to take discrete values,

$$P(X; \mathcal{H}) = \sum_{\theta} P(X \mid \theta; \mathcal{H})\ P(\theta; \mathcal{H}).$$

If we suppose that X is independent of $\mathcal{H}$, were we to know θ, that is if $P(X \mid \theta; \mathcal{H}) = P(X \mid \theta)$ abbreviated '$X \perp \mathcal{H}$, given θ', then the above expression becomes

$$P(X; \mathcal{H}) = \sum_{\theta} P(X \mid \theta)\ P(\theta; \mathcal{H}); \tag{2.3}$$

θ could be vector-valued, and, if it is continuous, an integral would replace the summation. When $P(X \mid \theta; \mathcal{H}) = P(X \mid \theta)$ for all values of θ, X is said to be conditionally (given θ) independent of $\mathcal{H}$.

The term $P(X \mid \theta)$ is known as a probability model for X with a parameter θ, and the term $P(\theta; \mathcal{H})$ is known as the prior distribution for θ. If the dimensions of θ are smaller than the dimensions of $\mathcal{H}$, then the specification of the probability model is a simpler task than the specification of $P(X; \mathcal{H})$. The assumption of independence is crucial, since it says that were we to know θ, we need not keep track of $\mathcal{H}$. It is important to note that, like probability, independence is also a judgment; it is always conditional (in our case, conditional on θ).

Probability models play a central role in statistic, both frequentist and Bayesian. There are no restrictions on the functional form that $P(X \mid \theta)$ can take, other than the fact that it must obey the rules of probability. The choice of a model is subjective, and from the point of view of personal probability, the notion of a 'true probability model' is as elusive as that of a 'right bet'. Often, our choice of the model is dictated by the nature of the problem that is being studied. For example, in reliability and biometry, where probability models are called failure models or lifelength models, the notion of 'aging' is used to specify a model, or sometimes the model is based on a probabilistic description of the physics of failure of the item (cf. Singpurwalla, 1988a; 2002). Often, models such as the Bernoulli, the multinomial, the Poisson, etc., are justified as 'chance distributions' that arise as a consequence of the judgment of 'exchangeability' (Chapter 3). More recently, the 'principle of indifference' (Barlow and Mendel, 1992) has been used to motivate the use of failure models such as the exponential, the gamma and the Weibull. The principle of indifference has its roots in the work of Laplace, who used the notion of 'insufficient reason' to convert ignorance to a uniform distribution of prior probabilities. Non-parametric analysis pertains to the scenario wherein one is reluctant to specify the probability model $P(X \mid \theta)$; Hollander and Wolfe (1972) give a readable account of such analysis.

Regarding the parameter θ, we have seen that it was introduced as a mathematical device with the aim of simplifying our probability specifications. However, it was understood that even though θ may not be observable, as typically is the case, it influences the observable X, upon which bets can be made and gambles settled. The above interpretation of a parameter, as a summarization of the background information $\mathcal{H}$, though simple to appreciate, is naive. A more satisfying interpretation of parameters (cf. Hill, 1993) comes from de Finetti's theorem about 'exchangeable sequences', a topic discussed in Chapter 3; there, parameters are seen as limits of some functions of observable sequences. Section 4.4 contains a discussion about the nature of the units of measurement for parameters in probability models.

Observe that in (2.3), the prior distribution $P(\theta; \mathcal{H})$ has arisen naturally, as a consequence of the laws of probability. Whereas subjectivity in the specification of probability models has not been the center of debate, the specification of the prior has been; it is the bone of contention for many frequentists. Consequently, they have chosen to drop the last term of (2.3) and to base their analysis on the probability model alone; see Singpurwalla (2001) for a discussion of the genesis of failure models in a frequentist's approach to reliability analysis. Several strategies for choosing the prior have been proposed (cf. Berger, 1985; or Kass and Wasserman, 1996) and these range from choosing the 'natural conjugate priors' (section 5.4.7) to priors that are 'objective', such as the reference priors, the indifference priors, robust priors and so on (sections 5.2 and 5.3). The dialogue with Bernardo (1997) provides an informative overview. Whereas such 'neutral' priors used by none other than Bayes, Laplace and Jeffreys are germane when dealing with issues pertaining to public policy and scientific inquiries, some Bayesians object to a choice of automated priors because they violate the spirit of a truly subjective approach. Instead, they advocate priors based on expert testimonies; see, for example, Lindley and Singpurwalla (1986), and the references therein.

In the above development, a probability model was motivated via a consideration of a single random quantity X. This can be extended for the case of two or more random quantities by following a parallel line of reasoning. To see how, consider two random quantities, X_1 and X_2, and suppose that we are interested in assessing our uncertainty about them in the light of $\mathcal{H}$. Then, for any random quantity θ, we have, by the law of the extension of conversation,

$$P(X_1, X_2; \mathcal{H}) = \sum_{\theta} P(X_1,\ X_2 \mid \theta; \mathcal{H})\, P(\theta; \mathcal{H}),$$

which by the multiplication rule of probability can be written as

$$P(X_1, X_2; \mathcal{H}) = \sum_{\theta} P_1(X_1 \mid X_2, \theta; \mathcal{H})\, P_2(X_2 \mid \theta; \mathcal{H})\, P(\theta; \mathcal{H}),$$

where the suffixes associated with the Ps reflect the fact that assessed probabilities could have different functional forms. If we assume, as is commonly done, that $X_2 \perp \mathcal{H}$, given θ, and that $X_1 \perp (X_2 \perp \mathcal{H})$, given θ, then simplification results. As a consequence, we have

$$P(X_1,\ X_2; \mathcal{H}) = \sum_{\theta} P_1(X_1 \mid \theta)\, P_2(X_2 \mid \theta)\, P(\theta; \mathcal{H}).$$

Observe that the role of θ has been to stand not only between X_2 and $\mathcal{H}$ and give them the property of independence, but also between X_2 and X_1 and make them independent. In essence, these assumptions imply that, were θ to be known, the uncertainty of any random quantity can be described without an appreciation of the background information $\mathcal{H}$ or the revealed values of the other random quantities.

A further simplification that is often made in practice is to assume that $P_i(X_i \mid \theta) = f(X_i \mid \theta)$, $i = 1, 2$; that is, X_1 and X_2 are 'identically distributed'. When this is done,

$$P(X_1, X_2; \mathcal{H}) = \sum_{\theta} f(X_1 \mid \theta)\, f(X_2 \mid \theta)\, P(\theta; \mathcal{H}),$$

and generalizing to the case on n random quantities $X^{(n)} = (X_1, X_2, \ldots, X_n)$, we have

$$P(X^{(n)}; \mathcal{H}) = \sum_{\theta} \prod_{i=1}^{n} f(X_i \mid \theta)\, P(\theta; \mathcal{H}) \tag{2.4}$$

which is a convenient starting point for discussing inference about θ and predictions about observables. An expansion of the $P(\theta; \mathcal{H})$ of (2.4) via the total law of probability produces what are known as hierarchical models (section 6.1). Such models have proved to be useful for addressing a large class of problems in reliability.

2.6.2 Statistical Inference and Probabilistic Prediction

With reference to (2.4), the frequentist approach to statistics focuses attention only on $\prod_{i=1}^{n} f(X_i \mid \theta)$ and generally emphasizes estimation and hypotheses testing about θ. By contrast, the Bayesian approach considers the entire right-hand side of (2.4), and emphasis is placed on $P(X_{n+1}; x^{(n)}, \mathcal{H})$, the **predictive distribution** of the future observable, in light of $x^{(n)} = (x_1, \ldots, x_n)$, where x_i is the observed value of X_i, $i = 1, \ldots, n$, and $\mathcal{H}$. The predictive distribution is a statement of uncertainty about a future observation, in light of all the previous information including the data $x^{(n)}$. However, in order to be able to assess $P(X_{n+1}; x^{(n)}, \mathcal{H})$ efficiently, we need to obtain $P(\theta; x^{(n)}, \mathcal{H})$, the **posterior distribution** of θ in the light of $x^{(n)}$ and $\mathcal{H}$. The posterior distribution of θ is a statement of uncertainty about θ, posterior to $x^{(n)}$, and in the Bayesian paradigm is indeed the estimation of θ. Furthermore, as we shall see later in section 2.7, $P(\theta; x^{(n)}, \mathcal{H})$ plays a key role for testing hypotheses about θ.

To see how the above works, we start with a consideration of the quantity $P(X_{n+1} \mid X^{(n)}; \mathcal{H})$ and note that by the law of the extension of conversation and the usual assumptions of independence

$$P(X_{n+1} \mid X^{(n)}; \mathcal{H}) = \sum_{\theta} f(X_{n+1} \mid \theta)\, P(\theta \mid X^{(n)}; \mathcal{H}) \tag{2.5}$$

where the first term on the right-hand side of (2.5) is a consequence of the assumption that $X_{n+1} \perp X^{(n)}$ given θ. In order to assess $P(\theta \mid X^{(n)}; \mathcal{H})$, we invoke Bayes' law so that

$$P(\theta \mid X^{(n)}; \mathcal{H}) \propto \prod_{i=1}^{n} f(X_i \mid \theta)\, P(\theta; \mathcal{H}). \tag{2.6}$$

When the X_i's get revealed as x_i, $i = 1, \ldots, n$, $X^{(n)}$ gets replaced by $x^{(n)}$, and the left-hand terms of (2.5) and (2.6) become $P(X_{n+1}; x^{(n)}, \mathcal{H})$ and $P(\theta; x^{(n)}, \mathcal{H})$, respectively. However, the product term on the right-hand side of (2.6) is no longer a probability. Rather, it is the likelihood $\mathcal{L}(\theta; x^{(n)})$, which, as a function of θ, is the likelihood function of the unknown θ for a fixed $x^{(n)}$. Thus, to summarize, the predictive distribution of X_{n+1} in light of $x^{(n)}$ and $\mathcal{H}$ is given as

$$P(X_{n+1}; x^{(n)}, \mathcal{H}) \propto \sum_{\theta} f(X_{n+1} \mid \theta)\, \mathcal{L}(\theta; x^{(n)})\, P(\theta; \mathcal{H}) \tag{2.7}$$

so that, once the prior distribution $P(\theta; \mathcal{H})$ and the likelihood function $\mathcal{L}(\theta; x^{(n)})$ are specified, the problem of inference about θ and prediction of X_{n+1} can be addressed in a unified manner. It has been a common practice to specify the function $\mathcal{L}(\theta; x^{(n)})$ via the functional form that is adopted for $f(X_{n+1} \mid \theta)$. For example, if $f(x \mid \theta) = \theta e^{-\theta x}$, then $\mathcal{L}(\theta; x^{(n)})$ is taken to be $\theta^n \exp(-\theta \sum_{i=1}^{n} x_i)$. However, there is nothing in the subjectivistic theme which requires that the likelihood be based on the probability model. That is, an adherence to the principle of conditionalization is not mandatory (section 2.4.3). Since the likelihood function describes the relative support that the data $x^{(n)}$ provides to the various values of θ (in the opinion of the assessor of probabilities), one is free to choose any functional form for $\mathcal{L}(\theta; x^{(n)})$. The symmetric form illustrated above is one of convenience. This flexibility, namely that of being able to divorce the likelihood from the model, is not available within the frequentist paradigm (Singpurwalla, 2002).

The development leading to (2.7) above can be easily generalized to cover the case of predicting m future observations, $X_{n+1}, \ldots, X_{n+m}$, in light of $x^{(n)}$ and $\mathcal{H}$. It is easy to verify – the details left as an exercise for the reader – that

$$P(X_{n+1}, \ldots, X_{n+m}; x^{(n)}, \mathcal{H}) \propto \sum_{\theta} \prod_{n+1}^{m} f(X_i \mid \theta)\ \mathcal{L}(\theta; x^{(n)})\ P(\theta; \mathcal{H}).$$

The predictive approach in Bayesian statistical inference has been comprehensively treated by Geisser (1993); also see San Martini and Spezzaferri (1974).

Before closing this section, we need to highlight two important matters. The first pertains to the fact that the predictivity of X_{n+1} given $X^{(n)}$ – see (2.5) – assumes that all the observations bear a relationship to each other, so that a knowledge of $X^{(n)}$ enhances an appreciation of X_{n+1}. The notion that makes this idea concrete is that of exchangeability; this notion, overviewed in Chapter 3, is due to de Finetti. For now it suffices to say that the random quantities $X_1, \ldots, X_{n+1}$ are judged exchangeable if they can be represented in the form of (2.5). Like the notion of independence, exchangeability is a judgment. The second matter pertains to the fact that in (2.6), if there was any value of θ, say θ^*, for which $P(\theta^*; \mathcal{H})$ was zero, then no amount of evidence provided by $x^{(n)}$ would be able to change the result that $P(\theta^*; x^{(n)}, \mathcal{H})$ should also be zero. To avoid this scenario (namely an inability to change one's opinion in light of glaring evidence), zero prior probabilities should not be assigned to discrete quantities – unless this is dictated by a logical argument – and the probability density should nowhere vanish for quantities that are continuous. Lindley (1982a) refers to this feature of an unwillingness to assign zero probabilities as an adherence to Cromwell's Rule.

2.7 TESTING HYPOTHESES: POSTERIOR ODDS AND BAYES FACTORS

The statistical testing of a hypothesis is usually done to verify a theory or a claim, be it in science, engineering, law or medicine, in light of evidence or data. We have said that reliability and risk analysis pertains to decision making under uncertainty. Then why should we be interested in the topic of testing hypotheses?

There could be many answers to the above question, but the one that immediately comes to mind pertains to the fact that associated with each action of a decision tree are consequences, and a consequence could have been entertained as the result of verifying a claim. For example, in the cholesterol drug problem of Chapter 1, we consider administering a drug because we believe that the drug has the potential of avoiding a heart attack. How did we arrive upon this belief? It is most likely that an experiment was conducted on several individuals, some of whom received the drug and some did not, and the results of this experiment provided evidence to certify the claim that the drug was effective in avoiding a heart attack. The certification of this claim could have been based on the test of a hypothesis. Similarly, the claim that the drug has minimal side effects could have been based on the test of suitable hypotheses. Other such examples of hypotheses testing in reliability and risk analyses are claims about the improved performance of an engine in a new design, claims that emergency diesel generators in nuclear power plants have a reliability that exceeds requirements (cf. Chen and Singpurwalla, 1996), claims that a piece of computer software is free of bugs (cf. Singpurwalla and Wilson, 1999) and so on. To summarize, the testing of statistical hypotheses is a part of decision making under uncertainty and as such plays an important role in reliability, risk and survival analysis. Indeed, one of the most visible exports of statistical reliability theory is the Military Standard 781-C, which is used worldwide for life-testing and acceptance sampling. Its theoretical basis is in the testing of hypothesis about the mean of an exponential distribution (cf. Montagne and Singpurwalla, 1985).

Because the testing of hypothesis is an important scientific problem, much has been written about it, starting from the days of Laplace. Lehmann (1950) gives an authoritative account of the frequentist treatment of this topic. However, Berger and Berry (1988) are critical of frequentist methods for testing hypotheses, their criticism centering upon a violation of the likelihood principle. A review article by Berger and Delampady (1987) proposes Bayesian alternatives. The book by Lee (1989), from which much of the material that follows is taken, provides a nice account of Bayesian hypothesis testing.

We begin with the following setup. Consider an unknown quantity X, discrete or continuous, and ignoring technicalities, suppose that $P(X \mid \theta)$ is a suitable probability model for X, where the parameter θ belongs to the set Θ; i.e. $\theta \in \Theta$. Let $P(\theta; \mathcal{H})$ be our prior distribution for θ, assuming that θ is discrete. Suppose that $\Theta = \Theta_0 \cup \Theta_1$, with $\Theta_0 \cap \Theta_1 = \emptyset$, where $\cup$, $\cap$ and $\emptyset$ denoting union, intersection and the empty (null) set, respectively. That is, the parameter set Θ is partitioned into two non-overlapping sets of Θ_0 and Θ_1. Suppose now that X has revealed itself as x. The problem that we wish to entertain is that, given x and $\mathcal{H}$, does $\theta \in \Theta_0$ or does $\theta \in \Theta_1$? In the context of any particular application, $\theta \in \Theta_0$ could correspond to the validity of a claim, and $\theta \in \Theta_1$, its negation, i.e. the falsity of the claim.

The premise behind a Bayesian approach to testing a hypothesis is that it is unreasonable, except in rare situations, that x gives us conclusive evidence about the disposition of θ. However, x could give us evidence that enhances our prior opinion about H_0, the null hypothesis, that $\theta \in \Theta_0$, or about H_1, the alternate hypothesis, that $\theta \in \Theta_1$. Bayes' law enables us to incorporate the evidence provided by x. To see how, let $\pi_0 = P(\theta \in \Theta_0; \mathcal{H})$ and $\pi_1 = 1 - \pi_0$ be our prior probabilities, and consider the quantity π_0/π_1, which is our prior odds on H_0 against H_1; the prior probabilities are obtained via our prior distribution $P(\theta; \mathcal{H})$. The notion of odds is useful because if the prior odds is close to one, we regard H_0 and H_1 to be equally likely, whereas if the ratio is large, then we regard H_0 to be more likely than H_1; vice versa if the ratio is small.

Given x, Bayes' law is used to compute the posterior probabilities $p_0 = P(\theta \in \Theta_0; x, \mathcal{H}) \propto \mathcal{L}(\theta \in \Theta_0; x, \mathcal{H})\pi_0$, and $p_1 = P(\theta \in \Theta_1; x, \mathcal{H}) \propto \mathcal{L}(\theta \in \Theta_1; x, \mathcal{H})\pi_1$, where $\mathcal{L}(\theta \in \Theta_0; x, \mathcal{H})$ is the likelihood that $\theta \in \Theta_0$, in light of x and $\mathcal{H}$. The posterior odds p_0/p_1 are analogously interpreted as the prior odds. It is easy to verify that

$$\frac{p_0}{p_1} = \frac{\mathcal{L}(\theta \in \Theta_0; x, \mathcal{H})}{\mathcal{L}(\theta \in \Theta_1; x, \mathcal{H})} \frac{\pi_0}{\pi_1}.$$

The above development was initiated by Jeffreys in the 1920s (Jeffreys, 1961) as an approach to testing hypotheses.

2.7.1 Bayes Factors: Weight of Evidence and Change in Odds

The absence of evidence is not evidence of absence.

Simple Hypotheses

Suppose that the two hypotheses are simple; i.e.$\Theta_0 = \{\theta_0\}$ and $\Theta_1 = \{\theta_1\}$, for some singletons $\theta_0 \neq \theta_1$. Then the posterior odds on H_0 against H_1 will be of the form

$$\frac{p_0}{p_1} = \frac{\mathcal{L}(\theta_0; x, \mathcal{H})}{\mathcal{L}(\theta_1; x, \mathcal{H})} \frac{\pi_0}{\pi_1},$$

implying that the posterior odds are the prior odds multiplied by the middle term, which is called the Bayes factor $\mathcal{B}$ in favor of H_0 against H_1. Thus $\mathcal{B}$ is simply the ratio of the likelihoods under H_0 and H_1. Alternatively,

$$\mathcal{B} = \frac{p_0/p_1}{\pi_0/\pi_1} = \frac{\text{Posterior odds (on } H_0 \text{ against } H_1)}{\text{Prior odds (on } H_0 \text{ against } H_1)};$$

this terminology is due to Good (1950), who attributed the method to Turing, in addition to and independently of Jeffreys.

If, $\pi_0 = \pi_1 = 1/2$, then $\pi_0/\pi_1 = 1$, and now $\mathcal{B} = p_0/p_1$, the posterior odds on H_0 against H_1. Since $\mathcal{B}$ is the ratio of likelihoods (when the hypotheses are simple), the Bayes factor gives the odds on H_0 against H_1, in light of the data x.

Composite Hypotheses

When θ is continuous, one (or both) of the two hypotheses H_0 and H_1 will be composite. Composite hypotheses are of interest in reliability and life-testing whenever concerns center around items satisfying requirements, like the mean time to failure should exceed a specified number, or the failure rate should not exceed a specified number, and so on. Indeed MIL-STD-781C mentioned before pertains to the testing of composite hypotheses.

Suppose that θ is continuous and let $f(\theta; \mathcal{H})$ denote its probability density at θ. Then, the prior probabilities π_0 and π_1, mentioned before, are:

$$\pi_0 = \int_{\theta \in \Theta_0} f(\theta; \mathcal{H}) \mathrm{d}\theta, \text{ and } \pi_1 = \int_{\theta \in \Theta_1} f(\theta; \mathcal{H}) \mathrm{d}\theta.$$

Let $\rho_0(\theta)$ and $\rho_1(\theta)$ denote the restriction of $f(\theta; \mathcal{H})$ on Θ_0 and Θ_1, respectively, re-normalized so that they are probability density functions. That is,

$$\rho_0(\theta) = \frac{f(\theta; \mathcal{H})}{\pi_0}, \text{ for } \theta \in \Theta_0, \text{ and}$$

$$\rho_1(\theta) = \frac{f(\theta; \mathcal{H})}{\pi_1}, \text{ for } \theta \in \Theta_1.$$

Given x, the posterior probability of H_0 is

$$\begin{aligned} p_0 = P(\theta \in \Theta_0; x, \mathcal{H}) &\propto \textstyle\int_{\theta \in \Theta_0} \mathcal{L}(\theta; x, \mathcal{H})\ f(\theta; \mathcal{H}) \mathrm{d}\theta \\ &= \textstyle\int_{\theta \in \Theta_0} \mathcal{L}(\theta; x, \mathcal{H})\ \rho_0(\theta) \pi_0 \mathrm{d}\theta. \end{aligned}$$

Similarly,

$$p_1 \propto \int_{\theta \in \Theta_1} \mathcal{L}(\theta; x, \mathcal{H}) \rho_1(\theta) \pi_1 \mathrm{d}\theta,$$

so that the posterior odds on H_0 against H_1 is

$$\frac{p_0}{p_1} = \frac{\pi_0}{\pi_1} \frac{\int_{\theta \in \Theta_0} \mathcal{L}(\theta; x, \mathcal{H}) \rho_0(\theta) \mathrm{d}\theta}{\int_{\theta \in \Theta_1} \mathcal{L}(\theta; x, \mathcal{H}) \rho_1(\theta) \mathrm{d}\theta}.$$

Thus in the case of composite hypotheses, the Bayes factor $\mathcal{B}$ in favor of H_0 against H_1 is

$$\mathcal{B} = \frac{p_0/p_1}{\pi_0/\pi_1} = \frac{\int_{\theta \in \Theta_0} \mathcal{L}(\theta; x, \mathcal{H}) \rho_0(\theta) \mathrm{d}\theta}{\int_{\theta \in \Theta_1} \mathcal{L}(\theta; x, \mathcal{H}) \rho_1(\theta) \mathrm{d}\theta}.$$

The above suggests that, in the case of composite hypotheses, the Bayes factor is the ratio of **weighted likelihoods**, with weights $\rho_0(\theta)$ and $\rho_1(\theta)$. This is in contrast to the structure of the Bayes factor when the hypotheses are simple; there, $\mathcal{B}$ was solely determined by the data (and the assumed model) irrespective of the prior. In the composite hypotheses case, the prior enters into the construction of the Bayes factor via the weights $\rho_0(\theta)$ and $\rho_1(\theta)$; it is not solely determined by the data.

Point (or Sharp) Null Hypotheses

Point null hypotheses are particularly useful when one wishes to test a theory or a claim, such as 'aspirin cures headaches'. Other examples were given at the beginning of this section. Point null hypotheses are characterized by the fact that even though θ can take continuous values, Θ_0 is simple (say $\Theta_0 = \{\theta_0\}$), and Θ_1 is the complement of Θ_0. The prior distribution of θ, $f(\theta; \mathcal{H})$, is such that a point mass $\pi_0 \geq 0$ is assigned to θ_0 and the rest, $\pi_1 = 1 - \pi_0$, is spread over the remaining values of $\theta(\neq \theta_0)$ according to a probability density function $\pi_1\rho_1(\theta)$, where $\rho_1(\theta)$ integrates to one. It is usual to choose $\pi_0 = 1/2$, and $\rho_1(\theta)$ to be uniform, so that all the values of θ, save θ_0, receive equal prior probability (cf. Lindley, 1957a).

Given x, the Bayesian approach for testing $H_0 : \theta = \theta_0$ versus the alternative $H_1 : \theta \neq \theta_0$, proceeds along the lines outlined before. However, in this case the Bayesian conclusions often differ quite substantially from those obtained by frequentist methods. This disparity has sometimes been referred to as the 'Jeffreys Paradox' (cf. Jeffreys, 1961) or as 'Lindley's Paradox' (cf. Shafer, 1982a). The discussion by Hill (1982) in Shafer (1982a) provides insights about the nature of this paradox.

Verify that the posterior probabilities for H_0 and H_1 are:

$$p_0 = \frac{\mathcal{L}(\theta_0; x, \mathcal{H})\pi_0}{\mathcal{L}(\theta_0; x, \mathcal{H})\pi_0 + \ell_1(x)\pi_1} = \frac{\mathcal{L}(\theta_0; x, \mathcal{H})\pi_0}{\ell(x)}, \text{ and}$$

$$p_1 = \frac{\pi_1\ell_1(x)}{\ell(x)}, \text{ where}$$

$$\ell_1(x) = \int \rho_1(\theta)\mathcal{L}(\theta; x, \mathcal{H})\mathrm{d}\theta \text{ and } \ell(x) = \mathcal{L}(\theta_0; x, \mathcal{H})\pi_0 + \ell_1(x)\pi_1.$$

Thus, for the case of a sharp null hypothesis, the Bayes factor $\mathcal{B}$ is of the form

$$\mathcal{B} = \frac{p_0/p_1}{\pi_0/\pi_1} = \frac{\mathcal{L}(\theta_0; x, \mathcal{H})}{\ell_1(x)}.$$

2.7.2 Uses of the Bayes Factor

Good (1950) refers to the logarithm of the Bayes factor as the weight of evidence. His motivation for considering logarithms is that, if we have several experiments pertaining to the testing of two simple hypotheses, then the Bayes factors multiply whereas the weights of evidence add. To see how, consider the two simple hypotheses of section 2.7.1, and suppose that the observed data is x_1. Then, the posterior odds corresponding to x_1 are given by

$$\frac{p_0}{p_1} = \mathcal{B}(x_1)\frac{\pi_0}{\pi_1},$$

where $\mathcal{B}(x_1)$ is the Bayes factor based on x_1. Now suppose that subsequent to observing x_1, we observe x_2. The posterior odds corresponding to x_2 will be $\mathcal{B}(x_2)\ p_0/p_1 = \mathcal{B}(x_2)\ \mathcal{B}(x_1)\ \pi_0/\pi_1$. The Bayes factors multiply, but by taking the logarithms of the prior and the posterior odds, the weights of evidence would add.

From the material of sections 2.7.1 and 2.7.2, we have seen that

$$\text{posterior odds} = \text{Bayes factor} \times \text{prior odds}.$$

The above feature has motivated some to claim that the Bayes factor is a summary of the evidence provided by the data in favor of one scientific theory (represented by a statistical model) as opposed to another (cf. Kass and Raftery, 1995). Indeed Jeffreys suggests using $\log_{10}(\mathcal{B}^{-1})$ as evidence against H_1 according to the following guidelines: (0 to 0.5) $\Rightarrow$ not worth more than a bare mention; (0.5 to 1) $\Rightarrow$ substantial; (1 to 2) $\Rightarrow$ strong; and (>2) $\Rightarrow$ decisive. In view of such guidelines, some Bayesians have declined to specify prior odds, and have chosen to use only Bayes factors as an alternative to the frequentist's significance probabilities. Whereas this strategy may be appropriate in the case of simple hypotheses (with $\pi_0 = \pi_1 = 1/2$), it is not so in the case of composite hypotheses because, here, the Bayes factor also depends on how the prior mass is spread over the two hypotheses see (section 2.7.1). In the latter case, the Bayes factor cannot be interpreted as a summary of the evidence provided by the data alone. Rather, the Bayes factor measures the change in the odds in favor of the hypothesis when going from the prior to the posterior; see Lindley (1997a), and Lavine and Schervish (1999) for a discussion of this and related matters. Bayes factors also play a role in the general area of model selection, model comparison and model monitoring; see the lecture notes by Chipman, George, and McCulloch (2001). Model selection and model comparison are commonly discussed issues in reliability and failure data analyses.

2.7.3 Alternatives to Bayes Factors

The foregoing discussion has assumed that the prior distributions used are proper, i.e. they integrate to one. When using Bayes factors for model choice or for tests of hypotheses, it is sometimes the case that the prior distributions used are improper. Improper distributions come into play when one wishes to adopt an impartial stance or when one claims to have little knowledge of the parameters. A consequence is that the Bayes factor contains an arbitrary ratio, say c_0/c_1, where the prior is $p_i(\theta) = c_i h_i(\theta)$, $i = 0, 1$, for a known $h_i(\theta)$, but an arbitrary, positive multiplier c_i. Thus, when calculating the Bayes factor the c_i's do not cancel, so that c_0/c_1 appears. Ways of overcoming this difficulty have been proposed by O'Hagan (1995) via his fractional Bayes factors, and by Berger and Pericchi (1996) via their intrinsic Bayes factors (IBF). A recent paper by Ghosh and Samanta (2002) provides a unified derivation of the fractional and intrinsic Bayes factors and concludes that these factors are close to each other and to certain Bayes factors based on proper priors.

It is of interest to note that when the probability model $P(X|\theta)$ has a density at x of the exponential form, namely $\theta^{-1}\exp(-x/\theta)$, and we wish to test a point (or sharp) null hypothesis at θ_0, the methodology of producing an IBF yields a proper prior, namely $f(\theta;\mathcal{H}) = \theta_0/(\theta_0+\theta)^2$ [personal communication with James Berger]. The fact that this prior depends on θ_0, the point at which the null hypothesis is positioned, makes $\theta_0/(\theta_0+\theta)^2$ non-subjective, and therefore impartial to all those whose interest centers around θ_0. In the context of estimation and prediction (section 5.3.2), this prior cannot be called objective, because using it involves anchoring it around a specific value of θ, namely θ_0.

2.8 UTILITY AS PROBABILITY AND MAXIMIZATION OF EXPECTED UTILITY

The idea of utilities and their relevance to decision making under uncertainty was introduced in the context of the cholesterol drug problem of Chapter 1, section 1.4. There, it was stated that utilities are typically functions of costs incurred as a consequence of a particular decision. In many standard business and engineering problems, it may be possible to assess meaningful costs and to transform these to utilities (section 2.8.1). However, there are many decision problems, especially those in medicine, public policy and safety, in which it is difficult to assign costs to the outcomes of a decision; examples are pain, suffering, quality of life, pleasure, etc. All the same, the insurance industry and modern legal practice have used some of these intangibles as a basis for specifying policies or for litigation, and have succeeded in assigning costs to them. The above goes to suggest that there needs to be a scale to quantify and measure preferences that are not monetary. Utilities are indeed such a scale, and they are measured in terms of the units called utiles. The aim of this section is to overview utility theory, and to argue that sensible decision making involves choosing the action that will maximize expected utility. Specifically, we shall point out that utilities are probabilities, that utility cannot be separated from probability, and that it is the laws of probability that lead us to the 'principal of maximization of expected utility'.

The theory of utility can be developed together with that of subjective probability, as was done by Ramsey (1926) and Savage (1972). A theory of utility that is separate from that of probability, as long as we are speaking of not too large sums of money, was developed by de Finetti. DeGroot (1970) gives an excellent account of the axiomatic development of utility theory; also see Hill (1978). A more recent contribution is by Rubin (1987), who sets up a weak system of axioms to argue the existence of utilities and point out that utility cannot be separated from the prior. An intuitive and readable account of the essence of utility theory is given by Lindley (1985), from which the following material has been abstracted. Readers not interested in foundational issues pertaining to the notion of utility and those who find the principle of the maximization of expected utility intuitive, may choose to skip the rest of this section and proceed directly to the next.

2.8.1 Utility as a Probability

Consider a situation of uncertainty involving m decisions $d_1, \ldots, d_m$, and n uncertain events, say $\mathcal{E}_1, \ldots, \mathcal{E}_n$. Let p_{ij} denote the probability that event $\mathcal{E}_i$ occurs when decision d_j is chosen, $i = 1, \ldots, n$, $j = 1, \ldots, m$. For example, in the cholesterol drug problem of Chapter 1, we had two decisions $\mathcal{A}$ and $(\mathcal{A})$, and the six uncertain events shown in Figure 1.1. The probability of the event $[\mathcal{HA}, \mathcal{SE}]$ when the decision chosen is $\mathcal{A}$ was denoted by p_1. Let C_{ij} be the consequence of decision d_i when outcome $\mathcal{E}_j$ occurs. An example of a consequence in the cholesterol drug problem, when the drug is administered and a heart attack occurs, together with the discomfort caused by the side effects of the drug, the cost of administering the drug, the pain of suffering a heart attack, and the difficulties due to the loss of life or a restricted lifestyle because of the attack. Clearly, this consequence is difficult to quantify in terms of cost alone because several intangibles are involved. To introduce the notion of a numerical value to the consequences, we need to have a standard and the means for coherent comparisons to it. The standard is given by two reference consequences, C and c, where C is such that no other consequence is preferable to it, and c such that no other consequence is worse than it.

Let us now focus on any consequence C_{ij} and inquire its relationship to C and c. For instance, is C_{ij} closer to C or to c? To address this issue, we consider an urn with N balls of which U are black and $N - U$ are white. Suppose that a ball is drawn from the urn at random, and if it is black then C results, and if it is white then c results. Thus we will receive C with a 'chance' (not probability) $U/N = u$, say, and c with a chance $1 - u$. We now ask, 'how does C with chance u compare with C_{ij}?' Clearly, if $u = 1$, then C with chance u is preferable (denoted $>^*$) to C_{ij}, and if $u = 0$, C with chance u, is not preferable (denoted $<^*$) to C_{ij}. As u increases, the gamble gets better, and so there is a unique value of u, say u^*, at which the two gambles (C with chance u^* and C_{ij}) are equally preferable (denoted $=^*$). To summarize, associated with every consequence C_{ij}, there is a unique number $u^* \in [0, 1]$ such that $C_{ij} =^*$ (C with chance u^* and c with chance $1 - u^*$). The number u^* [denoted $u(C_{ij})$] is called the **utility** of C_{ij}. Observe that $C_{ij} >^* C_{kl}$ implies that $u(C_{ij}) > u(C_{kl})$, and that $C_{ij} =^* C_{kl}$ implies that $u(C_{ij}) = u(C_{kl})$. Since u and u^* are like probabilities, i.e. they obey the rules of probability, we say that utilities are probabilities and follow the rules of probability. However, whereas probabilities measure our (personal) uncertainties about events, utilities measure our preferences about the consequences of events and, like probabilities, are personal and subjective. Furthermore, like the axioms of probability, the axioms of utility theory (not given here) prescribe normative behavior. Novick and Lindley (1979) describe a statistical approach for the coherent assessment of utilities (see also Singpurwalla and Wilson, 2006).

We may ask as to why it is useful to view utilities as probabilities? The answer is that now we are able to use the laws of probability which tell us how to combine utilities with probabilities and to arrive upon a paradigm for decision making under uncertainty. This is described next.

2.8.2 Maximization of Expected Utility

To see the role of utility and probability in decision making under uncertainty, consider what happens if we choose decision d_i and outcome $\mathcal{E}_j$ occurs. The consequence of this decision–outcome combination is C_{ij}, which, in line with our discussion of the previous section, suggests that $C_{ij} =^* [C$ with chance $u(C_{ij})]$. Thus, whatever be the outcome of choosing d_i, we can think about it in terms of C or c. To reflect this fact [namely that, if we choose d_i and $\mathcal{E}_j$ occurs, the probability of obtaining C is $u(C_{ij})$], we write

$$P(C \mid d_i, \mathcal{E}_j) = u(C_{ij}).$$

However, p_{ij} is the probability that $\mathcal{E}_j$ occurs when decision d_i is chosen. Thus, by the law of total probability, we have

$$P(C \mid d_i) = \sum_j^n P(C \mid d_i, \mathcal{E}_j)\, p_{ij} = \sum_j^n u(C_{ij})\, p_{ij}$$

and we must choose that decision for which the above is maximum. The term on the right-hand side of the above expression is called the expected utility of action d_i, and the decision problem is therefore to choose that decision for which the expected utility is maximum. This therefore, is the **principle of maximization of expected utility** (MEU).

2.8.3 Attitudes to Risk: The Utility of Money

Making decisions based on the principle of MEU may sometimes lead to decisions that are not intuitively appealing. This is typically the case when utility is assumed to be linear in monetary units. As an example, consider the decision to choose between one of the following two lotteries: in Lottery 1, you win \$30 with probability .5 and loose \$1 with probability .5; in Lottery 2, you win \$2000 with probability .5 and loose \$1900 with probability .5. Assuming that our utility is a linear function of the amounts won or lost, the MEU principle leads to Lottery 2, because its expected utility is $(2000 \times .5 - 1900 \times .5) = 50$, whereas that of Lottery 1 is 14.5. However, most people would prefer Lottery 1 over Lottery 2 because the latter would be viewed as being more 'risky' than the former. Such individuals are said to be risk averse, and this attitude toward risk is best captured by a utility function that is concave in money. If x denotes a monetary unit, say a dollar, then some commonly used examples of utility functions $\mathcal{U}(x)$ that are concave are $\mathcal{U}(x) = \log x$, $\mathcal{U}(x) = \sqrt{x}$ or $\mathcal{U}(x) = 1 - \exp(-x/\theta)$, for some $\theta > 0$, where θ is known as the **risk tolerance**. The logarithmic form of the utility function was considered by Daniel Bernoulli in 1738 in his resolution of the St. Petersburg paradox. Not everyone displays risk averse behavior, and so utility curves need not always be concave. Convex utility functions describe risk seeking or risk proneness, an example of which is participants in state lotteries which typically cost \$1 and have an expected value of approximately \$0.5. Individuals whose utility function is linear in x are known as risk neutral. For a general discussion on attitudes to risk and the assessment of utility functions, see section 4.9 of Keeney and Raiffa (1976).

2.9 DECISION TREES AND INFLUENCE DIAGRAMS FOR RISK ANALYSIS

We have said before that reliability and risk analyses are the ingredients for decision making in the face of uncertainty. The cholesterol drug problem of Chapter 1 serves as an illustrative example. There are several approaches to decision making under uncertainty, some of which have a statistical foundation (frequentist or Bayesian), and some which are more grounded in deterministic mathematics. Of the latter, 'The Analytical Hierarchy Process' of Saaty (cf. Forman and Dyer, 1991) seems to have the upper hand. The foundation for a frequentist theory of decision making was laid out by Wald (1950). Bayesians are critical of this theory because it violates the likelihood principle. The Bayesian attitude to decision making under uncertainty is based on two themes: the first is that all uncertainty should be described only by probability, and the second is that we choose only those actions that maximize expected utility. A tool that graphically portrays the various steps that are involved in implementing the above is the decision tree, introduced in Chapter 1. The decision tree originally appeared in Wald's frequentist decision theory, but since the late 1950s has emerged as the best known practical Bayesian technique. Besides its attractiveness as a visual device, the decision tree is easy to codify for computer applications, making an implementation of the Bayesian risk analysis methodology feasible for actual use. A more recent development is the **influence diagram** of Howard and Matheson (1984); it is a generalization of the fault tree mentioned in Chapter 1. Whereas the decision tree graphically portrays the steps in implementing statistical decision theory, the influence diagram graphically displays the relationships and dependencies among the random and decision nodes in a decision problem. Both these tools complement each other and should be of interest to those who wish to model decision problems and implement Bayesian decision theory for risk analysis.

2.9.1 The Decision Tree

A decision tree consists of nodes and branches (Figure 1.1). There are two kinds of nodes: a decision node and a random node. The decision nodes are indicated by rectangles and the random nodes by circles. The branches that emanate from a decision node show the possible decisions (or actions) that a decision maker can take. The branches that emanate from a random node indicate the various outcomes (also known as states of nature) that follow a particular action. Effectively, at each decision node, the decision maker acts whereas, at each random node, nature acts. Associated with each branch of a random node are probabilities. These are personal probabilities of the decision maker and reflect the probability of occurrence of the outcome associated with the branch. At the terminus of the tree are the utilities of the decision maker. Each utility represents the consequence of an action–outcome combination. For example, the utility associated with the topmost branch of Figure 1.1 comprises of the cost of administering the drug and the discomfort of suffering from both a heart attack and the drug's side effects. At each random node, we compute the **expected utility** at that node by multiplying each utility associated with that node by its probability, and summing these products. At each decision node, we choose that decision which has the largest expected utility. Our final decision is the one we make at the decision node that is the leftmost in the tree. Thus, to solve a decision problem, we commence from the right-hand end (the terminus) of the tree and work our way back to the left-hand end (the root) of the tree, by taking expectations at each random node and choosing the maximum at each decision node. One way to think of this process is that at each random node, we bundle up all the branches and at each decision node, we prune the branches. Finally, we are left with only one branch and the decision we choose is that decision associated with the last branch. Even though the decision tree of Figure 1.1 shows only one decision node $\mathcal{D}$ and two random nodes $\mathcal{R}_1$ and $\mathcal{R}_2$, decision trees in general can have several such nodes.

To illustrate the above process, suppose that, in the decision tree of Figure 1.1, we first concentrate on the random node $\mathcal{R}_1$ and denote by p_1, p_2, p_3 and p_4, the physicians' probabilities for the events $[\mathcal{HA}, \mathcal{SE}]$, $[\mathcal{HA}, (\mathcal{SE})]$, $[(\mathcal{HA}), \mathcal{SE}]$ and $[(\mathcal{HA}), (\mathcal{SE})]$, respectively, when the action taken is $\mathcal{A}$, to administer the drug; note that $\sum_i p_i = 1$. The abbreviations used here have been explained in section 1.4. Let the utilities associated with the above events be U_1, U_2, U_3 and U_4, respectively. That is, U_1 is the utility to the patient suffering both a heart attack and the drug's side effects when the drug is administered. We would expect to have $U_1 < U_2 < U_3 < U_4$, and that $p_3 > p_4 > p_1 > p_2$, though not necessarily. Similarly, at random node $\mathcal{R}_2$, let q_1 and q_2 be the probabilities of the events $[\mathcal{HA}]$ and $[(\mathcal{HA})]$, respectively, when the action taken is $(\mathcal{A})$, to not administer the drug; also $q_1 + q_2 = 1$. Finally, let U_5 and U_6 be the utilities associated with the above two events; we would expect to see $U_5 < U_6$. The expected utility of action $\mathcal{A}$, administering the drug, calculated at the random node $\mathcal{R}_1$ is $\sum_{i=1}^{4} p_i U_i = \mathcal{U}_1$, say. Similarly, the expected utility of $(\mathcal{A})$, not administering the drug, calculated at $\mathcal{R}_2$ is $q_1 U_5 + q_2 U_6 = \mathcal{U}_2$, say. At the decision node $\mathcal{D}$, we would choose action $\mathcal{A}$ if $\mathcal{U}_1 > \mathcal{U}_2$ and action $(\mathcal{A})$ if $\mathcal{U}_1 \leq \mathcal{U}_2$.

Figure 2.1, abstracted from the National Research Council's Report (1998), is another illustration of the use of decision trees. It pertains to an engineering design issue involving aircraft safety; the role of reliability and associated techniques, like fault tree analysis, failure data analysis, the use of expert judgments, etc., is shown.

In Figure 2.1, $\mathcal{D}_1$ pertains to the decision of installing a smoke detector in the hold of all cargo planes, and $\mathcal{D}_2$ to forgo such an illustration. The outcomes O_1, O_2, O_3 and O_4 pertain to the events 'no fire in the hold', 'fire in the hold and the smoke detector functions reliably', 'fire in the hold and the smoke detector fails to function' and 'no inflammable material in the hold', respectively. With these decisions and outcomes in place, consequences for each decision–outcome pair are calculated. Thus, for example, $\mathcal{U}(\mathcal{D}_1, O_3)$ is the consequence of having installed a failed detector when there is a fire in the hold. Once these consequences are assessed, the next task is to calculate the probabilities associated with each consequence. Thus $P(\mathcal{D}_1, O_3)$ is the probability of a fire in the hold and the failure of an installed detector. To assess this probability, we need to assess the reliability of the detector, for which techniques such as failure data analysis, expert judgments and fault tree analysis would play a role. The assessed probabilities and consequences result in expected utilities, which in turn give us a means for choosing $\mathcal{D}_1$ or $\mathcal{D}_2$.

2.9.2 The Influence Diagram

Whereas decision trees depict the many scenarios of a decision problem (including the time sequence of events and decisions), they suffer from the disadvantage that they become very large, even for modestly sized situations. The size of a decision tree, as measured by the number of its terminal branches, increases exponentially with the number of variables in the problem.

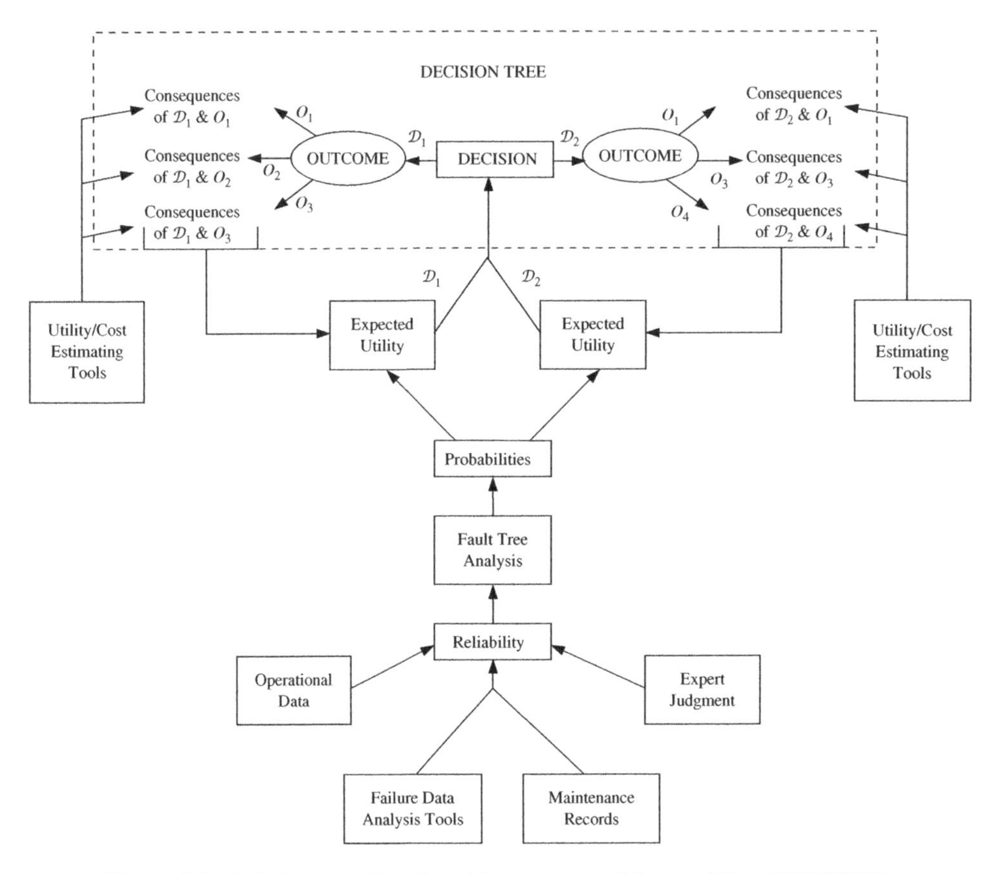

Figure 2.1 A decision tree for safety risk management (abstracted from NRR (1998)).

Another limitation of decision trees is that they do not explicitly depict the interdependencies between the various nodes that appear in a decision problem. Influence diagrams overcome these limitations of decision trees. They depict the decision problem as a compact graph, whose size, as measured by the number of its nodes, grows linearly with the number of variables. All the same, decision trees and influence diagrams are isomorphic; that is, any properly constructed influence diagram can be converted to a decision tree, and vice versa. A common strategy is to start off by using an influence diagram to help understand the major elements of a problem and its interdependencies, and then to convert it to a decision tree for its systematic solution. It has been the experience of many practitioners that communicating the elements of a decision problem via influence diagrams is more effective than doing so via decision trees. However, it is also the experience of many that drawing an influence diagram is much more involved than drawing a decision tree. An expository tutorial on influence diagrams, together with several examples of their applications, is in Clemen (1991); a more mathematical treatment is in Barlow (1988) and in Barlow and Pereira (1990). The material which follows has been abstracted from these sources.

As is done with decision trees, the elements of a decision problem (namely the decisions to be made), the uncertain outcomes and the consequences of a decision–outcome combination show up in the influence diagram as different shapes. These shapes, called nodes, are then linked up by arrows to show the interrelationships between the elements. A decision node is indicated by a rectangle and a random node by a circle; the consequence node, also called the value or payoff, is denoted by two concentric rectangles with rounded corners (Figure 2.2). The arrows connecting the nodes are called arcs, and the entire display is known as a graph. The node at the beginning of an arc is called a predecessor, and the node at the end of an arc is called a successor. A node with no adjacent predecessors is called a root node, and a node with no adjacent successors is called a sink node. A path between two nodes, say $\mathcal{R}_1$ and $\mathcal{R}_2$, is a collection of arcs that leads one from $\mathcal{R}_1$ to $\mathcal{R}_2$ via intermediate connecting nodes. In Figure 2.2, the node labeled $\mathcal{D}$ is a root node, whereas the node labeled 'Value' is a sink node. Also, node R is a successor to the node labeled $\mathcal{D}$, which is a predecessor to $\mathcal{R}$.

The simplest decision problem is one in which there is only one decision to make, one uncertain event and one outcome that is determined by both the decision and the uncertain event. Thus, for example, consider a simplified version of the cholesterol drug problem considered before, in which the only decision to be made at node $\mathcal{D}$ is $\mathcal{A}$ or $(\mathcal{A})$ and the only uncertain event of interest is $\mathcal{HA}$ or $(\mathcal{HA})$ at node R; that is, we ignore the issue of the drug's side effects. An influence diagram for this decision problem is shown in Figure 2.2. Observe that, since both the decision node and the random node precede the value node and also influence it, there are arcs going from these nodes to the value node. Also, since the decision to administer the drug would influence the occurrence or not of a heart attack, there is an arc going from the decision node to the random node. The absence of an arc from the random node to the decision node reflects the fact that, when the decision is made, we do not know if the patient will suffer a heart attack or not. Any arc going from a random node to a decision node indicates the fact that, when the decision is made, the outcome of the (predecessor) random node is known; such arcs are usually denoted by dotted lines. Also denoted by dotted lines are arcs going from one decision node to another; these indicate the fact that the first decision is made before the second. Finally, a well-constructed influence diagram should have no cycles of all solid lines; that is, once we leave a node we cannot get back to it, and the diagram must have at least one root node and one sink node.

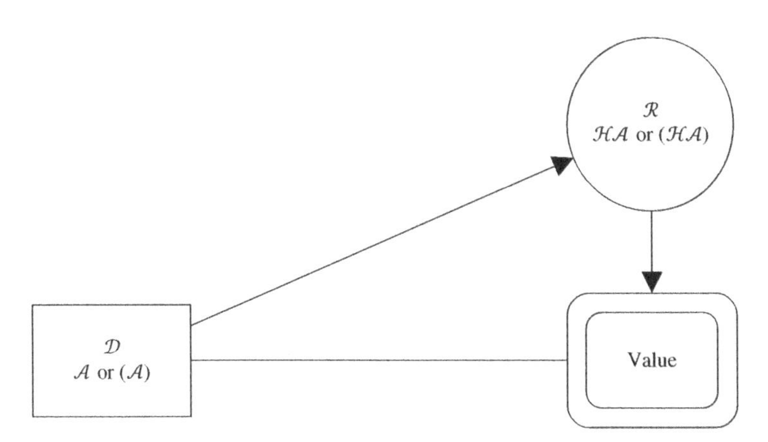

Figure 2.2 Influence diagram for a simplified version of the cholesterol drug problem.

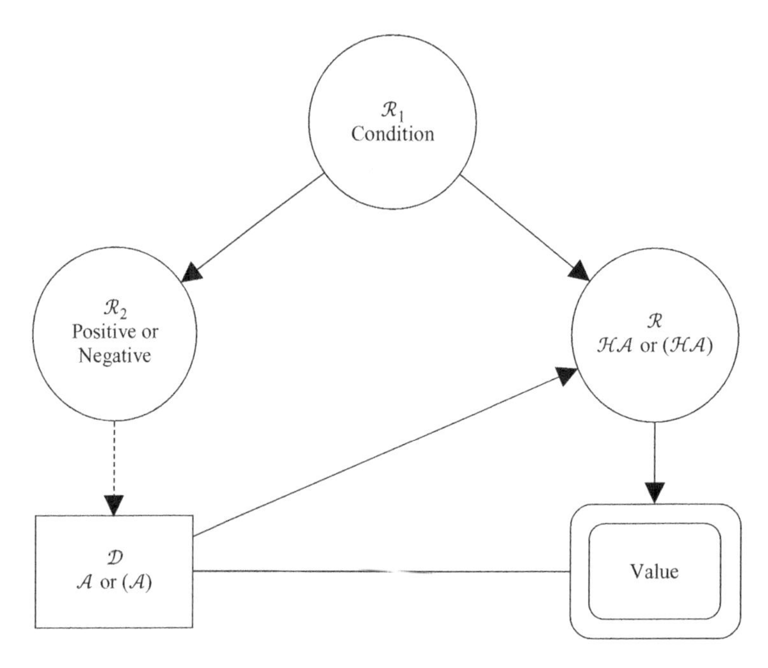

Figure 2.3 Influence diagram for the cholesterol drug problem with imperfect information.

The influence diagram of Figure 2.2 can be expanded to include the commonly occurring situation of imperfect information which decision makers often have prior to making a decision. With respect to the cholesterol drug problem of Chapter 1, suppose that we have the benefit of an inexpensive medical test on the general health of the patient which gives us added but inconclusive information about the patient's susceptibility to a heart attack. In Figure 2.3, the

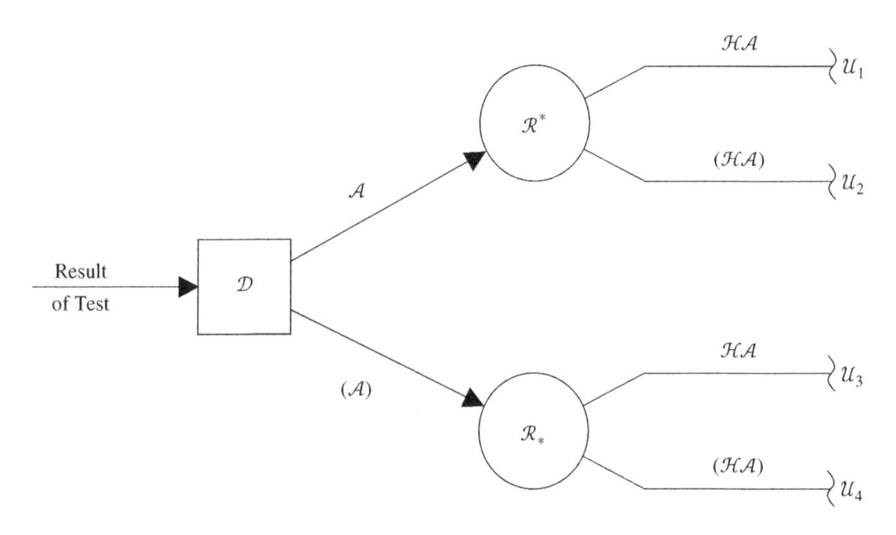

Figure 2.4 Decision tree for the cholesterol drug problem with imperfect information.

general health of the patient is indicated by a random node $\mathcal{R}_1$, and the additional test by a random node $\mathcal{R}_2$. This latter node is a random node because the test may reveal whether the patient is risking a heart attack (test positive) or not (test negative). The results of the test would of course depend on the true condition of the patient, and thus we have an arc going from $\mathcal{R}_1$ to $\mathcal{R}_2$. Furthermore, when the decision to administer the drug is made, the result of the test is known. Thus, we have a dotted arc going from $\mathcal{R}_2$ to $\mathcal{D}$ (Figure 2.3). Note that the situation described above is different from that in which we have the option of deciding whether to order the test or not. In the latter case, we would have an additional decision node pertaining to the test and preceding node $\mathcal{D}$.

To see how an influence diagram displays features of interdependence between the nodes that the decision tree does not, we show in Figure 2.4 a decision tree that is isomorphic to the influence diagram of Figure 2.3. Observe that, unlike the arc going from $\mathcal{D}$ to $\mathcal{R}$ in the influence diagram, there is nothing in the decision tree which graphically portrays the fact that our decision influences the outcome.$^{\infty}$ However, node $\mathcal{R}$ of the influence diagram corresponds to nodes $\mathcal{R}^*$ and $\mathcal{R}_*$ of the decision tree to indicate the fact that our decision could influence the outcome. Similarly, the decision tree has no analog for the arc from $\mathcal{R}_1$ to R of the influence diagram, which reflects the fact that the outcome of the test depends on the health of the patient.

Lauritzen and Spiegelhalter (1988) discuss the use of influence diagrams in medical expert systems, whereas Good (1961) has used a similar vehicle to illustrate notions of causality. The examples given here do not illustrate the fact that influence diagrams are generally more compact than decision trees. This feature is better appreciated in the context of sequential decision problems which have a tendency to grow exponentially. The fact that influence diagrams are useful aids for decision making vis-à-vis communication dependencies needs no further elaboration. However, the fact that they are generally difficult to construct and that they do not supplant decision trees in their entirety makes them less of a panacea than what their proponents have us believe.

Probabilistic influence diagrams

A **probabilistic influence diagram** is a special influence diagram in which all the nodes are random and, as before, the arcs between the nodes indicate their possible dependencies. If there is no arc connecting two nodes, then these nodes are judged to be conditionally independent, given the states of their adjacent predecessor nodes. Also, any two root nodes in a probabilistic influence diagram are independent. Associated with each node is a conditional probability for the node, and this probability depends on the states of the adjacent predecessor nodes, if any. Probabilities associated with the root nodes are conditioned on the background information. Given a probabilistic influence diagram, there exists a unique joint probability function corresponding to the random quantities represented by the nodes. This joint probability is the product of the probabilities associated with all the nodes in the diagram. Consequently, in addressing practical problems, it may be easier to use an influence diagram to assess the joint probability distribution by multiplying the node probabilities, as opposed to a direct probability assessment.

Probabilistic influence diagrams are in essence a pictorial depiction of the calculus of probability, and as such have been used by some to ensure that the laws of probability are observed. That is, probabilistic influence diagrams are isomorphic with the calculus of probability and serve as an aid for ensuring coherence. This isomorphism is achieved by introducing the three operations, 'node splitting', 'node merging' and 'arc reversal', all of which are derived from the addition and multiplication rules and Bayes' Rule. The following example in forensic science, taken from Barlow and Pereira (1990), is illustrative. Scenarios involving medical diagnosis and machine maintenance can be seen as alternate versions of this example.

An archetypal problem in forensic science goes as follows. A robbery has been committed by breaking a window and, in the process, a blood stain has been left by the robber. An individual with the same blood type as that on the window stain has been charged with the crime. Based on this evidence, we need to assess the probability of the individual's guilt. Let $\mathcal{E}_1$ ($\mathcal{E}_2$) denote the blood type of the individual (window stain), and let $\mathcal{E}$ denote the 'culpability' of the individual, with $\mathcal{E} = 1(0)$ implying guilt (innocence). Let $\mathcal{E}_i = 1(0)$, $i = 1, 2$, if the blood type is $\mathcal{A}$ (not $\mathcal{A}$). We need to evaluate $P(\mathcal{E} = 1 | \mathcal{E}_1 = \mathcal{E}_2 = 1)$. The probabilistic influence diagram of Figure 2.5 describes the probability model for this problem.

Observe that the diagram does not portray the actual values of the $\mathcal{E}_i$s that are known to us. It merely describes the dependence relationships among the quantities and the probabilities to be used. Specifically, if p represents the proportion (chance) of persons in the population having blood type A, and q our prior probability that the suspect is guilty of the crime, where q has been assessed before we learn of the blood type evidence, then $P(\mathcal{E}_1 = 1 | \ p) = p$, and $P(\mathcal{E} = 1 | \ q) = q$. The probabilities p and q are assessed based on background history $\mathcal{H}$ alone, and thus $\mathcal{E}$ and $\mathcal{E}_1$ go to define the root nodes $\mathcal{R}_1$ and $\mathcal{R}_2$; note that $p \neq (1 - q)$. A knowledge of p helps us assess the conditional probability of $\mathcal{E}_2$. Specifically, we can see that

$$
\begin{aligned}
P(\mathcal{E}_2 | \mathcal{E}_1, \mathcal{E}) &= p, && \text{if } \mathcal{E} \neq \mathcal{E}_2 = 1; \\
&= (1 - p), && \text{if } \mathcal{E} = \mathcal{E}_2 = 0; \\
&= 1, && \text{if } \mathcal{E} = 1 \text{ and } \mathcal{E}_1 = \mathcal{E}_2; \text{ and} \\
&= 0, && \text{otherwise.}
\end{aligned}
$$

Thus all the probabilities associated with the nodes of Figure 2.5 can be assessed.

Because $\mathcal{R}_1$ and $\mathcal{R}_2$ are root nodes, the events $\mathcal{E}$ and $\mathcal{E}_1$ are judged independent, and thus

$$P(\mathcal{E}, \mathcal{E}_1, \mathcal{E}_2) = P(\mathcal{E}_2 | \mathcal{E}, \mathcal{E}_1) \ P(\mathcal{E}_1) \ P(\mathcal{E})$$

by the multiplication rule. Therefore, we see that the joint probability of all the events in a probabilistic influence diagram is simply the product of probabilities associated with the nodes.

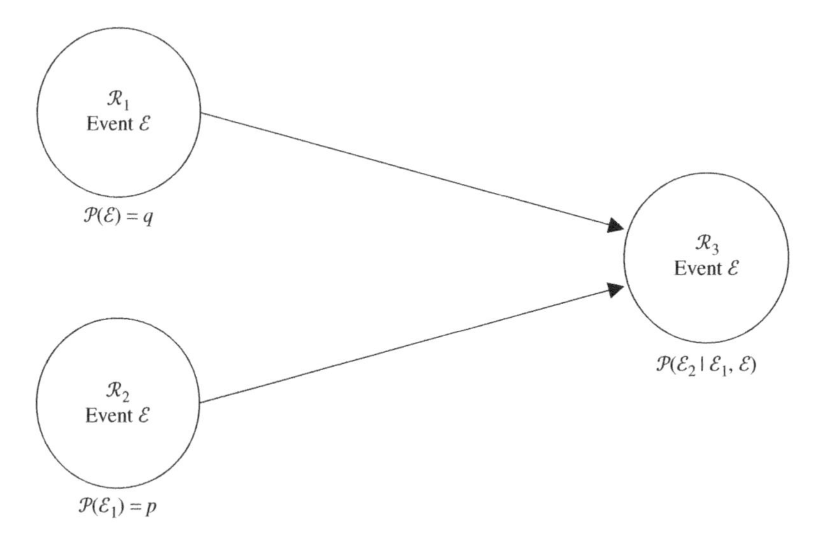

Figure 2.5 Probabilistic influence diagram for a problem in forensic science.

To obtain $P(\mathcal{E}=1 | \mathcal{E}_1=\mathcal{E}_2=1)$ we use the multiplication rule where

$$P(\mathcal{E}|\mathcal{E}_1, \mathcal{E}_2) = \frac{P(\mathcal{E}, \mathcal{E}_1, \mathcal{E}_2)}{P(\mathcal{E}_1, \mathcal{E}_2)},$$

and proceed to evaluate $P(\mathcal{E}_1, \mathcal{E}_2)$. The evaluation of this latter quantity gives us an opportunity to illustrate the **node splitting** operation of influence diagrams. Since

$$P(\mathcal{E}_1, \mathcal{E}_2) = P(\mathcal{E}_1)\ P(\mathcal{E}_2|\mathcal{E}_1) = P(\mathcal{E}_2)\ P(\mathcal{E}_1|\mathcal{E}_2),$$

there are two ways in which the node representing the event $(\mathcal{E}_1,\ \mathcal{E}_2)$ can be split (Figure 2.6).

The choice of how to split a node depends on our ability to assess the ensuing probabilities. For example, if we choose to go with the third box of Figure 2.6, then we are required to assess $P(\mathcal{E}_1|\mathcal{E}_2)$, the probability that an individual having blood type A is liable to commit a robbery and, in so doing, get hurt. That is, we need the probability that persons of blood type A are sloppy thieves. Similarly, the node representing the event $(\mathcal{E},\ \mathcal{E}_1,\ \mathcal{E}_2)$ can be split in six different ways, two of which are shown in Figure 2.7.

The second probabilistic influence diagram operation is **node merging**. This operation is the reverse of node splitting, and a simple way to appreciate this operation is to look at the first two boxes of Figure 2.7 in reverse order. That is, we can go from the two nodes of the second box to the single node of the first box. The same is also true of the third box. However, it is not always possible to merge any two adjacent nodes in a probabilistic influence diagram. In general, two nodes, say $\mathcal{E}_1$ and $\mathcal{E}_2$, can be merged into a single node $(\mathcal{E}_1,\ \mathcal{E}_2)$ only if there is a list ordering of the nodes such that $\mathcal{E}_1$ is an immediate predecessor or successor of $\mathcal{E}_2$ in the list. For example, in the diagram of Figure 2.8, the two nodes $\mathcal{E}$ and $\mathcal{E}_2$ cannot be merged because the only list

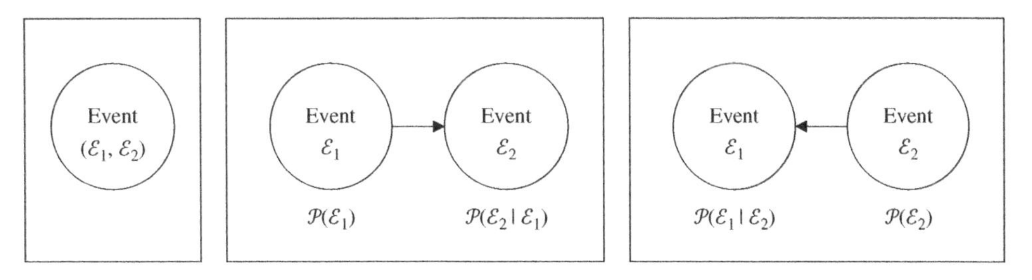

Figure 2.6 An illustration of node splitting.

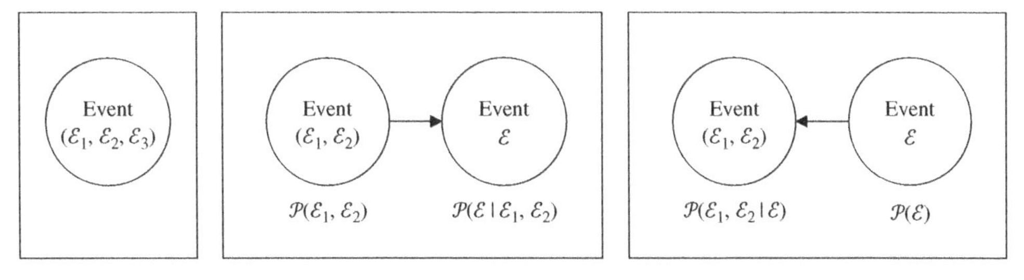

Figure 2.7 Another illustration of node splitting.

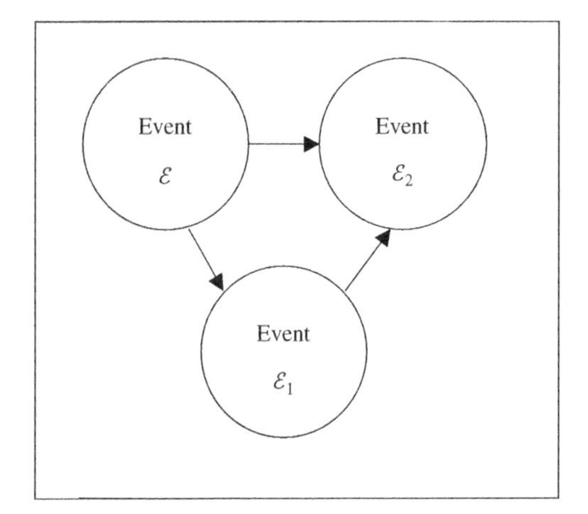

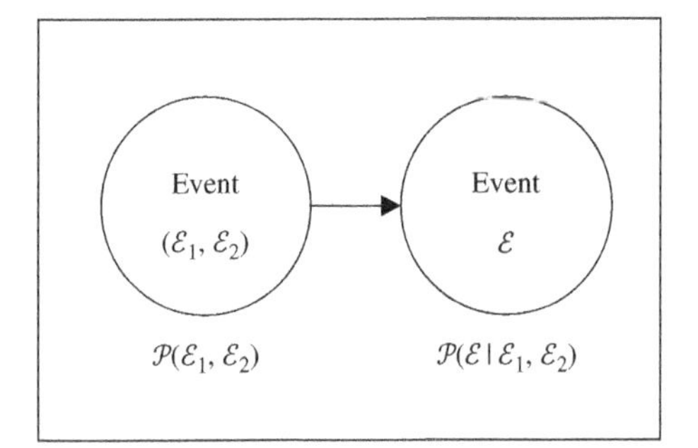

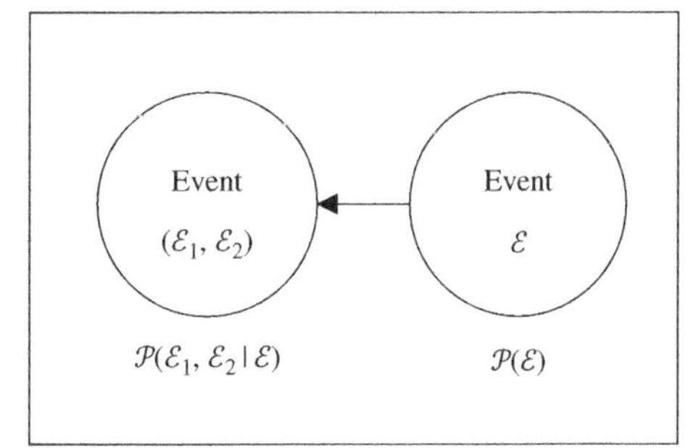

Figure 2.8 Illustration of node merging.

ordering here is $\mathcal{E} < \mathcal{E}_1 < \mathcal{E}_2$, and $\mathcal{E}$ and $\mathcal{E}_2$ are not neighbors on the list. We are, of course, able to merge $\mathcal{E}$ and $\mathcal{E}_1$ into $(\mathcal{E}, \mathcal{E}_1)$ and $\mathcal{E}_1$ and $\mathcal{E}_2$ into $(\mathcal{E}_1, \mathcal{E}_2)$.

The third probabilistic influence diagram operation is **arc reversal**. This operation corresponds to Bayes' formula, which pertains to inverting probabilities. To see how this works, consider the first box of Figure 2.9, which contains two nodes $\mathcal{E}_1$ and $\mathcal{E}_2$ with an arc from $\mathcal{E}_1$ to $\mathcal{E}_2$. Using the node merging operation, we merge the nodes corresponding to the events $\mathcal{E}_1$ and $\mathcal{E}_2$ to obtain the single node box in the center of Figure 2.9. We then apply the node splitting operation to

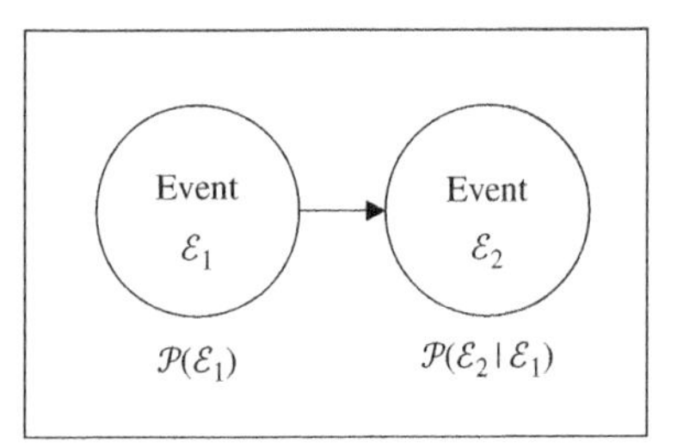

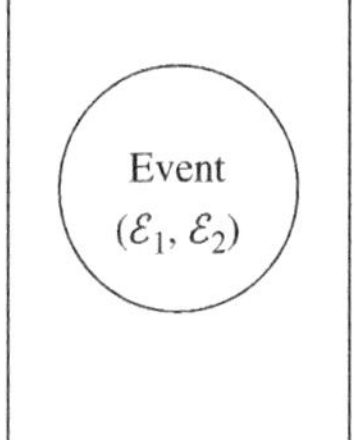

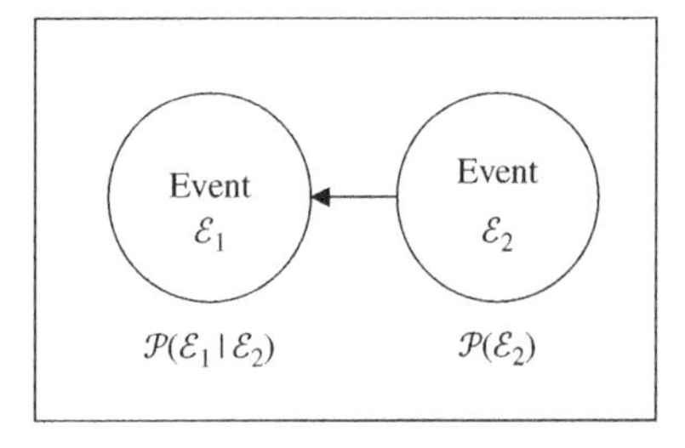

Figure 2.9 Illustration of arc reversal operation (Bayes' law).

the single node of the center box to obtain the third box of Figure 2.9, which has two nodes $\mathcal{E}_1$ and $\mathcal{E}_2$ with an arc going from $\mathcal{E}_2$ to $\mathcal{E}_1$. The arcs in the first and the third boxes are reversed. In order to obtain the probabilities $P(\mathcal{E}_1|\, \mathcal{E}_2)$ and $P(\mathcal{E}_2)$ we use Bayes' law and the law of total probability, respectively. We may thus interpret the law of total probability and Bayes' law as algebraic operations that enable us to go from the first box to the third without going through the middle box, the first and third boxes entailing a reversal of arcs.

Theorems pertaining to the conditions under which node merging and arc reversal can be undertaken are given by Barlow and Pereira (1990), who also describe the visual force of the probabilistic influence diagrams to explain the notions of conditional independence. We find them interesting, but leave it up to the reader to decide on their usefulness. Our inclusion of probabilistic influence diagrams is for the sake of completeness.

Chapter 3

Exchangeability and Indifference

Here lies Fermat: This page is too small to contain a proper epitaph.

3.1 INTRODUCTION TO EXCHANGEABILITY: DE FINETTI'S THEOREM

In section 2.6.1 of Chapter 2, the notion of a probability model (or a failure model) was introduced and are saw that it was subjectively specified. It was mentioned that the principle of indifference has sometimes been used to facilitate the subjective specification of failure models. The principle of indifference derives from the judgment of exchangeability. The aim of this chapter is to introduce the ideas of indifference and exchangeability, and to explore their ramifications for specifying failure models. First-time readers and those whose main interested is in methodology may choose to skip this chapter and proceed directly to Chapter 4. However, I strongly feel that the material here is at the heart of the theory and the practice of reliability and survival analysis, and is fundamental to a deeper appreciation of what the subject is all about. This is true for both probabilistic modeling and life-testing.

Exchangeability is an important notion in probability theory; Kingman (1978) surveys its features. It has also played a useful role in reliability theory vis-à-vis the mathematical characterization of positively dependent lifelengths (Shaked, 1977). Exchangeability was introduced by de Finetti in 1937, and has played a key role in both the theory and the practice of Bayesian statistics. Philosophically, exchangeability has had an impact on the positivist interpretation of the degree of belief notion of probability. In survival analysis, it has provided a justification for the predictability of random quantities given observations on similar quantities (Lindley and Novick, 1981). In reliability, it has facilitated the development of models for failure and life-data analysis (Barlow and Mendel, 1992). Exchangeability is also the basis for hierarchical models which have proved to be useful for addressing questions pertaining to the failure of technological systems, such as emergency diesel generators in nuclear reactors (Chen and Singpurwalla, 1996), and software for telecommunication systems (Singpurwalla and Soyer, 1992). In addition to the above, exchangeability has provided the probability theorists fertile territory for research in an area that is of interest to statisticians [(Diaconis and Freedman, 1980), Diaconis (1988)].

Reliability and Risk: A Bayesian Perspective N.D. Singpurwalla
© 2006 John Wiley & Sons, Ltd

3.1.1 Motivation for the Judgment of Exchangeability

A need for the judgment of exchangeability is motivated by restricting attention to a collection of random events. Consider the following scenario: A subject is administered a drug today; at the end of the month, the subject is to be observed for its 'response' or 'non-response'. The event 'response' is denoted by 1, and the event 'non-response' by 0. Suppose that 10 such subjects, all judged to be similar to each other, are administered the drug and our interest today focuses on the $2^{10} = 1024$ possible outcomes that can occur at the end of the month. What can we say about the probability of occurrence of each of these 1024 outcomes? Our scenario is generic, and applies equally well to many other situations ranging from coin tossing to acceptance sampling and life-testing.

From a personalistic point of view, we should be able to think hard about the subjects and the drug and coherently assign probabilities to each of the 1024 outcomes. There are no restrictions on how we make our assessments, the only requirement being that the assigned probabilities must add up to one. Even with so much latitude, it is a difficult task to come up with a sensible assignment of so many probabilities. de Finetti suggested exchangeability as a way to simplify this situation. Specifically, his idea was that all sequences of length 10, comprising of 1s and 0s, be assigned the same probability if they have the same total number of 1s. That is, all sequences that have a total of five 1s should be assigned the same probability; similarly, all sequences of eight 1s. This simply means that the probability assignment is 'symmetric', or 'invariant', or 'indifferent' under the changes in order. Thus, as far as the probability assignment is concerned, what matters in a sequence is the total number of 1s and not their position and order in the sequence. If exchangeability is to be believed, then the total number of probabilities to be assigned reduces from 1024 to 11. This simplification has been possible because we have made a judgment. Our judgment is that all the ten subjects are similar or indistinguishable, so that it does not matter to us which particular subject produces a response; all that matters to us is how many of the ten subjects responded to the drug. Thus exchangeability is a subjective judgment that we have made in order to simplify our assignment of probabilities. A precursor to exchangeability is Laplace's principle of 'insufficient reason'. In actuality, judgments of exchangeability should be supported by the science of the problem, and there could be situations where exchangeability cannot be assumed. To alleviate concerns such as these, de Finetti (1938) generalized his idea of exchangeability and introduced the notion of partial exchangeability. This is a topic that I do not plan to address, save for the material of section 6.2.1 wherein the simulation of exchangeable sequences is discussed. We are now ready to define exchangeability.

Exchangeability

Random quantities $X_1, X_2, \ldots, X_n$, discrete or continuous, are said to be exchangeable if their $n!$ permutations, say $(X_{k_1}, \ldots, X_{k_n})$ have the same n-dimensional probability distribution $P(X_1, \ldots, X_n)$. The elements of an infinite sequence $X_1, X_2, \ldots$ are exchangeable, if $X_1, X_2, \ldots, X_m$ are exchangeable for each m. The n-dimensional probability distribution $P(X_1, \ldots, X_n)$ is said to be exchangeable if it is invariant under a permutation of the coordinates.

3.1.2 Relationship between Independence and Exchangeability

It has been pointed out that exchangeability was introduced by de Finetti as a way of simplifying the assignment of probabilities. Are there other attributes of exchangeability that make it attractive for use? Is there a relationship between independence and exchangeability? Does independence imply exchangeability or is it the other way around? These are some of the questions that arise, and the aim of this section is to attempt to answer them. We start off by considering the following situation which commonly occurs in life-testing and industrial quality control.

Several identically manufactured items are tested for success or failure, with the uncertain event X_i taking the value one if the i-th item tests successfully, and zero otherwise. Given the results of the test, interest generally centers around either accepting or rejecting a large lot from which these items have been selected at random.

A frequentist approach to a problem of this type generally starts with the assumption that the sequence of zero-one events $X_1, X_2, \ldots$ are independent and have a common Bernoulli distribution with the parameter θ; i.e. $P(X_i = 1 \mid \theta) = \theta,\ i = 1, 2, \ldots$, and $P(X_i = 0 \mid \theta) = (1 - \theta)$. Experiments which lead to such events are called **Bernoulli trials**, and the sequence X_is, a **Bernoulli sequence**. Once this assumption is made, standard machinery can be applied, provided that the 'stopping rule' (cf. Lindley and Phillips, 1976) is known. For example, we can obtain point estimates and interval estimates of θ, test hypotheses about θ, and so on. Based on the outcome of such procedures, we can then decide to either accept or to reject the lot. In fact, procedures like this are so common that they have been codified for use by the United States government as MIL-STD-105D (MIL-STD-105D, 1963), and have become the standard by which many organizations, all over the world, do business.

The subjective Bayesian approach, however, proceeds differently. To a subjective Bayesian, when any of the X_is, say X_1 reveals itself, opinion about the uncertainty of the remaining X_is, namely, $X_2, X_3, \ldots$, changes; similarly, when X_2 reveals itself, opinion about the remaining X_is changes, and so on. By contrast, the assumption of independence of the X_is, which implies that observing X_1 has no impact on $X_2, X_3, \ldots$, is unjustifiably severe. The starting point for a Bayesian analysis would therefore be something which reflects the dependence between the X_is, and one such vehicle is the assumption that the zero-one sequence $X_1, X_2, \ldots$ is exchangeable. But why would an exchangeable sequence reflect dependence? The answer was provided by de Finetti who, after introducing the notion of exchangeability, went on to prove a famous theorem that bears his name. This theorem formalizes what was said at the end of section 2.6.2 namely that the defining representation for an exchangeable sequence is an equation of the type (2.4). That is, an exchangeable sequence can be generated by a mixture of conditionally independent identically distributed sequences.

Thus a Bayesian can proceed along one of two lines of reasoning, both leading to the same formulation. The first would be along the lines outlined in sections 2.6.1 and 2.6.2, and the second based on the judgment of exchangeability. With the first, the driving premise was the law of total probability, and the assumption that the X_is are conditionally independent, conditioned on knowing θ. Given the parameter θ the X_is may then be assumed to have a common probability model, which in this case is a Bernoulli with $0 \leq \theta \leq 1$. Since θ is an unknown quantity, the laws of probability require that for making statements of uncertainty about the X_is we average over all values θ. This line of development would bring us to the following special versions of (2.5) and (2.6); namely that, for $x = 1$ or 0,

$$P\left(X_{n+1} = x \mid X^{(n)}; H\right) = \int_0^1 \theta^x (1-\theta)^{1-x}\, P(\theta \mid X^{(n)}; H)\mathrm{d}\theta, \tag{3.1}$$

where

$$P\left(\theta \mid X^{(n)}; H\right) \propto \prod_{i=1}^{n} \theta^{x_i} (1-\theta)^{1-x_i}\, P(\theta; H), \tag{3.2}$$

and $P(\theta; H)$ is our prior density at θ. Since $0 < \theta < 1$ is continuous, the prior at θ is typically chosen to be a **beta density** on (0,1); i.e. $P(\theta; \alpha, \beta) = (\Gamma(\alpha+\beta)/(\Gamma(\alpha)\Gamma(\beta)))\theta^{\alpha-1}(1-\theta)^{\beta-1}$, where α and β are parameters of this density, and $\alpha, \beta > 0$; $\Gamma(\beta) = \int_0^\infty x^{\beta-1} e^{-x}\, dx$ is the **gamma function**. The notion of a density will be clarified later in section 4.2.

With the second line of reasoning, a Bayesian would start with the judgment of exchangeability, and then appeal to de Finetti's theorem (given below) to arrive at (3.1) and (3.2). Thus exchangeability and the theorem can be viewed as providing a foundation for the Bayesian development. It is important to emphasize that, like all probability judgments, the judgment of exchangeability is made before the sequence is observed. Once all this is done, the decision to accept or to reject the lot would be based on the assessed utilities and the principle of maximization of expected utilities; more details on how this can be done are given in Chapter 5.

It is easy to see, from the definition of exchangeability, that a sequence of independent random quantities is also an exchangeable sequence. However, exchangeability is a weaker condition than independence, in the sense that exchangeable random quantities are not necessarily independent. To verify this – details left as an exercise – consider an urn containing three balls, two of which are marked '1' and the remaining ball marked '0'. If these balls were to be drawn from the urn, one at a time without replacement, and if X_k were to denote the digit on the k-th ball drawn, then the sequence of random quantities X_1, X_2, X_3 would be exchangeable, but not independent.

3.1.3 de Finetti's Representation Theorem for Zero-one Exchangeable Sequences

We have seen before that the theory of probability does not concern itself with the interpretation of probability, neither does it concern itself with how the probabilities of the elementary events are assigned. The aim of the theory is to develop methods by which the probability of complicated events can be obtained from the probabilities of the elementary events. Thus, as a result in probability theory, de Finetti's theorem is striking. Controversy about it arises when in a particular application, the judgment of exchangeability is made and the theorem is then invoked to justify a Bayesian analysis. Since exchangeability as a judgment is personal, it follows therefore that any application of the theorem is liable to be the subject of debate. However, we have seen that exchangeability as a judgment is more realistic than independence, at least for zero-one sequences. The following version of the theorem and the proof has been taken from Heath and Sudderth (1976).

Theorem 3.1 (finite form), de Finetti (1937). *Let $X_1, \ldots, X_m$ be a sequence of exchangeable random quantities taking values 0 or 1. Then,*

$$P(X_1=1,\ldots,X_k=1,X_{k+1}=0,\ldots,X_n=0)=\sum_{r=0}^{m}\left(\frac{(r)_r(m-r)_{n-k}}{(m)_n}\right)q_r,\ \text{for } 0\le k\le n\le m,$$

where $q_r=P(\sum_{j=1}^{m} X_j=r)$, and $(x)_k=\prod_{j=0}^{k-1}(x-j)$.

Proof. From the exchangeability of the sequence $X_1, \ldots, X_m$ it follows that given $\sum_{j=1}^{m} X_j=r$, all possible arrangements of the r ones among the m places is equally likely. The situation is analogous to drawing from an urn containing r ones and $(m-r)$ zeros.

Thus

$$P\left(X_1=1,\ldots,X_k=1,X_{k+1}=0,\ldots,X_n=0 \mid \sum_{j=1}^{m} X_j=r\right)=\frac{\binom{r}{k}\binom{m-r}{n-k}}{\binom{m}{n}}\div\binom{n}{k},$$

and so

$$P(X_1=1,\ldots,X_k=1,X_{k+1}=0,\ldots,X_n=0)=\sum_{r=0}^{m}\left\{\frac{\binom{r}{k}\binom{m-r}{n-k}}{\binom{m}{n}}\div\binom{n}{k}\right\}q_r,$$

which, when simplified, reduces to the statement of the theorem. Note that division by $\binom{n}{k}$ is necessary because we are looking for a particular pattern of ones and zeros.

The finite form of the theorem holds for all sequences of length m that cannot be 'extended' to a length larger than m. Thus, for example, if we are to sample without replacement from an urn with m balls in it, we cannot have $(m+1)$ trials. A similar situation arises in sample surveys where we have to draw a sample from a finite population. However, in the drug testing scenario considered by us, the number of subjects to whom we can administer the drug can, in principle, be extended to a very large number. The same could also be true of the item testing scenario; if the lot size is very large, the number of items that we can test can, in principle, be very large. For such situations, the infinite form of de Finetti's theorem, given below, would apply.

The infinite form of de Finetti's theorem is simply the limiting case, as $m \to \infty$ and $r/m \to \theta$, of the above theorem. The proof involves convergence of distribution functions and approximations by integrals (Heath and Sudderth, 1976); it is therefore omitted.

Theorem 3.2 (infinite form), de Finetti (1937). *For every exchangeable probability assignment that can be extended to a probability assignment on an infinite sequence of zeros and ones, there corresponds a unique probability distribution function F concentrated on [0,1] such that, for all n and* $0 \leq k \leq n$.

$$P(X_1 = 1, \ldots, X_k = 1, X_{k+1} = 0, \ldots, X_n = 0) = \int_0^1 \theta^k (1-\theta)^{n-k} F(\mathrm{d}\theta),$$

where $F(\mathrm{d}\theta) = f(\theta)\mathrm{d}\theta$, *and* $f(\theta) = \frac{\mathrm{d}}{\mathrm{d}\theta}F(\theta)$ *is the derivative of* $F(\theta)$ *with respect to* θ; $F(\theta)$ *is the distribution function of* θ.

Comments on the Theorem

The infinite form of de Finetti's theorem suggests that exchangeable probability assignments of ones and zeros which can be extended have the special form of mixtures of (the independent and identically distributed) Bernoulli trials. The mixing distribution F may be regarded as the prior for the unknown parameter θ. Note the similarity between the right-hand side of the statement of the theorem and the right-hand side of (3.1). The theorem also implies that for many problems, specifying a probability assignment on the number of ones and zeros in any sequence of size n is equivalent to specifying a prior probability distribution on (0,1).

It is important to note that the infinite form of the theorem fails to hold for finite sequences. To see this, consider the case of two exchangeable random variables X_1 and X_2, with $P(X_1 = 1, X_2 = 0) = P(X_1 = 0, X_2 = 1) = 1/2$. Thus $P(X_1 = 1, X_2 = 1) = P(X_1 = 0, X_2 = 0) = 0$. Now, if there exists a probability distribution F such that $0 = P(X_1 = 1, X_2 = 1) = \int \theta^2 F(\mathrm{d}\theta)$, then F must put mass 1 at 0, making impossible $P(X_1 = 0, X_2 = 0) = \int (1-\theta)^2 F(\mathrm{d}\theta) = 1$. Thus there cannot exist a probability distribution F concentrated on (0,1) that will satisfy the statement of the theorem.

Finally, de Finetti's result generalizes so that every sequence of exchangeable random variables, not just the zero or one variables, is a mixture of sequences of independent, identically distributed variables. Thus equation (2.5), which is quite general and which was developed via a direct argument, can also be motivated via the judgment of exchangeability. However, as we shall see in section 3.2, a practical implementation of this idea for getting specific versions of (2.5) is difficult, and conditions that are more restrictive than exchangeability are sought to cut the problem to a manageable size.

3.1.4 Exchangeable Sequences and the Law of Large Numbers

The law of large numbers was mentioned in section 2.3.1 as Bernoulli's great theorem which bound up fair price, belief and frequency. Specifically, in the context of a large sequence of Bernoulli trials (that is, random quantities whose outcome is either a 1 or a 0 and all having the same θ as the chance of observing a 1), Bernoulli proved that the relative frequency with which a 1 occurs approximates θ. Stated formally, if S_m is the total number of 1s that occur in a large number, say m, of Bernoulli trials with a parameter θ, then according to the **weak law of large numbers** (for Bernoulli trials),

$$P\left\{\left|\frac{S_m}{m}-\theta\right|<\varepsilon\right\}\rightarrow 1,\ \text{ for } \varepsilon>0 \text{ arbitrarily small but fixed.}$$

Recall that, even to Bernoulli, a relative frequency was not a probability. Thus both θ and the probability statement of the theorem were regarded by Bernoulli as 'ease of happening', not relative frequencies. **The strong law of large numbers** (first stated by Cantelli in 1917) asserts that, with probability 1, $|S_m/m-\theta|$ becomes small and remains small.

Relative Frequency of Exchangeable Bernoulli Sequences

After proving his theorem for exchangeable sequences, de Finetti went on to prove a second result which is of great importance. In what follows, I shall adhere to the notation used in the infinite version of de Finetti's theorem. He proved that, for exchangeable sequences, the (personal) probability that the $\lim_{m\rightarrow\infty}(k/m)$ exists is 1 and, if this limit were to be denoted by θ, then the (personal) probability distribution for θ is F. If F is such that all its mass is concentrated on a single value, then the weak law of large numbers follows as a special case of de Finetti's theorem.

It is important to note that again θ is not to be interpreted as a probability, because to a subjectivist a probability is the amount you are willing to bet on the occurrence of an uncertain outcome. Lindley and Phillips (1976) view θ as a property of the real world and refer to it as a **propensity** or a **chance**, and the quantity $\theta^k(1-\theta)^{n-k}$ as a **chance distribution**. Some authors on subjective probability, such as Kyburg and Smokler (1964, pp. 13–14), have suggested that the above result of de Finetti bridges the gap between personal probability and physical probability. However, de Finetti (1976) disagrees, and Diaconis and Freedman (1979) also have trouble with this type of synthesis. Hill (1993) attempts to clarify this misunderstanding by saying that even though there does not pre-exist a 'true probability' θ, one could implicitly act as though there were one. Similar things could have also been said about the law of large numbers; however, it is likely that Bernoulli too would have disagreed.

3.2 DE FINETTI-STYLE THEOREMS FOR INFINITE SEQUENCES OF NON-BINARY RANDOM QUANTITIES

Most subjectivists find the de Finetti representation theorem for Bernoulli sequences very satisfactory, because specifying a reasonable prior distribution F on (0,1) is a manageable task. Here, a natural choice for F would be a beta distribution. For example, Laplace in his famous calculation of the probability that the sun will rise tomorrow given that it has risen in the previous n days, used the uniform distribution, a special case of the beta distribution, and arrived at the answer $n/(n+1)$.

However, de Finetti's representation theorem for sequences that are not zero-one is not straightforward to implement. It requires that we specify a prior distribution on the class of all probability distributions on the real line. This class is so large that it is difficult, if not literally impossible, for a subjectivist to describe a meaningful personal prior on this set. Ferguson (1974) contains some examples of the various attempts to choose such a prior. Thus it appears that, for real-valued random quantities, the judgment of exchangeability may not provide the justification for much of the practical Bayesian analysis that is currently being done. Conditions more restrictive than exchangeability are necessary for pinning down the specific cases (cf. Diaconis and Freedman, 1979). All the same, the general approach is still in the style of a de Finetti theorem; that is, to characterize models in terms of an 'invariance' property. By invariance, I mean 'equiprobable' or 'indifference', just like what we have done in the case of exchangeable zero-one sequences. The idea is to begin with observables, postulate symmetries or summary statistics (section 3.2.1), and then find a simple description of all models with the given symmetries.

The aim of this section is to explore the above symmetries and to produce versions of (2.5) other than that of the mixture of Bernoulli sequences. Before we do this, we must first gain additional insight about what we have already done with zero-one exchangeable sequences, particularly an insight into sufficiency and invariance. Informally speaking, by 'sufficiency' I mean a summarization of information in a manner that essentially preserves some needed characteristics.

3.2.1 Sufficiency and Indifference in Zero-one Exchangeable Sequences

We begin by obtaining an equivalent formulation of an exchangeable, infinitely extendible, zero-one sequence $X_1, X_2, \ldots$. Define the partial sum $S_n = X_1 + X_2 + \cdots + X_n$, and let $S_n = t$. Then, given t, exchangeability implies that the sequence $(X_1, \ldots, X_n)$ is uniformly distributed over the $\binom{n}{t}$ sequences having t ones and $(n-t)$ zeros. That is, each of the $\binom{n}{t}$ sequences has probability $1/\binom{n}{t}$. But infinitely exchangeable zero-one sequences are, by de Finetti's theorem, mixtures of Bernoulli sequences. Thus we have the following:

Indifference Principle for Zero-one Sequences
An infinite sequence of zero-one variables can be represented as a mixture of Bernoulli sequences if and only if, for each n, given $S_n = t$, the sequence $(X_1, \ldots, X_n)$ is 'uniformly' distributed over the $\binom{n}{t}$ sequences having t ones and $(n-t)$ zeros.

This equivalence formulation for the exchangeability of zero-one sequences of infinite length paves the way for questions about other kinds of sequences (section 3.3.2), but it also brings to surface two points: the relevance of the sufficiency of the partial sums S_n, and the uniform distribution of the sequence $(X_1, \ldots, X_n)$. In other words, if zero-one sequences are judged invariant (i.e. uniformly distributed) under permutation given the sufficient statistic S_n, then they are also exchangeable, and exchangeability, in turn, gives us the mixture representation. The 'indifference principle' refers to the act of judging invariance (under permutation in the present case).

3.2.2 Invariance Conditions Leading to Mixtures of Other Distributions

Prompted by the observation that mixtures of Bernoulli sequences arise when we assume invariance under permutation, we look for conditions (i.e. judgments about observables) that lead to mixtures of other well-known chance distributions that commonly arise in practice. This may enable us to better interpret specific forms of (2.5) that are used. In what follows, some of the classical as well as newer results on this topic are summarized. They pertain to mixtures of both discrete and continuous distributions. We start off with one of the best known distribution, the Gaussian (known as the 'normal distribution').

Scale Mixtures of Gaussian Distributions
When can a sequence of real valued random quantities $\{X_i\}$, $1 \le i < \infty$, be represented as a scale mixture of Gaussian $(0, \sigma^2)$ variables (or chance distributions)? That is, when is

$$P\{X_1 \le x_1, \ldots, X_n \le x_n\} = \int_0^\infty \prod_{i=1}^{n} \Phi(\sigma x_i)\, F(\mathrm{d}\sigma), \tag{3.3}$$

where, $P(X_1 \leq x_1) = \Phi(x) = (1/\sqrt{2\pi}) \int_{-\infty}^{x} e^{-u^2/2} du$, is the **Gaussian** (0,1) distribution function, and F a unique probability distribution on $(0, \infty)$?

Freedman (1962) (also Schoenberg (1938), and Kingman (1972)) has shown that a necessary and sufficient condition for the above to hold is that, for each n, the joint distribution of $X_1, \ldots, X_n$ be rotationally symmetric. Note that an $(n+1)$ vector of random quantities, say $X = (X_1, \ldots, X_n)$, is rotationally symmetric (or spherically symmetric, or orthogonally invariant) if the joint distribution of X is identical to that of MX for all $(m \times n)$ orthogonal matrices M.

An equivalent characterization based on sufficiency and invariance (cf. Diaconis and Freedman, 1981) is that, for every n, given the sufficient statistic $(\sum_{i=1}^{n} X_i^2)^{1/2} = t$, the joint distribution of $(X_1, \ldots, X_n)$ should be 'uniform' on the $(n-1)$ sphere of radius t in $\mathcal{R}^n$. Note that an $(n \times 1)$ vector of random quantities X has a uniform distribution on the unit n sphere if the i-th element of the vector is defined as

$$X_i = \frac{U_i}{(U_i^2 + \cdots + U_n^2)^{1/2}}, \quad i = 1, \ldots, n,$$

where the U_is are independent and identically distributed as the Gaussian (0, 1) distribution.

Location Mixtures of Gaussian Distributions

When can a sequence of real valued random quantities $\{X_i\}$, $1 \leq i \leq \infty$, be represented as a location mixture of Gaussian (θ, σ^2) variables with σ^2 known? That is, when is

$$P\{X_1 \leq x_1, \ldots, X_n \leq x_n \mid \sigma\} = \int_0^\infty \prod_{i=1}^{n} \Phi(\theta; \sigma x_i)\, F(\mathrm{d}\sigma), \tag{3.4}$$

where $P(X \leq x) = \Phi(\theta; x) = (1/\sqrt{2\pi}) \int_{-\infty}^{x} e^{-(u-\theta)^2/2} du$ is the Gaussian $(\theta, 1)$ distribution function?

The necessary and sufficient conditions (cf. Diaconis and Freedman, 1981) for the above to hold are the following:

(i) $X_1, X_2, \ldots$ is an exchangeable sequence, and
(ii) Given $X_1 + \cdots + X_n = t$, the joint distribution of $(X_1, \ldots, X_n)$ is a multivariate Gaussian distribution (Anderson, 1958; pp. 11–19) with mean vector t/n and covariance matrix with all diagonal terms σ^2 and off diagonal terms $(\sigma^2/n) - 1$.

Location and Scale Mixtures of Gaussian Distributions

When can a sequence of real-valued random quantities $\{X_i\}$, $1 \leq i \leq \infty$, be presented as a mixture of Gaussian (θ, σ^2) variables? That is, when is

$$P\{X_1 \leq x_1, \ldots, X_n \leq x_n\} = \int_0^\infty \prod_{i=1}^{n} \Phi(\theta; \sigma x_i)\, F(\mathrm{d}\theta, \mathrm{d}\sigma), \tag{3.5}$$

where $P(X \leq x) = \Phi(\theta; x) = (1/\sqrt{2\pi}) \int_{-\infty}^{x} e^{-(u-\theta)^2/2} du$ is the Gaussian $(\theta, 1)$ distribution function?

Smith (1981) has shown that a necessary and sufficient condition for the above to hold is that, for every n, given the two sufficient statistics $U_n = X_1 + \cdots + X_n$ and $V_n = (X_1^2 + \cdots + X_n^2)^{1/2}$, the joint distribution of $(X_1, \ldots, X_n)$ is 'uniform' over the $(n-2)$ – sphere in $\mathcal{R}^n$ with center at U_n and radius V_n.

Problems involving location and scale mixtures of Gaussian distributions are often encountered in many applications of Bayesian statistics, though not necessarily in reliability and survival analysis. The result given here is important because it shows that, whenever we take location and scale mixtures of a Gaussian distribution, we are de facto making a judgment of indifference of the type shown above.

Mixtures of Uniform Distributions

When is $X_1, X_2, \ldots$ a mixture over θ of sequences of independent uniform $[0, \theta]$ variables? That is, when is

$$P\{X_1 \leq x_1, \ldots, X_n \leq x_n\} = \int_0^\infty \prod_{i=1}^n \Phi(x_i|\theta)\, F(\mathrm{d}\theta)? \tag{3.6}$$

The necessary and sufficient condition is that, for every n, given $M_n = \max(X_1, \ldots, X_n)$, the X_is are independent and 'uniform' over $[0, M_n]$ (Diaconis and Freedman, 1981).

This elementary result provides an opportunity to illustrate the practical value of the de Finetti-style theorems we are discussing here. To see this, suppose that the X_is represent the lifetimes of items, and we are to be given only M_n, the largest of n lifetimes. If (upon receiving this information) our judgment about the other $(n-1)$ lifetimes was that each could be anywhere between 0 and M_n with equal probability, and if the knowledge about any other lifetime did not change this judgment, then a mixture over θ of uniform variables would be a suitable model for the n lifetimes.

Mixtures of Poisson Distributions

Let $\{X_i\}$, $1 \leq i \leq \infty$, take integer values. Freedman (1962) has shown that a necessary and a sufficient condition for the X_is to have a representation as a mixture of Poisson (λ) variables, i.e. as

$$P\{X_1 = x_1, \ldots, X_n = x_n\} = \int_0^\infty \prod_{i=1}^n \frac{e^{-\lambda}\lambda^{x_i}}{x_i!}\, F(\mathrm{d}\lambda) \tag{3.7}$$

is that for every n, the joint distribution of $(X_1, \ldots, X_n)$, given $S_n = \sum_{i=1}^n X_i$, is a multinomial on n-tuples of nonnegative integers whose sum is S_n with 'uniform' probabilities $(1/n, \ldots, 1/n)$. That is, for any integer valued a_i, $i = 1, 2, \ldots, n$, such that $S_n = \sum_{i=1}^n a_i$,

$$P\{X_1 = a_1, \ldots, X_n = a_n \mid S_n\} = \frac{S_n!}{a_1! \times \ldots \times a_n!} \prod_{i=1}^n (1/n)^{a_i}. \tag{3.8}$$

The distribution (3.8) given above is that of S_n balls dropped at random into n boxes, and is also known as the 'Maxwell–Boltzman' distribution.

Much of the more recent work in survival analysis (cf. Andersen *et al.*, 1993) and in reliability (cf. Asher and Fiengold, 1984) deals with count data. Modeling counts by point process models, like the Poisson, has turned out to be quite useful. The Bayesian analysis of a Poisson process model would involve expressions like (3.7). For example, Campodónico and Singpurwalla (1995) analyze the number of fatigue defects in railroad tracks and encounter a version of (3.7). The result above says that if we are to use the mixture model (3.7), we must be prepared to make the judgment implied by (3.8). Alternatively, we may start off in the true spirit of a de Finetti-style theorem and first elicit expert judgment on counts; if this happens to be of the form (3.8), we must use the mixture model (3.7).

Mixtures of Geometric Distributions

Let $\{X_i\}$, $1 \leq i \leq \infty$, take integer values. Diaconis and Freedman (1981) have shown that a necessary and a sufficient condition for the X_is to have a representation as a mixture of geometric (θ) variables, i.e.

$$P\{X_1 = x_1, \ldots, X_n = x_n\} = \int_0^1 \prod_{i=1}^n (1-\theta)^{x_i - 1} \theta F(d\theta) \tag{3.9}$$

is that for every n, the joint distribution of $(X_1, \ldots, X_n)$, given $S_n = \sum_{i=1}^n X_i$, is a 'uniform' distribution over all nonnegative n-tuples of integers whose sum is S_n. That is,

$$P\{X_1 = x_1, \ldots, X_n = x_n \mid S_n\} = \frac{1}{k}, \tag{3.10}$$

where k is the total number of n-tuples whose sum is S_n. For example, if $S_n = 1$, then the total number of n-tuples whose sum is 1 is n; i.e. $k = n$. The distribution (3.10) is called the 'Bose–Einstein' distribution.

Mixtures of Exponential Distributions

When is $X_1, \ldots, X_n$ a mixture over θ of sequences of independent exponential chance variables with parameter θ? That is, when is

$$P\{X_1 \leq x_1, \ldots, X_n \leq x_n\} = \int_0^\infty \prod_{i=1}^n (1 - e^{-x_i/\theta}) F(d\theta), \tag{3.11}$$

where $P(X \leq x \mid \theta) = (1 - e^{-x/\theta})$, an exponential distribution function?

A necessary and sufficient condition (cf. Diaconis and Freedman, 1987) for the above to hold is that, for each n, given the sufficient statistic $S_n = \sum_{i=1}^n X_i$, the joint distribution of $(X_1, \ldots, X_n)$ is uniform on the simplex $\{X_i \geq 0, \text{ and } S_n\}$.

The exponential chance distribution is one of the most frequently used failure models in reliability and survival analysis. Its popularity is attributed to its simplicity, and also to the fact that it characterizes non-aging or lack of wear; that is, its failure rate (section 4.3) is a constant function of time. In the mathematical theory of reliability, its importance stems from the fact that the exponential distribution function provides bounds on the distribution function of a large family of failure models (cf. Barlow and Proschan, 1965). Because of this central role, many papers dealing with a Bayesian analysis of the exponential failure model have been written. Much of this focuses on suitable choices for $F(d\theta)$. Martz and Waller (1982) is a convenient source of reference. The starting point for all such analyses are specific versions of (3.11). The de Finetti-style result given above says that underlying the use of (3.11) is a judgment of indifference, namely that, were we given only the sum of the n lifetimes, we would judge all the lifetimes to be equiprobable over a region which is defined by a simplex. Alternatively, we could have (in the spirit of de Finetti) started with the judgment of indifference, presumably provided by an expert's opinion, and then would be lead to (3.11). This would be a different motivation for using the exponential failure model.

Mixtures of Gamma and Weibull Distributions

Barlow and Mendel (1992) describe conditions under which $X_1, X_2, \ldots$ would be a mixture of gamma chance distributions. Starting with a finite population of N items with lifetimes X_i, $i = 1, \ldots, N$, and guided by the view that the easiest way to make probability judgments is via the principle of 'insufficient reason', they assume indifference of W_is (the transformed values of the X_is) on the simplex $\{X_i \geq 0, \sum_{i=1}^N X_i\}$ with $\sum_{i=1}^N X_i$ known. In the special case when $W_i = X_i^\beta$ and $\lim_{N\to\infty} \sum_{i=1}^N X_i/N = \lambda$, they obtain a mixture of gamma distributions with shape β and scale β/λ. The density function at x of a gamma-distributed variable X having scale λ and shape β is given as $\exp(-\lambda x)\lambda^\beta x^{\beta-1}/\Gamma(\beta)$, $x \geq 0$, where $\Gamma(\bullet)$ is the gamma function. Thus, parameters of failure models are functions of observable lifetimes. Mixtures of Weibull chance distributions with shape β and scale λ arise when all of the above hold except that the $\sum_{i=1}^N X_i$ in the simplex is replaced by $\sum_{i=1}^N X_i^\beta$. A random quantity X is said to have a **Weibull distribution** with shape β and scale λ if $P(X \geq x \mid \lambda, \beta) = \exp(-\lambda x^\beta)$, for $x \geq 0$.

Results that are analogous to the above, but pertaining to mixtures of the inverse binomial and the binomial are given by Freedman (1962). I think that these results are important to subjective Bayesians working on practical problems, because they provide a foundation for the starting point of their work, namely their choice of a model; also see Spizzichino (1988). All of the above results explain the consequences of the judgment of indifference on observables, given a statistic. In practice, since it is much easier to assess and to agree upon a 'uniform' distribution than upon any other distribution, indifference plays a key role. For this reason, I have placed the word 'uniform' within quotes.

3.3 ERROR BOUNDS ON DE FINETTI-STYLE RESULTS FOR FINITE SEQUENCES OF RANDOM QUANTITIES

There is one disturbing aspect of the material given in the previous section. It pertains to sequences of random quantities that are infinitely extendable. We have seen before that not all sequences of length k can be extended to length $n > k$. Also, the requirement of infinite extendability is not in keeping with de Finetti's general program of restricting attention to finite samples. Yet the results of the infinite case are simple and have been used since the time of Laplace. Thus a question arises as to how much of an error is committed in using the infinite case results for sequences that are not infinitely extendable. This question has been addressed by Diaconis and Freedman in a series of papers; in what follows, I reproduce a few of these for cases that are of interest to us.

3.3.1 Bounds for Finitely Extendable Zero-one Random Quantities

I start with the simplest case of zero-one variables. Recall, from the infinite version of de Finetti's theorem and the notation of section 3.2, that for every infinitely exchangeable sequence,

$$P(j\text{ones in } k \text{ trials}) = \binom{k}{j} \int_0^1 \theta^j (1-\theta)^{k-j} F(d\theta) \tag{3.12}$$

for a unique F and for every k with the same F.

Now consider a zero-one exchangeable sequence of length k which is extendable to an exchangeable sequence of length n, where n is finite but greater than or equal to k. Clearly, (3.12) will not hold for this sequence (see the comments at the end of section 3.1.3), but how wrong will we be if we use it? Diaconis and Freedman (1980) show that (3.12) almost holds in the sense that there exists an F such that, for any set $A \subset \{0, 1, 2, \ldots, k\}$,

$$\left|P\{\#\text{ of ones in } k \text{ trials} \in A\} - \sum_{j\in A} \binom{k}{j} \int_0^1 \theta^j (1-\theta)^{k-j} F(d\theta)\right| \leq 2k/n$$

uniformly in n, k and A.

Thus, in the case of the acceptance sampling and life-testing problem, testing 10 items from a lot of 1000 and using (3.12) instead of its finite version would imply an error of at most 0.02 in absolute value. For any fixed k, this error goes to zero as the lot size increases to infinity.

3.3.2 Bounds for Finitely Extendable Non-binary Random Quantities

Bounds analogous to the one above have also been obtained for non-binary random quantities. These bounds have their origin in a theorem due to Borel (1914) pertaining to the 'total variation distance' (see below) between the joint distribution of a finite number of random quantities and the joint distribution of the same number of independent Gaussian quantities. But before presenting these results, I introduce some convenient abbreviation. The Gaussian (θ, σ^2) distribution will be denoted as $\mathcal{N}(\theta, \sigma^2)$, the exponential (θ) as $\mathcal{E}(\theta)$, the Poisson (λ) as $P(\lambda)$, and the geometric (θ) as (θ).

The Gaussian Case

Suppose that the vector $(X_1, \ldots, X_n)$ is judged rotationally symmetric, and let P_k be the joint distribution function (also known as the **law**) of $(X_1, \ldots, X_k)$. Let $\zeta_1, \ldots, \zeta_k$ be independent and identically distributed as $\mathcal{N}(0, 1)$, and let $\mathcal{N}^k(\sigma)$ be the law of $\sigma\zeta_1, \ldots, \sigma\zeta_k$. Finally, let $\mathcal{N}_{Fk} = \int \mathcal{N}^k(\sigma)F(\mathrm{d}\sigma)$ (the mixture over σ of $\mathcal{N}^k(\sigma)$), where F is a probability on $[0, \infty)$. Then, the following theorem provides a (reasonably sharp) bound on the total variation distance between P_k and $\mathcal{N}_{Fk}$.

Theorem 3.3 (Diaconis and Freedman, 1987). *If $(X_1, \ldots, X_n)$ is rotationally symmetric, then there is a probability F on $[0, \infty)$ such that, for $1 \leq k \leq n-4$*

$$||P_k - \mathcal{N}_{Fk}|| \leq 2(k+3)/(n-k-3).$$

Loosely speaking, if P and Q are two probability distributions defined on all sets $A \in \mathcal{A}$, then their **total variation distance** is denoted by $||P - Q||$ and is defined as

$$||P - Q|| = 2 \sup_{A \in \mathcal{A}} |P(A) - Q(A)|.$$

The Exponential Case

Analogous to the above, suppose that the vector $(X_1, \ldots, X_n)$ is judged uniform on the simplex $\{X_i \geq 0, S_n\}$, given $\sum_{i=1}^n X_i = S_n$. Let P_k be the law of $(X_1, \ldots, X_n)$, and let $\zeta_1, \ldots, \zeta_k$, be independent and identically distributed as $\mathcal{E}(\lambda)$. Let $\mathcal{E}^k(\lambda)$ be the law of $(\zeta_1, \ldots, \zeta_k)$, and $\mathcal{E}_{Fk} = \int \mathcal{E}^k(\lambda)F(\mathrm{d}\lambda)$ be the mixture over λ of $\mathcal{E}^k(\lambda)$, where F is a probability on $[0, \infty)$. Then,

Theorem 3.4 (Diaconis and Freedman, 1987). *Given $\sum_{i=1}^n X_i = S_n$, if $(X_1, \ldots, X_n)$ is uniform on the simplex $\{X_i \geq 0 \text{and } S_n\}$, then there exists a probability F on $[0, \infty)$ such that, for all $1 \leq k \leq n-2$,*

$$||P_k - \mathcal{E}_{Fk}|| \leq 2(k+1)/(n-k+1).$$

The US Government's Military Standard MIL-STD-781C (MIL-STD-781C, 1977) codifies acceptance sampling methods based on life tests. These methods, used worldwide, assume exponentially distributed lifelengths. The result given here indicates the error of this assumption when sampling from finite-sized lots.

The Geometric Case

Given $\sum_{i=1}^n X_i = S_n$, suppose that $(X_1, \ldots, X_n)$ is judged uniform on the simplex $\{X_i \in I, S_n\}$, where I is the set of nonnegative integers. Let P_k be the law of $(X_1, \ldots, X_n)$, and let $\zeta_1, \ldots, \zeta_k$ be independent and identically distributed as $\mathcal{GM}(\theta)$. Let $(\mathcal{GM})^k(\theta)$ be the law of $\theta\zeta_1, \ldots, \theta\zeta_k$. Finally, let $(\mathcal{GM})_{Fk} = \int (\mathcal{GM})^k(\theta)F(\mathrm{d}\theta)$ be the mixture over θ of $(\mathcal{GM})^k(\theta)$, where F is a probability on $[0, \infty)$. Then, analogous to the above results we have

Theorem 3.5 (Diaconis and Freedman, 1987). *Given $\sum_{i=1}^{n} X_i = S_n$, if $(X_1, \ldots, X_n)$ is judged uniform on the simplex $\{X_i \in I,\ S_n\}$, then there exists a probability F on $[0, \infty)$ such that, for all $1 \leq k \leq n-3$,*

$$||P_k - (\mathcal{GM})_{Fk}|| \leq 2\left\{\frac{n^2}{(n-k-1)(n-k-2)} - 1\right\}.$$

The Poisson Case

Let $\{X_i\}$, $1 \leq i \leq \infty$ take integer values. Suppose that the joint distribution of $(X_1, \ldots, X_n)$, given $S_n = \sum_{i=1}^{n} X_i$, is judged a **multinomial** on n-tuples of nonnegative integers whose sum is S_n, with uniform probabilities $(1/n, \ldots, 1/n)$. Let P_k be the law of $X_1, \ldots, X_k$, and let $\zeta_1, \ldots, \zeta_k$ be independent and identically distributed as $P(\lambda)$. Let $(P)^k(\lambda)$ be the law of $\lambda\zeta_1, \ldots, \lambda\zeta_k$. Finally, let $(P)_{Fk} = \int (P)^k(\lambda)F(\mathrm{d}\lambda)$ be the mixture over θ of $(P)^k(\lambda)$ where F is a probability on $[0, \infty)$. Then,

Theorem 3.6 (Diaconis and Freedman, 1987). *Let $\sum_{i=1}^{n} X_i = S_n$, and $\{X_i\}$, $1 \leq i \leq \infty$, take integer values. Given S_n, if $(X_1, \ldots, X_n)$ is judged multinomial on n-tuples of nonnegative integers whose sum is S_n with uniform probabilities $(1/n, \ldots, 1/n)$, then there is a probability F on $[0, \infty)$ such that, for all $k \leq n/2$,*

$$||P_k - (P)_{Fk}|| \leq 2k/n.$$

In a recent report, Iglesias, Pereira and Tanka (1994) obtain results similar to the above for finite sequences of uniformly (over $[-\theta, \theta]$) distributed variables; their bound is of the order $4k/n$. Thus in most cases of interest, the law for finite sequences (of length k) which can be extended to length n is within $2k/n$ of the law of mixtures of independent identically distributed sequences. In view of such results, the search for failure models based on finite population sizes appears to be of limited value.

Chapter 4

Stochastic Models of Failure

Here lies Henri Lebesgue: Talent beyond measure.

4.1 INTRODUCTION

We have seen that problems of risk analysis entail an assessment of uncertainty associated with the occurrence of undesirable events, such as the failure of an item or the onset of a disease. Consequently, probability models for failure play an important role. Indeed much of what is done in the statistical theory of reliability, and in survival analysis, pertains to the selection of suitable failure models, assessing a model's goodness-of-fit to observed data [e.g. Chandra, Singpurwalla and Stephens (1981)], inference about the parameters of a chosen model, and predictions based on a chosen model.

In section 2.6.1 of Chapter 2, I stated that there are three general approaches for selecting a failure model. One is based on an assessment of the aging characteristics of an item, the other based on exchangeability and indifference, and the third based on a probabilistic modeling of the physics of failure. Chapter 3 pertained to the role of exchangeability and the principle of indifference in selecting some of the commonly used models of failure. Many analysts find this approach to model selection lacking with respect to the incorporation of knowledge about the causes and the nature of failure. The more traditional approaches for selecting failure models have been the other two, namely, those based on an appreciation of the aging properties of an item, and those based on an analysis of its physics of failure [cf. Gertsbach and Kovdonsky (1969)]. The purpose of this chapter is to describe features of these two traditional approaches. But first it is helpful to overview some notation and terminology introduced in sections 2.1 and 2.2, and to outline preliminaries which set the stage for a better discussion of the ensuing text.

4.2 PRELIMINARIES: UNIVARIATE, MULTIVARIATE AND MULTI-INDEXED DISTRIBUTION FUNCTIONS

Suppose that at some reference time τ, an item is to be commissioned for use, and suppose that T denotes its unknown time to failure; then T is said to be the **lifelength** of the item. Suppose that the possible values that T can take are denoted by $t \geq 0$. Let $\mathcal{H}$ be the background information that is available to an analyst (henceforth us) at time τ, and let $F(t; \mathcal{H}) = P(T \leq t; \mathcal{H})$ be our

Reliability and Risk: A Bayesian Perspective N.D. Singpurwalla
© 2006 John Wiley & Sons, Ltd

assessment of the uncertainty about T. Let $R(t;\mathcal{H}) = P(T \geq t;\mathcal{H}) = 1 - F(t;\mathcal{H}) = \overline{F}(t;\mathcal{H})$. In reliability theory, $R(t;\mathcal{H})$ is known as the **survival function** of T for a **mission** of duration $[0, t]$. In general, T need not stand for time; alternate candidates could be the number of cycles to failure, or the number of miles to failure, etc. Furthermore, T need not be a lifelength, and could denote things such as the time to the onset (or the remission) of a disease subsequent to the exposure to a hazardous material (or a medical intervention). No matter what T represents, it is important to bear in mind that like the distribution function $F(t;\mathcal{H})$, the survival function $R(t;\mathcal{H})$ is assessed at some reference time τ, so that $R(t;\mathcal{H})$ encompasses two kinds of time: the reference time τ and the mission time t. Whereas the latter is explicit in the expression $R(t;\mathcal{H})$, the former is implicit in $\mathcal{H}$; indeed, for clarity, $\mathcal{H}$ should really be read as $\mathcal{H}(\tau)$.

Depending, among other things, on the values t that T is able to take, $F(t;\mathcal{H})$ can be absolutely continuous (section 2.2), discontinuous with jumps at one or more values of t, or singular.[1] A celebrated result in probability theory is the **Lebesgue decomposition formula**, which says that any distribution function can be written as the weighted sum of an absolutely continuous, a discontinuous and a singular distribution function, (for example, Kolmogorov and Fomin, 1970, p. 341). This decomposition formula is germane to me; it has made an appearance in reliability theory via Marshall and Olkin's (1967) multivariate exponential (chance) distribution.

When $F(t;\mathcal{H})$ is absolutely continuous (Figure 4.1(a)) it is differentiable almost everywhere, and $F(t;\mathcal{H}) = \int_0^t f(s;\mathcal{H})ds$, the **Lebesgue integral**; $f(s;\mathcal{H})$ is known as the **probability density** generated by $F(t;\mathcal{H})$ at the point s. It is instructive to think of $f(t;\mathcal{H})ds$ as, approximately, the probability that T is between s and $s + ds$, where ds is a small increment of T in the vicinity of s. In Chapter 3, the quantity $f(x;\mathcal{H})ds$ was denoted $F(ds)$.

When $F(t;\mathcal{H})$ is discontinuous in t, that is, $F(t;\mathcal{H})$ consists of jumps (Figure 4.1(b)) then the derivative of $F(t;\mathcal{H})$ will not exist everywhere, and T will not have a probability density at the points of discontinuity of $F(t;\mathcal{H})$. A jump in $F(t;\mathcal{H})$ will occur at say $t = t^*$, if it were known in advance (i.e. at time τ) that at t^* there is a nonzero probability of failure. This type of a situation can arise in engineering reliability under scenarios wherein an item is known to experience a shock or some other type of a disturbance at a pre-specified time t^*; the same is also true if T is able to take only discrete values. Whenever $F(t;\mathcal{H})$ contains jumps of the type mentioned above, $F(t;\mathcal{H}) = \int_0^t dF(s;\mathcal{H})$, the **Lebesgue–Stieltjes integral**. Here $\int dF(s;\mathcal{H})$ is taken to be $\int_0^t f(s;\mathcal{H})ds$ whenever $F(s;\mathcal{H})$ is differentiable at s, and it is the size of the jump

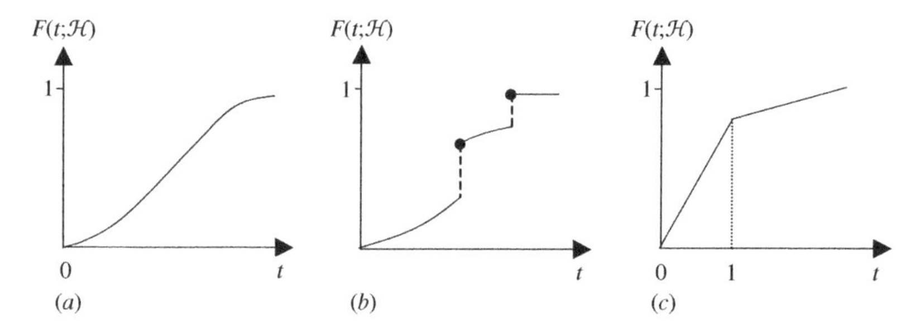

Figure 4.1 Absolutely continuous, discontinuous and continuous distribution functions.

[1] A distribution function is said to be 'singular' if it is continuous, not identically zero, and has derivatives that exist but vanish, almost everywhere. In the single variable case, an example of a singular distribution is the 'Cantor distribution'; this is a continuous distribution which does not have a probability density (cf. Feller, 1966, pp. 35–36). Singular distributions arise in reliability in the context of multi-component systems having dependent lifelengths. The multivariate exponential of Marshall and Olkin (1967) has a singular component. Singular components pose difficulties with inference (section 4.7.4).

in $F(t; \mathcal{H})$ at all s where $F(t; \mathcal{H})$ takes a jump. A discontinuous function is not absolutely continuous with respect to the Lebesgue integral.

Figure 4.1(c) illustrates the distribution function $F(t; \mathcal{H}) = t/2$ for $t \in (0, 1]$ and $F(t; \mathcal{H}) = 1/2 + (t-1)/4$ for $t \in (1, 3)$. Clearly, $F(t; \mathcal{H})$ is continuous but not differentiable at $t = 1$. Specifically, $F(t; \mathcal{H})$ has a derivative $f(t; \mathcal{H}) = 1/2$ for $t \in (0, 1)$, and $f(t; \mathcal{H}) = 1/4$ for $t \in (1, 3)$; thus $F(t; \mathcal{H})$ is differentiable almost everywhere, i.e. for all $t \in (0, 3)$, except $t = 1$. Furthermore, since $F(t; \mathcal{H}) = \int_0^t f(s; \mathcal{H})\mathrm{d}s$, the distribution function $F(t; \mathcal{H})$ is also absolutely continuous. Note that every absolutely continuous function is the indefinite (Lebesgue) integral of its derivative (cf. Royden, 1968, p. 107). The above technicalities are important; they are germane to the material of section 4.7.

When $F(t; \mathcal{H})$ is absolutely continuous, the quantity $\int_0^\infty t f(t; \mathcal{H})\mathrm{d}t$ is known as the **expected value** of $F(t; \mathcal{H})$, or since T denotes a lifelength, it is also known as the **mean time to failure** (MTTF); it is denoted by $E(T)$. When $F(t; \mathcal{H})$ has discontinuities, $E(T) = \int_0^\infty t\mathrm{d}F(t; \mathcal{H})$. With either case, since $T \geq 0$, it is easy to verify that $E(T) = \int_0^\infty R(t; \mathcal{H})\mathrm{d}t$. In general, if $g(T)$ is some function of T, then $E[g(T)] = \int_0^\infty g(t) f(t; \mathcal{H})\mathrm{d}t$ or if $F(t; \mathcal{H})$ has discontinuities, then $E[g(T)] = \int_0^\infty g(t)\mathrm{d}F(t; \mathcal{H})$. The quantity $E(T^2) - (E(T))^2$ is the **variance** of $F(t; \mathcal{H})$; it is denoted $V(T)$. The quantity $\sqrt{V(T)}$ is known as the ***standard deviation*** of $F(t; \mathcal{H})$ and $\sqrt{V(T)}/E(T)$ is known as the ***coefficient of variation*** of $F(t; \mathcal{H})$. The mode of T is that value of T at which $f(t; \mathcal{H})$ attains its maximum, and the **median** is that T for which $F(t; \mathcal{H}) = 1/2$.

The remaining lifetimes of a unit of age t is known as the **residual life**, and its expected value, the **mean residual life** (MRL). The latter plays a useful role in planning for maintenance and replacement, and in the context of health care, issues pertaining to the quality of life. Using the notation given above, but suppressing the $\mathcal{H}$, the residual life is $((T-t) \mid T \geq t)$ and $E((T-t) \mid T \geq t)$ is the MRL. Since $P((T-t) \geq u \mid T \geq t) = \overline{F}(t+u)/\overline{F}(t)$, the MRL is given as

$$E((T-t) \mid T \geq t) = \frac{\int_t^\infty \overline{F}(u)\,\mathrm{d}u}{\overline{F}(t)}.$$

Our discussion thus far has been restricted to the case of the lifelength T of a single item (or a single system) that is to be commissioned for use at time τ. In practice, interest can also center around the lifelengths of a collection of n items to be commissioned for use at τ. For example, we may be interested in the lifelengths of a system of n components (or a network of n nodes) with component (node) i having lifelength T_i, $i = 1, \ldots, n$. If $t_i \geq 0$ denotes the possible values that T_i can take, and if $\mathcal{H}$ is the background information about all the n components (nodes) available to us at time τ, then $P(T_1 \leq t_1, \ldots, T_n \leq t_n; \mathcal{H})$, our uncertainty about $T_1, \ldots, T_n$ at τ, will be denoted by $F(t_1, \ldots, t_n; \mathcal{H})$. As before, $R(t_1, \ldots, t_n; \mathcal{H}) = P(T_1 \geq t_1, \ldots, T_n \geq t_n; \mathcal{H})$ is the joint survival function of $T_1, \ldots, T_n$, for mission times $t_1, \ldots, t_n$, respectively. It is important to note that in this multivariate (i.e. $n \geq 2$) scenario, $R(t_1, \ldots, t_n; \mathcal{H}) \neq 1 - F(t_1, \ldots, t_n; \mathcal{H})$ where $F(t_1, \ldots, t_n; \mathcal{H})$ is the joint probability distribution function of $T_1, \ldots, T_n$. As a function of $t_1, \ldots, t_n$, $F(t_1, \ldots, t_n; \mathcal{H})$ can be absolutely continuous, discontinuous or singular; also, by the Lebesgue decomposition formula, $F(t_1, \ldots, t_n; \mathcal{H})$ can be written as a weighted sum of the above three components. As in the univariate (i.e. $n = 1$) case, if $F(t_1, \ldots, t_n; \mathcal{H})$ is absolutely continuous, then it is differentiable at (almost) all values of $t_1, \ldots, t_n$, and $f(s_1, \ldots, s_n; \mathcal{H})$, its partial derivative at $s_1, \ldots, s_n$ (known as the joint probability density function generated by $F(t_1, \ldots, t_n; \mathcal{H})$ at $s_1, \ldots, s_n$) is such that $F(t_1, \ldots, t_n; \mathcal{H}) = \int_0^{t_1} \ldots \int_0^{t_n} f(s_1, \ldots, s_n; \mathcal{H})\mathrm{d}s_1 \ldots \mathrm{d}s_n$; specifically, $f(s_1, \ldots, s_n; \mathcal{H}) = \partial^n F(s_1, \ldots, s_n; \mathcal{H})/\partial s_1 \ldots \partial s_n$.

Given $R(t_1, \ldots, t_n; \mathcal{H})$ it is easy to see that for any $i, i = 1, \ldots, n$, $R(t_i; \mathcal{H}) = P(T_i \geq t_i; \mathcal{H})$ is given by $R(0, \ldots, 0, t_i, 0, \ldots, 0; \mathcal{H})$; i.e. by setting all the t_js, $j \neq i$, in $R(t_1, \ldots, t_n; \mathcal{H})$ equal to zero. Once $R(t_i; \mathcal{H})$ is obtained, $F(t_i; \mathcal{H}) = P(T_i \leq t_i; \mathcal{H})$, the marginal distribution function of T_i, is simply $F(t_i; \mathcal{H}) = 1 - R(t_i; \mathcal{H})$. Finally, the lifelengths $T_1, \ldots, T_n$ are judged to be **mutually independent** if the $R(t_i; \mathcal{H})$, $i = 1, \ldots, n$, are such that

$$R(t_1, \ldots, t_n; \mathcal{H}) = \prod_{i=1}^{n} R(t_i; \mathcal{H}),$$

and the above type of relationship holds for all subsets of $(T_1, \ldots, T_n)$ of size $k \geq 2$, and all permutations of the indices of T_i.

The judgment of independence, commonly made in reliability, is rarely meaningful; it is often an idealization. The common environment in which several components are commissioned to function suggests that it is more realistic to judge their lifelengths as being positively dependent (for example, Marshall, 1975 or Lindley and Singpurwalla, 1986). Another argument that motivates positive dependence stems from the fact that often the failure of one component induces added stresses on the surviving components, and this in turn exacerbates their susceptibility to failure (cf. Freund (1961)). In view of such considerations, several failure models for dependent lifelengths have been proposed in the literature; a few are discussed in section 4.7. Also discussed in this chapter (section 4.8) is the notion of what are known as causal and cascading failures, and models describing such failures are given.

The topic of multi-indexed failure distributions, or distributions with multiple scales, though introduced by Mercer as early as 1961, and also alluded to by Cox (1972), is relatively new. Such distributions are to be distinguished from multivariate failure distributions in the sense that they pertain to the stochastic failure of a single item, where an item's failure is recorded with respect to two or more scales, say age (or time) and usage. A classic example is the failure of an automobile under warranty; when the automobile fails one records both its chronological age since purchase, and its mileage. Often, there is a positive dependence between the two scales. For example, the mileage accumulated by an automobile increases with its age, and often a meaningful relationship between the two scales can be proposed (cf. Singpurwalla and Wilson, 1998). In principle, multi-indexed failure distributions are no different from multivariate failure distributions; the distinction between the two lies mainly in the context of their use and the manner in which they are constructed. Some preliminary results on distributions with multiple scales and other examples of their plausible use are in Singpurwalla and Wilson (1998); these are overviewed in section 4.9.

4.3 THE PREDICTIVE FAILURE RATE FUNCTION OF A UNIVARIATE PROBABILITY DISTRIBUTION

A concept that plays a key role in reliability and survival analysis is that of the failure rate function, also known as the hazard rate function, or the force of mortality. The origins of the notion of a hazard rate function can be traced to the earliest work in actuarial sciences though much of the recent work on it has been spawned by problems in reliability. In view of the vast literature in applied probability that has been devoted to the nature and the crossing properties of the failure rate function, it is appropriate to make the claim that the notion of the failure rate is perhaps the main contribution of reliability to probability. In this section, I introduce and articulate the idea of the failure rate function and show its relationship to the survival function. In the sequel, I also make the distinction between the predictive failure rate and the model failure rate. In what follows, I omit the background history argument $\mathcal{H}$, so that the $F(t; \mathcal{H})$ of the previous section is simply $F(t)$, and similarly $R(t; \mathcal{H})$ is $R(t)$; furthermore, following a common convention, I denote the quantity $R(T) = 1 - F(t)$ by $\overline{F}(T)$.

Suppose that T is discrete, taking values $0, \Delta, 2\Delta, \ldots, k\Delta, \ldots$, for some $\Delta > 0$, and suppose that at time τ, the uncertainty about T is assessed via $F(k\Delta)$, for $k = 0, 1, 2, \ldots$,. Then, the **predictive failure rate** of $F(k\Delta)$ at some future time $k\Delta$, and as assessed at time τ (the now time) is defined as

$$\begin{aligned} r(k\Delta) &= \frac{P(k\Delta < T \leq (k+1)\Delta)}{P(T > k\Delta)} \\ &= \frac{\overline{F}(k\Delta) - \overline{F}((k+1)\Delta)}{\overline{F}(k\Delta)}. \end{aligned} \tag{4.1}$$

Even though the failure rate is a property of the distribution function $F(k\Delta)$, it is common to refer to it as the failure rate of T. Since $r(k\Delta)$ can be defined for all values of k, $r(k\Delta)$ as a function of $k\Delta$, for $k = 0, 1, 2, \ldots$, is known as the **predictive failure rate function** of $F(k\Delta)$. It is easy to verify that

$$1 - r(k\Delta) = \frac{\overline{F}((k+1)\Delta)}{\overline{F}(k\Delta)}; \tag{4.2}$$

thus, if we are able to specify $r(k\Delta)$, for $k = 0, 1, 2, \ldots$, then because of the fact that $\overline{F}(0) = 1$ we are able to obtain $\overline{F}(k\Delta)$ – the survival function – for $k = 0, 1, 2, \ldots$, via the relationship

$$\overline{F}(k\Delta) = \prod_{i=0}^{k-1} (1 - r(i\Delta)). \tag{4.3}$$

The above relationship, known as the multiplicative formula of reliability and survival, shows that there exists a one-to-one relationship between the failure rate function $r(k\Delta)$ and the survival function $\overline{F}(k\Delta)$ for $k = 0, 1, \ldots$ This one-to-one relationship can be advantageously exploited in the manner discussed below.

We first note that $r(k\Delta)$ is a conditional probability (assessed at time τ) that the item in question fails at the time point $(k+1)\Delta$, were it to be surviving at $k\Delta$; both $k\Delta$ and $(k+1)\Delta$ are future (subsequent to τ) time points, and $r(t) = 0$, for all $t \neq 0, \Delta, 2\Delta, \ldots$ Since probability is subjective, and since conditional probabilities are generally easier to assess than the unconditional ones, $r(k\Delta)$ as a function of k encapsulates one's judgment about the aging or the non-aging features of an item. By 'aging' I mean the degradation of an item with use; for example, humans and mechanical devices experience some form of a deterioration and wear over time. The characteristic of non-aging is opposite to that of aging. Non-aging as a physical or a biological phenomenon is rare, and occurs for instance, in the context of work-hardening of materials, or with the development of a newborn's immune system. With aging one would assess an increasing sequence of conditional failure probabilities over equal intervals of time. That is, $r(k\Delta)$ will be judged increasing in k. Conversely, $r(k\Delta)$ will be judged decreasing if the item experiences non-aging (section 4.4.2). If the physical state of an item does not change with age then $r(k\Delta)$ is assessed as being constant over k. For example, it has been claimed that electronic components neither degrade nor strengthen with age. For such devices $r(k\Delta)$ is constant in k. Thus to summarize, the failure rate as a conditional probability can be subjectively specified by a subject matter specialist based on the aging characteristics of an item, and once this is done the survival function $\overline{F}(k\Delta)$ can be induced via (4.3). As an aside, suppose that we are to assume that the item has failed at $k\Delta$ and are required to assess $r(k\Delta)$. Alternatively put, we are asked 'what is the failure rate of a failed item?' One is tempted to give zero for an answer, but this is not correct! To see why, we note from (4.2) that assuming failure at $k\Delta$, $1 - r(k\Delta) = 0/0$ which is undefined; this is to be expected since probabilities are meaningful only for those events whose disposition is not known. Thus it does not make sense to talk about the failure rate of a failed item.

For T continuous, one is tempted to define the failure rate as the limit of the right-hand side of (4.1) as $\Delta \downarrow 0$, and $k\Delta \to t$. However, such a temptation does not lead us to a satisfactory limiting operation. Instead for T continuous, the **instantaneous predictive failure rate** at t (when the following limit exists) is defined as

$$r(t) = \lim_{\mathrm{d}t \downarrow 0} \frac{P(t < T \leq t + \mathrm{d}t \mid T > t)}{\mathrm{d}t} \tag{4.4}$$

and $r(t)$ as a function of t is called the predictive failure rate function of the distribution function $F(t) = P(T \leq t)$, or equivalently of T. Informally, $r(t)$ is a measure of the risk that a failure will occur at t.

Note that for small values of $\mathrm{d}t$, $r(t)\mathrm{d}t$ can be interpreted as, approximately, the probability that an item of age t will fail in $(t, t + \mathrm{d}t)$. Thus, like $r(k\Delta)$, the instantaneous failure rate is a conditional probability, the probability assessment made at time τ. It is important to note that the above interpretation of $r(t)\mathrm{d}t$ is not appropriate if $P(T \leq t)$ is singular, or is otherwise not absolutely continuous. Since $F(t) = 1 - \overline{F}(t)$, We may rewrite (4.4) as

$$r(t) = \lim_{\mathrm{d}t \downarrow 0} \frac{F(t + \mathrm{d}t) - F(t)}{\overline{F}(t)\mathrm{d}t}, \tag{4.5}$$

provided that $\overline{F}(t) > 0$.

Clearly, the above limit will exist if $F(t)$ is absolutely continuous, in which case

$$r(t)\overline{F}(t) = \frac{\mathrm{d}}{\mathrm{d}t}F(t) = -\frac{\mathrm{d}}{\mathrm{d}t}\overline{F}(t). \tag{4.6}$$

If $\mathrm{d}F(t)/\mathrm{d}t = f(t)$, the probability density generated by $F(t)$ at t, then (4.6) leads us to the relationship

$$r(t) = \frac{f(t)}{\overline{F}(t)}, \tag{4.7}$$

which with $\overline{F}(t) > 0$ is often taken as the definition of the predictive failure rate and is the starting point of a mathematical theory of reliability (Barlow and Proschan (1965)).

Since the limit in (4.5) need not always exist for all $t \geq 0$, the predictive failure rate at t is sometimes also defined as

$$r(t) = \frac{\mathrm{d}F(t)}{\overline{F}(t)}, \tag{4.8}$$

where $\mathrm{d}F(t)$ is $f(t)\mathrm{d}t$ wherever $F(t)$ is differentiable, and is otherwise $F(t^+) - F(t^-)$, the jump in $F(t)$ at t. When $F(t)$ is the distribution function of Figure 4.1(c), its failure rate at $t = 1$ does not exist.

Equation (4.6), which comes into play when $F(t)$ is absolutely continuous, is a differential equation with $\overline{F}(0) = 1$ as the initial condition; indeed $\overline{F}(0) = 1$ is a defining feature of failure time distributions. The solution of this differential equation is the famous exponentiation formula of reliability and survival (not to be confused with the product integral formula which will be described soon). Specifically, for $t \geq 0$

$$\overline{F}(t) = \exp(-\int_0^t r(u)\mathrm{d}u). \tag{4.9}$$

The quantity $H(t) = \int_0^t r(u)\mathrm{d}u$ is important. As a function of t, it is known as the **cumulative hazard function** of $F(t)$ (or equivalently of T). Since $H(t) = -\log \overline{F}(t)$ – a classic relationship in reliability and survival analysis – it is easy to see that $f(t)$ the probability density generated by $F(t)$ is given as

$$f(t) = \exp(-\int_0^t r(u)\mathrm{d}u)\ r(t), \qquad \text{for } t \geq 0. \tag{4.10}$$

Finally, using the fact that since $H(t) = H(t_1) + (H(t) - H(t_1))$, for any t_1, $0 \leq t_1 \leq t$, we have

$$\overline{F}(t) = \mathrm{e}^{-H(t_1)}\mathrm{e}^{-(H(t)-H(t_1))},$$

from which the well-known fact that

$$\overline{F}(t) = P(T \geq t) = P(T \geq t_1)\ P(T \geq t \mid T \geq t_1)$$

follows.

The one-to-one relationship between the failure rate function and the survival function of (4.9) parallels that given by (4.3) for T discrete. To see this parallel, we may rewrite (4.3) as

$$\overline{F}(k\Delta) = \exp\left(-\sum_{i=0}^{k-1} \ln\left(\frac{1}{1 - r(i\Delta)}\right)\right).$$

4.3.1 The Case of Discontinuity

When $F(t)$ is not absolutely continuous, it does not have a probability density at all values of t, and thus it is (4.8) that comes into play. We may use this equation to define the cumulative hazard function as

$$H(t) = \int_0^t \mathrm{d}F(s)/\overline{F}(s),$$

Note that $H(t)$ will be non-decreasing and right-continuous on $[0, \infty)$; it will be interpreted as

$$\mathrm{d}H(s) = P(T \in [s, s + \mathrm{d}s) \mid T \geq s) = \mathrm{d}F(s)/\overline{F}(s).$$

Thus

$$\mathrm{d}F(s) = \mathrm{d}H(s)\overline{F}(s),$$

so that for any $0 \leq a \leq b < \infty$

$$\int_a^b \mathrm{d}F(s) = \int_a^b \mathrm{d}H(s)\overline{F}(s),$$

or that $F[a, b) = \int_a^b \mathrm{d}H(s)\overline{F}(s)$.

The solution to the above differential equation is the famous **product integral formula**; namely, for $t \geq 0$

$$\overline{F}(t) = \prod_{[0,t]} (1 - \mathrm{d}H(s));$$

see (4.3) for a parallel. It can be shown (Hjort (1990), p. 1268) that the right-hand side of the above equation equals $\exp[-h[a, b)]$ if and only if H is continuous. This means that the classic relationship $H(t) = -\log \overline{F}(t)$ is true only when F is absolutely continuous.

4.4 INTERPRETATION AND USES OF THE FAILURE RATE FUNCTION – THE MODEL FAILURE RATE

No single concept in probability appears to be more associated with reliability and survival analysis than that of the failure rate function. Engineers are attracted to the concept of a failure rate because it captures their intuitive notions about wear, in the sense that items that wear out should have a failure rate function that increases with time, and vice versa. By wear, we mean the depletion with use of material or resources so that an item's ability to perform its intended functions gets diminished. Similarly, biometricians find the failure rate function a useful device for expressing their opinions about aging, which again is a deterioration in the ability to perform specified tasks over time. From a mathematical point of view, the failure rate function is attractive because of formulae (4.3) and (4.9), which provide a one-to-one relationship between the failure rate function and the survival function. However, there exist circumstances under which the formulae (4.3) and (4.9) will not hold. These circumstances are discussed in Singpurwalla and Wilson (1995) (section 9.5.2). Thus, in practice, the general form of the failure rate function could be specified by a subject matter specialist, from which the reliability function can be automatically deduced. In effect, as discussed next, the failure rate is a useful device for making a connection between the physics or the biology of failure, and the probability of survival.

Suppose that an item is judged to neither age (wear) nor improve with use, then the failure rate function of its time to failure T, should be a constant, say $\lambda > 0$. But often λ cannot be specified, and so our judgment about the constant failure rate of T is conditional on λ being known. That is, the failure rate of $(T \mid \lambda)$ is $r(t \mid \lambda)$, i.e. a constant λ. From (4.9) it then follows that the survival function of $(T \mid \lambda)$ is $\overline{F}(t \mid \lambda) = \exp(-\lambda t)$ for $t \geq 0$; this is an exponential distribution with scale λ. Barlow (1985) refers to $r(t \mid \lambda)$ as the model failure rate of T, because this is the failure rate that generates a probability model (or a failure model, or a lifelength model) for T. Since $\overline{F}(T \mid \lambda) = \exp(-\lambda t)$ also appears in (3.11) as a chance distribution, $r(t \mid \lambda)$ can also be seen as the chance failure rate of T.

The model failure rate is to be distinguished from the **predictive failure rate** of T, which is the unconditional failure rate of T obtained by averaging out our uncertainty about λ. It will be pointed out in section 4.4.2 that the predictive failure rate of T is decreasing in t even though its model failure rate is constant in t. The key distinction between the model and the predictive failure rate is that the former is the failure rate of T conditioned on parameters, whereas the latter is its unconditional failure rate. In the current literature, it is the model failure rate that has been predominantly specified and discussed.

The exponential distribution, with a specified scale parameter λ, is a suitable failure model for items that do not deteriorate nor improve with time. For items that are judged to be otherwise, a commonly used model failure rate is $r(t \mid \lambda, \beta) = \beta \lambda^{\beta} t^{\beta - 1}$, for some $\lambda, \beta > 0$, and $t \geq 0$. Observe that $r(t \mid \lambda, \beta)$ increases (decreases) in t, when $\beta > (<) 1$, so that conditionally, both the judgments of aging and improvement of ability to perform can be represented. When $\beta = 1$, the model failure rate is a constant (Figure 4.2). The above form of $r(t \mid \lambda, \beta)$ results, via formula

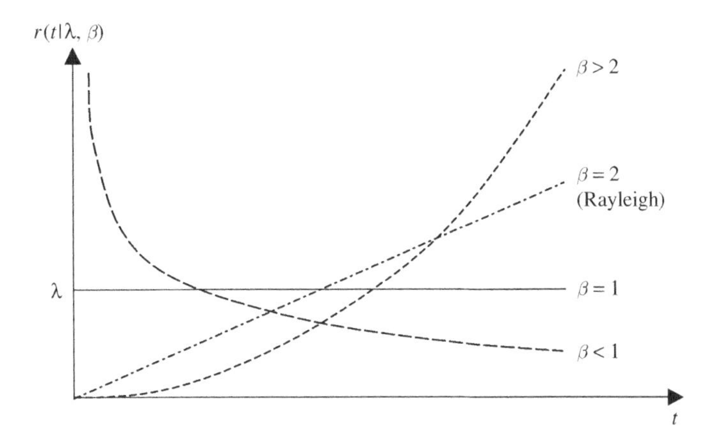

Figure 4.2 The model (chance) failure rate function of a Weibull distribution.

(4.9), in the famous Weibull distribution with scale parameter λ and shape parameter β; its survival function is $\overline{F}(t \mid \lambda, \beta) = \exp(-(\lambda t)^{\beta})$ for $t \geq 0$. With $\beta = 2$, the Weibull distribution is often referred to as the Rayleigh distribution. The scale parameter λ is sometimes also referred to as the characteristic life. The reason behind this is not clear, save for the fact that when $t = \lambda$, $\overline{F}(t \mid \lambda, \beta) = e^{-1} \approx 0.37$, suggesting that about 63% of the items will experience a failure by time $t = \lambda$.

The Weibull is a useful failure model in both biomedical and engineering applications, though Gavrilov and Gavrilova (2001) claim that for biological organisms, the *Gompertz Law* with a model failure rate function of the form $r(t \mid A, B, \alpha) = A + B\exp(\alpha t)$, $A, B, \alpha > 0$ is more appropriate. The Gompertz Law has been popular in the study of human aging and longevity. Other forms of the model failure rate and their corresponding survival functions are given in Chapter 4 of Mann, Schafer and Singpurwalla (1974) and Chapter 2 of Singpurwalla and Wilson (1999); these include the gamma, the lognormal, the extreme value, the Pareto, the truncated normal, a family of distributions proposed by Birnbaum and Saunders (1969), etc.

The lognormal with parameters $\mu \geq 0$ and $\sigma > 0$ is noteworthy because its failure rate function need not be monotone in t. For certain combinations of values of μ and σ, $r(t \mid \mu, \sigma)$ can increase in t and then decrease in t, so that the failure rate function is a reverse U-shape. The lognormal distribution function $P(T > t \mid \mu, \sigma)$ has a probability density at t of the form $f(t \mid \mu, \sigma) = \frac{1}{\sqrt{2\pi}\sigma t} \exp\left(-\frac{1}{2}\left(\frac{\log t - \mu}{\sigma}\right)^2\right)$, $t \geq 0$. The lognormal distribution is related to the Gaussian distribution via the relationship $T = \log_e X$, where X is Gaussian with parameters μ (mean) and σ^2 (variance); its mean $E(T)$ is $e^{\mu+\sigma^2/2}$, its median is e^{μ}, its mode $e^{\mu+\sigma^2}$ and its variance $e^{2\mu+\sigma^2}(e^{\sigma^2} - 1)$.

The extreme value distribution, also known as Gumbel's distribution – after Gumbel (1958) who used it extensively to describe extremal phenomena, like floods and gusts – bears a relationship to the Weibull distribution in a manner that is akin to the normal and lognormal. Specifically, if X has a Weibull distribution, then $T = \log_e X$ has an extreme value distribution of the form $\overline{F}(t \mid \gamma, \eta) = \exp(-\exp(t - \gamma)/\eta))$, for $-\infty < t < \infty$, γ, η. It can be easily seen that with $\gamma = 0$ and $\eta = 1$, the model failure rate of the extreme value distribution is e^t, $-\infty < t < \infty$; this is an exponentially increasing function of t, taking the value 1 when $t = 0$. The parameter γ is a location parameter and η a scale parameter. Because the support of T is the real line $(-\infty, +\infty)$, the extreme value distribution cannot be an appropriate choice for a failure model. Its inclusion here is for completeness.

A two-parameter family of distributions was proposed by Birnbaum and Saunders (1969) as a model for describing failures due to fatigue caused by cyclical loading. Here

$$P(T > t) \mid \omega, \delta) = \Phi\left[\frac{1}{\omega}\left\{\left(\frac{t}{\delta}\right)^{\frac{1}{2}} - \left(\frac{\delta}{t}\right)^{\frac{1}{2}}\right\}\right], \quad 0 < t < \theta, \ \omega, \delta > 0;$$

ω is a shape parameter and δ a scale parameter. The quantity $\Phi(x) = \int_{-\infty}^{x} \exp(-s^2/2)\mathrm{d}s$ is the cumulative distribution function of a standard Gaussian distribution; i.e. a Gaussian distribution with mean zero and variance one. The distribution given above is absolutely continuous with a density function that is unimodal. The model failure rate of the Birnbaum–Saunders family of distributions is not monotonic; however, it has been conjectured that it is U-shaped (Figure 4.6).

The Inverse Gaussian distribution has often been proposed as an alternative to the lognormal for a failure distribution function (cf. Chhikara and Folks (1977)). Here, given the parameters μ and ϕ, $\mu, \phi > 0$, $P(T > t \mid \mu, \phi)$ has density at t of the form

$$\left(\frac{\phi\mu}{2\pi}\right)^{\frac{1}{2}} t^{-\frac{3}{2}} \mathrm{e}^{\phi} \exp\left\{-\frac{\phi}{2}\left(\frac{t}{\mu} + \frac{\mu}{t}\right)\right\},$$

where μ, a scale parameter, is the mean of T, and $1/\sqrt{\phi}$ is the coefficient of variation; ϕ is a shape paramter. The model failure rate of this distribution is non-monotonic; it initially increases and then decreases to a non-zero value as t becomes large. More about this distribution is said later in section 5.2.6.

Barlow and Proschan (1965) describe a mathematical theory of reliability based on monotone model (or chance) failure rate functions and the manner in which these functions cross the constant failure rate function of an exponential distribution. Multivariate versions of some of the univariate distributions mentioned above will be discussed in section 4.7.

In reliability theory, survival functions of the kind $\overline{F}(t \mid \lambda) = \exp(-\lambda t)$ and $\overline{F}(t \mid \lambda, \beta) = \exp(-\lambda t^{\beta})$ are referred to as the **reliability** of T for a **mission time** t. In actuality, of course, these are survival functions of T conditional on λ and β. Since such functions also appear as chance distributions in the de Finetti-type representation theorems of Chapter 3 (for example equation (3.11)) I make the claim that

reliability is a chance: not a probability!

In contrast, the survival function of T when $\overline{F}(t \mid \lambda) = \exp(-\lambda t)$ is given as

$$\overline{F}(t) = P(T > t) = \int_{0}^{\infty} \exp(-\lambda t)\ F(\mathrm{d}\lambda);$$

similarly for $\overline{F}(t \mid \lambda, \beta) = \exp(-\lambda t^{\beta})$. Thus to summarize, in this book I make the distinction between reliability of T for a mission of duration t, and the survival function of T. The former entails unknown parameters; the latter does not. In practice, it is the survival function that is of interest to engineers and biometricians. The reliability function is a convenient step toward getting to the survival function.

Units of Measurement for Parameters of Chance Distributions

For purposes of discussion consider the exponential chance distribution with parameter $\lambda > 0$. Here $P(T > t \mid \lambda) = \exp(-\lambda t)$ and its probability density at t, $f(t)$ is $\lambda\exp(-\lambda t)$. What are the units of measurement for λ, or is λ, like a probability, a unitless quantity? To answer this question we start by recalling the definition of a probability density function, namely, that

$$f(t) = \lim_{h \downarrow 0} \frac{F(t+h) - F(t)}{h},$$

for some $h > 0$. The quantities $F(t+h)$ and $F(t)$ being probabilities are unitless, and h is measured in the same units as t. If t is measured in units of time (seconds, minutes, hours, etc.), then so is h. Consequently, the units in which $f(t)$ is measured is $(\text{time})^{-1}$. Similarly, since the failure rate $r(t) \stackrel{\text{def}}{=} f(t)/\overline{F}(t)$, the units in which $r(t)$ is measured is also $(\text{time})^{-1}$. Thus no matter what $\overline{F}(t)$ is, both $f(t)$ and $r(t)$ are measured in terms of the reciprocal of the units in which the random variable T is measured.

With the above in place, recall that for the exponential as a chance distribution of time to failure, the failure rate $r(t \mid \lambda) = \lambda$. Consequently, the units in which λ, the scale parameter would be measured is $(\text{time})^{-1}$. In the case of the Weibull as a chance distribution $P(T \geq t \mid \lambda, \beta) = \exp(-(\lambda t)^{\beta})$, $\lambda, \beta > 0$, and its failure rate $r(t \mid \lambda, \beta) = \beta \lambda^{\beta} t^{\beta - 1}$. Since the unit of measurement for $r(t \mid \lambda, \beta)$ is also $(\text{time})^{-1}$ – recall that for $\beta = 1$, the Weibull becomes an exponential – the shape parameter β must be a unitless quantity.

4.4.1 The True Failure Rate: Does it Exist?

We have stated that the failure rate, be it model or predictive, is a conditional probability and that probability, being personal, does not exist outside an analyst's mind. The same is therefore also true of the failure rate (cf. Singpurwalla, 1988a; 1995a); however, such a position may be contrary to the views of many who tend to think of the failure rate as an objective entity, so that any item has a 'true failure rate'. Furthermore, to most engineers, the notion of a failure rate is synonymous with its naive estimate, namely, the number of failures in an interval divided by the number of items that are surviving at the start of the interval. This parallel between the failure rate and its estimate is of course a misconception. In our development, the failure rate – be it model or predictive – is an assessment of our uncertainty about the failure of an item in the time interval $[t, t + dt)$ were we to assume that the item is surviving at t. If, based on the physics or the biology of failure, we judge the item to age or to deteriorate with time, so that the capability of the item to perform its intended function degrades, then our probability for the above event should increase with t (Figure 4.2). If in our opinion the capability of the item to perform its intended function is enhanced with time, and there would be two reasons for this (section 4.4.2 below) then the said probability should decrease with t.

4.4.2 Decreasing Failure Rates, Reliability Growth, Burn-in and the Bathtub Curve

Rationale for a Decreasing Failure Rate

Whereas the judgment that the failure rate of an item is increasing with time is easy to support, namely, that the item deteriorates with use, reasons for the judgment of a decreasing failure rate are more subtle. There are two possibilities. The first is based on the physics of failure, namely, that certain items exhibit an improvement (over time) in their ability to perform their intended tasks. Examples are reinforced concrete that strengthens with exposure, drill bits that sharpen with use, materials that experience work hardening, and immune systems which mature with time. The second, and probably the more commonly occurring reason, is a psychological one. It comes into play when we change our opinion, for the better, about an item's survivability. For example, our opinion of the credibility of computer software that is thoroughly tested and debugged keeps improving with use; thus we may judge the software as having a predictive failure rate that is decreasing. Since the software does not deteriorate with use, its model failure rate should be a constant, but since we do not know this constant, its predictive failure rate is decreasing; (Singpurwalla and Wilson, 1999, p. 77). Mathematically, the above type of argument is captured by a celebrated theorem in reliability which says that scale mixtures of certain chance distributions result in distributions having a decreasing failure rate. For example, in Barlow and Proschan (1975, p. 103), it is shown that $\overline{F}(t) = \int \exp(-t\lambda)F(d\lambda)$, which is a scale mixture of exponentials with a parameter λ, has a decreasing failure rate. In particular, if λ above has a gamma distribution with a scale parameter ψ and a shape parameter α (section 3.2.2), then $\overline{F}(t; \psi, \alpha) = (\psi/(\psi + t))^{\alpha}$, which is a **Pareto distribution**. Its density at t is of the form $\alpha\psi^{\alpha}/(t+\psi)^{\alpha+1}$, $t \geq 0$, and its failure rate is $\alpha/(t+\psi)$, which is a decreasing function of t. The mean of this Pareto is $\psi/(\alpha - 1)$; it exists only when $\alpha > 1$. Note that a scale mixture is merely the law of total probability of section 2.4.2. Scale mixtures of distributions can be motivated by the following two, related, scenarios.

The first scenario is a tangible one, involving the physical act of putting together several items to form a batch, with each item having its own model failure rate λ_i. Suppose that the λ_is are known. We are required to assess the failure rate of an item picked at random from the batch. We assume that in forming the batch, the identity of each item is lost so that we do not know the λ_i for the selected item, whose predictive failure rate we are required to assess. Such physical mixtures arise in practice when items coming from different sources are stored in a common bin, and the proportion of each λ_i determines the mixing distribution $\mu(\lambda)$.

The second scenario under which scale mixtures arise is less natural, because it models the mathematics of our thought process. In the case of exponentially distributed lifelengths, if we are uncertain about λ – as we will be – and if our uncertainty is described by $\mu(\lambda)$, then our unconditional distribution of lifelengths would be a scale mixture of exponentials. The mixing now goes on in our minds and is therefore purely psychological.

An intuitive explanation as to why a scale mixture of exponentials results in a predictive distribution with a decreasing failure rate is easy to see if we bear in mind the notion that the failure rate is an individual's judgment, and that judgments change with added information. In our case the judgment is made at some reference time τ, and it pertains to the uncertainty about failure at a future time t, assuming some added knowledge. The added knowledge is the supposition that the item has not failed at t. It is important to note that when the failure rate is assessed at time τ, for any future time t, we are not saying that in actuality the item is surviving at t; rather, we inquire as to how we would assess our probability were we to suppose that the item is surviving at t. Thus we may start off with a poor opinion of the item's reliability, and upon the supposition of its continued survival, change our opinion for the better, resulting in a decreasing failure rate; more details are in Barlow (1985).

To summarize, there are two possible reasons for the judgment of a decreasing failure rate. The first is motivated by the physics of failure and generally pertains to the model failure rate. The second pertains to the predictive failure rate and is due to the psychology of altering our opinion about survivability; this is accomplished by mixtures of distributions.

Often reliability growth and burn-in have been given as the reasons for assuming decreasing failure rates. This is incorrect because both the above operations change the physical characteristics of an item, so that the changed item has a failure rate which may be different from its failure rate before the change.

Reliability Growth and Debugging

Reliability growth pertains to an enhancement of reliability due to changes in the design of the item or due to an elimination of its identified defects (cf. Singpurwalla, 1998); for example, debugging a piece of software. To describe reliability growth, we must talk about the concatenated (or adjoined) failure rates – model or predictive – of different versions of the same item, each version having a failure rate function that is dominated by the failure rate function of its preceding

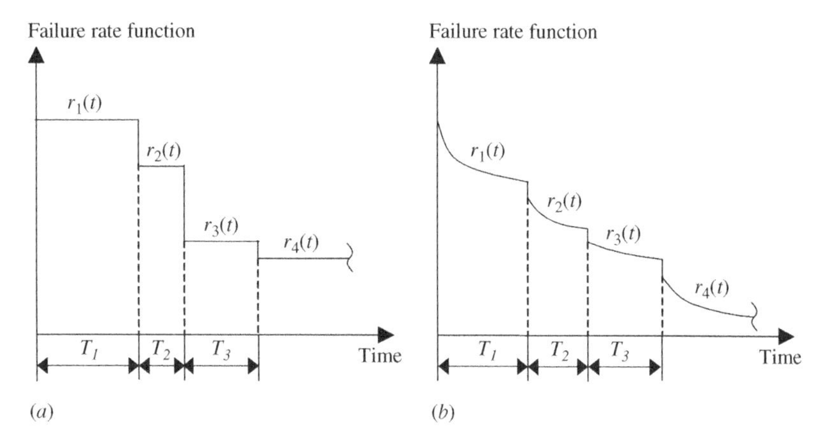

Figure 4.3 Concatenated failure rate functions for items undergoing reliability growth.

version. An example is the reliability growth of computer software that undergoes several stages of testing and debugging (Chen and Singpurwalla, 1997; or Al-Mutairi, Chen, and Singpurwalla, 1998). In Figure 4.3(a) and (b), we see concatenated failure rate functions that are composed of constant and decreasing failure rate segments, respectively. Concatenated failure rate functions are often encountered in the literature on software reliability and have been erroneously cited as examples of items having a decreasing failure rate.

Burn-in and Screening Tests

Burn-in pertains to the elimination of weak items by subjecting each item to a life test, called a **screening test**, for a specified period of time (for example Block and Savits, 1997). The hope is that the weak and defective items will drop out due to early failures so that what remains are items that have proven themselves in a life test. In practice, burn-in appears to be almost always done on a one-of-a-kind item that is to be used in life-critical systems, such as spacecraft. The bathtub curve, discussed later, is often given as a reason for undertaking burn-in. Because of an accumulation of age during the screening test, the failure rate of an item that survives the test will be different from its failure rate before the test. Thus, like reliability growth, burn-in changes the physical characteristics of the item tested. In particular, for items that are judged to have an increasing failure rate, a burn-in test will make the surviving items inferior to what they were before the test. Thus, theoretically, burn-in is advantageous only when lifetimes are judged to have a decreasing predictive failure rate. Why is it that in practice all life-critical items, even those that are known to deteriorate with use, are subjected to a burn-in? Are the engineers doing something they should not be doing, or is it so that even items which deteriorate with use could be judged as having a failure rate which is initially decreasing and then increasing in time? It turns out that when we are uncertain about the precise form of an increasing model failure rate function for a deteriorating device, the predictive failure rate would initially decrease and then increase (Figure 4.4) and thus the engineer's hunch to always burn-in makes sense. The intuition underlying the above phenomenon is analogous to the one used to explain the decreasing failure rate of mixtures of exponentials. That is, mixtures of distributions are a mathematical description of the psychological process of altering one's opinion about the reliability of an item. Figure 4.4 shows, for different values of a and b, a U-shaped predictive failure rate of T when its model failure rate is increasing and is of the form $r(t \mid \alpha, \beta) = \alpha t + \beta$ with α assumed known and the uncertainity of β described by a gamma distribution having a shape parameter a and a scale

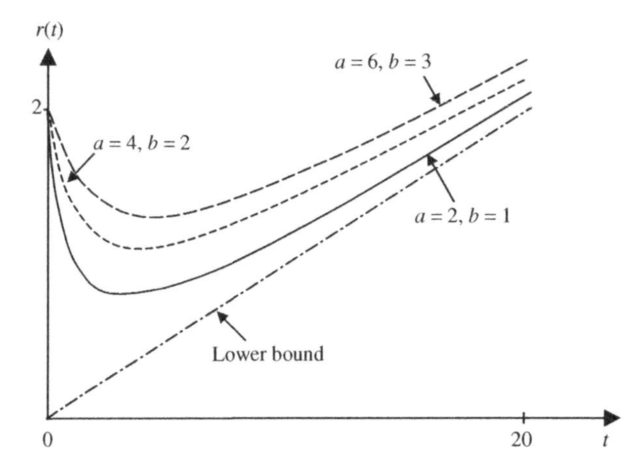

Figure 4.4 A U-shaped predictive failure rate generated by increasing model failure rates.

parameter b (for details, Lynn and Singpurwalla, 1997). Gurland and Sethuraman (1995) describe an analogous phenomenon in which increasing model failure rates could result in monotonically decreasing predictive failure rates.

Determining an optimum burn-in time is a problem in decision making under uncertainty wherein we are trading-off one attribute for another (for example Clarotti and Spizzichino, 1990). There are two separate cases that need to be discussed. In the case of items whose failure rate is always decreasing we are trading-off the cost of a burn-in versus having an item with the lowest possible failure rate. The more time we spend under burn-in, the lower will be the failure rate of an item that survives the burn-in test. Thus ideally, for items having a decreasing failure rate, the burn-in period should be indefinite, but then we would never commission a surviving item for use; thus the trade-off. In the case of items that are known to deteriorate with use, but whose model failure rate we are unable to specify, the predictive failure rate will be judged to initially decrease and then increase, and now we are trading off the cost of testing plus the depletion of useful life for the added knowledge about the specific form of the model failure rate. With the above perspective on burn-in testing, the prevailing argument that the purpose of burn-in is the elimination of defective items can be justified if we are to view defective items as those having large values for the model failure rate.

The Bathtub Curve

Many complex technological systems, and also humans, are judged to have a failure rate function which is in the form of a bathtub curve (Figure 4.5(a)). In fact the bathtub curve has become one of the hallmarks of engineering reliability. Observe that there are three segments to the bathtub curve (which is assessed at a reference time τ). The initial decreasing segment is referred to as the **infant mortality** phase, the middle constant segment as the **random** phase, and the final increasing segment as the **aging** or the **wear-out** phase. The rationale behind a choice of the bathtub curve for the failure rate function is as follows: Typically, a newly developed system may contain design and manufacturing faults, known as ***birth defects***, that would trigger an early failure; thus the initial failure rate is described by a decreasing function of time. Were we to be told that the system does not experience a failure due to birth defects, then it is likely that its subsequent failure is due to causes that cannot be explained, and so the middle phase of the failure rate is constant over time. Should we suppose that item survives this random phase, then

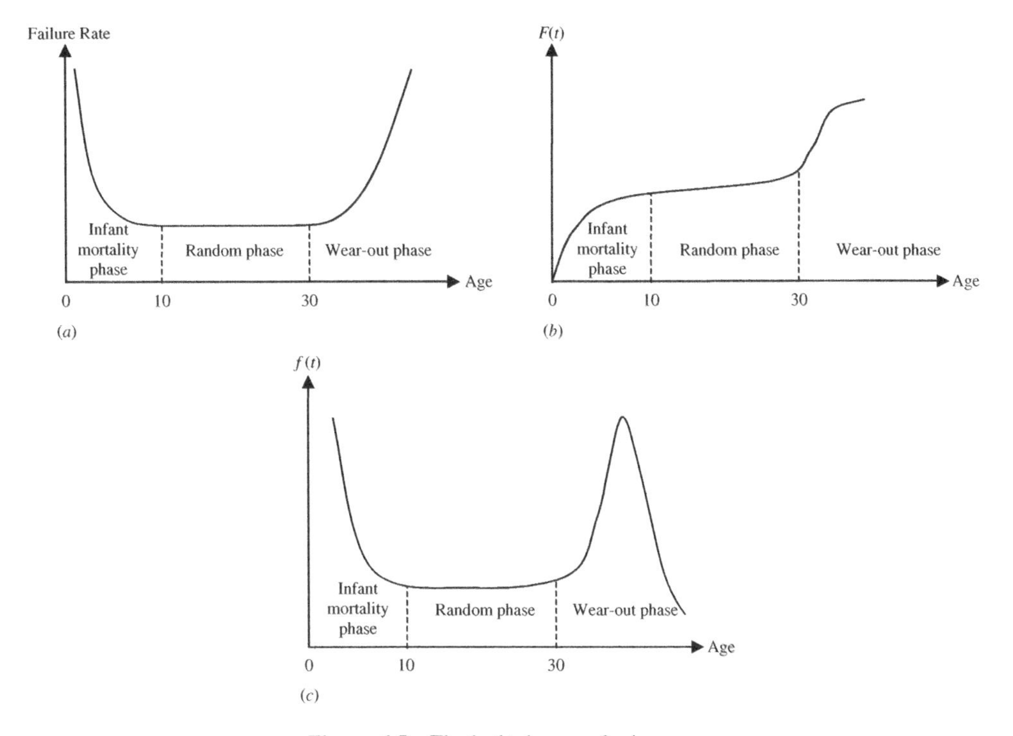

Figure 4.5 The bathtub curve for humans.

its subsequent failure is predominantly due to deterioration or wear, and so the final phase of the failure rate is an increasing function of time.

The bathtub form of the failure rate is used by actuaries to establish insurance premiums. The infant mortality period used by actuaries is from birth to age 10 years, the random phase from 10 to 30 years, and the wear-out phase commences at age 30, when premiums increase. The failure rate for humans during the initial period of the infant mortality phase is very high; that is, there is a high risk of death immediately after birth. Consequently, many insurance policies become effective after 15 days following birth. The causes of death between the ages of 10 and 30 years are assumed to be random and are due to events such as epidemics, wars, etc.; aging is assumed to commence at age 30. It is important to bear in mind that the bathtub curve is a specific form of the failure rate function (specified at time τ), and like all other forms of the failure rate function is the opinion or the judgment of an assessor about the survivability of a unit. Under our interpretation of probability, the bathtub curve does not have a physical reality, and could be chosen as a model for a single one-of-a-kind item, or for any member of a population of items. Figures 4.5(b) and (c) illustrate the nature of the distribution function $F(t)$ and its corresponding density function $f(t)$ for the bathtub curve of Figure 4.5(a).

Even though the bathtub form of the failure rate function is a reasonable idealization for the failure rate of humans, it may not be appropriate for certain engineering and biological systems. For many systems, the judgment of a constant failure rate, which implies that the system neither improves nor deteriorates with use, may not be meaningful. For such systems, a U-shaped failure rate function may be a more reasonable description of their failure behavior (Figure 4.6(a)). For systems that do not experience wear, a strictly decreasing or an L-shaped failure rate function may be appropriate, whereas for systems that do not experience infant mortality, a failure rate

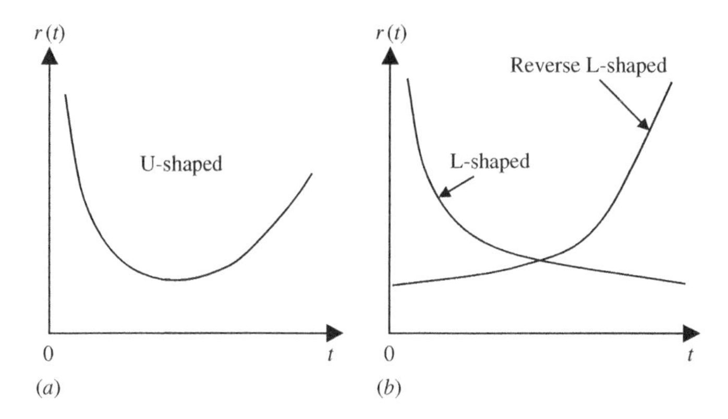

Figure 4.6 U-shaped and L-shaped failure rate functions.

function that is strictly increasing, or is a constant, or is initially constant and then increasing may be reasonable (Figure 4.6(b)).

Before closing this section it is desirable to comment on the infant mortality phase of the bathtub curve. We also need to comment on whether the failure rate function of a bathtub curve represents the model failure rate or the predictive failure rate. The decreasing form of the failure rate function does not necessarily imply that the system is indeed improving with use. As I have said before, this is a rare phenomenon, restricted to scenarios such as the setting of cement, or in the case of humans, the building-up of an immune system. Rather, the decreasing form of the failure rate function typically describes our improving opinion about the survivability of the item. In actuality, a system may experience a gradual wear and/or destruction due to randomly occurring events, as soon as it is put to use; but these are not judged as being the predominant causes of initial failure. It is our lack of precise knowledge about the presence or the absence of manufacturing (or birth) defects which causes us to judge a decreasing failure rate function during the initial phase of a system's life. Thus it appears that in most instances the infant mortality phase of the bathtub curve should pertain to the predictive failure rate. Since the other two phases of the bathtub curve are a constant and an increasing function of time, it appears (in the light of our discussions on decreasing failure rates), that these two phases must pertain to the model failure rate. Thus, in most instances the bathtub curve is more likely to be a concatenation of a predictive and a model failure rate. If the bathtub curve is U-shaped, then it could possibly pertain to the predictive failure rate, but most likely it will be the concatenation of a predictive and a model failure rate.

To conclude, it is our view that the bathtub curve that is commonly discussed and used by practitioners has an interpretation that is not as elementary as it is made out to be. In the framework that has been put forth, it is a representation of the opinion of an individual, or a collection of individuals, and this could be a concatenation of predictive and model failure rate functions, or just a predictive failure rate function. More on the shape of the failure rate function can be found in Aalen and Gjessing (2001).

4.4.3 The Retrospective (or Reversed) Failure Rate

In retrospective studies, like the postmortem of an observed failure, or the (unknown) time of the onset of a disease, the notion of a retrospective failure rate is useful. The retrospective failure rate is also known as the 'reversed hazard rate', and in many respects its behavior parallels that of the failure rate. Specifically, were we told that an item is in a failed state at t, we may be interested in assessing its actual time to failure. For example, in forensic studies involving a homicide, an investigator may want to know the actual time at which the crime was committed. Such scenarios are duals of the ones we have discussed so far, in the sense that the direction of time gets reversed from some time t toward zero, rather from t to infinity.

Following the notation of section 4.3, let T be a continuous random variable with an absolutely continuous distribution function $F(t)$ and probability density $f(t)$. Let $a = \inf\{t : F(t) > 0\}$ and $b = \sup\{t : F(t) < 1\}$; roughly speaking, a is the smallest value that T can take for which $F(a) > 0$. Similarly, b is the largest value that T can take for which $F(b) < 1$. The interval (a, b) with $-\infty \leq a < b < \infty$ is called the **interval of support** of $F(t)$. Then for any $t > a$, $\widetilde{r}(t)$, the **instantaneous retrospective predictive failure rate** of T at t is defined as:

$$\widetilde{r}(t) = \lim_{\mathrm{d}t \downarrow 0} \frac{P(t - \mathrm{d}t \leq T < t \mid T < t)}{\mathrm{d}t}; \quad \text{thus}$$

$$\widetilde{r}(t) = \lim_{\mathrm{d}t \downarrow 0} \frac{F(t) - F(t - \mathrm{d}t)}{F(t)\, \mathrm{d}t} = \frac{f(t)}{F(t)},$$

provided that $F(t) > 0$. The above expression for $\widetilde{r}(t)$ parallels that for the failure rate $r(t)$ given by (4.7). If the limit given above does not exist, then $\widetilde{r}(t) = \mathrm{d}F(t)/F(t)$ (equation (4.8)). Our definition for $\widetilde{r}(t)$ suggests that for small values of $\mathrm{d}t$, $\widetilde{r}(t)\mathrm{d}t$ is approximately the probability that an item that was recorded as being failed at t, actually failed in $(t - \mathrm{d}t, t)$.

As a function of t, the retrospective failure rate function could be monotonic (increasing or decreasing) or non-monotonic on the interval of support of $F(t)$. Block, Savits and Singh (1998) show that distributions whose failure rate is decreasing will necessarily have a decreasing retrospective failure rate, and that there does not exist a nonnegative random variable with an upper bound for which the retrospective failure rate is increasing. Thus distributions like the Weibull or the gamma, which have an increasing failure rate, will have a decreasing retrospective failure rate. The above feature makes the retrospective failure rate function unattractive vis-á-vis its ability to discriminate between failure models. Thus for example, whereas the failure rate function for a gamma distribution is increasing and that for a Pareto distribution is decreasing, the retrospective failure rate function of both these distribution functions is decreasing. Finally, Block, Savits and Singh (1998) show that if the retrospective failure rate function is increasing on the interval of support of some distribution, then the interval must have a finite upper point.

A relationship similar to the one involving the cumulative hazard function and the survival function (equation (4.9)) is available for the **cumulative retrospective hazard function**, defined as

$$\widetilde{H}(t) \stackrel{\text{def}}{=} \int_t^b \widetilde{r}(u)\, \mathrm{d}u,$$

and $F(t)$, the distribution function. Specifically,

$$F(t) = \exp(-\widetilde{H}(t));$$

see Chandra and Roy (2001).

Solving for $F(t)$ and $f(t)$ in (4.8) and the definition of $\widetilde{r}(t)$ given above, we can see that

$$F(t) = \frac{r(t)}{r(t) + \widetilde{r}(t)},$$

and that

$$f(t) = \frac{r(t)\widetilde{r}(t)}{r(t) + \widetilde{r}(t)}.$$

A notion that is converse to that of residual life (section 4.2) is that of a **dormant life**, also known as the 'inactivity time' (cf. Chandra and Roy, 2001). The dormant life of an item that is recorded as being failed at t is $(t-T \mid T \leq t)$; its expected value $E(t-T \mid T \leq t)$ is the mean dormant life (MDL). The notion of an MDL parallels that of MRL, the mean residual life. Verify that

$$P((t-T) \leq u \mid T \leq t) = (F(t) - F(t-u))/F(t),$$

so that the dormant life has density at u of the form $f(t-u)/F(t)$. It now follows that the MDL is given as

$$E((t-T) \mid T \leq t) = \frac{1}{F(t)} \int_a^t F(u) \, du.$$

4.5 MULTIVARIATE ANALOGUES OF THE FAILURE RATE FUNCTION

Multivariate analogues of the failure rate function have been considered, among others, by Basu (1971), Cox (1972), Johnson and Kotz (1975), Marshall (1975) and by Puri and Rubin (1974). One such analogue is the 'hazard gradient' discussed by Marshall (1975); this will be considered first.

4.5.1 The Hazard Gradient

A good starting point for introducing the hazard gradient is the cumulative hazard function

$$H(t) = -\ln \overline{F}(t) = \int_0^t r(u) du, \qquad t \geq 0 \tag{4.11}$$

mentioned in section (4.3).

Consider $R(t_1, \ldots, t_n) = P(T_1 \geq t_1, \ldots, T_n \geq t_n)$, the joint survival function of n lifelengths $T_1, \ldots, T_n$, where, for convenience, we have omitted the $\mathcal{H}$. Let $\mathbf{t} = (t_1, \ldots, t_n)$ be such that $R(t_1, \ldots, t_n) > 0$. Then, analogous to (4.11), we define $H(\mathbf{t})$, the **multivariate cumulative hazard function** as $H(\mathbf{t}) = -\ln R(t_1, \ldots, t_n)$. If the n-dimensional function $H(\mathbf{t})$ has a gradient, say $\mathbf{r}(\mathbf{t}) = \nabla H(\mathbf{t})$, where $\mathbf{r}(\mathbf{t}) = (r_1(\mathbf{t}), \ldots, r_n(\mathbf{t}))$, with

$$r_i(\mathbf{t}) = \frac{\partial}{\partial t} H(\mathbf{t}), \qquad i = 1, \ldots, n, \tag{4.12}$$

then $\mathbf{r}(\mathbf{t})$ is called the **hazard gradient** of $R(t_1, \ldots, t_n)$.

Johnson and Kotz (1975) interpret $r_i(\mathbf{t})$ as the conditional failure rate of T_i evaluated at t_i, were it to be so that $T_j > t_j$, for all $j \neq i$. That is

$$r_i(\mathbf{t}) = \frac{f_i(t_i \mid T_j > t_j, \text{ for all } j \neq i)}{P(T_i > t_i \mid T_j > t_j, \text{ for all } j \neq i)}, \tag{4.13}$$

where $f_i(t_i \mid T_j > t_j, \text{ for all } j \neq i)$ is the conditional probability density function of T_i, given that $T_j > t_j$, for all $j \neq i$.

From elementary calculus, it now follows that

$$H(\mathbf{t}) = \int_0^{\mathbf{t}} \mathbf{r}(\mathbf{u}) d\mathbf{u}, \tag{4.14}$$

from which we have as a multivariate analogue of (4.9) the result that

$$P(T_1 \geq t_1, \ldots, T_n \geq t_n) = \exp(-\int_{\mathbf{0}}^{\mathbf{t}} \mathbf{r}(\mathbf{u})d\mathbf{u}); \tag{4.15}$$

the role of the hazard gradient is now apparent. Equation (4.14) holds as long as $\mathbf{r}(\mathbf{u})$ exists almost everywhere and that along the path of integration $(\mathbf{0}, \mathbf{t})$, $H(\mathbf{t})$ is absolutely continuous.

The following (additive) decomposition of $H(\mathbf{t})$ given by Marshall (1975) is noteworthy; it parallels the (additive) decomposition of $H(t)$ in the univariate case. Specifically,

$$H(\mathbf{t}) = \int_0^{t_1} r_1(u_1, 0, \ldots, 0)du_1 + \int_0^{t_2} r_2(t_1, u_2, 0, \ldots, 0)du_2 + \ldots + \int_0^{t_n} r_n(t_1, \ldots, t_{n-1}, u_n)du_n, \tag{4.16}$$

where $r_1(u_1, 0, \ldots, 0)$ is the failure rate of T_1 at u_1, and $r_i(t_1, \ldots, t_{i-1}, u_i, 0, \ldots, 0)$ is the conditional failure rate of T_i at u_i, were it be such that $T_1 \geq t_1, T_2 \geq t_2, \ldots, T_{i-1} \geq t_{i-1}$. An interpretation of the decomposition of (4.16) will be given in section 4.6.2; however, a consequence of this decomposition is the well-known (multiplicative) decomposition of $R(t_1, \ldots, t_n)$. Specifically,

$$P(T_1 \geq t_1, \ldots, T_n \geq t_n) = P(T_1 \geq t_1)\, P(T_2 \geq t_2 \mid T_1 \geq t_1) \ldots\ldots P(T_n \geq t_n \mid T_1 \geq t_1, \ldots, T_{n-1} \geq t_{n-1});$$

see Marshall (1975).

The univariate analogues of the additive and the multiplicative decompositions given above appear in section 4.3, following Equation (4.10).

4.5.2 The Multivariate Failure Rate Function

A more natural, though perhaps less useful, multivariate analogue of the univariate failure rate function is the multivariate failure rate function introduced by Basu (1971). I give below its bivariate version, and state some of its interesting characteristics.

Suppose that an absolutely continuous bivariate distribution function $F(t_1, t_2)$ with $F(0, 0) = 0$ generates a probability density function $f(t_1, t_2)$ at (t_1, t_2). Then the bivariate failure rate of $F(t_1, t_2)$ at some (τ_1, τ_2) is, for $P(T_1 \geq \tau_1, T_2 \geq \tau_2) > 0$, given as

$$r(\tau_1, \tau_2) = \frac{f(\tau_1, \tau_2)}{P(T_1 \geq \tau_1, T_2 \geq \tau_2)} \tag{4.17}$$

$$= \frac{f(\tau_1, \tau_2)}{1 + F(\tau_1, \tau_2) - F(\tau_1, \infty) - F(\infty, \tau_2)}.$$

Clearly, when T_1 and T_2 are mutually independent, $r(\tau_1, \tau_2) = r(\tau_1)r(\tau_2)$, where $r(\tau_1)$ and $r(\tau_2)$ are the corresponding univariate failure rates at τ_1 and τ_2, respectively (equation (4.7)). The relationship of (4.17) can be extended to the multivariate case.

As seen before, in section 4.4, the only univariate distribution whose failure rate is a constant is the exponential. We now ask if there is an analog to this result when the $r(\tau_1, \tau_2)$ of (4.17) is a constant, say $\lambda > 0$. Basu (1971) shows that the only absolutely continuous bivariate distribution with marginal distributions that are exponential and whose bivariate failure rate function is a constant, is the one that is obtained when T_1 and T_2 are exponentially distributed. The search for multivariate distributions with exponential marginals is motivated by the fact that univariate exponential distributions, because of their simplicity and ease of use, are very popular in engineering reliability. The attractiveness of absolute continuity stems from the fact that statistical inference involving such distributions is relatively easy (cf. Proschan and Sullo, 1974). If Basu's (1971) requirement of exponential marginals is relaxed, then according to a theorem by Puri and Rubin (1974), the only absolutely continuous multivariate distribution whose multivariate failure rate (in the sense of (4.17)) is a constant is the one given by a mixture of exponential distributions.

Thus to summarize, it appears that the main value of the multivariate analogue of the failure rate function considered in this subsection is a characterization of the multivariate failure rate function that is a constant.

4.5.3 The Conditional Failure Rate Functions

Another multivariate analogue of the univariate failure rate function is the one proposed by Cox (1972) in his classic paper, and under the subheading 'bivariate life tables'. In the bivariate case involving lifelengths T_1 and T_2 we define, analogous to (4.4) of section 4.3, the following four univariate conditional failure rate functions:

$$r_{p0}(t)=\lim_{dt\downarrow 0}\frac{P(t\le T_p\le t+dt \mid T_1\ge t, T_2\ge t)}{dt}, \qquad \text{for } p=1,2;$$

$$r_{21}(t\mid u)=\lim_{dt\downarrow 0}\frac{P(t\le T_2\le t+dt \mid T_2\ge t, T_1=u)}{dt}, \qquad \text{for } u<t, \quad \text{and}$$

$$r_{12}(t\mid u)=\lim_{dt\downarrow 0}\frac{P(t\le T_1\le t+dt \mid T_1\ge t, T_2=u)}{dt}, \qquad \text{for } u<t. \tag{4.18}$$

In the latter two expressions, $T_i=u,\ i=1,2$, is to be interpreted as $T_i\in[u,u+du)$, where du is infinitesimal.

A motivation for introducing the conditional failure rate functions parallels that for introducing the failure rate function of (4.4) and (4.5). Namely, that a subject matter specialist can subjectively specify the functions $r_{p0}(t)$, for $p=1,2$, based on the aging characteristics of an item, and the functions $r_{ij}(t\mid u),\ i,j=1,2$ based on the aging as well as the load sharing features of the surviving item; for example, Freund (1961). If $F(t_1,t_2)$ denotes the joint distribution function of T_1 and T_2, and $f(t_1,t_2)$ the probability density at (t_1,t_2) generated by $F(t_1,t_2)$, then using arguments that parallel to those leading to (4.10), it can be shown that for $t_1\le t_2$

$$f(t_1,t_2)=\exp[-\int_0^{t_1}\{r_{10}(u)+r_{20}(u)\}du-\int_{t_1}^{t_2}r_{21}(u\mid t_1)du]r_{10}(t_1)r_{21}(t_2\mid t_1); \tag{4.19}$$

analogously for the case $t_2\le t_1$.

Consequently, as was the case with the univariate failure rate function, a specification of the conditional failure rate functions of (4.18) enables us to obtain $f(t_1,t_2)$ and hence $F(t_1,t_2)$. Alternatively, if we are to know $R(t_1,t_2)=P(T_1\ge t_1, T_2\ge t_2)$, then we may obtain $r_{10}(t)$ and $r_{ij}(t\mid u),\ i,j=1,2$, via relationships of the type:

$$r_{10}(t)=-\frac{1}{R(t,t)}\left[\frac{\partial}{\partial t}R(t,u)\right]_{u=t},$$

and

$$r_{12}(t\mid u)=-\frac{\partial^2 R(t,u)}{\partial t\,\partial u}\Big/\frac{\partial R(t,u)}{\partial u}.$$

Finally, it is relatively easy to show that T_1 is independent of T_2, if and only if $r_{12}(t\mid u)=r_{10}(t)$, and $r_{21}(t\mid u)=r_{20}(t)$.

To conclude this section, we note that $R(t_1, \ldots, t_n)$ can be obtained through (4.15) via the hazard gradient $\mathbf{r}(\mathbf{u})$ or through (4.19) via the conditional failure rate function. The former is of mathematical interest, especially for the additive decomposition of (4.16); the latter has an operational appeal since it upholds the original spirit of introducing the failure rate function.

4.6 THE HAZARD POTENTIAL OF ITEMS AND INDIVIDUALS

In sections 4.3 and 4.5, we have seen the useful role played by the cumulative hazard function $H(t)$, and its multivariate analog $H(\mathbf{t})$. In the univariate case, we have seen that when $F(t)$ is absolutely continuous

$$\overline{F}(t) = \mathrm{e}^{-H(t)}, \tag{4.20}$$

where $H(t) = \int_0^t r(u)\mathrm{d}u$; note that for $r(u) > 0$, for all $u \geq 0$, $H(t)$ monotonically increases in t. The aim of this section is to interpret the quantity $H(t)$ in a manner that provides some insight into the relationship $\overline{F}(t) = \exp(-H(t))$.

We start with the remark that since $r(u)\mathrm{d}u$ is approximately the conditional probability that an item of age u will fail in the time interval $(u, u + \mathrm{d}u)$, the quantity $\int r(u)\mathrm{d}u$ is a sum of conditional probabilities. However, the conditioning events change over time, so that taking their sum is not a meaningful operation within the calculus of probability. Consequently, the cumulative hazard function $H(t)$ cannot be a probability, and therefore lacks a probabilistic interpretation. What then does the quantity $H(t)$ represent? In what follows I endeavor to address this question; however, I am cognizant of the possibility that there could be other interpretations.

It is noted that the conditional probability $r(u)\mathrm{d}u$ is subjectively specified, at least in the framework that I have espoused, and as such it incorporates an assessor's judgment about both the inherent quality of an item and the environment under which it is scheduled to operate. By 'quality' I mean a resistance to failure-causing agents, such as crack growth, wear, a weakening of the immune system, etc. Consequently, the failure rate function of an item of poor quality that operates in a benign environment could be smaller than that of a high-quality item operating in a harsh environment. In effect, the quantity $r(u)\mathrm{d}u$ encapsulates an assessor's judgment about the manner in which an item and its environment interact, vis-à-vis the item's lifelength. Keeping the above in mind we now turn our attention to (4.20), and note that its right-hand side $\mathrm{e}^{-H(t)}$ is also the distribution function of an exponentially distributed random variable, say X, with a scale parameter one, evaluated at $H(t)$. That is

$$P(T \geq t) = \mathrm{e}^{-H(t)} = P(X \geq H(t)). \tag{4.21}$$

Alternatively said, corresponding to every nonnegative random variable T, taking values t, for $t \in [0, \infty)$, and having a distribution function $\overline{F}(t)$, there exists a random variable X, taking values $H(t)$, for $0 < H(t) < \infty$, whose distribution function is an exponential with scale parameter one. The distribution function of T is said to be **indexed** by t, whereas that of X is indexed by $H(t)$, where $H(t) = \int_0^t \mathrm{d}F(t)/\overline{F}(t)$.

From (4.21), we see that if T represents the time to failure of an item, then the item will fail when $H(t)$ reaches a **random threshold**, X (Figure 4.7). We call this threshold the **hazard potential** (HP) of the item, and propose the view that when conceived, every item possesses some kind of an unknown resource X, and that the effect of its use is a depletion of this resource according to the dictates of $H(t)$. This unknown resource is the hazard potential of the item, and like potential energy, which reflects the amount of work that an item can do, the hazard potential represents an item's resistance to failure. Our uncertainty about the unknown resource is described by an exponential distribution with scale parameter one, and the function $H(t)$ (being

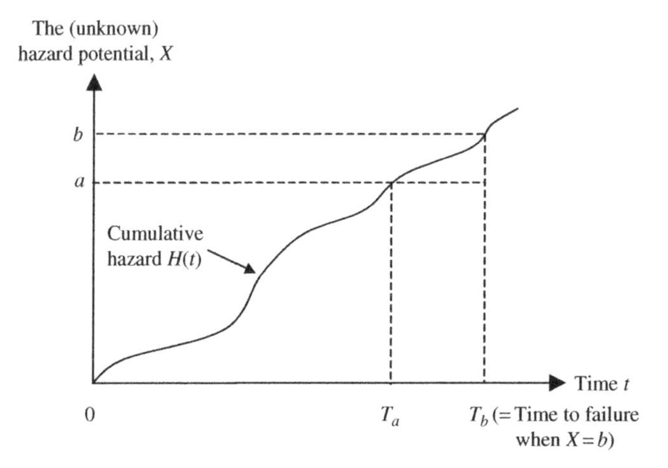

Figure 4.7 Illustration of resource depletion by the cumulative hazard function.

nondecreasing in t) depicts the rate at which the resource gets consumed (Figure 4.7). This is one interpretation of $H(t)$; its role in explaining lifelengths having a decreasing failure rate function has been articulated by Kotz and Singpurwalla (1999).

An alternate interpretation of $H(t)$ is motivated by the fact that the exponential distribution of X is indexed by $H(t)$. That is, $H(t)$ may be viewed as a change of time scale from the natural clock time t to a transformed time $H(t)$. Under this interpretation of $H(t)$, we may, de facto, make the claim that the lifelengths of any and all items are always exponentially distributed (with a scale parameter one) on a suitably chosen scale, $H(t)$. The choice of the scale $H(t)$ is subjective, and is determined by an assessment of the item's inherent quality and its operating environment. For example, two different components operating in the same environment may not necessarily have the same $H(t)$; similarly, changing the environment from say E_1 to E_2, will generally change the cumulative hazard function from $H_1(t)$ to $H_2(t)$ (Figure 4.8). In general $H(t) > t$ would reflect a harsh environment (e.g. an accelerated test) whereas $H(t) < t$ would

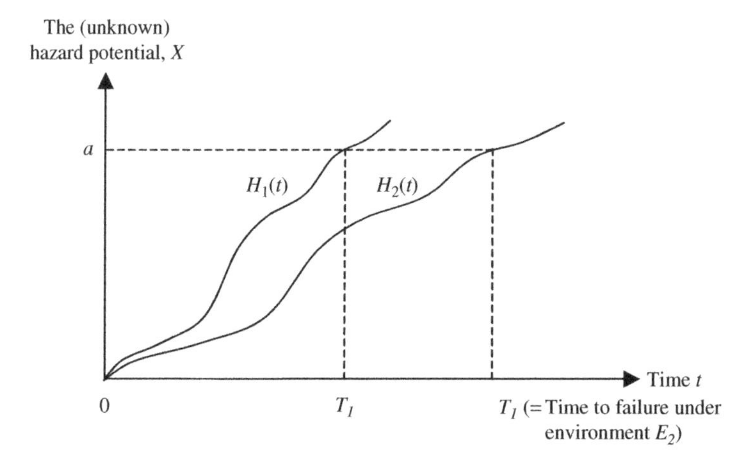

Figure 4.8 Effect of changing the environment on the cumulative hazard function.

correspond to a benign environment. The change of a time scale interpretation of $H(t)$ has also been recognized by Çinlar and Ozekici (1987).

The above material can be summarized via the following two assertions, the second of which may be interpreted as a type of an indifference principle of survival.

Theorem 4.1. *Corresponding to every nonnegative random variable T having distribution function F(t), there exists an exponentially distributed random variable X (with a scale parameter one) that is indexed by* $H(t)=\int_0^t \mathrm{d}F(t)/\overline{F}(t)$. *Equivalently,*

Corollary 4.1. *All lifelengths can be regarded as having an exponential distribution (with a scale parameter one) on a suitably chosen time scale.*

Regarding the exponentially distributed random variable X, it is evident that, in general, it will not possess the lack of memory property unless $H(t)=t$. A distribution function $F(x)=1-\overline{F}(x)$ is said to possess the lack of memory property if for all $x,\ t\geq 0,\ \overline{F}(x+t)=\overline{F}(x)\overline{F}(t)$; the only univariate distribution function possessing this property is the exponential distribution indexed by $H(t)=t$. Continuing along this vein, it is also easy to verify that the entropy of the exponentially distributed random variable X is one, only when $H(t)=t$. The entropy of a random variable X having an absolutely continuous distribution function $F(x)$ is $-\int_0^\infty f(x)\log(f(x))\mathrm{d}x$, where $f(x)$ is the probability density generated by $F(x)$ at x.

We close this discussion by noting that three quantities have been introduced here, a lifelength T that can be observed (i.e. measured), its cumulative hazard function $H(t)$ and the hazard potential X; both X and $H(t)$ are not observable. Given any two of these three entities we can obtain the third. Finally, we are able to interpret X as a resource, and $H(t)$ as the rate at which this resource gets consumed.

4.6.1 Hazard Potentials and Dependent Lifelengths

The likes of Figure 4.8 suggest that T_1 and T_2, the lifelengths of two items having a common (unknown) hazard potential X, will be dependent. This is because a knowledge of T_1 (or T_2) tells us something about X, and this in turn changes our assessment of T_2 (or T_1). The notion of dependent lifelengths is further articulated in section 4.7. Similarly, if the hazard potentials X_1 and X_2 are dependent, then their lifelengths T_1 and T_2 will be dependent, but only if we can specify both $H_1(t)$ and $H_2(t)$, for $t\geq 0$, or if only one of these can be specified, then we should be able to say something about the relationship between $H_1(t)$ and $H_2(t)$. Such assertions are best summarized via the following two theorems (and also Theorem 4.4, of section 4.6.2).

Theorem 4.2. *Lifetimes T_1 and T_2 are independent if and only if their hazard potentials X_1 and X_2 are indepenent, and if $H_1(t),\ H_2(t),\ t\geq 0$ are assumed known.*

Proof. When X_1 and X_2 are indepenent,

$$P(X_1\geq H_1(t_1), X_2\geq H_2(t_2))=P(X_1\geq H_1(t_1))\cdot P(X_2\geq H_2(t_2)),$$

for any $H_1(t_1)$ and $H_2(t_2)$. Consequently,

$$\begin{aligned}
&P(T_1\geq t_1, T_2\geq t_2; H_1(t), H_2(t), t\geq 0)\\
&\quad =P(X_1\geq H_1(t_1), X_2\geq H_2(t_2))\\
&\quad =P(X_1\geq H_1(t_1))\cdot P(X_2\geq H_2(t_2))\\
&\quad =P(T_1\geq t_1; H_1(t), t\geq 0)\cdot P(T_2\geq t_2; H_2(t), t\geq 0).
\end{aligned}$$

Thus knowing $H_1(t)$ and $H_2(t)$, T_1 and T_2 are independent. Similarly the reverse.

When $H_i(t)$, $i=1,2$ or both $i=1$ and 2, for $t \geq 0$ cannot be specified, Theorem 4.2 gets weakened in the sense that only the 'if' part of the theorem will hold. Specifically, T_1 and T_2 will be independent even when X_1 and X_2 are dependent. The intuitive argument proceeds as follows. Observing T_1 provides no added insight about X_1 since $H_1(t)$ is not specified. Consequently, there is no added insight about X_2 either, and thence about T_2. Similarly, when T_2 is observed and $H_2(t)$ not specified. Thus T_1 and T_2 are independent (unless of course one is willing to make other assumptions about $H_1(t)$ and $H_2(t)$, $t \geq 0$). Mathematically, without knowing $H_i(t)$, $i=1,2$, we are unable to relate $P(T_1 \geq t_1, T_2 \geq t_2)$ with the distribution of X_1 and X_2. The above is summarized via

Corollary 4.2.1. *Lifetimes T_1 and T_2 are independent whenever $H_1(t)$ and/or $H_2(t)$, $t \geq 0$ are unknown.*

The converse of Theorem 4.2 is

Corollary 4.2.2. *Lifetimes T_1 and T_2 are dependent if and only if their hazard potentials X_1 and X_2 are depenent, and if $H_1(t)$ and $H_2(t)$ are specified.*

Corollary 4.2.2 puts aside the often expressed view that the lifetimes of items sharing a common environment are necessarily dependent (cf. Marshall, 1975; Lindley and Singpurwalla, 1986). That is, it is a common environment that causes dependence among lifetimes. Corollary 4.2.2 asserts that it is the commanilities in the (HP)s, that is the cause of interdependent lifetimes. Dependent (HP)s are a manifestation of similarities in design, manufacture, or a genetic make-up. In the language of probabilistic causality of Suppes (1970), the common environment can be interpreted as a spurious cause of dependent lifetimes, whereas dependent (or identical) (HP)s as their prima facie (or genuine) cause.

Theorem 4.2 and Corollary 4.2.1 pertain to the two extreme cases wherein the $H_i(t)$s, $i=1,2$ are either known or are unknown. An intermediate case is wherein one of the $H_i(t)$s, say $H_1(t)$, $t \geq 0$ is known, and the other unknown, save for the fact that $H_1(t) > H_2(t)$. For such scenarios we are able to show that

Theorem 4.3. *Suppose that $H_1(t) > (\leq H_2(t))$, and either $H_1(t)$ or $H_2(t)$, $t \geq 0$ is specified; then X_1 and X_2 dependent implies that T_1 and T_2 are also dependent.*

Proof. The proof is by contradiction. For this, suppose that X_1 and X_2 have a bivariate exponential distribution of Marshall and Olkin (1967). Specifically, for λ_1, λ_2 and $\lambda_{12} > 0$,

$$P(X_1 \geq x, X_2 \geq y) = \exp(-\lambda_1 x - \lambda_2 y - \lambda_{12} \max(x, y)),$$
$$= \exp(-(\lambda_1 + \lambda_{12})x - \lambda_2 y), \text{ if } x > y.$$

The marginal distribution of X_i, $P(X_i \geq x) = \exp(-(\lambda_i + \lambda_{12})x)$, $i=1,2$. For the X_is to be dependent on (HP)s we need to have $(\lambda_i + \lambda_{12}) = 1$, for $i=1,2$, and $\lambda_{12} > 0$; this would imply that $\lambda_1 = \lambda_2 = \lambda$. Thus

$$P(X_1 \geq x, X_2 \geq y) = \exp(-(x + \lambda_2 y)).$$

If we set $x = H_1(t_1)$ and $y = H_2(t_2)$, for some $t_1, t_2 \geq 0$, then $x > y$ would imply that $H_2(t_2) = H_1(t_2) - \delta$, for some unknown $\delta > 0$. Consequently

$$P(X_1 \geq x, X_2 \geq y) = P(X_1 \geq H_1(t_1), X_2 \geq H_1(t_2) - \delta)$$
$$= \exp(-(H_1(t_1) + \lambda_2(H_1(t_2) - \delta))).$$

Given the above we need to show that T_1 and T_2 are dependent. Suppose not; then

$$\begin{aligned}
&P(T_1 \geq t_1, T_2 \geq t_2; H_1(t_1), H_2(t_2), t_1, t_2 \geq 0) \\
&= P(T_1 \geq t_1; H_1(t_1), t_1 \geq 0)\, P(T_2 \geq t_2; H_2(t_2), t_2 \geq 0) \\
&= P(X_1 \geq H_1(t_1))\, P(X_2 \geq H_2(t_2)) \\
&= \exp(-H_1(t_1)) \exp(-(H_1(t_2) - \delta)) \\
&= P(X_1 \geq H_1(t_1), X_2 \geq H_1(t_2) - \delta),
\end{aligned}$$

since the first term of the above equation does not entail elements of the second. Thus I have

$$P(X_1 \geq H_1(t_1), X_2 \geq H_1(t_2) - \delta) = \exp(-(H_1(t_1) + H_1(t_2) - \delta)).$$

The expressions $P(X_1 \geq x, X_2 \geq y)$ and $P(X_1 \geq H_1(t_1), X_2 \geq H_1(t_2) - \delta)$ will agree with each other if $\lambda_2 = 1$. However, $\lambda_2 = 1$ would imply that $\lambda_{12} = 0$, which would contradict the hypothesis that X_1 and X_2 are dependent. The proof when $H_1(t) \leq H_2(t)$ will follow along similar lines.

A consequence of Corollary 4.2.2 is that we are able to generate families of dependent lifelengths using a multivariate distribution with exponential marginals as a seed. The multivariate exponential distribution of Marshall and Olkin (1967) can be one such seed; others could be the one proposed by Singpurwalla and Youngren (1993) and those referred to in Kotz and Singpurwalla (1999). Details are in Singpurwalla (2006a).

4.6.2 The Hazard Gradient and Conditional Hazard Potentials

In this section, we obtain a converse to Theorem 4.3. Specifically, we start with dependent lifelengths and explore the nature of their dependence on the hazard potentials. The hazard gradient plays a role here, and in the sequel, it motivates the introduction of another notion, namely that of the 'conditional hazard potential'. The main result of this section is that a collection of dependent lifelengths can be replaced by a collection of independent exponentially distributed random variables indexed on suitably chosen scales. The independent random variables are the conditional hazard potentials.

To put matters in perspective, we recall from section 4.5.1 that if $H(\mathbf{t}) = -\ln R(\mathbf{t})$, where $R(\mathbf{t}) = P(T_1 \geq t_1, \ldots, T_n \geq t_n)$, then $\mathbf{r}(\mathbf{u})$ is the hazard gradient of $H(\mathbf{t})$. Furthermore $H(\mathbf{t}) = \int_{\mathbf{0}}^{\mathbf{t}} \mathbf{r}(\mathbf{u})\, d\mathbf{u}$, and that $H(\mathbf{t})$ has an additive decomposition given by (4.16). The first term of this decomposition is the integral of $r_1(u_1, 0, \ldots, 0)$ – the failure rate of T_1 at u_1. This integral, which is the cumulative hazard rate function of T_1 at t_1 will be denoted by $H_1(t_1)$; that is

$$H_1(t_1) = \int_0^{t_1} r_1(u_1, 0, \ldots, 0) du_1.$$

Similarly, the second term of the decomposition is denoted by

$$H_2(t_2 \mid t_1) = \int_0^{t_2} r_2(t_1, u_2, 0, \ldots, 0) du_2,$$

where the integrand $r_2(t_1, u_2, 0, \ldots, 0)$ is the conditional failure rate of T_2 at u_2 given that $T_1 \geq t_1$. Continuing in this vein, the last term of the decomposition is

$$H_n(t_n \mid t_1, \ldots, t_{n-1}) = \int_0^{t_n} r_n(t_1, \ldots, t_{n-1}, u_n) du_n.$$

Thus we have

$$H(\mathbf{t}) = H_1(t_1) + H_2(t_2 \mid t_1) + \ldots + H_n(t_n \mid t_1, \ldots, t_{n-1}). \tag{4.22}$$

Since the cumulative hazard function has no interpretive content within the calculus of probability, $H(\mathbf{t})$ and its components will also not have any interpretive content. However, since $R(\mathbf{t}) = \exp(-H(\mathbf{t}))$, we may write

$$\begin{aligned} P(T_1 \geq t_1, \ldots, T_n \geq t_n) &= e^{-(H_1(t_1)+H_2(t_2 \mid t_1)+\ldots+H_n(t_n \mid t_1,\ldots,t_{n-1}))} \\ &= e^{-H_1(t_1)}\, e^{-H_2(t_2 \mid t_1)} \ldots e^{-H_n(t_n \mid t_1,\ldots,t_{n-1})}. \end{aligned} \tag{4.23}$$

Clearly, $\exp(-H_1(t_1)) = P(T_1 \geq t_1)$, since $H_1(t_1)$ is the cumulative hazard function of T_1. Similarly, by following arguments analogous to those leading to (4.9), we can see that $\exp(-H_2(t_2 \mid t_1)) = P(T_2 \geq t_2 \mid T_1 \geq t_1)$, and in general

$$P(T_n \geq t_n \mid T_1 \geq t_1, \ldots, T_{n-1} \geq t_{n-1}) = e^{-H_n(t_n \mid t_1,\ldots,t_{n-1})}. \tag{4.24}$$

As a consequence of the above analogies we have, from (4.23), the relationships

$$\begin{aligned} P(T_1 \geq t_1, \ldots, T_n \geq t_n) &= e^{-H_1(t_1)}\, e^{-H_2(t_2 \mid t_1)} \ldots e^{-H_n(t_n \mid t_1,\ldots,t_{n-1})} \\ &= P(T_1 \geq t_1)\, P(T_2 \geq t_2 \mid T_1 \geq t_1) \ldots \\ &\quad \ldots P(T_n \geq t_n \mid T_1 \geq t_1, \ldots, T_{n-1} \geq t_{n-1}), \end{aligned}$$

the latter equality also being true by virtue of the multiplicative decomposition of $P(T_1 \geq t_1, \ldots, T_n \geq t_n)$.

Let $X_1, \ldots, X_n$ be the hazard potentials corresponding to the lifelengths $T_1, \ldots, T_n$, and the cumulative hazard functions $H_1(t_1), \ldots, H_n(t_n)$, respectively. Then a consequence of (4.24) is the result that

$$\begin{aligned} P(T_n \geq t_n \mid T_1 \geq t_1, \ldots, T_{n-1} \geq t_{n-1}) &= P(X_n \geq H_n(t_n) \mid X_1 \geq H_1(t_1), \ldots, \\ &\qquad \ldots X_{n-1} \geq H_{n-1}(t_{n-1})) \\ &= e^{-H_n(t_n \mid t_1,\ldots,t_{n-1})}. \end{aligned} \tag{4.25}$$

Since $T_1, \ldots, T_n$ are not assumed to be independent, the hazard potentials $X_1, \ldots, X_n$ are, by virtue of Theorem 2, not independent. However, $\exp(-H_n(t_n \mid t_1, \ldots, t_{n-1}))$ is the distribution function of an exponentially distributed random variable, say X_n^*, with a scale parameter one, evaluated at $H_n(t_n \mid t_1, \ldots, t_{n-1})$. Thus, a consequence of (4.25) is the result that for all $n \geq 2$

$$P(X_n \geq H_n(t_n) \mid X_1 \geq H_1(t_1), \ldots, X_{n-1} \geq H_{n-1}(t_{n-1})) = P(X_n^* \geq H_n(t_n \mid t_1, \ldots, t_{n-1})).$$

The random variable X_n^* is called the **conditional hazard potential** of the n-th item; i.e. the item whose lifelength is T_n. Its (unit) exponential distribution function is indexed by $H_n(t_n \mid t_1, \ldots, t_{n-1})$. By contrast, X_n, the hazard potential of the n-th item has a (unit) exponential distribution that is indexed by $H_n(t_n)$.

Similarly, corresponding to each term of (4.23), save the first, there exist random variables $X_2^*, X_3^*, \ldots, X_{n-1}^*$, independent of each other and also of X_n^* such that

$$P(T_1 \geq t_1, \ldots, T_n \geq t_n) = P(X_1 \geq H_1(t_1))\ P(X_2^* \geq H_2(t_2 \mid t_1)) \ldots$$
$$\ldots\ P(X_n^* \geq H_n(t_n \mid t_1, \ldots, t_{n-1})).$$

We now have, as a multivariate analogue to Theorem 4.1,

Theorem 4.4. *Corresponding to every collection of nonnegative random variables $T_1, \ldots, T_n$ having a survival function $R(t_1, \ldots, t_n)$, there exists a collection of n independent and exponentially distributed random variables $X_1, X_2^*, \ldots, X_n^*$, (with scale parameter one), with X_1 indexed on $H_1(t)$, and X_i^* indexed on $H_i(t_i \mid t_1, \ldots, t_{i-1})$, for $i = 2, \ldots, n$ and $n \geq 2$.*

4.7 PROBABILITY MODELS FOR INTERDEPENDENT LIFELENGTHS

The notion of independent events was introduced in section 2.2.1, and that of independent lifelengths in section 4.2. Within the framework of betting, independence is made operational by asserting that for two events $\mathcal{A}$ and $\mathcal{B}$, a knowledge of the occurrence (or not) of $\mathcal{A}$, does not change one's bets on $\mathcal{B}$, and vice versa. Lifelengths that are not independent are said to be **interdependent**, or simply, **dependent**. Dependent lifelengths are also referred to as **interacting lifelengths**. An example of dependent lifelengths is a sequence of exchangeable lifelengths, assuming that the sequence is not entirely comprised of independent random variables (section 3.1.2). We have seen that both independence and exchangeability are judgments, and that such judgments simplify the assignment of probabilities. Consequently, dependence, as the negation of independence, is also a judgment. However, to be useful, judgments of dependence demand more of a user, namely, a specification of the exact manner in which the bets on one lifelength change as knowledge about the disposition of the remaining lifelengths were to be given. Probability models for dependent lifelengths provide such specifications, and the purpose of this section is to highlight some such models. The models described here are motivated by scenarios that involve an underlying physical or biological structure. However, there do exist in the literature several models for dependence that have been motivated by mathematical considerations alone; an encyclopedic review is in Kotz, Balakrishnan and Johnson (2000). To gain a deeper appreciation of the mathematical structure of dependence I recommend the research monograph edited by Block, Sampson and Savits (1990), and the book by Arnold, Castillo, and Sarabia (1992). For purposes of discussion I focus attention on the bivariate case, so that the models proposed in sections 4.7.1 and thereafter pertain to two lifelengths. But before doing so, I introduce some preliminaries.

4.7.1 Preliminaries: Bivariate Distributions

Let T_1 and T_2 be the lifelengths of two items, and following the notation of section 4.2, let $P(T_1 \geq t_1, T_2 \geq t_2; \mathcal{H}) = R(t_1, t_2; \mathcal{H}) = \overline{F}(t_1, t_2; \mathcal{H})$. Similarly, let $P(T_1 \leq t_1, T_2 \leq t_2; \mathcal{H}) = F(t_1, t_2; \mathcal{H})$. In what follows we shall omit $\mathcal{H}$. It is easy to verify that

$$\overline{F}(t_1, t_2) = 1 - F_1(t_1) - F_2(t_2) + F(t_1, t_2), \tag{4.26}$$

where $F_i(t_i)$ is the marginal distribution function of T_i, obtained via the operations $F(t_1, \infty) = F_1(t_1)$ and $F(\infty, t_2) = F_2(t_2)$. Also, $F_1(t_1, 0) = F_2(0, t_2) = F(0, 0) = 0$, $F(\infty, \infty) = 1$, and the expected value of $F_i(t_i)$ is, $E(T_i) = \int_0^\infty \overline{F}_i(t_i)\mathrm{d}t_i$, where $\overline{F}_i(t_i) = 1 - F_i(t_i)$, $i = 1, 2$.

Suppose that the partial derivative $\partial^2 F(t_1, t_2)/\partial t_1 \partial t_2$ exists (almost) everywhere, and let $f(t_1, t_2) = \partial^2 F(t_1, t_2)/\partial t_1 \partial t_2$. Then, the bivariate distribution function $F(t_1, t_2)$ is said to have a **bivariate density** $f(t_1, t_2)$, and

$$F(t_1, t_2) = \int_0^{t_1} \int_0^{t_2} f(s_1, s_2) \mathrm{d}s_1 \mathrm{d}s_2.$$

It is important to note that a bivariate distribution function $F(t_1, t_2)$ is not uniquely determined by its marginal distribution functions $F_1(t_1)$ and $F_2(t_2)$. There are an infinite number of solutions to the problem of determining $F(t_1, t_2)$ from $F_1(t_1)$ and $F_2(t_2)$. One such solution is given by the **Fréchet bounds** (Fréchet, 1951).

$$\max[F_1(t_1) + F_2(t_2) - 1, 0] \leq F(t_1, t_2) \leq \min[F_1(t_1), F_2(t_2)]; \tag{4.27}$$

the bounds are themselves bivariate distributions whose marginals are $F_1(t_1)$ and $F_2(t_2)$. An exception to the above occurs when for all $t_1, t_2 \geq 0$, the events $(T_1 \leq t_1)$ and $(T_2 \leq t_2)$ are judged independent, because now

$$F(t_1, t_2) = F_1(t_1)\ F_2(t_2),$$

and so the marginals provide a unique joint distribution.

Another way to construct $F(t_1, t_2)$ from $F_1(t_1)$ and $F_2(t_2)$ is due to Morgenstern (1956). Specifically, for some α, $0 \leq \alpha \leq 1$,

$$F(t_1, t_2 \mid \alpha) = F_1(t_1)\ F_2(t_2)\ [1 + \alpha(1 - F_1(t_1))(1 - F_2(t_2))]. \tag{4.28}$$

A bivariate exponential distribution of Gumbel (1960) (section 4.7.2) is based on this construction. Yet another approach is the method of copulas described below. Singpurwalla and Kong (2004) have used this approach to construct a new family of bivariate distributions with exponential marginals (section 4.7.7).

The Method of Copulas

A **copula** C is a bivariate distribution on $[0, 1] \times [0, 1]$, whose marginal distributions are uniform. Copulas join (i.e. couple) univariate distribution functions to form multivariate distribution functions (Nelson, 1995). This feature is encapsulated in the following theorem.

Theorem 4.5. (Sklar, 1959). *Let F be a two-dimensional distribution function with marginal distribution functions F_1 and F_2. Then there exists a copula C such that $F(t_1, t_2) = C[F_1(t_1), F_2(t_2)]$. Conversely, for any univariate distribution functions F_1 and F_2 and any copula C, the function $F(t_1, t_2) = C[F_1(t_1), F_2(t_2)]$ is a two-dimensional function with marginal distribution functions F_1 and F_2.*

The notion of copulas and Sklar's theorem generalize to the multivariate case. Sklar's theorem enables us to generate copulas, and copulas can be used to characterize certain properties of sequences of random variables. Specifically, $F(t_1, t_2) = C[F_1(t_1), F_2(t_2)]$ implies that for $0 \leq u,\ v \leq 1$

$$C(u, v) = F\left[F_1^{-1}(u), F_2^{-1}(v)\right]$$

so that knowing F, F_1 and F_2, we are able to generate C. Thus, for example, if T_1 and T_2 are independent random variables with distribution functions F_1 and F_2, respectively, then $F(t_1, t_2) = F_1(t_1) F_2(t_2)$ from which it follows that $C(u, v) = uv$. Conversely, by Sklar's theorem, T_1 and T_2 are independent if $C(u, v) = uv$. Thus T_1 and T_2 are independent if, and only if, $C(u, v) = uv$. In a similar vein, we can argue that since T_1 and T_2 are exchangeable implies that the vectors (T_1, T_2) and (T_2, T_1) have the same distribution, T_1 and T_2 are exchangeable if, and only if, $F_1 = F_2$ and $C(u, v) = C(u, v)$.

Sklar's theorem also enables us to obtain bounds on a copula $C(u, v)$ via the Fréchet bounds of (4.27); this is because the Fréchet bounds are themselves distributions. Specifically, for $0 \leq u,\ v \leq 1$,

$$\max(u + v - 1, 0) \leq C(u, v) \leq \min(u, v);$$

$\max(u + v - 1, 0)$ is a copula for $\max(F_1(t_1) + F_2(t_2) - 1, 0)$ and $\min(u, v)$ is a copula for $\min(F_1(t_1), F_2(t_2))$. Finally, Morgenstern's distribution of (4.28) gives rise to the copula

$$C(u, v) = uv(1 + \alpha(1 - u)(1 - v)).$$

An alternate method of generating copulas is due to Genest and MacKay (1986).

Whereas a copula C joins univariate distribution functions to form a multivariate distribution, a survival copula $\widehat{C}$ joins univariate survival functions to form a multivariate survival function. Thus in the bivariate case, we have $\overline{F}(t_1, t_2) = \widehat{C}(\overline{F}_1(t_1), \overline{F}_2(t_2))$, where it can be easily seen that

$$\widehat{C}(u, v) = u + v - 1 + C(1 - u, 1 - v).$$

Like C, $\widehat{C}$ is a copula, but $\widehat{C}$ is not to be confused with $\overline{C}(u, v)$, the bivariate survival function of two uniformly distributed random variables. This is because

$$\overline{C}(u, v) = 1 - u - v + C(u, v) = \widehat{C}(1 - u, 1 - v).$$

Interest in survival copulas stems from the fact that for some multivariate distributions, such as Marshall and Olkin's (1967) multivariate exponential, survival copulas take simpler forms than their corresponding copulas.

Measures of Interdependence

The dependence of T_1 on T_2 (or vice versa) is best described via the **conditional survival function** $P(T_1 \geq t_1 \mid T_2 = t_2)$, and/or the **conditional mean** $E(T_1 \mid T_2 = t_2)$, where

$$P(T_1 \geq t_1 \mid T_2 = t_2) = \frac{P(T_1 \geq t_1,\ T_2 = t_2)}{P(T_2 = t_2)}, \qquad \text{and} \qquad (4.29)$$

$$E(T_1 \mid T_2 = t_2) = \int_0^\infty (1 - P(T_1 \geq t_1 \mid T_2 = t_2)) \mathrm{d}t_1.$$

It follows from the above that when $(T_1 \geq t_1)$ and $(T_2 = t_2)$ are judged independent, $P(T_1 \geq t_1 \mid T_2 = t_2) = P(T_1 \geq t_1)$, and $E(T_1 \mid T_2 = t_2) = E(T_1)$. The conditional mean $E(T_1 \mid T_2 = t_2)$ is also known as the **regression** of T_1 on T_2.

If T_1 and T_2 are judged dependent, then the extend to which T_1 and T_2 experience a linear relationship is measured by the product moment $E(T_1 T_2)$, where (assuming that $f(t_1, t_2)$ exists),

$$E(T_1 T_2) = \int_0^\infty \int_0^\infty t_1 t_2 f(t_1, t_2) \mathrm{d}t_1 \mathrm{d}t_2, \tag{4.30}$$
$$= \int_0^\infty t_1 E(T_2 \mid T_1 = t_1) f_1(t_1) \mathrm{d}t_1;$$

$f_1(t_1)$ is the probability density generated by $F_1(t_1)$.

A normalization of the product moment to yield values between -1 and $+1$ results in Pearson's **coefficient of correlation** $\rho(T_1, T_2)$, where

$$\rho(T_1, T_2) = \frac{E(T_1 T_2) - E(T_1)E(T_2)}{(V(T_1)V(T_2))^{1/2}};$$

$V(T_i)$ is the variance of $F_i(t_i)$, $i = 1, 2$. The numerator, $E(T_1 T_2) - E(T_1)E(T_2)$, is known as the **covariance** of T_1 and T_2; it is denoted $\mathrm{Cov}(T_1, T_2)$.

It can be verified that $-1 \leq \rho(T_1, T_2) \leq +1$, and that $\rho(T_1, T_2) = 0$ if T_1 and T_2 are independent. However, since $\rho(T_1, T_2)$ provides an assessment of only the extent of a linear relationship between T_1 and T_2, $\rho(T_1, T_2) = 0$, does not necessarily imply that T_1 and T_2 are independent; one exception is the case wherein T_1 and T_2 have a bivariate Gaussian distribution. Another exception is the BVE of section 4.7.4. A stronger result pertaining to independence under uncorrelatedness is due to Joag-Dev (1983), who shows that when T_1 and T_2 are 'associated', $\rho(T_1, T_2) = 0$ implies that T_1 and T_2 are independent. The notion of association, as a measure of dependence, is due to Esary, Proschan and Walkup (1967). It says that a random vector $T = (T_1, \ldots, T_n)$ is associated if for every (co-ordinatewise) non-decreasing function f and g, $Cov[f(T), g(T)] \geq 0$.

Values of $\rho(T_1, T_2) > (<) 0$ suggest that the events $(T_1 \geq t_1)$ and $(T_2 \geq t_2)$ are positively (negatively) dependent, for all values of t_1 and t_2. Under positive dependence the failure of one component increases the probability of failure of the surviving components. Positive dependence manifests itself when components share a common load or when components operate in a common environment. With negative dependence, the failure of one component increases (decreases) the probability of survival (failure) of the other component. For physical systems, the judgment of negative dependence is difficult to foresee. With certain biological systems, negative dependence is justified on grounds that for entities that compete for limited resources, such as food, the failure of one unit increases the available resources for the surviving unit, and thus its probability of failure decreases.

The departure of T_1 from the regression of T_1 on T_2 is measured by the **squared correlation ratio**

$$\eta^2(T_1 \mid T_2) = \frac{1}{V(T_2)} \int_0^\infty (E(T_1) - E(T_1 \mid T_2 = t_2))^2 f_2(t_2) \mathrm{d}t_2.$$

Like the conditional mean and the coefficient of correlation, the squared correlation ratio is also a measure of the dependence of T_1 on T_2.

By making the transformations $u = F_1(t_1)$, $v = F_2(t_2)$ and $c(u, v) = F(F_1^{-1}(u), F_2^{-1}(v))$, it can be seen that in terms of copulas, the coefficient of correlation takes the form

$$\rho(T_1, T_2) = (V(T_1)\; V(T_2))^{-\frac{1}{2}} \int_0^1 \int_0^1 [C(u, v) - uv] \mathrm{d}F_1^{-1}(u) \mathrm{d}F_2^{-1}(v),$$

so that $\rho(T_1, T_2) = 0$, if T_1 and T_2 are independent.

The above representation of $\rho(T_1, T_2)$ suggests some alternate measures of dependence of T_1 on T_2. These are **Spearman's Rho**, given by

$$r(T_1, T_2) = 12\int_0^1\int_0^1 [C(u, v) - u\, v]\,du\,dv,$$

and **Kendall's Tau** given by

$$\tau(T_1, T_2) = 4\int_0^1\int_0^1 C(u, v)dC(u, v) - 1, \text{ for } 0 \leq u, v \leq 1.$$

Finally, there is a measure of positive dependence that is suitable for lifelengths sharing a common environment. This is a **positive quadrant dependence**, wherein for all t_1 and t_2, $P(T_1 \leq t_1, T_2 \leq t_2) \geq P(T_1 \geq t_1)\,P(T_2 \geq t_2)$. According to this definition, T_1 and T_2 are positively quadrant dependent if, and only if, $C(u, v) \geq uv$, for all $0 \leq u, v \leq 1$, or equivalently $\widehat{C}(u, v) \geq uv$. Geometrically, the above inequality suggests that for a positive quadrant dependence, the graph of $C(u, v)$ must lie on or above that line of uv. Thus copulas provide a graphical way to portray the positive dependence between two lifetimes T_1 and T_2, the strength of dependence being a function of the deviation of $C(u, v)$ from uv. Since $C(u, v)$ changes with u and v, the dependence portrayed by $C(u, v) \geq uv$ is 'local'. Spearman's Rho, $r(T_1, T_2)$, provides a more global measure of positive quadrant dependence.

4.7.2 The Bivariate Exponential Distributions of Gumbel

Gumbel (1960) has proposed a system of bivariate distributions whose marginal distributions are exponential – thus the name 'bivariate exponential'. This system may be viewed as the very first family of probability models for describing dependent lifelengths. A drawback of Gumbel's system is that the proposed models lack a physical motivation, and that a generalization to the multivariate case is not obvious. Its advantages are that unlike many of the other multivariate lifelength distributions, both positive and negative dependence can be represented; also the marginal distributions are unit exponential (i.e. an exponential distribution with a scale parameter one). This latter feature makes the system a natural choice for generating other families of distributions for dependent lifelengths via hazard potentials (Theorem 4.3). This is one of the main reasons for introducing the Gumbel family.

Under Gumbel's system, there are two versions of the bivariate exponential distribution. Under the first version the correlation is always negative, and the largest value that it can take is -0.4036. Under the second version, both positive and negative values of the correlation can be had, and the maximum value of the correlation is ± 0.25. I overview below each of the two versions.

Version I of Gumbel's Bivariate Exponential Distribution

Motivated by the lower Fréchet bound, Gumbel has proposed, for a parameter δ, where $0 \leq \delta \leq 1$, the bivariate survival function

$$P(T_1 \geq t_1,\ T_2 \geq t_2 \mid \delta) = \exp(-(t_1 + t_2 + \delta t_1 t_2)), \quad \text{for } t_1, t_2 \geq 0.$$

The marginal distribution functions are unit exponential; i.e. $P(T_i \geq t_i \mid \delta) = \exp(-t_i), i = 1, 2$. This suggests that the parameter δ plays a role only with respect to the joint distribution function, and that T_1 and T_2 are independent when $\delta = 0$. Supressing δ, the conditional mean is of the form

$$E(T_1 \mid T_2 = t_2) = \frac{1 + \delta + \delta t_2}{(1 + \delta t_2)^2},$$

implying that for $\delta > 0$, the regression of T_1 on T_2 decreases from $(1+\delta)$ when $t_2 = 0$, to zero when t_2 becomes infinite. When $\delta = 0$, T_1 and T_2 are independent; thus $E(T_1 \mid T_2 = t_2) = 1$, for all values of t_2. Analogous to the conditional mean, we may also obtain the conditional variance. It has been shown (cf. Gumbel, 1960) that the conditional variance

$$V(T_1 \mid T_2 = t_2) = \frac{(1+\delta+\delta t_2)^2 - 2\delta^2}{(1+\delta t_2)^4}.$$

Thus like the conditional mean, the conditional variance decreases in t_2, ranging from one when $\delta = 0$, to $(2+4t_2+t_2^2)/(1+t_2)^4$, when $\delta = 1$.

An expression for the coefficient of correlation involves the integral logarithm and is therefore cumbersome to write out. However, $\rho(T_1, T_2)$ depends on δ alone, and is always negative ranging from zero when $\delta = 0$, to -0.4036 when $\delta = 1$. Because of this negative correlation, the bivariate exponential given here is a meaningful model only when the dependence between T_1 and T_2 can be judged as being negative.

Version II of Gumbel's Bivariate Exponential Distribution

If in (4.28) the marginals are chosen to be unit exponentials, that is, if $F_i(t_i) = (1 - \exp(-t_i))$, for $t_i \geq 0, i = 1, 2$, then it can be seen that

$$P(T_1 \geq t_1,\ T_2 \geq t_2 \mid \alpha) = \mathrm{e}^{-(t_1+t_2)}(1 + \alpha(1 - \mathrm{e}^{-t_1} - \mathrm{e}^{-t_2} + \mathrm{e}^{-(t_1+t_2)})), \quad \text{for } -1 \leq \alpha \leq +1;$$

this is Version II of Gumbel's system. It leads to the copula

$$C(u, v) = \exp\left\{-[(-\log u)^\alpha + (-\log v)^\alpha]^{\frac{1}{\alpha}}\right\}.$$

Here the regression of T_1 on T_2 (with α supressed) is of the form

$$E(T_1 \mid T_2 = t_2) = 1 + \frac{\alpha}{2} - \alpha \mathrm{e}^{-t_2},$$

suggesting that for $\alpha > 0$, the conditional mean of T_1 increases with t_2 from $(1 - \alpha/2)$ to $(1 + \alpha/2)$. For $\alpha < 0$, the conditional mean decreases. In either case, the regression curves are exponential functions of t_2 (illustrated in Figure 4.15). Similarly, for $\alpha > (<)0$, the conditional variance increases (decreases) with t_2. This is because

$$V(T_1 \mid T_2 = t_2) = 1 + \frac{\alpha}{2}(1 - 2\mathrm{e}^{-t_2}) - \frac{\alpha^2}{4}(1 - 2\mathrm{e}^{-t_2})^2.$$

Finally, it can be shown (Gumbel, 1960), that the coefficient of correlation $\rho(T_1, T_2)$ is $\alpha/4$, so that $\rho(T_1, T_2)$ is restricted to take values between $-1/4$ and $+1/4$. When $\alpha = 0$, $\rho(T_1, T_2) = 0$; however $\alpha = 0$ also implies that T_1 and T_2 are independent. Thus the bivariate exponential distribution discussed here provides another example of the case wherein $\rho(T_1, T_2) = 0$, implies that T_1 and T_2 are independent. The same is also true of Version I of Gumbel's bivariate exponential distribution; namely, that when $\delta = 0$, $\rho(T_1, T_2) = 0$, and T_1 and T_2 are independent. To summarize, the bivariate Gaussian and Gumbel's bivariate exponential distributions provide examples of the case wherein $\rho(T_1, T_2) = 0$, implies that T_1 and T_2 are independent; another example is the BVE of section 4.7.4.

System Reliability Considerations

With $P(T_1 \geq t_1, T_2 \geq t_2 \mid \cdot)$ specified, it is easy to obtain the reliability of a series or a parallel (redundant) system having interdependent lifelengths described by Gumbel's bivariate exponential distributions. Specifically, let T_s and T_p denote the times to failure or a series and a parallel-redundant two-component system whose lifetimes are T_1 and T_2. Then, for any $t \geq 0$, where t is the mission time, the reliability of the series system is

$$P(T_s \geq t \mid \cdot) = P(T_1 \geq t, T_2 \geq t \mid \cdot)$$

and that of the parallel redundant system is

$$P(T_p \geq t \mid \cdot) = P(T_1 \geq t \mid \cdot) + P(T_2 \geq t \mid \cdot) - P(T_1 \geq t, T_2 \geq t \mid \cdot);$$

the second equality is a consequence of (4.26). The marginals $P(T_i > t)$, $i = 1, 2$, are unit exponentials.

4.7.3 Freund's Bivariate Exponential Distribution

In contrast to Gumbel's bivariate exponential distributions which do not appear to have a physical motivation, Freund (1961) proposed a bivariate extension of the exponential distribution that is based on the following construction.

Consider a two-unit **parallel redundant** system; i.e. a system wherein all components are simultaneously put to use, but all that is needed for the system to function is one working component. Examples are paired organs like eyes, kidneys and lungs, or a twin-engine aircraft like Boeing's 777. The system is considered to be functioning (albeit in a degraded state) even when one of the two units has failed. However, the failed unit increases the stress on the surviving unit, and in so doing increases the latter unit's propensity of failure. Freund (1961) proposed a simple model for encapsulating such situations. He assumed that, when both the components are functioning, the lifelength of component i, T_i, $i = 1, 2$, has a constant (model) failure rate λ_i. Upon the failure of the first component of the (model) failure rate of the surviving component increases from λ_i to λ_i^*, $i = 1, 2$. Figure 4.9 shows the failure rate function of T_2 assuming that component 1 has failed first, at $T_1 = t_1$.

For the set-up described above, T_1 and T_2 are not independent, since the failure of one component changes the parameter of the surviving one. Consequently, for $0 < t_1 < t_2 < \infty$,

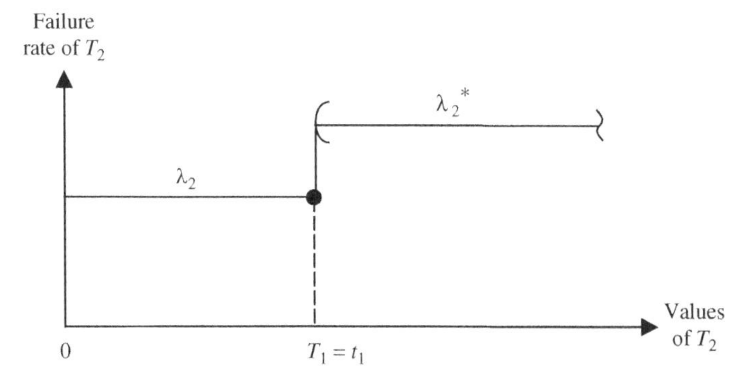

Figure 4.9 Failure rate function of the second to fail component (Freund's model).

$$\begin{aligned} P(t_1 \le T_1 \le t_1 + \mathrm{d}t_1, t_2 \le T_2 \le t_2 + \mathrm{d}t_2) &= P(t_2 \le T_2 \le t_2 + \mathrm{d}t_2 \mid T_1 = t_1) \cdot \\ &\quad P(t_1 \le T_1 \le t_1 + \mathrm{d}t_1) \\ &= (\mathrm{e}^{-\lambda_2 t_1})(\lambda_2^* \mathrm{e}^{-\lambda_2^*(t_2 - t_1)} \mathrm{d}t_2)(\lambda_1 \mathrm{e}^{-\lambda_1 t_1} \mathrm{d}t_1) \\ &= \lambda_1 \lambda_2^* \mathrm{e}^{-\lambda_2^* t_2 - (\lambda_1 + \lambda_2 - \lambda_2^*) t_1} \mathrm{d}t_1 \mathrm{d}t_2; \end{aligned}$$

similarly for $0 < t_2 < t_1 < \infty$, *mutatis mutandis*. Thus, suppressing all parameters on the left-hand side, the joint density function of $P(T_1 \ge t_1, T_2 \ge t_2)$ at (t_1, t_2) is

$$f(t_1, t_2) = \begin{cases} \lambda_1 \lambda_2^* \mathrm{e}^{-\lambda_2^* t_2 - (\lambda_1 + \lambda_2 - \lambda_2^*) t_1}, & 0 < t_1 < t_2 < \infty \\ \lambda_2 \lambda_1^* \mathrm{e}^{-\lambda_1^* t_1 - (\lambda_1 + \lambda_2 - \lambda_1^*) t_2}, & 0 < t_2 < t_1 < \infty. \end{cases} \tag{4.31}$$

Since $\lim_{t_2 \downarrow t_1} f(t_1, t_2)$ does not equal $\lim_{t_2 \uparrow t_1} f(t_1, t_2)$, Freund's model does not have a probability density at $t_1 = t_2$; i.e. $f(t, t)$ is not defined.

The probability that component 1 is the first to fail is $P(T_1 < T_2)$. Clearly

$$\begin{aligned} P(T_1 < T_2) &= \int_0^\infty P(T_2 > t_1 \mid T_1 = t_1)\, f(t_1)\, \mathrm{d}t_1 \\ &= \int_0^\infty \mathrm{e}^{-\lambda_2 t_1}\, \lambda_1\, \mathrm{e}^{-\lambda_1 t_1}\, \mathrm{d}t_1 = \frac{\lambda_1}{\lambda_1 + \lambda_2}, \end{aligned}$$

provided that the events $(T_2 > t_1)$ and $(T_1 = t_1)$ are judged independent. Similarly, $P(T_2 < T_1) = \lambda_2/(\lambda_1 + \lambda_2)$.

Properties: Moments, Marginals and Memory

Using moment-generating functions (for details see Freund, 1961) it can be shown that

$$\begin{aligned} E(T_1) &= \frac{(\lambda_1^* + \lambda_2)}{\lambda_1^*(\lambda_1 + \lambda_2)}, \\ V(T_1) &= \frac{((\lambda_1^*)^2 + 2\,\lambda_1 \lambda_2 + \lambda_2^2)}{(\lambda_1^*)^2 (\lambda_1 + \lambda_2)^2}, \quad \text{and} \\ \mathrm{Cov}(T_1, T_2) &= \frac{(\lambda_1^* \lambda_2^* - \lambda_1 \lambda_2)}{\lambda_1^* \lambda_2^* (\lambda_1 + \lambda_2)^2}; \end{aligned}$$

similarly, $E(T_2)$ and $V(T_2)$.

The marginal density of T_1 at t_1, obtained by integrating (4.31) over t_2, is for $\lambda_1 + \lambda_2 - \lambda_1^* \ne 0$, given as

$$f(t_1) = \frac{(\lambda_1 - \lambda_1^*)(\lambda_1 + \lambda_2)\, \mathrm{e}^{-(\lambda_1 + \lambda_2) t_1}}{\lambda_1 + \lambda_2 - \lambda_1^*} + \frac{\lambda_1^* \lambda_2\, \mathrm{e}^{-\lambda_1^* t_1}}{\lambda_1 + \lambda_2 - \lambda_1^*}; \tag{4.32}$$

similarly for $f(t_2)$. If $\lambda_1 + \lambda_2 - \lambda_1^* = 0$, then

$$f(t_1) = (\lambda_1 + \lambda_1^* \lambda_2\, t_1)\, \mathrm{e}^{-\lambda_1^* t_1}. \tag{4.33}$$

The above equations imply that, unlike the marginal distributions of Gumbel (1960), those of Freund (1961) are not exponential. In the case of $\lambda_1 + \lambda_2 - \lambda_1^* \ne 0$, the marginal of T_1 is a mixture of two exponentials, one with scale parameter $(\lambda_1 + \lambda_2)$, and the other with a scale parameter λ_1^*. The mixing coefficients are $(\lambda_1 - \lambda_1^*)/(\lambda_1 + \lambda_2 - \lambda_1^*)$ and $\lambda_2/(\lambda_1 + \lambda_2 - \lambda_1^*)$, respectively. When $\lambda_1^* > \lambda_1$ one of the two mixing coefficients will be negative and so the failure rate function of the distribution of T_1 will be increasing; similarly, with T_2.

When $\lambda_i^* = \lambda_i$, $i = 1, 2$, $E(T_i) = \lambda_i$ and $f(t_i) = \lambda_i e^{-\lambda_i t_i}$. Furthermore $\text{Cov}(T_1, T_2) = 0$ and $f(t_1, t_2) = \lambda_1 \lambda_2 e^{-(\lambda_1 t_1 + \lambda_2 t_2)}$; thus T_1 and T_2 are independent. Finally, the expression for $\text{Cov}(T_1, T_2)$ suggests that when $\lambda_1^* \lambda_2^* < \lambda_1 \lambda_2$, T_1 and T_2 can be negatively dependent. This can happen if the failure of the first component decreases the propensity of failure of the surviving component. Such scenarios can arise with resource-sharing systems, which operate under limited resources; for example, a multi-unit power-generating facility supported by on-line preventive maintenance. The failure of a unit makes more resources available for the maintenance of the surviving units.

It is well known that the exponential (chance) distribution $P(X \geq x \mid \lambda) = \overline{F}(x \mid \lambda) = e^{-\lambda x}$ enjoys the **lack of memory** property, i.e. $P(X \geq x + s \mid X \geq x, \lambda) = P(X \geq s \mid \lambda)$. Indeed, it is only the univariate distribution for continuous lifelengths having this property. The lack of memory property is appropriate for items that do not deteriorate (or age) with use, since the essence of this property is that our assessment of the item's future survivability is not influenced by its past use. The bivariate analogue of the lack of memory property is of the form $P(T_1 \geq t_1 + s, T_2 > t_2 + s \mid T_1 \geq t_1, T_2 \geq t_2) = P(T_1 \geq s, T_2 \geq s)$; where, for convenience, the conditioning parameters have been suppressed. Block and Basu (1974) have shown that Freund's bivariate exponential distribution is absolutely continuous and that it possesses the bivariate lack of memory property. Indeed, besides a bivariate distribution based on independent exponential marginals, Freund's distribution is one of the few absolutely continuous bivariate distributions which possess the bivariate lack of memory property.

System Reliability Considerations

It is relatively straightforward to verify that when $\lambda_1 = \lambda_2 = \lambda$ and $\lambda_1^* = \lambda_2^* = 2\lambda$, T, the time to failure of the paired-organ system has density at t given by

$$f(t \mid \lambda) = 4t\lambda^2 \exp(-2\lambda t), \tag{4.34}$$

with survival function (or reliability) of the form

$$P(T > t \mid \lambda) = 2\lambda t \exp(-2\lambda t) + \exp(-2\lambda t); \tag{4.35}$$

thus, the system's model failure rate function is

$$r(t \mid \lambda) = \frac{4\lambda^2 t}{2\lambda t + 1}. \tag{4.36}$$

More about this system will be said later, in section 4.8 on models for cascading failures.

4.7.4 The Bivariate Exponential of Marshall and Olkin

Freund's bivariate distribution is absolutely continuous and possesses the bivariate lack of memory property; however, it does not allow for the simultaneous failure of both components. Also, it does not generalize easily to the multivariate case. To account for scenarios that involve the simultaneous failure of two (or more) components, Marshall and Olkin (1967) introduced a bivariate distribution with exponential marginals, which have the bivariate lack of memory property. They termed this distribution the 'BVE' and its multivariate version the 'MVE'. As we shall soon see, the BVE has many attractive features; its main disadvantage is that it is not absolutely continuous, a consequence of which is that statistical methods based on densities cannot be easily invoked. To appreciate the structure of the BVE, it is necessary for us to overview the notion of a Poisson counting process.

The Poisson Counting Process – An Overview

A **point process** is, roughly speaking, a countable random collection of points on the real line. Let $N(t)$ be the number of points in $[0, t]$. Then $N(t)$, as a function of $t \geq 0$, counts the events of the point process, and is hence called a **counting process**. To see why point processes are of interest to us in reliability, consider a scenario wherein certain events (like shocks) occur over time $t \geq 0$, according to the postulates of a special and an important kind of a point process, namely a Poisson process. The postulates of a Poisson process are:

(i) The probability of an event occuring in an interval of time $[t, t+h] = \lambda h + o(h)$, where λ is a specified constant.
(ii) The probability of two or more events occurring in $[t, t+h] = o(h)$;

$o(h)$ is a function of h, say $g(h)$, for $h > 0$ with the property that $\lim_{h \downarrow 0}[g(h)/h] = 0$.

As before, let $N(t)$ denote the (unknown) number of events that have occurred in the time interval $[0, t]$, with the proviso that $N(0) = 0$. Then the sequence of random variables $\{N(t);\ t \geq 0\}$ is called a **homogeneous Poisson counting process with intensity** λ. The following properties of this process are well-known (for example, Ross, 1996).

(a) For time point s and t with $s < t$, $N(t) - N(s)$, the number of events that occur in the interval $[s, t]$ has a **Poisson distribution** with parameter $\lambda(t-s)$; i.e.

$$P[(N(t) - N(s)) = k \mid \lambda] = \frac{e^{-\lambda(t-s)}[\lambda(t-s)]^k}{k!};$$

consequently, taking $s = 0$,

$$P((N(t) = k \mid \lambda) = \frac{\exp(-\lambda t)(\lambda t)^k}{k!}$$

with $E(N(t) \mid \lambda) = \lambda t$.

(b) If $T_1, T_2, \ldots,$ denote the times between the arrivals of the events, i.e. the inter-arrival times, then given λ, the T_is are independent and identically exponentially distributed. Thus $P(T_i \geq t \mid \lambda) = e^{-\lambda t},\ i = 1, 2, \ldots$

(c) The process $\{N(t);\ t \geq 0\}$ has stationary **independent increments**, i.e. for $0 \equiv t_0 < t_1 < t_2 < \cdots < t_h < t_{h+1} < \cdots$, where $t_i = t_0 + i\Delta$, for $\Delta > 0$, the random variables $(N(t_1) - N(t_0)), (N(t_2) - N(t_1)), \ldots, (N(t_{h+1}) - N(t_h))$ are independent and identically distributed.

The sequence of random variables $\{N(t);\ t \geq 0\}$ is called a **non-homogeneous Poisson process with a mean value function** $\Lambda(t)$, if the process has independent increments – see (c) above – and if for all s and t, with $s < t$,

$$P[(N(t) - N(s)) = k \mid \Lambda(t)] = \frac{e^{-[\Lambda(t) - \Lambda(s)]}[\Lambda(t) - \Lambda(s)]^k}{k!};$$

here $E(N(t)) = \Lambda(t)$ and $d[\Lambda(t)]/dt \stackrel{\text{def}}{=} \lambda(t)$ is called the **intensity function of the nonhomogenous Poisson process**.

Structure of the Bivariate Exponential Distribution – BVE

Consider a system comprising of two components, connected in series or in parallel. The system operates in an environment in which three types of shocks occur, each occurring per the postulates of a homogenous Poisson process. Let $\{N_i(t);\ t \geq 0\}$, $i = 1, 2, 3$, denote the associated counting processes, with intensities λ_1, λ_2 and λ_{12}, respectively. Suppose that the above three sequences of random variables are contemporaneously independent. The shocks associated with λ_i have an effect on component i alone, $i = 1, 2$, and those associated with λ_{12} have an effect on both components. Whenever a component receives a shock that is associated with it, the component fails. The set-up given here is very general and one can conceive of many scenarios for which it is reasonable. A simple example is an electro-mechanical system wherein certain shocks are pertinent to the mechanical part alone; certain pertinent to the electrical part alone, and certain, like an earthquake, pertinent to both. Probability models based on scenarios wherein shocks lead to failure are called 'shock models'. Let T_i denote the time to failure of component i, and U_j, $j = 1, 2, 3$, the time to occurrence of the first shock in the process $\{N_j(t);\ t \geq 0\}$. Since the three Poisson processes mentioned above are independent, i.e., for any t, the random variables $N_1(t)$, $N_2(t)$ and $N_3(t)$ are independent, the corresponding U_js are also independent. Furthermore, by property (b) of homogenous Poisson processes, each U_j has an exponential distribution.

Thus for the 'fatal shock model' described above

$$T_i = \min(U_i, U_3),\ \ i = 1, 2,\ \text{ and so}$$

$$\begin{aligned} P(T_1 \geq t_1, T_2 \geq t_2 \mid \lambda_1, \lambda_2, \lambda_{12}) &= P(U_1 \geq t_1, U_2 \geq t_2, U_3 \geq \max(t_1, t_2)) \\ &= \exp(-\lambda_1 t_1 - \lambda_2 t_2 - \lambda_{12} \max(t_1, t_2)). \end{aligned} \tag{4.37}$$

A generalization of the fatal shock model is the 'non-fatal shock model', wherein each shock is fatal to its associated component with a certain probability. Specifically, let p_i be the probability that a shock generated by the process $\{N_i(t);\ t \geq 0\}$ destroys component i, $i = 1, 2$, and let p_{10} be the probability that a shock generated by the process $\{N_3(t);\ t \geq 0\}$ destroys component 1 but keeps component 2 intact. Similarly, p_{01}, p_{00} and p_{11}, where $p_{00} + p_{10} + p_{01} + p_{11} = 1$. Then, using property (b) of the homogenous Poisson process, it can be verified that for $0 \leq t_1 \leq t_2$, and $\underline{p} = (p_1, p_2, p_{00}, p_{10}, p_{01}, p_{11})$,

$$\begin{aligned} &P(T_1 \geq t_1, T_2 \geq t_2 \mid \lambda_1, \lambda_2, \lambda_{12}, \underline{p}) = \\ &\qquad \exp\{-t_1(\lambda_1\ p_1 + \lambda_{12}\ p_{10}) - t_2(\lambda_2\ p_2 + \lambda_{12}(1 - p_{00} - p_{10}))\}; \end{aligned}$$

the details can be found in Marshall and Olkin (1967). Similarly, for $0 \leq t_2 \leq t_1$,

$$\begin{aligned} &P(T_1 \geq t_1, T_2 \geq t_2 \mid \lambda_1, \lambda_2, \lambda_{12}, \underline{p}) = \\ &\qquad \exp\{-t_1(\lambda_1\ p_1 + \lambda_{12}(1 - p_{00} - p_{01})) - t_2(\lambda_2\ p_2 + \lambda_{12}\ p_{01})\}. \end{aligned}$$

If we set $\Lambda_1 = \lambda_1 p_1 + \lambda_{12} p_{10}$, $\Lambda_2 = \lambda_2 p_2 + \lambda_{12} p_{01}$ and $\Lambda_{12} = \lambda_{12} p_{11}$, we obtain

$$P(T_1 \geq t_1, T_2 \geq t_2 \mid \Lambda_1, \Lambda_2, \Lambda_{12}) = \exp(-\Lambda_1\ t_1 - \Lambda_2\ t_2 - \Lambda_{12} \max(t_1, t_2)). \tag{4.38}$$

To gain insight about the arguments that lead to the above equation, consider the case of a single component with lifelength T, which experiences shocks per the postulates of a homogenous Poisson process with intensity λ. Suppose that each shock is fatal to the component with probability p. Then the component survives to time t if all the shocks it experiences in $[0, t]$ are non-fatal. Thus

$$P(T \geq t) = \sum_{j=0}^{\infty} \frac{e^{-\lambda t}(\lambda t)^j}{j!}(1-p)^j = e^{-\lambda p t}.$$

If every shock is a fatal shock, then $p = 1$, and $P(T \geq t) = e^{-\lambda t}$, an exponential chance distribution with parameter λ.

Similarly, when $p_1 = p_2 = p_{11} = 1$, the non-fatal shock model of (4.38) reduces to the fatal shock model of (4.37). In deriving (4.38) we have assumed that there are no after-effects of each shock that is not fatal to its component(s). When T_1 and T_2 have a survival function of the form given by (4.37) or (4.38), they are said to have a **bivariate exponential distribution**, abbreviated the BVE. The distribution easily generalizes to the n-variate case for $n > 2$. However, its number of parameters grows exponentially. For example with $n = 3$, the number of parameters is seven: $\lambda_1, \lambda_2, \lambda_3, \lambda_{12}, \lambda_{13}, \lambda_{23}$ and λ_{123}. Thus unless one is prepared to assume that some of the λs, particularly those having multiple indices such as λ_{123}, are zero, the MVE as a model for failure is unwieldy. Engineers refer to such models as being non-scalable, and models with several λs set to zero are said to have a loss of granularity. Thus in using the MVE as a model for failure one may have to trade off between scalability and granularity; this is perhaps the MVE's biggest disadvantage. However, the model has many attractive features, some of which serve to illustrate the finer aspects of modeling interdependency. These are described below with the bivariate case as a point of discussion. In the interest of giving a broad coverage, many of the results given below are without proof; the details can be found in Marshall and Olkin (1967) or in Barlow and Proschan (1975, pp. 127–138).

Moments, Marginals and Memory of the BVE

Consider the joint survival function of (4.37); here again, in all that follows, we suppress the conditioning parameters λ_1, λ_2, and λ_{12}, to write the BVE as

$$\overline{F}(t_1, t_2) = P(T_1 \geq t_1, T_2 \geq t_2) = \exp(-\lambda_1\ t_1 - \lambda_1\ t_1 - \lambda_{12}\max(t_1, t_2)). \tag{4.39}$$

Claim 4.1 below is a natural consequence of the construction of the BVE.

Claim 4.1. *If T_1 and T_2 have the BVE, then there exist independent exponentially distributed random variables U_1, U_2 and U_3 such that $T_1 = \min(U_1, U_3)$ and $T_2 = \min(U_2, U_3)$.*

By setting t_1 (or t_2) equal to zero, we see that for any $t_i > 0$, $P(T_i \geq t_i) = \exp(-(\lambda_i + \lambda_{12})t_i)$, $i = 1, 2$. Thus we have

Claim 4.2. *The marginal distributions of the BVE are exponential. Consequently, $E(T_i) = (\lambda_i + \lambda_{12})^{-1}$, and $V(T_i) = (\lambda_i + \lambda_{12})^{-2}$, $i = 1, 2$.*

For any $t \geq 0$, (4.39) leads to the result that for all $s_1, s_2 \geq 0$

$$P(T_1 \geq t + s_1, T_2 \geq t + s_2 \mid T_1 \geq s_1, T_2 \geq s_2) = P(T_1 \geq t, T_2 \geq t). \tag{4.40}$$

Thus we have

Claim 4.3. *The BVE enjoys the bivariate lack of memory property.*

Indeed, it can be shown (cf. Barlow and Proschan, 1975, p. 130) that, besides a bivariate distribution that is based on independent exponential marginals, the BVE is the only bivariate distribution having exponential marginals which satisfies the bivariate lack of memory property. A consequence of this feature, plus (4.40) is

Claim 4.4. *The probability of survival of a two-component series system is independent of the ages of each component if, and only if, the joint lifelengths have the BVE.*

The Nature of Dependence in the BVE

To investigate the dependence of T_2 on T_1, we consider the quantity $P(T_2 \geq t_2 \mid T_1 = t_1) \stackrel{\text{def}}{=} \overline{F}(t_2 \mid t_1)$. This quantity, evaluated via an evaluation of $\lim_{\Delta t_1 \downarrow 0} P(T_2 \geq t_2 \mid t_1 \leq T_1 < t_1 + \Delta t_1)$, leads to the result that

$$\overline{F}(t_2 \mid t_1) = \begin{cases} \exp(-\lambda_2 t_2), & t_2 \leq t_1 \\ \frac{\lambda_1}{\lambda_1 + \lambda_{12}} \exp(-\lambda_{12}(t_2 - t_1) - \lambda_2 t_2), & t_2 > t_1. \end{cases} \tag{4.41}$$

The regression of T_2 on T_1, obtained by integrating $\overline{F}(t_2 \mid t_1)$ over 0 to ∞, is

$$E(T_2 \mid T_1 = t_1) = \frac{1}{\lambda_2} - \frac{\lambda \lambda_{12} e^{-\lambda_2 t_1}}{\lambda_2(\lambda_1 + \lambda_{12})(\lambda_2 + \lambda_{12})},$$

where $\lambda = \lambda_1 + \lambda_2 + \lambda_{12}$. Thus the regression of T_2 on T_1 is an exponentially increasing function of t_1 that is bounded by $1/\lambda_2$ (Figure 4.15). This behavior parallels that of $E(T_2 \mid T_1 = t_1)$ in the case of Gumbel's bivariate exponential distribution, Version II, for $\alpha > 0$; here the conditional mean is bounded by $(1 + \alpha/2)$.

Positive dependence in the case of the BVE can be asserted via the covariance. Specifically, it can be seen (cf. Barlow and Proschan, 1975, p. 135) that

$$\text{Cov}(T_1, T_2) = \frac{\lambda_{12}}{\lambda(\lambda_1 + \lambda_{12})(\lambda_2 + \lambda_{12})},$$

so that $\rho(T_1, T_2)$, Pearson's coefficient of correlation is λ_{12}/λ. When $\lambda_{12} = 0$, $\rho(T_1, T_2) = 0$, and (4.37) together with Claim 4.2 imply that T_1 and T_2 are independent. Thus for the BVE, $\rho(T_1, T_2) = 0$ implies that T_1 and T_2 are independent. Another distribution which shares this property, namely that $\rho(T_1, T_2) = 0$ implies that T_1 and T_2 are independent, is the bivariate Gaussian.

The BVE of (4.39) generates the survival copula

$$\widehat{C}(u, v) = uv \min(u^{-\alpha}, v^{-\beta}),$$

where $\alpha = \lambda_{12}/(\lambda_1 + \lambda_{12})$ and $\beta = \lambda_{12}/(\lambda_2 + \lambda_{12})$ (Nelson, 1999, p. 47). Since $C(u, v) \geq uv$, the lifelengths T_1 and T_2 are positively quadrant dependent.

Whereas $\overline{F}(t_2 \mid t_1)$ is well defined at $t_2 = t_1$ – see equation (4.41) – it does take a downward jump of size $[\lambda_{12}/(\lambda_1 + \lambda_{12})]\exp(-\lambda_2 t_1)$ at that point. The size of the jump depends on both λ_1 and λ_2, and is greater than zero, even if $\lambda_1 = \lambda_2$; the jump vanishes when $\lambda_{12} = 0$. I am able to show, details omitted, that the size of the jump is (approximately) equal to $P(T_2 = t_1 \mid T_1 = t_1)$, (Figure 4.10).

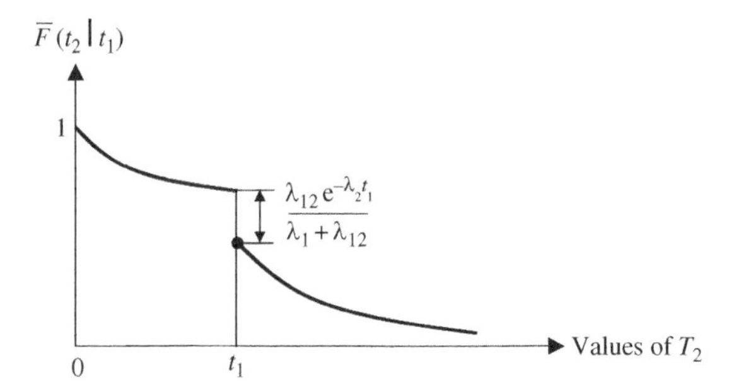

Figure 4.10 Jump in the conditional survival function of the BVE.

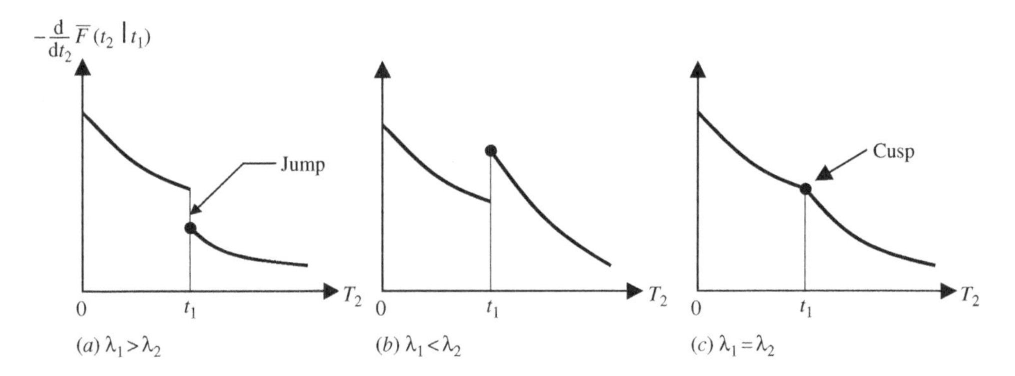

Figure 4.11 The conditional density of the BVE.

Corresponding to the jump of the conditional survival function of Figure 4.10 is a jump in its conditional probability density $-\mathrm{d}[\overline{F}(t_2 \mid t_1)]/\mathrm{d}t_2$, which exists for all $t_2 \neq t_1$; for $t_1 = t_2$, the density does not exist. It is easy to verify that the size of the jump in the density, at $t_2 = t_1$, is $[\lambda_{12}/(\lambda_1 + \lambda_{12})](\lambda_1 - \lambda_2)\mathrm{e}^{-\lambda_2 t_1}$. The jump is upward when $\lambda_1 < \lambda_2$, and the density has a cusp when $\lambda_1 = \lambda_2$; see Figures 4.11 (*b*) and (*c*), respectively.

Even though the probability density of $\overline{F}(t_2 \mid t_1)$ is not defined at the point $t_2 = t_1$, its failure rate, by virtue of (4.8), is defined everywhere. Specifically, if

$$r(t_2 \mid t_1) \stackrel{\text{def}}{=} -\mathrm{d}[\overline{F}(t_2 \mid t_1)/\overline{F}(t_2 \mid t_1)]/\mathrm{d}t_2,$$

then it is easy to verify that

$$r(t_2 \mid t_1) = \begin{cases} \lambda_2, & t_2 < t_1, \\ \lambda_2 + \lambda_{12}, & t_2 > t_1, \text{ and} \\ \frac{\lambda_{12}}{\lambda_1 + \lambda_{12}}, & t_2 = t_1; \end{cases}$$

and the last equation is a consequence of the jump in $\overline{F}(t_2 \mid t_1)$. Figure 4.12 shows a plot of $r(t_2 \mid t_1)$. It is instructive to compare Figures 4.9 and 4.12; recall that the former pertains to the

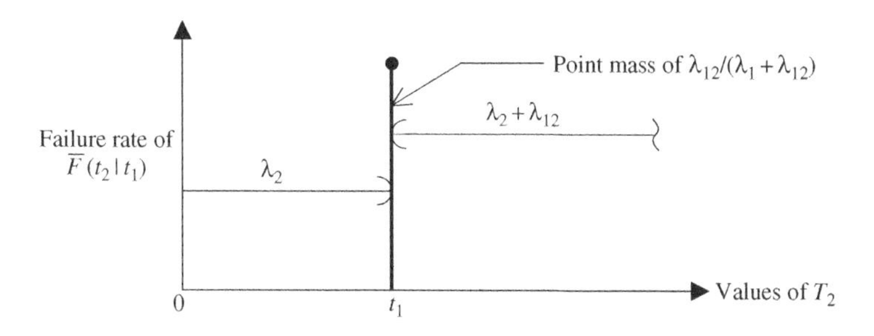

Figure 4.12 Failure rate function of the second to fail component in the BVE.

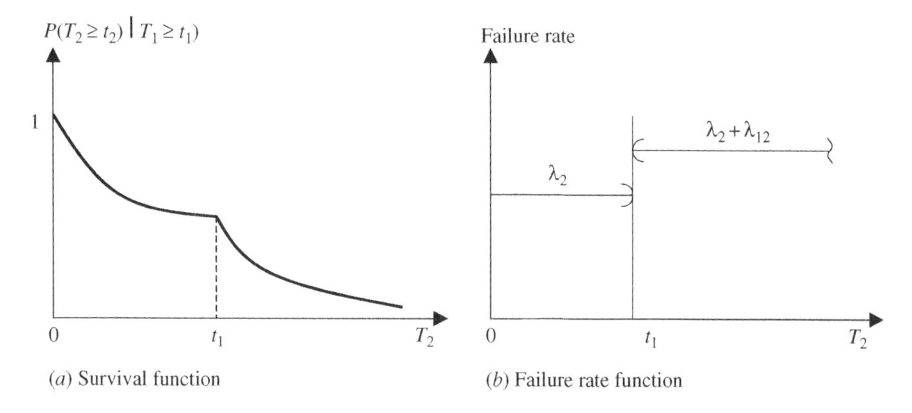

Figure 4.13 The conditional survival function $\mathcal{P}(T_2 \geq t_2 \mid T_1 \geq t_1)$ of the BVE and its failure rate function.

failure rate of the second component to fail in Freund's model. Both the failure rate functions experience a jump at $t_2 = t_1$, but the BVE has a point mass of size $\lambda_{12}/(\lambda_1 + \lambda_{12})$ at $t_2 = t_1$. The point mass could be smaller than λ_2, between λ_2 and $\lambda_2 + \lambda_{12}$, or greater than $\lambda_2 + \lambda_{12}$. In Figure 4.12, we show only the last case.

Our discussion thus far, pertaining to the nature of dependence of T_2 on T_1, has been based on a consideration of $P(T_2 \geq t_2 \mid T_1 = t_1)$. An analogous discussion based upon $(T_1 \geq t_1)$ as a conditioning event can also be conducted. The major difference between the cases $P(T_2 \geq t_2 \mid T_1 = t_1)$ and $P(T_2 \geq t_2 \mid T_1 \geq t_1)$ is that whereas the former experiences a jump at $t_2 = t_1$ (Figure 4.10), the latter experiences a cusp at $t_2 = t_1$. That is $P(T_2 \geq t_2 \mid T_1 \geq t_1)$ is continuous, but not differentiable at $t_2 = t_1$ (Figure 4.13 (a)). A consequence is that the failure rate of $P(T_2 \geq t_2 \mid T_1 \geq t_1)$ is not defined at $t_2 = t_1$ (Figure 4.13 (b)). Thus, when investigating the dependence of T_2 on T_1, the nature of the conditioning event is to be borne in mind.

A comparison of Figures 4.9, 4.12 and 4.13(b) is instructive.

Survival Function of the BVE and its Decomposition

The BVE has other noteworthy features, many of which are of mathematical interest alone. Some of these are given below; they help us gain a deeper appreciation of this remarkably interesting most referenced but least used joint distribution.

We start by noting that for $t_1 > t_2 > 0$,

$$\lim_{t_2 \uparrow t_1} \frac{\partial F(t_1, t_2)}{\partial t_1} \neq \lim_{t_2 \downarrow t_1} \frac{\partial F(t_1, t_2)}{\partial t_1},$$

for $t_2 > t_1 > 0$. Thus we have

Claim 4.5. *The BVE has a singularity along the line $t_1 = t_2$; the singularity disappears when $\lambda_{12} = 0$.*

We next note that

$$P(T_1 = T_2) = \int_0^\infty e^{-\lambda_1 t} e^{-\lambda_2 t} \lambda_{12} e^{-\lambda_{12} t} dt$$

$$= \frac{\lambda_{12}}{\lambda_1 + \lambda_2 + \lambda_{12}}.$$

The integral term is a consequence of the fact that the lifelengths of the two components can only be equal at the time of occurrence of U_3, but provided that U_3 precedes both U_1 and U_2. Thus we have

Claim 4.6. *The BVE has a probability mass along the line $t_1 = t_2$, unless of course $\lambda_{12} = 0$.*

After some tedious but routine calculations, it is evident that $\partial^2[\overline{F}(t_1, t_2)]/\partial t_1 \partial t_2$ exists for both $t_1 > t_2 > 0$ and $t_2 < t_1 < 0$; however, it does not exist for $t_1 = t_2$. Since $t_1 = t_2$ has a two-dimensional Lebesgue measure zero, we claim that $\partial^2[\overline{F}(t_1, t_2)]/\partial t_1 \partial t_2$ exists almost everywhere. However, it can be seen – again after some tedious calculations – that

$$\int_{t_1} \int_{t_2} \frac{\partial^2 \overline{F}(t_1, t_2)}{\partial t_1 \partial t_2} \neq \overline{F}(t_1, t_2);$$

thus $\partial^2[\overline{F}(t_1, t_2)]/\partial t_1 \partial t_2$ cannot be regarded as a density. We therefore have

Claim 4.7. *The BVE is not absolutely continuous; it does not have a probability density with respect to the two-dimensional Lebesgue measure.*

A consequence of Claim 4.7 is that one is unable to use methods based on densities because writing out the likelihood poses difficulties (Bemis, Bain and Higgins, 1972; and section 5.4.7). Since the BVE is not absolutely continuous, its Lebesgue decomposition (section 4.2) will entail discontinuities and/or singularities. However, as shown in Barlow and Proschan (1975, p. 133), the BVE has an absolutely continuous part and only a singular part; it has no discrete part. This is summarized via

Claim 4.8. *For $\lambda = \lambda_1 + \lambda_2 + \lambda_{12}$, the Lebesgue decomposition of $\overline{F}(t_1, t_2)$ yields*

$$\overline{F}(t_1, t_2) = \frac{\lambda_1 + \lambda_2}{\lambda} \overline{F}_a(t_1, t_2) + \frac{\lambda_{12}}{\lambda} \overline{F}_s(t_1, t_2)$$

where the absolutely continuous part is

$$\overline{F}_a(t_1, t_2) = \frac{\lambda}{\lambda_1 + \lambda_2} e^{-\lambda_1 t_1 - \lambda_2 t_2 - \lambda_{12} \max(t_1, t_2)} - \frac{\lambda_{12}}{\lambda_1 + \lambda_2} e^{-\lambda \max(t_1, t_2)},$$

and the singular part is

$$\overline{F}_s(t_1, t_2) = e^{-\lambda \max(t_1, t_2)}.$$

$\overline{F}_a(t_1, t_2)$ is the absolutely continuous part because $\partial^2[\overline{F}_a(t_1, t_2)]/\partial t_1 \partial t_2$ exists almost everywhere, and because $\int_{t_1}\int_{t_2} \partial^2[\overline{F}_a(t_1, t_2)]/\partial t_1 \partial t_2 = \overline{F}_a(t_1, t_2)$. Similarly $\overline{F}_s(t_1, t_2)$ is the singular part because $\partial^2[\overline{F}_s(t_1, t_2)]/\partial t_1 \partial t_2 = 0$ for $0 < t_1 < t_2 < \infty$ and for $0 < t_2 < t_1 < \infty$; it does not exist when $t_1 = t_2$. A verification of these statements entails several laborious steps.

Let $f_a(t_1, t_2) = \partial^2[\overline{F}_a(t_1, t_2)]/\partial t_1 \partial t_2$ be the probability density function of $\overline{F}_a(t_1, t_2)$. Then, it is evident that

$$f_a(t_1, t_2) = \begin{cases} \dfrac{\lambda(\lambda_1 + \lambda_{12})\lambda_2}{\lambda_1 + \lambda_2} e^{-(\lambda_1+\lambda_{12})t_1 - \lambda_2 t_2}, & t_1 > t_2 > 0 \\ \dfrac{\lambda(\lambda_2 + \lambda_{12})\lambda_1}{\lambda_1 + \lambda_2} e^{-\lambda_1 t_1 - (\lambda_2+\lambda_{12})t_2}, & t_2 > t_1 > 0; \end{cases}$$

it is undefined for $t_1 = t_2$.

The marginals of $\overline{F}_a(t_1, t_2)$, $\overline{F}_a(t_1)$ and $\overline{F}_a(t_2)$, obtained by setting in $\overline{F}_a(t_1, t_2)$, $t_2 = 0$ and $t_1 = 0$, respectively, turn out to be mixtures of exponential distributions. Specifically,

$$\overline{F}_a(t_1, 0) = \overline{F}_a(t_1) = \frac{\lambda}{\lambda_1 + \lambda_2} e^{-(\lambda_1+\lambda_{12})t_1} - \frac{\lambda_{12}}{\lambda_1 + \lambda_2} e^{-\lambda t_1}, \quad t_1 > 0, \text{ and}$$

$$\overline{F}_a(0, t_2) = \overline{F}_a(t_2) = \frac{\lambda}{\lambda_1 + \lambda_2} e^{-(\lambda_2+\lambda_{12})t_2} - \frac{\lambda_{12}}{\lambda_1 + \lambda_2} e^{-\lambda t_2}, \quad t_2 > 0.$$

Interestingly, if in $f_a(t_1, t_2)$ we set

$$\alpha = \lambda_1 + \frac{\lambda_1 \lambda_{12}}{\lambda_1 + \lambda_2}, \quad \beta = \lambda_2 + \frac{\lambda_2 \lambda_{12}}{\lambda_1 + \lambda_2},$$

$$\alpha' = \lambda_1 + \lambda_{12}, \qquad \beta' = \lambda_2 + \lambda_{12},$$

then the resulting expression takes the form of (4.31) with parameters α, β, α' and β'. This, we recall, is the probability density function of Freund's bivariate exponential distribution. We thus have

Claim 4.9. *Freund's bivariate exponential distribution can also be derived via a shock model. The survival function of Freund's distribution is $\overline{F}_a(t_1, t_2)$, the absolutely continuous part of the BVE, and its marginals are mixtures of exponential distributions.*

In $\overline{F}_a(t_1)$ the weights assigned to the exponential components have opposite signs; thus its failure rate is increasing.

Block and Basu (1974) derive $\overline{F}_a(t_1, t_2)$ via the bivariate lack of memory property, and call $\overline{F}_a(t_1, t_2)$ an 'absolutely continuous BVE', abbreviated ACBVE. Because of Claim 4.9, Block and Basu (1974) state that the ACBVE is a variant of Freund's distribution, which we recall has the bivariate lack of memory property; thus the ACBVE $\overline{F}_a(t_1, t_2)$ also possesses the lack of memory property.

The marginals of $\overline{F}_s(t_1, t_2)$, the singular component of the BVE, are $\overline{F}_s(t_i) = e^{-\lambda t_i}$, $t_i > 0$, $i = 1, 2$. Analogous to Claim 4.8, which is a decomposition of $\overline{F}(t_1, t_2)$, is a decomposition of $\overline{F}(t_1)$ and $\overline{F}(t_2)$, the marginals of the BVE. This decomposition turns out to be a mixture of $\overline{F}_a(t_i)$ and $\overline{F}_s(t_i)$, $i = 1, 2$, the marginals of the absolutely continuous and the singular components of the BVE. Specifically

Claim 4.10. *The marginals of the BVE are mixtures of the marginals $\overline{F}_a(t_i)$ and $\overline{F}_s(t_i)$, respectively. That is, for $i=1,2$,*

$$\overline{F}(t_i) = \frac{\lambda_1+\lambda_2}{\lambda}\overline{F}_a(t_i) + \frac{\lambda_{12}}{\lambda}\,e^{-\lambda t_i}, \quad t_i > 0.$$

Systems Having Interdependent Lifelengths Described by a BVE

Let T_s and T_p be the times to failure of a series and a parallel system, respectively, for a two-component system whose lifelengths T_1 and T_2 have the BVE of (4.39). Clearly,

$$P(T_s \geq t) = P(T_1 \geq t, T_2 \geq t) = \exp(-\lambda t),$$

where $\lambda = \lambda_1 + \lambda_2 + \lambda_{12}$. Thus the survival function of the series system is exponential, and its failure rate is constant, λ.

The case of the parallel system is more interesting. For one, it brings into play the problem of identifiability of the parameters λ_1, λ_2 and λ_{12}. Specifically, should the first failure experienced by the system be an individual failure (i.e. the failure of component 1 or component 2, but not both), then the cause of the second failure cannot be identified; it could be due to a component-specific shock or a common shock. With the matter of identifiability, it is difficult to use the observed time of the second failure, say t_2, to estimate λ_2 and λ_{12}. Identifiability also manifests itself in the form of identical survival functions for the fatal and the non-fatal shock models; ((4.37) and (4.38)).

It is easy to verify (section 4.7.4) that since

$$\begin{aligned} P\left(T_p \geq t\right) &= 1 - P(T_p \leq t) \overset{\text{def}}{=} \overline{F}_p(t) \\ &= \exp(-(\lambda_1+\lambda_{12})\,t) + \exp(-(\lambda_2+\lambda_{12})\,t) - \exp(-\lambda t), \end{aligned}$$

the probability density of $\overline{F}_p(t)$ exists everywhere and that $\overline{F}_p(t)$ is absolutely continuous. The failure rate of $\overline{F}_p(t)$, $r_p(t)$ therefore exists, and can be shown to be

$$r_p(t) = \lambda_{12} + \lambda_1\frac{e^{-(\lambda_1+\lambda_{12})t}(1-e^{-\lambda_2 t})}{\overline{F}_p(t)} + \lambda_2\frac{e^{-(\lambda_2+\lambda_{12})t}(1-e^{-\lambda_1 t})}{\overline{F}_p(t)}.$$

Figure 4.14 is a plot of this failure rate function for $\lambda_1 = 1$, $\lambda_2 = 2$, and $\lambda_{12} = 1.5$.

Figure 4.14 reveals some interesting features. The first is that $r_p(0) = 1.5$, suggesting that the two-component system can fail instantly, as soon as it is commissioned into operation. The second is that $r_p(t)$ initially increases and then decreases (albeit slightly), asymptoting to a constant 2.5. This feature of $r_p(t)$ is reminiscent of the failure rate function of a gamma distribution, which starting from zero increases and then asymptotes to a constant (cf. Figure 3.5.2 of Barlow and Proschan 1975, p. 75). The analogy is not surprising, because when $\lambda_1 = \lambda_2 = \gamma$ and $\lambda_{12} = 0$, T_p has a gamma distribution with scale γ and shape 2. The decrease in $r_p(t)$ prior to its asymptoting to 2.5 is a consequence of the fact that $\lambda_1 \neq \lambda_2$; the constant 2.5 can be identified with $\lambda_{12} + \min(\lambda_1, \lambda_2)$.

The above behavior of $r_p(t)$ motivates us to consider its decomposition. The components of this decomposition are shown by the dotted graphs of Figure 4.14; their interpretation – given below – is instructive.

We start by noting that for the two-component parallel redundant system $P(T_1 < t \mid T_p > t)$ is the conditional probability that component 1 has failed by t given that the system has not. Since

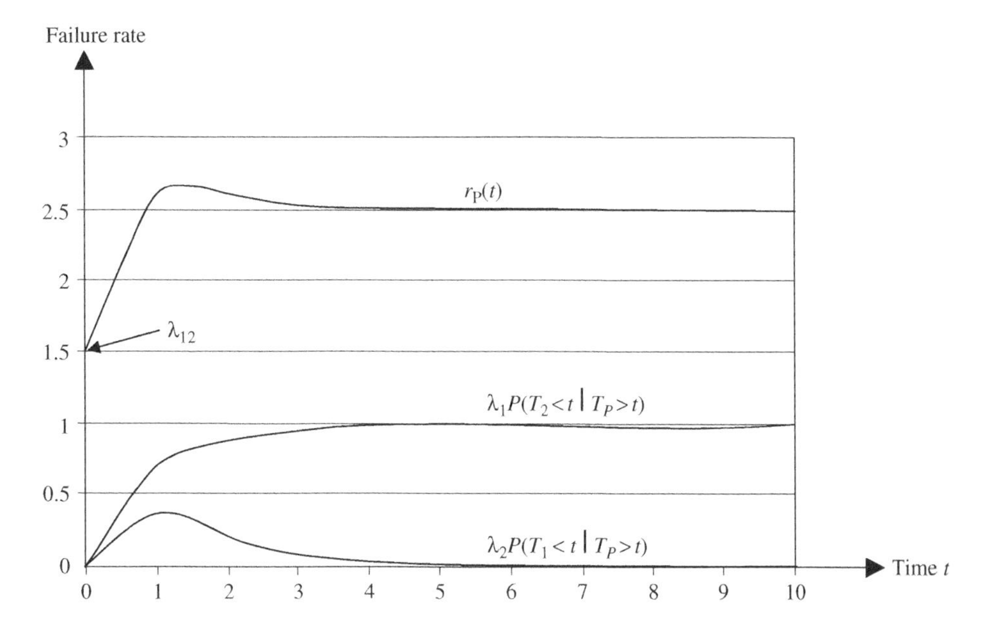

Figure 4.14 The Failure rate function for a two-component parallel system whose lifelengths are a BVE.

$$P\left(T_1 < t \mid T_\mathrm{p} > t\right) = \frac{P(T_1 < t, T_\mathrm{p} > t)}{\overline{F}_\mathrm{p}(t)}, \quad \text{and since}$$

$$P\left(T_1 < t,\ T_\mathrm{p} > t\right) = P(N_1(t) > 0, N_2(t) = N_3(t) = 0),$$

$$P\left(T_1 < t \mid T_\mathrm{p} > t\right) = \frac{(1 - \exp(-\lambda_1 t))\exp(-(\lambda_2 + \lambda_{12})t)}{\overline{F}_\mathrm{p}(t)}.$$

Similarly, we can find $P(T_2 < t \mid T_\mathrm{p} > t)$, *mutatis mutandis*. Thus, we may write the following as a decomposition of $r_\mathrm{p}(t)$:

$$r_\mathrm{p}(t) = \lambda_{12} + \lambda_1 P(T_2 < t \mid T_\mathrm{p} > t) + \lambda_2 P(T_1 < t \mid T_\mathrm{p} > t).$$

Since $r_\mathrm{p}(t)\mathrm{d}t \approx P(t \leq T_\mathrm{p} < t + \mathrm{d}t \mid T_\mathrm{p} \geq t)$, and since $\lambda_i \mathrm{d}t$ is approximately the probability that the i-th shock occurs in an interval $\mathrm{d}t$, the above decomposition has a probabilistic interpretation in terms of λ_i and $r_\mathrm{p}(t)$.

Generalizations and Extensions of the BVE

Besides a generalization to the n-variate case, a consequence of which is the problem of scalability, the structure and the manner of construction of the BVE suggests some natural ways to build upon it. Some of these are overviewed below.

One possibility is to allow the shocks in a fatal shock model to have after-effects. Thus, for example, we may assume that each component can withstand exactly k shocks, be they of a component-specific type or a common type; k is assumed to be known. When such is the case, the bivariate survival function $\overline{F}(t_1, t_2)$ will have gamma distributions for their marginals (cf. Barlow and Proschan, 1975, p. 138). Such survival functions do not posses the bivariate lack of memory property. The above theme can be extended by supposing that component i can withstand exactly k_i shocks, $i = 1, 2$. Downton (1970) and Hawkes (1972) consider the case wherein k_1 and k_2 cannot be precisely specified, but their joint bivariate distribution $\pi(k_1, k_2)$ can. Downton assumes a bivariate geometric distribution for $\pi(k_1, k_2)$; Hawkes considers a generalization of the bivariate geometric. Neither Downton nor Hawkes provide a physical motivation for their particular choice for $\pi(k_1, k_2)$. However, in Downton's case, the regression of T_2 in T_1 is a linear function of t_1, whereas in Hawkes' case it is an exponentially increasing function of t_1 that is bounded as $t \to \infty$. A concave, increasing and bounded form of the regression is more realistic than the linear, increasing and unbounded version. As mentioned by Hawkes, an infinitely reliable engine in an automobile does not lead one to expect that the automobile's exhaust pipe will last forever.

A second strategy for building upon the construction of the BVE is to assume that the shock-generating processes are non-homogenous Poisson with mean value functions $\Delta_i(t)$, $i = 1, 2, 3$. In the case of a fatal shock model with $\Delta_1(t) = \lambda_1 t^{\beta}$, $\Delta_2(t) = \lambda_2 t^{\beta}$, and $\Delta_3(t) = \lambda_{12} t^{\beta}$, for some $\beta > 0$, the resulting survival function will have Weibull distributions for its marginals (cf. Marshall and Olkin, 1967). Since the intensity functions corresponding to $\Delta_i(t)$ are of the form $\lambda_i \beta t^{\beta-1}$, which also happens to be the failure rate of a Weibull distribution, the corresponding non-homogenous Poisson processes $\{N_i(t);\ t \geq 0\}$ are sometimes referred to as 'Weibull processes' (cf. Lee and Lee, 1978). This terminology is misleading because the finite dimensional distributions of the process $\{N_i(t);\ t \geq 0\}$ are not Weibull. However, were the model described above to be extended to the n-variate case, then the sequence of random variables $\{T_1, T_2, \ldots, T_n\}$ would have marginal distributions that are Weibull. Consequently, the process $\{T_i;\ i = 1, \ldots, n\}$ could legitimately be called a Weibull process; multivariate Weibull distributions have been discussed by Lee (1979).

There are other such strategies for building upon the structure of the BVE, each entailing the addition of new parameters, and in so doing exacerbating the problem of scalability. Thus it does not pay to pursue them further. However, there is one aspect of the BVE that has attracted much attention, and in so doing has motivated its many extensions. This aspect pertains to the matter of absolute continuity, which the survival function of the BVE is not. The ACBVE of Block and Basu (1974) can be viewed as one such extension. There are at least two other extensions, one by Sarkar (1987) and the other by Ryu (1993). Like Block and Basu, Sarkar and Ryu question the appropriateness of the BVE when the available data does not consist of simultaneous failures, or when the physical scenario is such that simultaneous failures are not possible. However, complex technological systems, like nuclear reactors and power system networks, often exhibit simultaneous failures and so the BVE has a useful role to play.

Sarkar's extension produces an absolutely continuous bivariate distribution with exponential marginals, which does not have the bivariate lack of memory property. Unlike the BVE, Sarkar's bivariate distribution lacks a physical motivation; like the ACBVE, its derivation is based on a characterization of certain independence properties. For specified constants λ_1, λ_2 and $\lambda_{12} > 0$, the survival function of Sarkar's distribution is of the form

$$\begin{aligned} P(T_1 \geq t_1, T_2 \geq t_2) &= \exp\{-(\lambda_2 + \lambda_{12})t_2\}\left\{1 - [A(\lambda_1 t_2)]^{-\gamma}[A(\lambda_1 t_1)]^{1+\gamma}\right\}, \text{ for } 0 < t_1 \leq t_2, \\ &= \exp\{-(\lambda_1 + \lambda_{12})t_1\}\left\{1 - [A(\lambda_2 t_1)]^{-\gamma}[A(\lambda_2 t_2)]^{1+\gamma}\right\}, \text{ for } 0 < t_2 \leq t_1; \end{aligned}$$

here $\gamma = \lambda_{12}/(\lambda_1 + \lambda_2)$ and $A(z) = 1 - \exp(-z)$, for $z > 0$.

Sarkar (1987) also obtains the coefficient of correlation for his distribution, and shows that it is bounded on the left by $\lambda_{12}/(\lambda_1 + \lambda_2 + \lambda_{12})$, which we recall is the coefficient of correlation for the BVE. Specifically, here

$$\frac{\lambda_{12}}{\lambda_1 + \lambda_2 + \lambda_{12}} \leq \rho(T_1, T_2) < 1,$$

which implies that the naure of linear dependence between T_1 and T_2 under Sarkar's distribution is stronger than that under the BVE. The fact that $\rho(T_1, T_2) \geq 0$ suggests a positive dependence between T_1 and T_2; consequently, an absence of the lack of memory property implies that with Sarkar's distribution, for any $t \geq 0$, and for all $s_1, s_2 \geq 0$,

$$P(T_1 \geq t + s_1, T_2 \geq t + s_2 \mid T_1 \geq s_1, T_2 \geq s_2) \geq P(T_1 \geq t, T_2 \geq t).$$

An extension of the BVE proposed by Ryu (1993) results in an absolutely continuous bivariate distribution whose marginal distributions have an increasing failure rate, and which does not have the bivariate lack of memory property. The exponential as a marginal distribution arises as a limiting case of the marginals of the Ryu's bivariate model. Unlike the ACBVE and Sarkar's bivariate distribution, Ryu's derivation has a physical motivation. Its construction is based on Marshall and Olkin's shock model, but the shocks are assumed to have an after-effect; i.e. the effect of the shocks is cumulative. The after-effect is reflected via the following construction of a **random** (or **stochastic**) **hazard function**.

Suppose that the failure rate of the lifelength of component 1 at time t is $d_1 N_1(t) + s_1 N_3(t)$, where $N_1(t)$ is the number of component-specific shocks received by component 1 at time t, and $N_3(t)$ the number of common shocks at time t; d_1 and s_1 are constants. The failure rate is random because when it is specified (at time 0), $N_1(t)$ and $N_3(t)$ are unknown. More about this useful idea of specifying random failure rate functions will be discussed later in Chapter 7, on reliability in dynamic environments. Similarly, $d_2 N_2(t) + s_2 N_3(t)$ is the failure rate of component 2 at time t. Because the $N_i(t)$, $i = 1, 2, 3$, are non-decreasing functions of t, the two failure rate functions are also non-decreasing. This failure will manifest itself in the fact that Ryu's bivariate distribution has marginal distributions that have an increasing failure rate. In what follows, we assume that d_1 and d_2 are very large, say infinite; this makes every component specific shock a fatal shock. With $d_1 = d_2 = \infty$, and $s_1, s_2, \lambda_1, \lambda_2$ and λ_{12} specified, Ryu's bivariate distribution has the joint survival function:

$$\begin{aligned} P(T_1 \geq t_1, T_2 \geq t_2) &= \exp\Big[-(\lambda_1 + \lambda_{12})\, t_1 - \lambda_2 t_2 + \tfrac{\lambda_{12}}{s_1}\left(1 - e^{-s_1(t_1 - t_2)}\right) \\ &\quad + \frac{\lambda_{12}}{s_1 + s_2}\left(e^{-s_1(t_1 - t_2)} - e^{-s_1 t_1 - s_2 t_2}\right)\Big], \quad \text{for } t_1 > t_2, \\ &= \exp\Big[-\lambda_1 t_1 - (\lambda_2 + \lambda_{12})\, t_2 + \tfrac{\lambda_{12}}{s_1}\left(1 - e^{-s_1(t_2 - t_1)}\right) \\ &\quad + \frac{\lambda_{12}}{s_1 + s_2}\left(e^{-s_1(t_2 - t_1)} - e^{-s_1 t_1 - s_2 t_2}\right)\Big], \quad \text{for } t_1 \leq t_2; \end{aligned}$$

for details see Appendix B of Ryu (1993).

The marginal distribution of T_i, $i = 1, 2$, is of the form

$$P(T_i > t) = \exp\left[-\lambda_i t - \lambda_{12} t + \frac{\lambda_{12}}{s_i}\left(1 - e^{-s_i t}\right)\right], \quad t \geq 0.$$

The marginals are akin to the exponential marginals of the BVE, save for the last term which vanishes when $s_i = 0$ or when s_i is large, say infinite. It is easy to verify that $h_i(t)$, the failure rate function of $P(T_i > t)$, is $h_i(t) = \lambda_i + \lambda_{12}(1 - e^{-s_i t})$, which is an increasing function of t, for any $s_i < \infty$; when s_i is very large, $h_i(t)$ is approximately a constant, $\lambda_i + \lambda_{12}$. The failure rate is a constant λ_i, when $s_i = 0$. It is important that we distinguish between $h_1(t)$, the failure rate of $P(T_1 > t)$, and the quantity $d_1 N_1(t) + s_1 N_3(t)$, which is the **random** failure rate function of $P(T_1 > t)$ conditional on $N_1(t)$ and $N_3(t)$. The function $h_1(t)$ is unconditional on $N_1(t)$ and $N_3(t)$; similarly for $h_2(t)$.

In general, it can be seen that when s_1 and s_2 become large, say infinite, then every common shock is also a fatal shock, and then Ryu's survival function reduces, albeit approximately, to the survival function of the BVE. Thus for $\lambda_{12} > 0$, we may claim that Ryu's construction is an extension of the BVE rather than claim that the BVE is a special case of Ryu's construction. This is because the exponential marginals in Ryu's bivariate distribution being the consequence of a limiting argument are only approximations.

Absolutely continuous distributions whose marginals reflect aging, such as the ones by Freund, Ryu and the ACBVE are attractive in the sense that they capture scenarios that are realistic. However, some of these distributions are more parsimonious (scalable) than the others. For example, the ACBVE has only three parameters, whereas Freund's distribution has four, and Ryu's has five. A disadvantage of the ACBVE is that it lacks a physical motivation, its genesis being the bivariate lack of memory property. The other two distributions have a physical motivation, but circumstances which lead to them, namely load sharing in the case of Freund, and shocks with after-effects in the case of Ryu, may not be meaningful for all applications. Thus the choice of which bivariate distribution to use should depend on considerations involving both the physical scenario at hand, and the properties of the chosen distribution. If simultaneous failures can occur, then the BVE or its generalizations that result in the gamma or the Weibull as marginal distributions would be suitable. If simultaneous failures are not possible, then any one of the absolutely continuous bivariate distributions discussed here could be an appropriate choice depending on the physical scenario at hand and/or the properties of the chosen distribution. Table 4.1 gives a summary of some of these properties; also included in this table are two other bivariate distributions, discussed later. In any particular application the choice of a distribution will depend on assessing which of the properties of Table 4.1 seem relevant, and then choosing that family which best captures the desired properties. Thus subjectivity plays a key role in one's initial choice of a distribution. The observed failure data, if available, would support this choice or negate it. The data will also facilitate an estimation of the parameters of the underlying distribution.

Table 4.1 A Comparison of the Properties of Some Bivariate Disributions

Family	Absolute continuity	Lack of memory	Marginals	Failure rate of marginals	Correlation
Independent exponentials	Yes	Yes	Exponential	Constant	Uncorrelated
Gumbel's Version II	Yes	No	Exponential	Constant	Positive and negative
Freund's load sharing	Yes	Yes	Mixture of exponentials	Increasing	Positive and negative
Marshall and Olkin's BVE	No	Yes	Exponential	Constant	Positive
Block and Basu's ACBVE	Yes	Yes	Mixture of exponentials	Increasing	Positive
Sarkar's extension	Yes	No	exponential	Constant	Positive
Ryu's extension	Yes	No	General	Increasing	Positive
Bivariate Pareto	Yes	No	Pareto	Decreasing	Positive
Single parameter bivariate exponential	Yes	No	Exponential	Constant	Positive
Bivariate Pareto copula-induced bivariate exponential	Yes	No	Exponential	Constant	Positive

4.7.5 The Bivariate Pareto as a Failure Model

An absolutely continuous bivariate distribution with Pareto marginals can be motivated as a failure model for certain types of components sharing a common (but unspecified) environment. The model generalizes to the multivariate case and it is much more scalable than the multivariate version of the BVE. Its dependence properties are different from those of the models considered earlier and the model is related to some well-known families of bivariate distributions that have been considered in other contexts. The physical scenario for which the model is proposed is quite general, and this generality enables us to consider its several extensions. Indeed, one such extension leads us to a single parameter family of bivariate distributions with exponential marginals (section 4.7.6).

Structure of the Bivariate Pareto Distribution

Consider a two-component, parallel redundant system that is required to operate in an environment whose effect on each component is unknown (in a sense to be described later). Let T_1 and T_2 be the lifelengths of each component. Our aim is to assess the reliability of the system for a mission of duration t, $t \geq 0$. Suppose that the model failure rate of the distribution of T_i is $\eta\lambda_i$, where λ_i encapsulates the inherent quality of component i, and η reflects the effect of the environment on λ_i, $i = 1, 2$. One way to interpret λ_i is to view it as the model failure rate of T_i when component i operates in some standard, carefully controlled test-bench environment; λ_i will be called the **intrinsic** model failure rate of the component. If the operating environment is judged to be harsher (milder) than the test-bench environment, then η would be greater (smaller) than one; otherwise $\eta = 1$.

Were we to assume that given η, λ_1 and λ_2, the lifelengths T_1 and T_2 are independent, then the reliability of the parallel-redundant system is $\exp(-\eta\lambda_1 t) + \exp(-\eta\lambda_2 t) - \exp(-\eta(\lambda_1 + \lambda_2)t)$. Suppose now that λ_1 and λ_2 can be specified to a very high degree of accuracy. This is plausible because the λ_is, being chance parameters, can be assessed by testing a large number of copies of component i in the test-bench environment. Since the operating environment can change from one user of the system to another, η will be treated as being unknown with a distribution function $G(\eta)$; this distribution function will involve parameters that have to be specified. With the λ_is assumed known and η having the distribution function $G(\eta)$, the reliability of the two-unit parallel redundant system becomes

$$G^*(\lambda_1 t) + G^*(\lambda_2 t) - G^*\left[(\lambda_1 + \lambda_2)t\right], \tag{4.42}$$

where $G^*(y) = \int_0^\infty \exp(-\eta y)\mathrm{d}G(\eta)$ is the **Laplace transform** of (the probability density function of) G. The above result easily generalizes to the case of multiple components, each new component introducing only one additional parameter, namely its λ. This feature gives the proposed model the feature of scalability.

As a special case of (4.42), suppose that $G(\eta)$ has a gamma distribution with parameters $a > -1$ and $b > 0$, where a is the shape parameter; note that here, it is $a = 0$ that results in an exponential distribution. Then, if T_p denotes the time to failure of the system, its reliability

$$P(T_\mathrm{p} \geq t) = \left(\frac{b}{\lambda_1 t + b}\right)^{a+1} + \left(\frac{b}{\lambda_2 t + b}\right)^{a+1} - \left(\frac{b}{(\lambda_1 + \lambda_2)\,t + b}\right)^{a+1}; \tag{4.43}$$

in writing the above, the conditioning parameters are suppressed.

Using arguments that mimic those used to obtain (4.43), we can easily obtain

$$P(T_1 \geq t_1, T_2 \geq t_2) = \left(\frac{b}{b + \lambda_1 t_1 + \lambda_2 t_2}\right)^{a+1}; \tag{4.44}$$

its joint density at (t_1, t_2) is of the form

$$f(t_1, t_2) = \frac{\lambda_1\lambda_2(a+1)(a+2)b^{a+1}}{(\lambda_1 t_1 + \lambda_2 t_2 + b)^{a+3}}. \tag{4.45}$$

Integrating the above with respect to t_2 gives the marginal density of T_1 at t_1 as

$$f(t_1) = \frac{\lambda_1(a+1)b^{a+1}}{(\lambda_1 t_1 + b)^{a+2}}.$$

This is a Pareto density of the first kind (cf. Johnson and Kotz, 1970, p. 234); it yields the survival function

$$P(T_1 \geq t_1) = \left(\frac{b}{b+\lambda_1 t_1}\right)^{a+1};$$

similarly, $P(T_2 \geq t_2)$. Since $P(T_1 \geq t_1, T_2 \geq t_2) \neq P(T_1 \geq t_1)P(T_2 \geq t_2)$, T_1 and T_2 are not independent. By contrast, when η is specified, T_1 and T_2 are independent – thus dependence here is attributed to our uncertainty about η; more about this is said later.

If, in (4.44), we set $t_1 = t_2 = t$, then $P(T_1 \geq t_1, T_2 \geq t_2)$ is the reliability of a series system of two components sharing a common environment. Specifically, if T_s denotes the lifelength of the series system, then

$$P(T_s \geq t) = \left(\frac{b}{b+t(\lambda_1+\lambda_2)}\right)^{a+1}.$$

Lindley and Singpurwalla (1986) discuss the relationship of the joint distribution (4.44) to other bivariate distributions like the Lomax and the logistic, making such distributions suitable candidates for failure models. Using the fact that the conditional density of T_1 at t_1, given $T_2 = t_2$, is of the form

$$f(t_1 \mid t_2) = \frac{\lambda_1(a+2)(b^*)^{a+2}}{(\lambda_1 t_1 + b^*)^{a+3}}, \tag{4.46}$$

where $b^* = b + \lambda_2 t$, it can be seen that the regression of T_1 on T_2 is of the form

$$E(T_1 \mid T_2 = t_2) = \frac{\lambda_2 t_2 + b}{\lambda_1(a+1)}; \qquad \text{also}$$

$$V(T_1 \mid T_2 = t_2) = \frac{1}{\lambda_1^2}\left[\frac{(a+2)(b+\lambda_2 t_2)^2}{a(a+1)^2}\right].$$

The conditional density of (4.46) is also a Pareto density, and the regression of T_1 on T_2 is linear with a positive regression coefficient $\lambda_2/\lambda_1(a+1)$. The latter is to be contrasted with the exponentially increasing and bounded regression of the Gumbel, the BVE and the Hawkes' models (Figure 4.15). Recall that for Downton's model, regression of T_1 and T_2 was linear, and this was a criticism of the model; the same criticism would therefore also apply to the model proposed here. The correlation between T_1 and T_2 is $(a+1)^{-1}$, which with $a > -1$

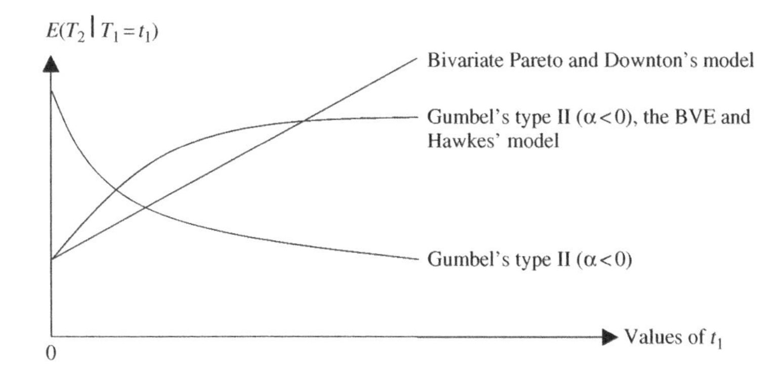

Figure 4.15 The regression of T_2 on T_1 for some models of interdependent lifetimes.

is always positive; the correlation depends only on a. Also, the probability that component 2 fails before component 1 is $\lambda_2/(\lambda_1 + \lambda_2)$, irrespective of η and its distribution. Finally, with $P(T_1 \geq t_1, T_2 \geq t_2)$ and $P(T_1 \geq t_1)$ known, it is obvious that

$$P(T_2 \geq t_2 \mid T_1 \geq t_1) = \left(\frac{b + \lambda_1 t_1}{b + \lambda_1 t_1 + \lambda_2 t_2} \right)^{a+1};$$

this expression enables us to assess the survivability of component 2, given the survivability of component 1.

Currit and Singpurwalla (1988) have investigated the crossing properties of $P(T_p \geq t)$ – equation (4.43) – with

$$\exp(-\lambda_1 t) + \exp(-\lambda_2 t) - \exp(-(\lambda_1 + \lambda_2)t) \stackrel{\text{def}}{=} P_I(T_p \geq t),$$

the reliability function of a parallel redundant system with independent exponentially distributed lifetimes. They show that when a and b are such that $(E(\eta))^2 + V(\eta) > 1$, $P(T_p \geq t)$ crosses $P_I(T_p \geq t)$ at least once, and that the crossing occurs from below. This means that the assumption of independence initially overestimates system reliability. The overestimation tantamounts to an underestimation of the risk of failure for small mission times, a feature whose consequences could be serious. A similar result is also true vis-à-vis the crossing of $P(T_s \geq t)$ and $P_I(T_s \geq t) \stackrel{\text{def}}{=} 1 - \exp(-(\lambda_1 + \lambda_2)t)$ – the reliability of a series system with independent exponentially distributed lifetimes – except that if a crossing occurs, it can occur at most once. Here again, the assumption of independence results in an overestimation of system reliability, at least initially. Chaudhuri and Sarkar (1998) strengthen the results of Currit and Singpurwalla (1988) by showing that when $E(\eta) > 1$, there is exactly one crossing of the type described above, for both series and parallel redundant systems.

There is a subjectivist angle from which the crossings mentioned above can be viewed. Specifically, the expressions for $P(T_p \geq t)$ and $P(T_s \geq t)$ are a consequence of our uncertainty about η. Their initial domination by $P_I(T_p \geq t)$ and $P_I(T_s \geq t)$ can be seen as the price that one has to pay for not being able to specify η with certainty. The price is an assessment of reliability that is smaller than that which would have been assessed were $\eta = 1$. More on the causes of dependence is given next.

The Cause of Interdependent Lifelengths

We have attributed the uncertainty about η as the cause of dependence between T_1 and T_2. Does this imply that dependent lifelengths are always a consequence of parameters that are not precisely specified? The answer to this question cannot be in the affirmative because, for the BVE, T_1 and T_2 are dependent even when λ_1, λ_2 and λ_{12} are assumed known. This is because with $\lambda_1 = \lambda_2 = 0$, a knowledge of $\lambda_{12} (\neq 0)$ does not tell us precisely when the common shock is going to occur. However, as soon as T_1 were to be observed as t_1, we do know that a common shock has occurred at t_1, and this causes us to revise our uncertainty assessment about T_2; indeed in the case of the fatal shock model, the uncertainty about T_2 vanishes. In the case of the bivariate Pareto model considered here, the occurrence of failure at t_1 tells us something about the operating environment, and hence about η. This in turn changes our belief about T_2.

To summarize, in the case of the BVE, a knowledge that $T_1 = t_1$, tells us when a common shock has occurred and this in turn enhances our knowledge about the common environment. The added knowledge about the common environment causes us to revise our uncertainty about T_2. Similarly, in the case of the bivariate Pareto, $T_1 = t_1$ influences our belief about η, and this in turn causes us to change our belief about T_2. From a subjective probability point of view, dependent lifelengths are a manifestation of changes in belief about one lifelength given the decomposition of the other.

Finally, there is one other issue that needs to be brought into the picture, namely that of hazard potentials and dependent lifelengths (section 4.6.1). He had seen there that dependence between lifetimes is a manifestation of dependence between the hazard potentials. This claim is still true; dependent hazard potentials are the genuine cause of interdependent lifelengths.

4.7.6 A Bivariate Exponential Induced by a Shot-noise Process

Implicit in the set-up of section 4.7.5 is the assumption that the operating environment, be it harsher or milder than the test-bed environment, is static. Thus η was taken to be an unknown constant. The more realistic scenario is the one wherein the operating environment changes over time so that its effect on component i, $i = 1, 2$, is to change λ_i to $\eta(t)\lambda_i$, where $\eta(t)$ is some unknown function of t; it is called the **modulating function**. When such is the case, the operating environment is said to be **dynamic**, and the topic of survivability under dynamic environments is interesting enough for an entire chapter to be devoted to it (Chapter 7). The purpose of this section, however, is to introduce an absolutely continuous bivariate distribution with exponential marginals that arises in the context of dynamic environments, and to compare the properties of this distribution with other bivariate distributions discussed here.

Since $\eta(t)$ is unknown, we start by introducing a probability model for describing our uncertainty about it. There are several approaches for doing so, a natural one being to suppose that $\eta(t)$ is a polynomial function of time with unknown coefficients. For example, $\eta(t) = a_0 + \sum_{i=1}^{n} a_i t^i$, where the a_i's are unknown. This kind of strategy is the essence of the approach used by Cox (1972) in his celebrated paper on 'proportional hazards'. Another approach is to assume that $\{\eta(t);\ t \geq 0\}$ is a meaningful stochastic process. That is, for every value of $t \in [0, \infty)$, $\eta(t)$ is a random variable with a specified distribution, and for every collection of k time points, $0 < t_1 < t_2 < \cdots < t_k$, $k \geq 1$, the joint distribution of the k random variables $\eta(t_1), \ldots, \eta(t_k)$ is also specified. In essence, a stochastic process is simply a collection of random variables whose marginal and joint distributions can be fully specified. The **sample path** of a stochastic process is the collection of values taken by the Poisson process. The homogenous and the non-homogenous Poisson counting processes of section 4.7.4 are examples of stochastic processes, and so are the other point processes discussed in Chapter 8.

The advantage of describing $\eta(t)$ by a stochastic process over assuming that $\eta(t)$ is some deterministic function of time is that the latter describes the effect of a systematically changing environment, whereas the former best encapsulates the features of a haphazard one. In Singpurwalla and Youngren (1993), two stochastic process models for $\eta(t)$ are considered. The first entails the assumption that $\left\{\int_0^t \lambda(u)\,\mathrm{d}Y(u);\ t\geq 0\right\}$ is a 'gamma process'; $\lambda(u)$ is the intrinsic failure rate function of a component, and $\mathrm{d}Y(u)=\eta(u)\,\mathrm{d}u$, whenever $\eta(u)$ exists. The gamma process will be formally defined in section 4.9.1. An overview of these processes and their generalizations is in Singpurwalla (1997) and in van der Weide (1997). With $\left\{\int_0^t \lambda(u)\,\mathrm{d}Y(u);\ t\geq 0\right\}$, $i=1,2,$ described by a gamma process, Singpurwalla and Youngren (1993) produce a generalization of the BVE whose essence boils down to the feature that in a non-fatal shock model, each shock induces its own probability of failure on its affiliated component. Recall that in Marshall and Olkin's (1967) set-up, the probability of failure from shock to shock (within a shock-generating process) is a constant. Whereas the above generalization of the BVE is of interest, the focus here is on the second process for $\{\eta(t);\ t\geq 0\}$, namely, a 'shot-noise process'.

The Shot-noise Process

Shot-noise processes have been used in the physical and the biological sciences to describe phenomenon having residual effects. Examples are surges of electrical power in a control system, or the after effects of a heart attack. Such residual effects are captured via an attention function, $h(t)$, which typically is a non-decreasing function of t (details are in Cox and Isham, 1980, p. 135). In the context considered here, suppose that the operating environment consists of shocks, or a series of events, whose effect is to induce stresses of unknown magnitudes $X_k>0,\ k=0,1,\ldots,$ on a component or components; X_k is the stress induced by the k-th shock. The shocks are assumed to occur according to a non-homogenous Poisson process with a specified intensity function $m(t),\ t\geq 0$. Suppose that when a stress of magnitude X is induced at some epoch of time t, then its contribution to the modulating function $\eta(t)$ at time $(t+u)$ is $\eta(t+u)=Xh(u)$. Specifically, if $0\equiv T_{(0)}<T_{(1)}<\ldots$ are the epochs of time at which stresses of magnitude $X_0, X_1, \ldots,$ respectively, are induced then

$$\eta(t)=\sum_{k=0}^{\infty} X_k\, h(t-T_{(k)});$$

$h(u)=0$, for $u<0$ (Figure 4.16). We are assuming here that time is measured from the instant the first shock occurs; thus $\eta(0)=X_0$. The process $\{\eta(t);\ t\geq 0\}$ is then a **shot-noise process.**

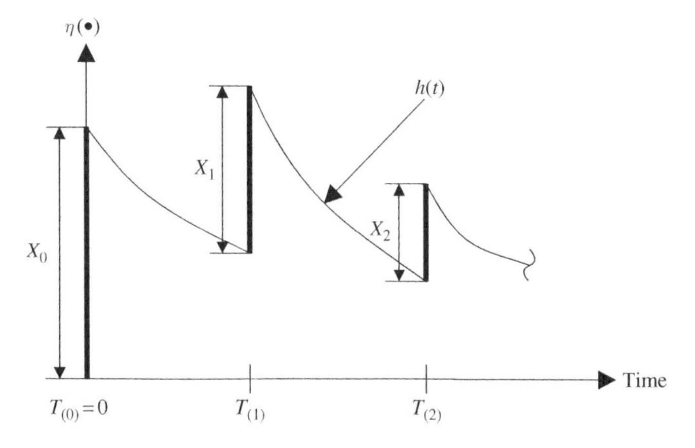

Figure 4.16 A shot-noise process for $\eta(\cdot)$.

When $h(t)$ is a constant, the effect of the induced stresses is cumulative; when $h(t)$ decreases in t, the unit reveals some form of healing or recovery.

If $\{\eta(t);\ t \geq 0\}$ is described by a shot-noise process, then the process $\{\lambda(t)\eta(t);\ t \geq 0\}$ is also a shot-noise process; $\lambda(t)$ is the intrinsic model failure rate of a component.

A Single-parameter Bivariate Distribution with Exponential Marginals

In what follows, we assume that the failure rate of component i is a constant λ_i, $i = 1, 2$, and that for all values of k, X_k is independent of $T_{(k)}$. Furthermore, the X_is are mutually independent and have a common distribution function G. Let G^* denote the Laplace transform of G, and let $M(t) = \int_0^t m(u)\,du$ and $H(t) = \int_0^t h(u)\,du$ be the cumulative intensity and attention functions, respectively. Then, in the case of a solo component experiencing the kind of environment described above, and having a constant failure rate λ, Lemoine and Wenocur (1986) give arguments which can be used to show that T, the time to failure of the component is such that

$$P(T \geq t) = G^*(\lambda H(t)) \exp\left[-M(t) + \int_0^t G^*(\lambda H(u))\, m(t-u)\, du\right];$$

as before, in writing out the above, the conditioning parameters have been suppressed. The proof is based on some well-known properties of a non-homogenous Poisson process. The details can be found in Singpurwalla and Youngren (1993), who also show that in the bivariate case, with $0 \leq t_1 \leq t_2$,

$$\begin{aligned} P(T_1 \geq t_1, T_2 \geq t_2) = {} & G^*(\lambda_1 H(t_1) + \lambda_2 H(t_2)) \cdot \\ & \exp\left[\int_0^{t_1} G^*(\lambda_1 H(t_1 - u_1) + \lambda_2 H(t_2 - u_1)) m(u_1) du_1\right] \cdot \\ & \exp\left[\int_{t_1}^{t_2} G^*(\lambda_2 H(t_2 - u_2) m(u_2)) du_2 - M(t_2)\right]; \end{aligned} \tag{4.47}$$

here again, the conditioning parameters have been suppressed.

As a special case of (4.47), suppose that $m(u) = m$, so that the shock generating process is a homogenous Poisson and that G is an exponential distribution with a scale parameter b. Furthermore, suppose that $h(u) = 1$, so that the effect of the imposed stresses is cumulative. Then, for $0 \leq t_1 \leq t_2$

$$\begin{aligned} P(T_1 \geq t_1, T_2 \geq t_2) = {} & \left(\frac{b}{b + \lambda_1 t_1 + \lambda_2 t_2}\right)\left(\frac{b + \lambda_2(t_2 - t_1)}{b + \lambda_1 t_1 + \lambda_2 t_2}\right)^{-mb/(\lambda_1+\lambda_2)} \\ & \cdot \left(\frac{b}{b + \lambda_2(t_2 - t_1)}\right)^{-mb/\lambda_2} \exp(-m t_2), \end{aligned} \tag{4.48}$$

and its marginal

$$P(T_i \geq t) = \left(\frac{b + \lambda_i t}{b}\right)^{\frac{mb}{\lambda_i} - 1} \exp(-mt), \tag{4.49}$$

$i = 1, 2$, is a Pareto distribution of the third kind (Johnson and Kotz, 1970, p. 234). With $m = \lambda_i / b$ the survival function (4.49) becomes an exponential distribution. It can be easily verified that when $\lambda_i / b > (<) m$, the model failure rate of $P(T_i \geq t)$ decreases (increases) in t from λ_i / b to m (Figure 4.17).

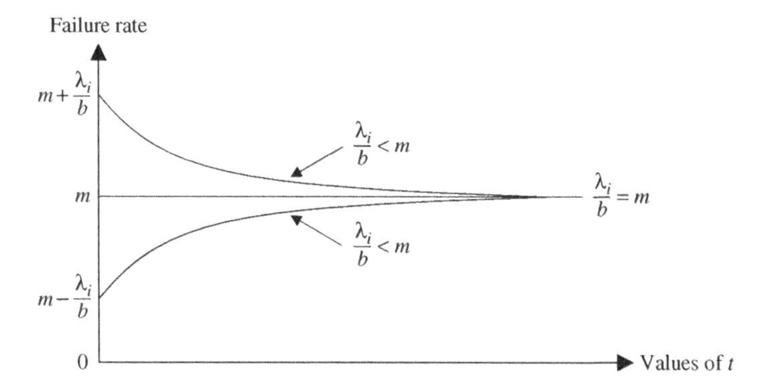

Figure 4.17 Model failure rate function of the marginal.

In (4.48), if we set $\lambda_1=\lambda_2=\lambda$, and $\lambda/b=m$, then we see that for $0\leq t_1\leq t_2$

$$P(T_1\geq t_1, T_2\geq t_2)=\sqrt{\frac{1-mt_1+mt_2}{1+mt_1+mt_2}}\exp(-mt_2);$$

similarly, its symmetric version holds for $0\leq t_2\leq t_1$.

Thus for $t_1, t_2>0$, $P(T_1\geq t_1, T_2\geq t_2)$

$$=\sqrt{\frac{1-m\min(t_1,t_2)+m\max(t_1,t_2)}{1+m(t_1+t_2)}}\exp(-m\max(t_1,t_2)), \tag{4.50}$$

$$=\sqrt{\frac{1-|t-s|}{1+t+s}}\exp(-\max(s,t)), \tag{4.51}$$

if $mt_1=t$ and $mt_2=s$.

This is a single parameter bivariate distribution with exponential marginals. The distribution generalizes easily to the multivariate case, and because it has only one parameter, namely m, it is highly scalable. Whereas the feature of scalability makes the distribution attractive, the question of realism needs to be resolved. Of particular concern here is equating the Poisson parameter m to the ratio of the component's intrinsic model failure rate λ and the scale parameter b of the exponential distribution for the inflicted stress.

Strength of Dependence in the Single-parameter Bivariate Distribution

It can be shown, details omitted, that the joint survival function $P(T_1\geq t_1, T_2\geq t_2)$ of (4.50) has a probability density at (t_1, t_2) for all $t_1, t_2>0$, and that it is absolutely continuous. Furthermore, T_1 and T_2 are positively quadrant dependent, and the distribution does not experience the bivariate lack of memory property. With $P(T_i>t)$, $i=1,2$, known, it is evident that the conditional survival function

$$P(T_1\geq t_1 \mid T_2\geq t_2)=\sqrt{\frac{1-mt_1+mt_2}{1+mt_1+mt_2}}, \qquad 0<t_1<t_2<\infty$$

$$=\sqrt{\frac{1+mt_1-mt_2}{1+mt_1+mt_2}}\exp(-m(t_1-t_2)), \quad t_1\geq t_2>0.$$

To investigate the regression of T_2 on T_1, we need to find $P(T_1 \geq t_1 \mid T_2 = t_2)$. After some routine, but laborious manipulations, this can be shown to be:

$$\begin{cases} \left(\dfrac{1+mt_1}{(1+m(t_1-t_2))(1+m(t_1+t_2))}\right)\sqrt{\dfrac{1+m(t_1-t_2)}{1+m(t_1+t_2)}}\,e^{-m(t_1-t_2)}, & \text{for } t_1 \geq t_2, \text{ and} \\ \left(1-\dfrac{mt_1}{(1+m(t_2-t_1))(1+m(t_1+t_2))}\right)\sqrt{\dfrac{1+m(t_2-t_1)}{1+m(t_1+t_2)}}, & \text{for } 0 < t_1 < t_2. \end{cases}$$

Integrating $P(T_1 \geq t_1 \mid T_2 = t_2)$ with respect to t_1 over $(0, \infty)$ would give us $E(T_1 \mid T_2 = t_2)$. This is analytically difficult to do. However, its numerical evaluation shows that $E(T_1 \mid T_2 = t_2)$ tends to be a convex function of t_2 becoming linear as t_2 gets large (Figure 4.18).

A comparison of Figures 4.15 and 4.18 indicates that the regression curves of the bivariate Pareto, the model of Downton, and the model of this section are generally similar. The regression of T_1 on T_2 suggests that T_1 and T_2 are positively correlated. Indeed Kotz and Singpurwalla (1999) show that the correlation between T_1 and T_2 is 0.4825, irrespective of the value of m. Thus m is purely a measure of location and scale.

Systems with Interdependent Lifelengths

If T_s denotes the time to failure of a series system of two components having lifelengths T_1 and T_2 described by the joint survival function of (4.50), then by setting $t_1 = t_2 = t$, we see that the reliability of the system is

$$P(T_s \geq t) = \left(\frac{1}{1+2mt}\right)^{\frac{1}{2}} \exp(-mt), \quad t \geq 0. \tag{4.52}$$

Contrast this expression with $\exp(-2mt)$, the reliability of the system assuming independent exponentially distributed lifelengths with a scale parameter m. Clearly, the assumption of independence underestimates the reliability of a series system functioning in a shot-noise environment.

Similarly, if T_p denotes the lifelength of a parallel redundant system, then its reliability is

$$P(T_p \geq t) = \left\{2 - \left(\frac{1}{1+2mt}\right)^{\frac{1}{2}}\right\} \exp(-mt), \quad t \geq 0. \tag{4.53}$$

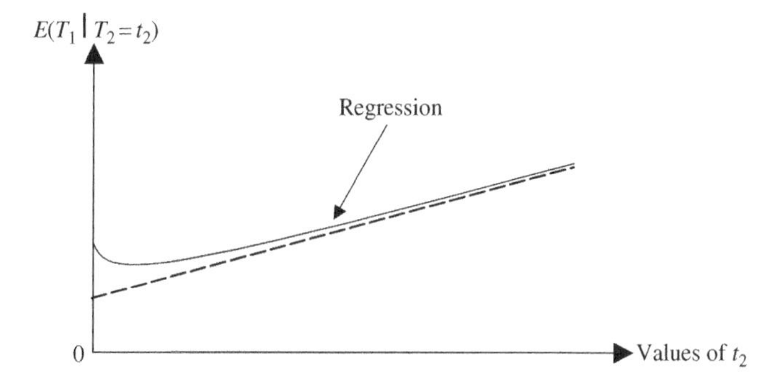

Figure 4.18 The Regression of T_1 on T_2 for the single parameter bivariate exponential.

Comparing the above with $(2-\exp(-mt))\exp(-mt)$, the reliability of the system assuming independent exponentially distributed lifelengths, suggests that independence overestimates system reliability of parallel redundant systems.

An intriguing aspect of (4.52) and (4.53) is the feature that as $t \to \infty$, the reliability of both the series system and the parallel system are approximated by $\exp(-mt)$. This suggests that the contribution of redundancy to the reliability of the system asymptotically diminishes, so that both the series system and the parallel system behave like a single unit having an exponentially distributed lifelength.

4.7.7 A Bivariate Exponential Induced by a Bivariate Pareto's Copula

In section 4.7.1, I introduced the notion of copulas and indicated that the 'method of copulas' can be used to generate new families of bivarate distributions. Motivated by the consideration that the regression function of the bivariate distributions discussed in sections 4.7.2 through 4.7.6 do not cover the convex increasing case, Singpurwalla and Kong (2004) develop a family of bivariate distributions with exponential marginals whose regression function fills the above void. They use the copula of a bivariate Pareto as a seed. Specifically, they consider the bivariate Pareto of (4.44) with $\lambda_1 = \lambda_2 = 1$; thus

$$P(T_1 \geq t_1, T_2 \geq t_2) = \left(\frac{b}{b+t_1+t_2}\right)^{a+1}, \quad \text{and}$$

$$P(T_i \geq t_i) = \left(\frac{b}{b+t_i}\right)^{a+1}, \quad \text{for } i = 1, 2.$$

By invoking Sklar's Theorem (section 4.7.1) on the above ingredients one obtains the bivariate copula

$$C_a(u, v) = u + v - 1 + \left((1-u)^{-(a+1)} + (1-v)^{-(a+1)} - 1\right)^{-(a+1)}.$$

Observe that in $C_a(u, v)$ the scale parameter b does not appear, suggesting that what matters is only the shape parameter a. If we set $u = 1 - \exp(-t_1)$ and $v = 1 - \exp(-t_2)$ in $C_a(u, v)$, and invoke Sklar's Theorem in reverse, then we are able to produce the following bivariate distribution with exponential (scale parameter 1) marginals:

$$P(T_1 \geq t_1, T_2 \geq t_2) = \left(\exp\left(\frac{t_1}{a+1}\right) + \exp\left(\frac{t_2}{a+1}\right) - 1\right)^{a+1}.$$

The regression function of the above bivariate distribution, namely, $E(T_1 \mid T_2 = t_2)$ does not exist in closed form. However, a numerical evaluation of the regression (with $a = 1$) suggests that $E(T_1 \mid T_2 = t_2)$ is convex and increases exponentially in t_2 (Singpurwalla and Kong, 2004).

4.7.8 Other Specialized Bivariate Distributions

The bivariate distributions discussed so far had marginals that belong to a common family, like the exponential, the Pareto and the Weibull. There could arise circumstances wherein the assumption of a common family for the marginals is unrealistic. With that in mind I give below some bivariate distributions whose marginals belong to different families. Such bivariate distributions provide flexibility of modeling and a means for describing different types of bivariate dependence.

The Gamma–Pareto Family
Here lifetimes T_1 and T_2 have a density at $t_1 > 0$ and $t_2 > 0$ of the form

$$f(t_1, t_2) = \frac{(\lambda t_1)^{\gamma} e^{-t_1(\lambda + t_2/\beta)}}{\Gamma(\gamma)\beta},$$

with gamma and Pareto marginal densities

$$f(t_1) = \frac{\lambda^{\gamma} t_1^{\gamma-1} e^{-\lambda t_1}}{\Gamma(\gamma)}, \quad t_1 > 0,$$

and

$$f(t_2) = \frac{\gamma(\lambda\beta)^{\gamma}}{(\lambda\beta + t_2)^{\gamma+1}}, \quad t_2 > 0,$$

respectively. The parameters β, λ, and γ are positive. Modeling interdependence is best conceptualized through conditional distributions. Here the conditional distribution of T_1 given t_2 is a gamma density of the form

$$f(t_1 \mid t_2) = \frac{t_1{}^{\gamma} \exp[-t_1((\lambda\beta + t_2)/\beta)]}{\Gamma(\gamma+1)\left(\frac{\beta}{\lambda\beta + t_2}\right)^{\gamma+1}}, \quad t_1, t_2 > 0,$$

and that of T_2 given t_1 an exponential of the form

$$f(t_2 \mid t_1) = \frac{t_1}{\beta} \exp\left(-\frac{t_1 t_2}{\beta}\right), \quad t_1, t_2 > 0.$$

The Gamma–Uniform Family
Here the joint density at $t_1 > 0$ and $t_2 > 0$ is given as

$$f(t_1, t_2) = \frac{\lambda^{\gamma} t_1^{\gamma-1} e^{-\lambda t_1}}{2\beta\Gamma(\gamma)},$$

with the marginal of T_1 being a gamma distribution with shape (scale) $\gamma(\lambda)$, so that

$$f(t_1) = \frac{\lambda^{\gamma} t_1^{\gamma-1} e^{-\lambda t_1}}{\Gamma(\gamma)}, \quad t_1 > 0,$$

and the marginal of T_2 a uniform

$$f(t_2) = (2\beta)^{-1}, \quad \mu - \beta < t_2 < \mu + \beta,$$

for some parameter $\mu \geq \beta$. Here again, the parameters β, λ, γ and μ are positive. The variables T_1 and T_2 are independent.

4.8 CAUSALITY AND MODELS FOR CASCADING FAILURES

Whereas the notion of dependence has been well incorporated and developed in the literature on reliability theory – the material of section 4.7 being a testament to this – one also often encounters in practice terms such as 'the cause of failure' and 'cascading failures'. These terms do not appear to have been well articulated within a mathematical framework. This section, based on the work Lindley and Singpurwalla (2002) and Swift (2001), sheds light on some issues of causal and cascading failures. I start by introducing the notion of probabilistic causality as enunciated by Suppes (1970). I then argue that failure models such as those of Freund, Marshall and Olkin, and perhaps some others (but not all) of section 4.7, are models of causal failure. I then claim that the notion of probabilistic causality with an added condition leads us to the notion of cascading failures. Consequently, a modified version of the model of Freund paves the way to developing a probabilistic characterization of cascading events.

4.8.1 Probabilistic Causality and Causal Failures

Deterministic causality has been a topic of discussion among philosophers since the times of Hobbes, Newton, and Hume. A probabilistic approach to causality has been suggested by Reichenbach, Good, and Suppes (cf. Salmon, 1980, p. 50). We consider here the version of Suppes (1970), according to which an event $\mathcal{D}$ is said to be a **prima facie probabilistic cause** of an event $\mathcal{E}$ if:

(i) $\mathcal{D}$ occurs before $\mathcal{E}$ (in time);
(ii) $P(\mathcal{D}) > 0$; and
(iii) $P(\mathcal{E} \mid \mathcal{D}) > P(\mathcal{E})$ and $P(\mathcal{E} \mid \overline{\mathcal{D}}) > P(\mathcal{E})$.

Condition *(iii)* above implies that a cause is a probability raising event. Contrast this with the notion of dependence wherein all that matters is a change (not necessarily an increase) in probability. The three conditions above label the cause to be a prima facie cause, because the cause so defined could only be an apparent cause. Suppes labels a cause to be a **genuine cause** if it is a prima facie cause that cannot be shown as being a 'spurious cause'. A prima facie cause $\mathcal{D}$ is said to be a **spurious cause** of $\mathcal{E}$ if and only if there exists, within a conceptual framework, an event $\mathcal{S}$ where:

(i) $\mathcal{S}$ occurs before $\mathcal{D}$;
(ii) $P(\mathcal{D}, \mathcal{S}) > 0$;
(iii) $\mathcal{D}$ is a prima facie cause of $\mathcal{E}$;
(iv) $P(\mathcal{E} \mid \mathcal{D}, \mathcal{S}) = P(\mathcal{E} \mid \mathcal{S})$; and
(v) $P(\mathcal{E} \mid \mathcal{D}, \mathcal{S}) \geq P(\mathcal{E} \mid \mathcal{D})$.

Thus a spurious cause is a prima facie cause that can be explained away by conditioning on an earlier event (or a common cause of $\mathcal{D}$ and $\mathcal{E}$) that accounts as well for the conditional probability of the effect (i.e. the event $\mathcal{E}$).

Marshall and Olkin's bivariate exponential distribution provides an example for illustrating the notions of probabilistic causality discussed above. With respect to the notation of section 4.7.4, suppose that $\mathcal{E}$ is the event $T_2 = t_2$, and $\mathcal{D}$ the event $T_1 = t_1$, where $t_1 < t_2$. Then it is easy to verify that for $t_1 > (\lambda_{12})^{-1} \ln((\lambda_1 + \lambda_{12})/\lambda_1)$, the three conditions of prima facie causality are satisfied, so that $\mathcal{D}$ is a prima facie probabilistic cause of $\mathcal{E}$. However, $\mathcal{D}$ is not a genuine cause, because if $\mathcal{S}$ denotes the event that the common shock of this distribution occurs at t_1, then $\mathcal{E}$ is independent of $\mathcal{D}$, given $\mathcal{S}$. Indeed, here $\mathcal{S}$ is a genuine probabilistic cause of $\mathcal{E}$. Consequently, for all t_1 greater than a constant, Marshall and Olkin's BVE is a model for a causal failure, with the event $T_2 = t_2$ as the effect, the event $T_1 = t_1$ as a prima facie cause, and $\mathcal{S}$ as the genuine cause. Similarly, we can see that in the case of Freund's model (section 4.7.3) if $T_1 = t_1$ denotes the event that the first component to fail, fails at t_1, and if $T_2 = t_2$, with $t_1 < t_2$, is the event where the surviving component fails at t_2, then the event $(T_1 = t_2)$ is a prima facie cause of the event $(T_2 = t_2)$ – the effect. Therefore Freund's model is also a description of causal failures.

Note that because of the time-ordering of causality, if $\mathcal{D}$ is the cause of $\mathcal{E}$, then $\mathcal{E}$ cannot be the cause of $\mathcal{D}$. Contrast this to the notion of dependence, wherein $\mathcal{D}$ dependent on $\mathcal{E}$ implies that $\mathcal{E}$ is dependent on $\mathcal{D}$. That is, dependence preserves the interchangeability of events; causality does not. Accordingly, the bivariate Pareto failure model of Lindley and Singpurwalla (1986) (section 4.7.5) is not a model for causal failures; it is a model for interdependent failures.

Finally, to some, such as Pearl (2000), the approach to causality discussed here is inadequate. Instead, Pearl introduces his notion of a 'causal mechanism', a lucid discussion of which is by Lindley (2002); also see Singpurwalla (2002). Hesslow (1976, 1981) finds Suppe's notion of probabilistic causality unsatisfactory on several grounds, one of which pertains to the view that for any effect there could be an inexhaustible number of causes and thus every conceivable cause is a spurious cause. However, Rosen (1978) rebuts Hesslow's arguments on grounds that assertions of causal relationships should be within some conceptual framework. In the context of probabilistic causality, Rosen's rebuttal re-affirms the importance of $\mathcal{H}$ (the background history) in probability assessments.

4.8.2 Cascading and Models of Cascading Failures

What are cascading failures and how are they different from causal failures? To distinguish between these two modes of failure, the first thing to note is that in a causal failure model, simultaneous failures are possible. However, in a model for cascading failures, there is a sequence of failures, one followed by the other, but within a specified period of time. Cascading is like a domino effect; the falling of a domino causes its neighbor to fall, but only if the neighbor is within striking distance of the falling domino. If the dominos are too far apart, a falling domino will not have any effect on its neighbors. Thus under cascading, there cannot be simultaneous failures. The failure of one component is followed by that of its neighbor, but within a specified time called the **critical time** or a **threshold**. If the failure of a component causes its neighbor to fail, but after the critical time has elapsed, then the failure mode is not one of cascading failures. The power outage of August 2003 in Canada and North-East United States is a classic example of cascading failures. The failure of a single transmission line due to an overload caused by an imbalance in the supply and demand of electrical power initiated a sequence (or a cascade) of failures which literally paralyzed a large region.

Freund's model (Figure 4.9) provides an architecture for developing the notion of a cascading failure. Suppose, for example, that Freund's model is modified so that the failure rate of the surviving component changes at t_1 from λ_1 to λ_2, but at some time $t_1 + \delta$, $\delta > 0$, it reverts back to λ_1 (Figure 4.19). The quantity δ is the critical time. The choice of δ is subjective, but one possible strategy would be to let δ be the time it takes to restore the failed component to an operational status. In actuality δ is best treated as a random quantity having a finite support. In Figure 4.19, I have chosen the parameters in Freund's model as $\lambda_1 = \lambda_2 = \theta$, and $\lambda_1^* = \lambda_2^* = 2\theta$.

For the set-up of Figure 4.19, it can be seen, details omitted, that the joint density function of T_1 and T_2, at t_1 and t_2, respectively, for $t_1 < t_2$, is of the form

$$\begin{cases} 2\theta^2 e^{-2\theta t_2}, & t_2 < \delta \\ 2\theta^2 e^{-2\theta t_2}, & t_1 < t_2 < t_1 + \delta \\ \theta^2 e^{-\theta(t_1 + t_2 + \delta)}, & t_2 > t_1 + \delta; \end{cases}$$

Similarly for $t_2 < t_1$, *mutatis mutandis*.

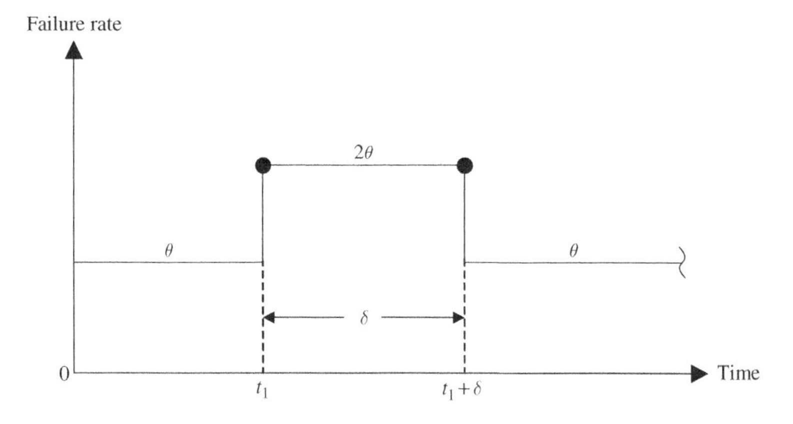

Figure 4.19 Failure rate of the surviving component in a model for cascading failures.

By comparison with the properties of Freund's model (section 4.7.3) we may verify that T_1 and T_2 are independent when $\delta = 0$, and are positively correlated when $\delta > 0$; the model reduces to Freund's model when $\delta \uparrow \infty$.

The above model can be generalized to the case of multiple components and can be extended in several possible directions. For example, we may set $\lambda_1^* = \lambda_2^* = c\theta$, where c is random. Another possibility is to make δ random. These and other related issues are discussed by Swift (2001).

System Reliability Under Cascading Failures

As before, if we let T_p denote the time to failure of a parallel redundant system experiencing cascading failures of the type described by the model of (4.54). Then it can be seen that the probability density of T_p at t is of the form

$$\begin{cases} 4\theta^2 t e^{-2\theta t}, & t < \delta \\ 4\theta^2 \delta e^{-2\theta t} + 2\theta e^{-\theta(t+\delta)} - 2\theta e^{-2\theta t}, & t \geq \delta, \end{cases}$$

and that

$$P(T_p \geq t) = \begin{cases} 2\theta t e^{-2\theta t} + e^{-2\theta t}, & t < \delta \\ 2\theta\delta e^{-2\theta t} + 2e^{-\theta(t+\delta)} - e^{-2\theta t}, & t \geq \delta. \end{cases} \tag{4.55}$$

Consequently, the failure rate function of $P(T_p \geq t)$ is

$$r_{T_p}(t) = \begin{cases} 4\theta^2 \dfrac{t}{2\theta t + 1}, & t < \delta \\ \dfrac{2\theta(2\theta\delta e^{-\theta t} + e^{-\theta\delta} - e^{-\theta t})}{2\theta\delta e^{-\theta t} + 2e^{-\theta\delta} - e^{-\theta t}}, & t \geq \delta. \end{cases} \tag{4.56}$$

The mean time to system failure $E(T_p)$ is $\theta^{-1} + (2\theta)^{-1} e^{-2\theta\delta}$, which is greater than θ^{-1}, the mean time to failure were the system to experience failure under Freund's set-up. Were the two-component parallel-redundant system to be described by independent exponentially distributed lifelengths, then the mean time to system failure would be $3/2\theta$.

Thus the mean time to failure of a two-component parallel-redundant system experiencing cascading failures is bounded from below (above) by the mean time to failure were it to experience causally dependent (independent) lifelengths. Results such as these plus those based on a comparison of the failure rate and the reliability functions, lead to the claim that in the case of parallel-redundant systems, it is the causal mode of failure that is more deleterious to system survivability than a cascading mode. The cascading mode is more deleterious than failures generated by independent lifelengths; (Lindley and Singpurwalla, 2002; Swift, 2001). The above conclusion should ease the often expressed concern of some engineers who feel that systems experiencing cascading failures are the ones that are the most prone to the risk of a total collapse.

4.9 FAILURE DISTRIBUTIONS WITH MULTIPLE SCALES

Our discussion so far has been limited to the development of failure models (univariate and multivariate) that are indexed by a single scale, namely, time. However, there are scenarios which require that the occurrence of failure be registered in terms of two (or more) scales; time and some other metric which may or may not be related to time. For example, when an automobile fails, we often note its chronological age as well as its mileage. This type of information is required for making claims against an automobile's warranty (cf. Singpurwalla and Wilson 1993). In the biological and medical contexts we would be interested in knowing the cumulative radiation from diagnostic radiotherapy as a function of age, in order to assess if the hazards of therapy outweigh its benefits. Often the two scales bear a strong relationship to each other, and this is what makes the topic of multiple scales interesting. The relationship could be a deterministic one, though most likely it tends to be stochastic. For example, the mileage accumulated by an automobile increases with its age, but its value depends on usage, and usage varies from one individual to the other.

Singpurwalla and Wilson (1998) propose a strategy for constructing failure models indexed by two scales, time and a time-dependent quantity such as usage. A time-dependent quantity is considered because most other measures of failure turn out to be functions of time; the approach generalizes to multiple scales. In what follows I describe the strategy proposed by Singpurwalla and Wilson (1998) since the emphasis there is on probabilistic modeling in a reliability setting. However, it is helpful to note that there have been other proposals for the treatment of multiple scales, namely, those by Nelson (1995), Oakes (1995), Kovdonsky and Gertsbach (1997), and Jewell and Kalbfleisch (1996). The work of Jewell and Kalbfleisch comes close in spirit to that described here; the works of Nelson and of Oakes boil down to combining scales so that the actual analysis is conducted on a single scale.

In section 4.9.1, I outline an overall strategy for developing failure distributions with multiple scales. Included here are candidate models for relating usage with chronological age, and an approach for capturing the effect of usage, which now becomes a **covariate** of the time to failure. A covariate of the time to failure is to be interpreted as a variable that influences the time to failure. For example, the operating environment, discussed in section 4.7.5, whose effect on the failure rate of the item is encapsulated by the parameter η, can be viewed as being a covariate. Our strategy for relating usage with age is based on Cox's (1972) celebrated proportional hazard model. The relationships between usage and time are prescribed by some well-known stochastic processes, such as the Poisson and its variants. In section 4.9.2, I summarize results on a specific model that is a consequence of the techniques of section 4.9.1.

4.9.1 Model Development

By way of notation, let T denote the time to failure of an item and $M(t)$ its cumulative usage (or dose) at time $t \geq 0$. Thus $M(T) \stackrel{\text{def}}{=} U$ is the cumulative usage at failure. Both T and U are random variables; furthermore, by construction they are dependent. Our aim is to specify the joint probability density function of T and U, at t and u, respectively, assuming that it exists; both t and u are non negative. Let $f_{T,U}(t, u)$ denote this joint density. Then

$$f_{T,U}(t, u) = f_T(t) f_{U \mid T}(u \mid t)$$

by the multiplication rule, where $f_{U \mid T}(u \mid t)$ is the density at u of the conditional distribution of U given T; we suppose that this conditional density also exists. Since $U = M(T)$, the above decomposition can also be written as

$$\begin{aligned} f_{T,U}(t, u) &= f_T(t) f_{M(T) \mid T}(u \mid t) \\ &= f_T(t) f_{M(t) \mid T}(u \mid t), \text{ since we are conditioning at } T = t, \\ &= f_{M(T)}(u) f_{T \mid M(t)}(t \mid u), \text{ by symmetry of the multiplication rule.} \end{aligned}$$

Observe that in $f_{T \mid M(t)}(t \mid u)$, t appears two times: as an argument of the variable $T \mid M(t)$, and as an indexing parameter of the random variable $M(t)$. Indeed as a function of t, $f_{T \mid M(t)}(t \mid u)$ is not a probability density function. Our next step is to prescribe meaningful forms for $f_{M(t)}(u)$ and $f_{T \mid M(t)}(t \mid u)$. This is done next.

Candidate Usage Processes

Since usage varies from unit to unit, a natural model for $M(t)$ is a non-decreasing stochastic process $\{M(t);\ t \geq 0\}$, with $M(0) \equiv 0$. For usage characterized by simple counts, such as the number of times a unit is turned on (and off), a suitable model for $\{M(t);\ t \geq 0\}$ would be a Poisson process with mean value function $\Lambda(t)$; (section 4.7.4). Thus, given $\Lambda(t)$, $t \geq 0$,

$$P(M(t) = u \mid \Lambda(t)) = \mathrm{e}^{-\Lambda(t)} (\Lambda(t))^u / u!, u = 0, 1, \ldots \tag{4.57}$$

For usage that manifests as damage due to shocks of random magnitude, such as the landing gear of an airplane, the 'compound Poisson process' is a meaningful model. Specifically, if $N(t)$, $t \geq 0$, denotes the number of shocks (or landings) inflicted on a unit by time t, and if $\{N(t);\ t \geq 0\}$ is described by a non-homogenous Poisson process with mean value function $\Lambda(t)$, then $M(t)$, the cumulative damage (or equivalently, the usage) at time t is $M(t) = \sum_{i=1}^{N(t)} X_i$, where X_i is the damage to the unit caused by the i-th shock. If all the X_is are assumed to be independent and have a distribution that is identical to that of X, where the distribution of X has a density at x given as $f_X(x)$, then the random variable $M(t)$ has density at u of the form

$$\mathrm{e}^{-\Lambda(t)} \sum_{j=1}^{\infty} \frac{(\Lambda(t))^j}{j!} f_X^{(j)}(u); \tag{4.58}$$

the stochastic process $\{M(t);\ t \geq 0\}$ is known as a **compound Poisson process**. The quantity $f_X^{(j)}(u)$ is the density at u of the random variable $(X_1 + X_2 + \cdots + X_j)$; it is known as the j-**fold convolution** of $f_X(u)$.

The Poisson and the compound Poisson processes have independent increments (section 4.7.4) and are therefore appropriate if the future of $M(t)$ is not influenced by its past. Furthermore the number of increments of a Poisson process in a finite interval of time are finite. Thus they are meaningful for describing usage such as damage due to shocks (which are intermittent). By contrast 'gamma processes' (section 4.7.6) have an infinite number of increments in a finite interval of time, and are therefore suitable for describing wear caused by continuous use. The structure of a gamma process is given below.

Suppose that $a(t)$ is non-decreasing and left continuous in t, with $a(0)=0$, and $b \neq 0$, a positive constant. Then the process $\{M(t);\ t \geq 0\}$ is said to be a **gamma process** with a shape function $a(t)$ and a scale parameter b, if for any $t \geq s \geq 0$, and $M(0) \equiv 0$,

(i) $M(t)$ has independent increments, and
(ii) $(M(t) - M(s))$ has a gamma distribution with a scale parameter $1/b$, and a shape parameter $(a(t) - a(s))$.

When $a(t)$ is linear in t, $\{M(t);\ t \geq 0\}$ will become a 'Lévy Process'; this process is appropriate for describing wear caused by a continuous use of the item. Lévy processes are discussed in section 7.2.2. Like damage, wear (which can be observed and measured) is a proxy for usage. When usage is intermittent so is the wear; when such is the case $\{M(t);\ t \geq 0\}$ can be described by Çinlar's (1972), 'Markov additive process' (Singpurwalla and Wilson, 1998). Markov additive processes are overviewed in section 7.3.4, wherein an illustration involving random usage is also given.

Describing the Effect of Usage on Time to Failure

Suppose that the item in question has a propensity to fail even when it does not experience any usage. This could happen due to a deterioration in the item's resistance to failure because of natural causes. Let $r_0(t)$ be the failure rate of the item were it not to be subjected to any usage; $r_0(t)$ is known as the **baseline failure rate**. Usage modifies $r_0(t)$ by increasing it; we assume that this modification is additive so that $r(t)$, the failure rate of the distribution of T, is of the form

$$r(t) = r_0(t) + \eta M(t), \tag{4.59}$$

where $\eta > 0$ is a constant. The model of (4.59) is known as an **additive hazards model**; it suggests that each unit of use increases the failure rate of T by the same amount. In actuality, such a model may or may not be true; all the same, it is considered here for illustrative purposes – also see Section 7.5.2. Let

$$\mathcal{M}(t) = \int_0^t M(u)\mathrm{d}u \text{ and } \mathcal{R}_0(t) = \int_0^t r_0(u)\mathrm{d}u.$$

Then, by the exponential formula of reliability (section 4.3) conditional on $M(t) = u$ and $\mathcal{M}(t) = \mathcal{U}$,

$$f_{T \mid M(t), \mathcal{M}(t)}(t \mid u, \mathcal{U}) = (r_0(t) + \eta u) \exp(-(\mathcal{R}_0(t) + \eta \mathcal{U}));$$

this is the conditional density of T at t, given $M(t)$ and $\mathcal{M}(t)$. Averaging out with respect to $\mathcal{M}(t)$ given $M(t)$, enables us to obtain the conditional density of T at t, given $M(t)$. The details are in (4.6) of Singpurwalla and Wilson (1998), who go on to show that when $\{M(t);\ t \geq 0\}$ is the Poisson process of (4.57), the conditional density of T at t, given $M(t) = u$ is of the form:

$$(r_0(t) + \eta u)\mathrm{e}^{-\mathcal{R}_0(t)} \left(\int_0^t \mathrm{e}^{-\eta(t-s)} \frac{\lambda(t)}{\Lambda(t)} \mathrm{d}s \right)^u, \tag{4.60}$$

where $\lambda(s) = \frac{\mathrm{d}}{\mathrm{d}s}\Lambda(s)$ is the intensity function of the process.

4.9.2 A Failure Model Indexed by Two Scales

In section 4.9.1, we discussed several possibilities for $f_{M(t)}(u)$ and $f_{T \mid M}(t \mid u)$, (4.60) being an example of the latter. The purpose of this section is to see an illustration of how the above two can be put together to arrive at $f_{T,U}(t, u)$, the joint density at (t, u) of T and U. Suppose now that the usage process is the non-homogenous Poisson process of (4.57), and that the additive hazards model of (4.59) is invoked. Then, $f_{T \mid M(t)}(t \mid u)$ will be given by (4.60) and with $f_{M(t)}(u)$ given by (4.57), we have

$$f_{T,U}(t, u) = \frac{r_0(t) + \eta u}{u!} \mathrm{e}^{-\mathcal{R}_0(t) - \Lambda(t)} \left(\int_0^t \lambda(s) \mathrm{e}^{-\eta(t-s)} \mathrm{d}s \right)^u,$$

for $t \geq 0$ and $u = 0, 1, 2, \ldots$

A simplification of the above is achieved if $r_0(t) = r$, a constant and $\Lambda(t) = \lambda t$, for some constant $\lambda > 0$; that is, $\{M(t);\ t \geq 0\}$ is a homogenous Poisson process. When such is the case our bivariate failure model, with time $t \geq 0$ as one scale, and usage $u = 0, 1, 2, \ldots,$ as the other scale becomes

$$f_{T,U}(t, u) = \frac{r + \eta u}{u!} \left(\frac{\lambda}{\eta} (1 - \mathrm{e}^{-\eta t}) \right)^u \mathrm{e}^{-(r+\lambda)t}. \tag{4.61}$$

The marginal distributions of T and U are

$$P(T > t) = \exp\left(-(r + \lambda)t + \frac{\lambda}{\eta}(1 - \mathrm{e}^{-\eta t}) \right), \text{ and}$$

$$P(U = u) = \frac{\lambda^u (r + \eta u)}{\prod_{i=0}^{u} (\lambda + r + \eta i)}.$$

Results analogous to (4.61) for other processes discussed in section 4.9.1 tend to be cumbersome; they entail approximations and simulations. More details, with an example involving an application of the ideas given here to a problem entailing the specification of a warranty for traction motors of electric locomotives are in Singpurwalla and Wilson (1998).

I conclude this section by citing the work of Lawless and his colleagues on the role of multi-indexed failure models in the context of statistical analysis of product warranty data; an overview is in Lawless (1998).

Chapter 5

Parametric Failure Data Analysis

Here lies Thomas Bayes: He rests on his posterior.

5.1 INTRODUCTION AND PERSPECTIVE

Much of the literature in reliability and survival analysis is devoted to the topics of life-testing and the analysis of failure data. Indeed, the journal *Lifetime Data Analysis* is exclusively devoted to research in these topics. In many respects life-testing, failure data analysis and the material of Chapter 4 on failure models constitute the core of a course in reliability. To appreciate the role of these three topics in reliability and survival analysis, it is helpful to put in perspective what has been covered so far.

I start by recalling the key message of Chapter 1, namely the claim that the presence of uncertain events is common to all problems of risk analysis. In the context of Chapter 4, one such uncertain event was $(T > t)$, where T is some index of measurement, typically, the time to failure of a unit; another event is $(N(t) = k)$, where $N(t)$ is the number of occurrences of some event of interest, like a shock. Chapter 2 was devoted to the quantification of uncertainty by probability, and thus $P(T > t)$ becomes an entity of interest. This entity was called the survival function, and a goal of reliability and survival analysis is to prescribe techniques for assessing $P(T > t)$. To facilitate this, the law of total probability (or the law of the extension of conversation) was invoked, so that for some parameter θ (vector or scalar)

$$P(T > t) = \int_{\theta} P(T > t \mid \theta) F(\mathrm{d}\theta), \tag{5.1}$$

where $F(\theta)$ is the distribution function of θ. The quantity $P(T > t \mid \theta)$ is known as the reliability of T for a mission time t; de facto, it is a failure model, or a chance distribution for T. The material of Chapter 4 pertained to considerations in choosing $P(T > t \mid \theta)$. In the absence of any information other than $\mathcal{H}$, which has been suppressed here, $F(\theta)$ represents one's best assessment of the uncertainty about θ; it is the prior distribution for θ. The quantity $P(T > t)$ is also known as the predictive distribution of T. Section 5.2 is devoted to the matter of assessing this predictive distribution assuming some commonly used chance distributions like the exponential, the Weibull and the Poisson.

Reliability and Risk: A Bayesian Perspective N.D. Singpurwalla
© 2006 John Wiley & Sons, Ltd

Suppose now that interest was centered around n items with T_i as the lifelength of the i-th item. Then following the line of reasoning that leads us to (2.4), namely conditional independence and identical distributions, we have as an extension of (5.1)

$$P(T_1 \geq t_1, \ldots, T_n \geq t_n) = \int_{\theta} \prod_{i=1}^{n} P(T_i \geq t_i \mid \theta) F(\mathrm{d}\theta). \tag{5.2}$$

Furthermore, if $(T_1 \geq t_1, \ldots, T_n \geq t_n)$ were denoted by $T^{(n)}$, then by the multiplication rule and the law of total probability, together with the assumptions of conditional independence and identical distributions, T_{n+1}, the lifelength of the $(n+1)$-th item, is

$$P(T_{n+1} \geq t_{n+1} | T^{(n)}) = \int_{\theta} P(T_{n+1} \geq t_{n+1} \mid \theta) F(\mathrm{d}\theta \mid T^{(n)}), \tag{5.3}$$

where by Bayes' law

$$F(\mathrm{d}\theta | T^{(n)}) \propto \prod_{i=1}^{n} P(T_i \geq t_i | \theta) F(\mathrm{d}\theta).$$

The above two equations are analogous to (2.5) and (2.6), respectively. Equations (5.2) and (5.3) can also be motivated by the judgment of exchangeability; see, for example, (3.11).

When the T_is get revealed as say $\tau^{(n)}$, $i = 1, \ldots, n$, $T^{(n)}$ gets replaced by $\tau^{(n)} = (\tau_1, \ldots, \tau_n)$, and (5.3) becomes

$$P(T_{n+1} \geq t_{n+1}; \tau^{(n)}) = \int_{\theta} P(T_{n+1} \geq t_{n+1} \mid \theta) F(\mathrm{d}\theta; \tau^{(n)}), \tag{5.4}$$

where $F(\mathrm{d}\theta; \tau^{(n)})$, the posterior for θ is given as

$$F(\mathrm{d}\theta; \tau^{(n)}) \propto L(\theta; \tau^{(n)}) F(\mathrm{d}\theta); \tag{5.5}$$

$L(\theta; \tau^{(n)})$ is the likelihood of θ for a fixed and known $\tau^{(n)}$. Ignoring for now considerations of the 'stopping rule' (section 5.4.2), it is usual to take $L(\theta; \tau^{(n)})$ as

$$L(\theta; \tau^{(n)}) = \prod_{i=1}^{n} f(\tau_i | \theta), \tag{5.6}$$

where $f(\tau_i|\theta)$ is the probability density at τ_i, generated by $P(T_i \geq t_i \mid \theta)$ – provided that the density exists. In principle, however, $L(\theta; \tau^{(n)})$ could have any suitable functional form (Singpurwalla, 2002).

The left-hand side of (5.4) is the predictive distribution of T_{n+1}, the lifelength of the $(n+1)$-th, in the light of $\tau^{(n)}$ the available failure data on the n lifelengths judged exchangeable with each other and also with T_{n+1}. Section 5.4 discusses an assessment of this predictive distribution supposing the commonly used chance distributions of section 5.2. Equation (5.4) generalizes, so that the joint predictive distribution of $T_{n+1}, T_{n+2}, \ldots, T_{n+m}$, lifelengths of m items judged exchangeable with each other and with n items whose lifetimes $\tau_1, \ldots, \tau_n$ have been observed. Specifically, we have

$$P(T_{n+1} \geq t_{n+1}, \ldots, T_{n+m} \geq t_{n+m}; \tau^{(n)}) = \int_\theta \prod_{i=1}^{m} P(T_{n+i} \geq t_{n+i} \mid \theta) L(\theta; \tau^{(n)}) F(\mathrm{d}\theta). \tag{5.7}$$

Such predictive life distributions play a useful role in system reliability analysis.

By **life-testing** I mean strategies or experimental designs for generating the failure data $\tau^{(n)}$. These strategies are important, in the sense that they may influence a specification of the likelihood (section 5.4.2). By **failure data analysis** I mean an assessment of the predictive distributions of the form given by (5.4) and (5.7). Failure data analysis should, in principle, also include some form of validating the assumption of exchangeability that is inherent to the results of (5.4), (5.6) and (5.7). The aim of this chapter is to articulate on the matter of assessing predictive distributions of the kind mentioned above, on the generation of failure data to facilitate these assessments, and on using the predictive distributions for making decisions such as lot acceptance/rejection. In section 5.3 I discuss methods for specifying the prior $F(\mathrm{d}\theta)$, either by elicitation or by standard rules.

Whereas the role of the parameter θ has been to facilitate an assessment of predictive distributions, it is sometimes the case that interest may center around θ itself. This can happen when one is able to attach some physical or scientific meaning to θ as opposed to viewing it as simply a Greek symbol. When such is the case, failure data analysis could also mean to include an assessment of the posterior distribution of θ (equation (5.5)).

5.2 ASSESSING PREDICTIVE DISTRIBUTIONS IN THE ABSENCE OF DATA

The archetypal problem I propose to address here is the assessment of $P(T > t)$ via (5.1), namely

$$P(T > t) = \int_\theta P(T > t \mid \theta) F(\mathrm{d}\theta).$$

It is important to bear in mind that implicit to the above is the role of $\mathcal{H}$ – history or background information. For convenience $\mathcal{H}$ has been suppressed; however $\mathcal{H}$ encapsulates all the engineering and scientific knowledge that an assessor of $P(T > t)$ has about an item whose lifelength is T. Such knowledge should play a role in specifying a failure model of the form $P(T > t \mid \theta)$ as well as the prior $F(\mathrm{d}\theta)$. What distinguishes the material here from that of section 5.3 is that in the latter, one has in addition to $\mathcal{H}$ the failure data $\tau^{(n)}$, so that an assessment of predictive distributions is made in the light of both $\tau^{(n)}$ and $\mathcal{H}$. The likes of (5.1) are germane when the item in question is one of a kind, so that it is not possible to obtain failure data from items judged exchangeable with the said item.

5.2.1 The Exponential as a Chance Distribution

Now suppose that based on $\mathcal{H}$, an assessor of $P(T > t)$ makes the judgment that the item is immune from aging and wear so that a reasonable choice for its failure model or chance distribution is an exponential with parameter λ; i.e. $\theta = \lambda$, $\lambda > 0$. Then $P(T > t \mid \theta = \lambda) = \exp(-\lambda t)$, for $t \geq 0$, and so

$$P(T > t) = \int_\lambda \exp(-\lambda t) F(\mathrm{d}\lambda).$$

Recall (section 4.2) that $F(\mathrm{d}\lambda) = f(\lambda)\mathrm{d}\lambda$, where $f(\lambda) = \mathrm{d}F(\lambda)/\mathrm{d}\lambda$ is the derivative of $F(\lambda)$, the prior distribution function of λ, provided that the derivative exists. What should $F(\lambda)$ be?

Since $\lambda > 0$, $F(\lambda)$ should have $(0, \infty)$ as its support. One possibility is to assume that $F(\lambda)$ has a gamma distribution with a scale parameter ψ and a shape parameter α (sections 3.2.2 and 4.4.2). Then it can be seen that

$$P(T > t) = (\psi/(\psi + t))^{\alpha} \tag{5.8}$$

a Pareto distribution. Since $E(\lambda) = \alpha/\psi$, $V(\lambda) = \alpha/\psi^2$, and $E(T \mid \lambda) = 1/\lambda$, we may pin down ψ and α by setting α/ψ equal to our best guess for the values of T, and $\sqrt{\alpha}/\psi$ equal to our standard deviation about this guess. In principle, one is free to choose any proper distribution as a prior for λ as long as there is a rationale and justification for the choice. More about this matter of proper distributions is said later in section 5.3.2.

5.2.2 The Weibull (and Gamma) as a Chance Distribution

Instead of the exponential, as a failure model for T, we may want to choose the Weibull distribution with scale parameter $\lambda > 0$, and shape parameter $\beta > 0$. Then $\theta = (\lambda, \beta)$ is a vector, and $P(T > t \mid \theta) = \exp(-(\lambda t)^{\beta})$. This choice has an advantage of providing an ability to incorporate the property of aging or wear with $\beta > 1$, and it reduces to an exponential distribution with $\beta = 1$. Thus using this distribution over the exponential makes the analysis robust to departures from exponentiality. Its disadvantage is that one now has to assess a two-dimensional joint prior distribution for $\theta = (\lambda, \beta)$, and this results in the unavailability of closed form expressions of the predictive distribution. Numerical approximations and simulation offers a way out, but the results so produced lack an interpretive flavor. For example, properties like the decreasing failure rate of the Pareto predictive distribution of (5.8) cannot be asserted.

The traditional approach for assigning a prior distribution for θ has been based on the assumption that λ and β are independent; an exception is Soland (1969). This cannot be meaningful because were we to anchor our best judgment about T via $E(T)$, its mean, or via $M(T)$, its median, then specifying either confounds λ and β. For example, were $P(T \geq t \mid \lambda, \beta)$ to be parameterized as $\exp(-(t/\gamma)^{\beta})$ with $\lambda = 1/\gamma$, then it can be seen that $M(T) = \gamma \exp(c/\beta)$, where $c = \ln\ln 2$, is a constant. Consequently, once $M(T)$ is specified, both γ and β cannot be freely chosen. Similarly, with $E(T)$. How then should a joint prior on γ and β be specified?

Since β characterizes aging or the absence of it, it is possible for subject matter specialists to offer expert testimonies on β. Large values of $\beta > 1$ suggest a rapid deterioration of the item with age; the converse with $\beta < 1$. Thus a gamma distribution with scale ψ and shape α would seem to be a reasonable choice for the prior distribution of β. The quantity α/ψ would encapsulate ones best guess about β and $\sqrt{\alpha}/\alpha$ the uncertainty about this guess. Suppose that $\pi(\beta; \alpha, \psi)$ denotes the probability density of this prior. However, eliciting from a subject matter specialist, a prior for γ conditioned on β poses a practical difficulty. Most subject matter specialists do not think in terms of parameters of failure models, and asking an expert to assess γ for some fixed value of β seems far fetched.

In Singpurwalla (1988b), a strategy for inducing a joint prior on γ and β based on $\pi(\beta; \alpha, \psi)$ and an elicited prior on the median $M(T)$ is proposed. Let $\pi(M(T); \mu, \sigma)$ denote the density of the prior distribution of $M(T)$; μ and σ are the parameters of this density (section 5.3.1). This prior density is based on the testimony of one or more subject matter specialists about the median life; it incorporates considerations such as expertise of the experts plus the experts' disposition to risk, etc. Details about the elicitation process and the functional form of $\pi(M(T); \mu, \sigma)$ are given in section 5.3.1, (5.21). However, there is a caveat to the proposed strategy in the sense that the elicitation of $M(T)$ is not conditional upon β; that is, $M(T)$ and β are judged independent. This seems reasonable – and the data analyzed by Singpurwalla (1988b) does not negate the assumption – because subject matter specialists are able to conceptualize and assess median lifetimes without any encumbrance from their assessment about aging. Even otherwise the elicitation of $M(T)$ conditioned on β is, from a practical point of view, more palatable than the assessment of γ conditioned on β. With the prior parameters α, ψ, μ and σ specified, the joint prior density at γ and β turns out to be of the form:

$$\frac{\kappa\alpha^{\psi}\beta^{\psi-1}}{\sqrt{2\pi}\sigma\Gamma(\psi)}\exp\left\{\left(\frac{c}{\beta}-\alpha\beta\right)-\frac{1}{2}\left(\frac{\gamma e^{c/\beta}-\mu}{\sigma}\right)^{2}\right\},$$

$$\text{where } \kappa=\left(\int_{-\mu/\sigma}^{\infty}(2\pi)^{-1/2}\exp\left(-\frac{u^{2}}{2}\right)du\right)^{-1}; \qquad (5.9)$$

$\kappa\approx 1$ if $\mu\geq 3\sigma$. Equation (5.9) though cumbersome in appearance has the virtue of being a proper density and one that has been reasonably well motivated. A gamma density for β has also been advocated by Martz and Waller (1982, p. 237). An advantage in eliciting expert testimony on the median $M(T)$ rather than the mean $E(T)$ is twofold. The first is that the relationship between $M(T)$ and the parameters γ and β takes a simple form, namely, $M(T)=\gamma\exp(c/\beta)$ where c is a known constant; by contrast $E(T)=\gamma\Gamma((\beta+1)/\beta)$, where $\Gamma(x)=\int_0^{\infty}u^{x-1}e^{-u}du$. This simplifies the task of inducing the joint prior on γ and β. The second, perhaps a more important reason is that eliciting expert testimonies on percentiles, like the median, has, according to empirical and psychological studies, proven to be more effective than that of the mean; see, for example, Peterson and Beach (1967) and Hogarth (1975).

The approach prescribed above, though focused on the special case of the Weibull, can be invoked for the other two parameter failure models. All that is required is that any measure of central tendency, such as the mean, the mode or the median, or for that matter some other percentile, be expressed as a simple function of the parameters, and that one of the two parameters be physically interpreted. For example, the mode of a gamma distribution with scale (shape) $\psi(\alpha)$ is $(\alpha-1)/\psi$, and for $\alpha>(<1)$ this distribution has an increasing (decreasing) failure rate, which for large amounts of the lifetime T asymptotes to ψ. Thus under the judgment of aging, the prior on α should have support $(1,\infty)$; otherwise its support should be $(0,1)$. The choice of gamma as a failure model has been argued for parallel systems in stand-by redundancy, or for items whose failure can be attributed to the accumulation of damage caused by Poisson shocks (Mann, Schafer and Singpurwalla, 1974, p. 125). The exact details on how to exploit the above features for inducing a proper prior distribution on ψ and α remain to be worked out; they are left as a project for the reader.

5.2.3 The Bernoulli as a Chance Distribution

The Bernoulli, as a chance distribution, was introduced in section 3.1.3 in connection with de Finetti-type representation theorems for zero-one exchangeable sequences. In the context of the set-up of this section, suppose that the event $(T>t)$ has as its proxy an event $(X=1)$, and that $(X=0)$ is a proxy for the event $(T\leq t)$. A motivation for introducing the binary X in place of the continuous T is simplification. Specifically, we are able to replace multi-parameter chance distributions by a single parameter chance distribution. To see why, we note from (5.1) that when the continuous T is replaced by the Bernoulli X with a parameter θ, we have

$$P(X=1)=\int_{\theta} P(X=1 \mid \theta)\, F(\mathrm{d}\theta)$$

$$=\int_{0}^{1} \theta F(\mathrm{d}\theta), \tag{5.10}$$

since θ has support (0, 1). Thus were $P(T>t \mid \bullet)$ to be a Weibull chance distribution of the form $\exp(-(t/\gamma)^{\beta})$, as was done in section 5.2.2, the representation of (5.10) would entail a chance distribution consisting of a single parameter θ. Whereas θ encapsulates the effect of both γ and β, there is a loss of granularity in going from the representation of (5.1) to (5.10). This is because with the former one can place bets on the event $(T>t)$ for a range of values of t, whereas the latter enables bets on only a single event defined by X.

In section 3.1.2, a beta distribution with parameters α and β was chosen as the prior for θ, $0<\theta<1$. For $\alpha, \beta>0$, its density at θ is of the form

$$\frac{\Gamma(\alpha+\beta)}{\Gamma(\alpha)\Gamma(\beta)}\theta^{\alpha-1}(1-\theta)^{\beta-1}, \tag{5.11}$$

where $\Gamma(\alpha)=\int_0^{\infty} x^{\alpha-1}\mathrm{e}^{-x}\mathrm{d}x$. The mean and variance of this beta distribution are $\alpha/(\alpha+\beta)$ and $\alpha\beta/((\alpha+\beta)^2(\alpha+\beta+1))$, respectively.

This distribution is attractive in the sense that by a suitable choice of α and β, various forms of subjective judgments about θ can be expressed. For example, $\alpha=\beta=1$ results in a uniform density for θ, whereas the choice $\alpha=2$, $\beta=1$ makes the density of the form 2θ – a triangle with an apex of 2 at $\theta=1$. Identical values of $\alpha>1$ and $\beta>1$ make the density symmetrical about $\theta=1/2$, which is where it also attains its maximum. Values of $\alpha<(>)1$ and $\beta>(<)1$ make the density L-shaped or J-shaped (Figure 5.1). The former is appropriate when one has a strong prior opinion about the outcome X being 0; the latter when the prior opinion that X when revealed will be 1 is strong. Values of α and β less than one enable us to describe a U-shaped density for θ (Figure 5.1). The U-shaped density for θ is appropriate when one opines that in repeated trials of the phenomenon that generates X, the values of the Xs when revealed will be a long sequence of either zeros or ones. When $\alpha<(=)1$ and $\beta=(<)1$ the density is L(J)-shaped, but unlike the illustrations of Figure 5.1, it does have a probability mass at $\theta=1(0)$. A hierarchical two-stage approach for specifying a prior for θ is in section 2 of Chen and Singpurwalla (1996).

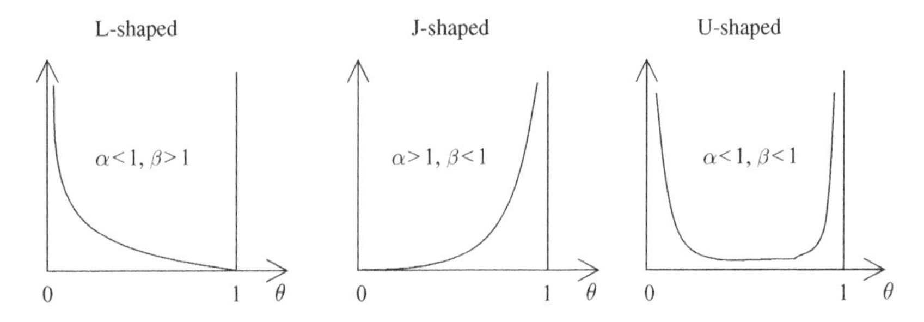

Figure 5.1 L, J, and U-shaped densities for the prior of θ.

A lot has been written about a suitable choice for the prior distribution of θ. Much of the discussion is philosophical and focused on densities that encapsulate little or no knowledge, or those that best allow the data – when available – to dominate the effect of the prior. Such priors are known under several names as 'objective', 'default', 'reference' and 'non-informative', and are used when one wishes to adopt an impartial stance. Geisser (1984), following Bayes and Laplace, advocates the uniform as a noninformative prior for θ; Laplace used the principle of indifference for justifying his use of this prior. Others who have written on this topic are Bernardo (1979a), Haldane (cf. Jeffreys, 1961; p. 123) and Zellner (1977). Bernardo and Zellner advocate (default) proper priors that are proportional to $\theta^{-1/2}(1-\theta)^{-1/2}$ and $\theta^{\theta}(1-\theta)^{(1-\theta)}$, respectively. However, some of the proposed priors for θ are improper, and in some cases they depend on the nature of the experiment that generates X; such priors are objectionable. For example, Haldane proposes that the density for θ be proportional to $(\theta(1-\theta))^{-1}$ which is improper because its integral is not one. Lindley (1997a) refers to the property of the failure of the density to integrate to one as **impropriety**. The desire to produce priors that are universal, and consequently non-subjective, has yielded distributions having the objectionable features mentioned above. More about this will be said in section 5.3.2.

A subjectivistic assessment of θ based on expert testimony has been proposed by Chaloner and Duncan (1983), much in the spirit of eliciting expert testimonies on Weibull lifetimes that result in (5.9). An alternative approach is to induce a prior distribution for θ using the elicited priors on the parameters of the chance distributions of sections 5.2.1 and 5.2.2. This task is easier in the case of the exponential than the Weibull. Specifically, for $P(T>t \mid \lambda)=\exp(-\lambda t)$, I have proposed the gamma distribution with parameters ψ and α elicited via expert testimonies on the mean time to failure $E(T \mid \lambda)$ as a suitable choice. Since

$$P(x=1 \mid \theta)=\theta=P(T>t \mid \lambda)=\exp(-\lambda t),$$

a prior on λ induces a prior on θ. It can be seen – details left as an exercise for the reader – that a gamma prior on λ with scale ψ and shape α results in the following as a prior density for θ, $0<\theta<1$:

$$\frac{\theta^{(\psi/t-1)}}{\Gamma(\alpha)}\left(\frac{\psi}{t}\right)^{\alpha}\left(\ln\left(\frac{1}{\theta}\right)\right)^{\alpha-1}. \tag{5.12}$$

It is important to bear in mind that the mission time t is a parameter of the above prior. This is to be expected since θ depends on t, taking the value $1(0)$ when $t=0(\infty)$. In the case of the Weibull with scale γ and shape β, $\theta=\exp(-(t/\gamma)^{\beta})$. This form makes the task of inducing a prior on θ based on a joint prior on γ and β cumbersome; it will have to be done by simulation. Similarly, for the case of a gamma chance distribution.

Since $P(X=1)=E(\theta)$ (equation (5.10)) we have the result that when the prior on θ is a beta density of the form given by (5.11), $P(X=1)=\alpha/(\alpha+\beta)$, the mean of θ. When the prior on θ is an induced density of the form given in (5.12), then $P(X=1)=(\psi/(\psi+t))^{\alpha}$, the mean of θ under this density; its variance is $(\psi/(\psi+2t))^{\alpha}-(\psi/(\psi+t))^{2\alpha}$. Note that the α in $\alpha/(\alpha+\beta)$ is not the same α in $(\psi/(\psi+t))^{\alpha}$; the former is a parameter of the beta density for θ, whereas the latter is the shape parameter of a gamma prior for λ. Since $(\psi/(\psi+t))^{\alpha}$ is also the survival function of a Pareto distribution (equation (5.8)) we have the interesting observation that the mean value function of the density of (5.12) is the survival function of a Pareto distribution.

Whereas (5.10) pertains to the special case of $X=1$, its more general version entailing $X=x$, for $x=1$ or 0, takes the form

$$P(X=x)=\int_0^1 \theta^{x}(1-\theta)^{1-x}F(\mathrm{d}\theta).$$

If $F(\theta)$ is taken to be the beta distribution with parameters α and β, then

$$P(X=x)=\int_0^1 \theta^x (1-\theta)^{1-x} \frac{\Gamma(\alpha+\beta)}{\Gamma(\alpha)\Gamma(\beta)} \theta^{\alpha-1}(1-\theta)^{\beta-1}\mathrm{d}\theta,$$

$$=\frac{\Gamma(\alpha+\beta)\Gamma(\alpha+x)\Gamma(\beta-x+1)}{\Gamma(\alpha+\beta)\Gamma(\alpha)\Gamma(\beta)}. \tag{5.13}$$

Verify that when $x=1$, the above expression becomes $\alpha/(\alpha+\beta)$, and that when $x=0$ it is $\beta/(\alpha+\beta)$. Furthermore, if the prior on θ were to be a uniform distribution, i.e. were $\alpha=\beta=1$, (5.13) would yield $P(X=1)=P(X=2)=1/2$. This result implies that under the uniform prior for θ, the reliability of the unit in question is $1/2$.

Series Systems with Bernoulli Chance Distributions

Now suppose that we have two such units and that they are connected to form a series system. What is the reliability of this system supposing that their associated X_is, $i=1, 2$, are conditionally – given θ – independent? The conventional answer, namely $\left(\frac{1}{2}\right)\cdot\left(\frac{1}{2}\right)=\frac{1}{4}$, would be incorrect! The correct answer is $1/3$. To see why, we start by considering, for $x_i=0$ or 1, $i=1, 2$,

$$P(X_1=x_1, X_2=x_2)=\int_0^1 P(X_1=x_1, X_2=x_2 \mid \theta)\, F(\mathrm{d}\theta)$$

$$=\int_0^1 P(X_1=x_1 \mid \theta)P(X_2=x_2 \mid \theta)\, F(\mathrm{d}\theta),$$

since X_1 and X_2 are conditionally independent, given θ. Thus

$$P(X_1=x_1, X_2=x_2)=\int_0^1 \theta^{x_1}(1-\theta)^{1-x_1}\theta^{x_2}(1-\theta)^{1-x_2}F(\mathrm{d}\theta),$$

$$=\int_0^1 \theta^{x_1+x_2}(1-\theta)^{2-x_1-x_2}\mathrm{d}\theta, \tag{5.14}$$

since $F(\theta)$ is a uniform distribution on (0, 1). Simplification of the above yields,

$$P(X_1=x_1, X_2=x_2)=\frac{\Gamma(x_1+x_2+1)\Gamma(3-x_1-x_2)}{\Gamma(4)},$$

which for $x_1=x_2=1$ becomes $\frac{\Gamma(3)\Gamma(1)}{\Gamma(4)}=\frac{1}{3}$.

The incorrect answer is a consequence of invoking the assumption of independence after averaging out θ. This is flawed because X_1 and X_2 are independent conditional on θ. Unconditionally, they are not independent; they are exchangeable. This example highlights two features: the assumption of unconditional independence that is commonly made in practice underestimates the reliability of series systems and the importance of when to average out with respect to the unknown parameters. The latter point is also illustrated in Singpurwalla (2000), who considers the prediction of events in a Poisson process with an unknown parameter.

Note that if $F(\theta)$ were to be a beta distribution with parameters α and β, then (5.14) would simplify as

$$P(X_1 = x_1, X_2 = x_2) = \frac{\Gamma(\alpha+\beta)\Gamma(\alpha+x_1+x_2)\Gamma(\beta-x_1-x_2+2)}{\Gamma(\alpha+\beta+2)\Gamma(\alpha)\Gamma(\beta)}.$$

5.2.4 The Poisson as a Chance Distribution

The exponential and the Weibull, as chance distributions, were considered in sections 5.2.1 and 5.2.2 on the premise that the former encapsulated an immunity to aging whereas the latter was able to incorporate aging. A choice of these distributions can also be justified via de Finetti's theorems on exchangeable sequences. In this section, I consider the case of the Poisson as a chance distribution. The appropriateness of this choice can be argued on two grounds. The first is that the Poisson distribution can be seen as the genesis for the exponential, the Weibull, and the gamma (an several other) distributions, in a sense to be made clear soon. The second is that the Poisson also appears in the context of a de Finetti-style representation of nonnegative integer valued exchangeable sequences (section 3.2.2). Many scenarios in reliability and survival analysis entail count data of the type $X_1, X_2, \ldots,$ where each $X_i = 0, 1, 2, \ldots$ For example, X_i could denote the number of times the i-th machine experiences a failure, assuming that it gets repaired subsequent to every failure. The X_is here are a generalization of the Bernoulli case of section 5.2.3 wherein each X_i took only two values 0 and 1.

To see the relationship between the Poisson distribution and the others considered here, suppose that a unit is scheduled to operate in an environment wherein damage-causing shocks occur according to the postulates of a Poisson process (section 4.7.4). Suppose that the unit is able to sustain no shocks, so that it fails upon the occurrence of the first shock. If T denotes the time to failure of the unit and $N(t)$ the number of shocks in the time interval $[0, t]$, then

$$P(N(t) = k \mid \Lambda(t)) = \exp(-\Lambda(t))(\Lambda(t))^k / k!,$$

for $k = 0, 1, 2$; $\Lambda(t)$ is the mean value function of the process and its derivative $\lambda(t)$ is the intensity function of the process. This is the Poisson distribution with $E[N(t)] = \Lambda(t)$. Since the unit is unable to sustain any shocks

$$P(T > t) = P(N(t) = 0) = \int_{\Lambda(t)} P(N(t) = 0 \mid \Lambda(t)) F(\mathrm{d}\Lambda(t))$$

by the law of the extension of conversation. But $P(N(t) = 0 \mid \Lambda(t)) = \mathrm{e}^{-\Lambda(t)}$, and thus

$$P(T > t) = \int_{\Lambda(t)} \mathrm{e}^{-\Lambda(t)} F(\mathrm{d}\Lambda(t)). \tag{5.15}$$

If $\Lambda(t) = \lambda t$, that is, the Poisson process is homogenous then

$$P(T > t) = \int_{\theta} \mathrm{e}^{-\lambda t} F(\mathrm{d}\lambda),$$

a relationship considered by us in section 5.2.1. If $\Lambda(t) = \exp(-(t/\gamma)^\beta)$, then

$$P(T > t) = \int_{\gamma, \beta} \mathrm{e}^{-(t/\gamma)^\beta} F(\mathrm{d}\gamma, \mathrm{d}\beta) \tag{5.16}$$

the scenario considered by us in section 5.2.2.

Suppose now, that the unit in question can withstand at most K shocks, where K takes values $k=0, 1, 2, \ldots$, then

$$P(T>t)=\sum_{k=0}^{\infty} P(T>t \mid k)\pi(k), \tag{5.17}$$

where $\pi(k)=P(K=k)$. However,

$$\begin{aligned} P(T>t \mid k) &= \int_{\Lambda(t)} P(T>t \mid k, \Lambda(t))F(\mathrm{d}\Lambda(t)) \\ &= \int_{\Lambda(t)} \sum_{j=0}^{k} \frac{\mathrm{e}^{-\Lambda(t)}(\Lambda(t))^j}{j!} F(\mathrm{d}\Lambda(t)), \end{aligned} \tag{5.18}$$

since $P(T>t \mid k, \Lambda(t))=P(N(t)\leq k \mid \Lambda(t))$ and the latter is prescribed by a Poisson distribution. Thus

$$P(T>t)=\sum_{k=0}^{\infty} \int_{\Lambda(t)} \sum_{j=0}^{k} \frac{\mathrm{e}^{-\Lambda(t)}(\Lambda(t))^j}{j!} F(\mathrm{d}\Lambda(t))\pi(k). \tag{5.19}$$

When the Poisson process is homogeneous, $\Lambda(t)=\lambda t$, and $P(T>t \mid k, \Lambda(t))$ becomes

$$P(T>t \mid k, \lambda t)=\sum_{j=0}^{k} \frac{\mathrm{e}^{-\lambda t}(\lambda t)^j}{j!}, \tag{5.20}$$

which is a gamma distribution with scale λ and shape k. Its density at t is given as (also see section 3.2.2)

$$F(\mathrm{d}t \mid k, \lambda)=\frac{\mathrm{e}^{-\lambda t}(\lambda t)^{k-1}\lambda}{(k-1)!}.$$

Thus we can see that the three chance distributions in question, namely, the exponential, the gamma and the Weibull, have their genesis in the Poisson process and the Poisson distribution. Choosing different forms for $\Lambda(t)$ motivates chance distributions other than the three mentioned above. Recall that $\Lambda(t)$ must be non-decreasing in t. Its derivative $\lambda(t)$ could, however, take any meaningful form, including the possibility that $\{\lambda(t); t\geq 0\}$ be a stochastic process. When such is the case, the underlying Poisson process is also known as a **doubly stochastic Poisson process** or the **Cox process**, after David Cox who may have been the first to propose it, and $\{\lambda(t); t\geq 0\}$ is known as the **intensity process**. When $\lambda(t)$ is described by a Markov process (section 7.3.3), the resulting model for $N(t)$ is known as a hidden Markov model. The above processes are often used for describing software failures (cf. Singpurwalla and Wilson, 1999). More about these processes is detailed later, in Chapters 7 and 8 on point process models. For now I concentrate on the issue of specifying prior distributions for $\Lambda(t)$.

Prior for the Mean Value Function of a Poisson Process

In Campódonico and Singpurwalla (1995) a strategy for inducing a prior distribution on the parameters of the functional form which describes $\Lambda(t)$ based on an elicited joint prior on $\Lambda(t_i), i=1,\ldots,n$, is proposed. If the functional form of $\Lambda(t)$ entails two parameters then $n=2$;

in general, n equals the number of parameters used to specify $\Lambda(t)$. Details of the elicitation process are given in section 5.3.1. For $n=2$, with $\mu_i = \Lambda(t_i)$, $i = 1, 2$, and $\mu_1 \leq \mu_2$, the joint prior on μ_1, μ_2, $\pi(\mu_1, \mu_2; \bullet)$ can be obtained. It can be shown that this prior is proportional to

$$\prod_{i=1}^{2} \frac{\exp\left\{-\frac{1}{2}\left(\frac{\mu_i - l_i}{r_i}\right)^2\right\}}{1 - \Phi\left(\frac{n_i - \mu_i}{r_i}\right)} \cdot \frac{\exp\left\{-\frac{1}{2}\left(\frac{s_2 - s_1}{\mu_2 - \mu_1}\right)^2\right\}}{1 - \Phi\left(-\frac{s_1}{\mu_2 - \mu_1}\right)(\mu_2 - \mu_1)} \cdot \frac{\exp\left(-\frac{s_1}{\mu_2 - \mu_1}\right)}{\mu_2 - \mu_1}; \tag{5.21}$$

l_i, n_i, r_i and s_i, $i = 1, 2$, are parameters whose interpretation is given at the end of section 5.3.1. As before $\Phi(x) = \int_{-\infty}^{x} \frac{1}{\sqrt{2\pi}} e^{-u^2/2} du$.

Once $\pi(\mu_1, \mu_2; \bullet)$ is specified, we are able to address the likes of (5.15)–(5.19). However, to do so, we must write $\Lambda(t_i)$ in terms of μ_i, $i = 1, 2$. For example, in the context of (5.16), $\mu_i = (t_i/\gamma)^{\beta}$, $i = 1, 2$; this when solved gives

$$\gamma = T_1 \mu_1^{-(\ln(t_1/t_2)/\ln(\mu_1/\mu_2))} \quad \text{and} \quad \beta = \frac{\ln(\mu_1/\mu_2)}{\ln(t_1/t_2)},$$

and $\pi(\mu_1, \mu_2; \bullet)$ gets used in place of $F(d\gamma, d\beta)$.

The cumbersome nature of (5.21) makes computations difficult; this is a disadvantage of proper priors that are formally elicited. A computer code for numerically addressing the computations entailed here is in Campódonico (1993).

5.2.5 The Generalized Gamma as a Chance Distribution

Flexibility in the choice of a chance distribution can be achieved by considering the generalized gamma family of distributions. This family entails four parameters, two for the shape, one for the scale and one for the threshold. Here, the probability density at t is of the form

$$\frac{\beta}{\Gamma(\nu)} \frac{(t-\mu)^{\beta\nu - 1}}{\gamma^{\beta\nu}} \exp\left(-\left(\frac{t-\mu}{\beta}\right)^{\beta}\right)$$

for $t > \mu$, γ, β, $\nu > 0$ and $\mu > 0$.

The shape of this distribution is determined by both β and ν; γ is a scale parameter, and μ the threshold parameter. The model assumes that no failures can occur before μ; consequently μ is of interest when setting warranties. The generalized gamma reduces to an exponential when $\beta = \nu = 1$, a gamma with (without) a threshold when $\beta = 1$ (and $\mu = 0$), a Weibull with (without) a threshold when $\nu = 1$ (and $\mu = 0$), and a lognormal when $\nu \to \infty$.

The flexibility of this chance distribution is a consequence of it having several parameters. This, however, is also its weakness, because now we need to elicit a 4-variate prior distribution, a task that is not trivial. As a consequence of this difficulty, Upadhyay and Smith (1994) assume independent priors for each of the four parameters, a gamma for β and ν, a uniform for μ, and an improper prior that is proportional to $1/\gamma$ for the parameter γ. This choice of priors is arbitrary; however, the authors' objective was not to discuss the selection of a meaningful subjective prior. Rather, their aim was to describe the use of computer-intensive techniques for analyzing failure data that is meaningfully described by a generalized gamma distribution. More about the use of such (objective) priors is said later in section 5.3.2, and the computational technique alluded to above will be overviewed in Appendix A, in the context of the material of section 5.4.5.

5.2.6 The Inverse Gaussian as a Chance Distribution

The Inverse Gaussian (IG) has been introduced before, in section 4.4, as an alternative to the lognormal distribution. Its attractiveness stems from the fact that it is better able to describe situations involving early failures, and from a structural point of view, an IG is the distribution of the first passage (or hitting) time of a Brownian motion process to a threshold (or a barrier). More about the role of this process in modeling lifetimes is said later, in section 7.3.

Given the parameters μ and ϕ, the density at t of $P(T > t \mid \mu, \phi)$ has been given in section 4.4. Betró and Rotondi (1991) illustrate the behavior of this density for different values of the shape parameter ϕ; recall that μ is a scale parameter. For small values of $\phi \geq 1$, the density is skewed to the right, becoming more and more symmetric as ϕ increases to say 50 or so. For values of $\phi < 1$, the density tends to be sharply peaked near the origin making the IG a suitable chance distribution for scenarios involving the predominance of early failures.

A joint prior distribution for the parameters μ and ϕ that is proper has been proposed by Betrò and Rotondi (1991). Specifically, given ϕ, the prior density at μ is also an IG with parameters η and ω; it is of the form

$$\left(\frac{\eta\phi\omega}{2\pi}\right)^{\frac{1}{2}} \mu^{-\frac{3}{2}} e^{\phi\omega} \exp\left\{-\frac{\phi\omega}{2}\left(\frac{\eta}{\mu}+\frac{\mu}{\eta}\right)\right\}.$$

The prior of ϕ is taken to be a gamma distribution with scale α and shape $\gamma > 1$, so that the density at ϕ is of the form

$$\frac{\alpha^{\gamma}}{\Gamma(\gamma)}\phi^{\gamma-1}e^{-\alpha\phi}, \quad \text{for } \alpha > 0.$$

The joint prior density at μ and ϕ is the product of the above two density functions. Besides propriety (section 5.2.3) this choice for a prior has the attractive feature of resulting in closed form expressions for the mean and the variance of μ; specifically, the former turns out to be η and the latter $\eta^2\alpha/\omega(\gamma-1)$. More important, the posterior distribution of μ and ϕ also turns out to be of a computable form (section 5.4.6).

5.3 PRIOR DISTRIBUTIONS IN CHANCE DISTRIBUTIONS

From a purely subjectivistic Bayesian point of view, all prior distributions, be they on unobservable parameters or on the observable unknowns, should be proper and should truthfully reflect the informed opinion of the analyst and any subject matter specialists that the analyst may consult. Furthermore, the priors should not depend on the nature of the experiment that generates the observables. For example, in the context of Bernoulli trials, the prior on θ should not be based on whether the experiment to be performed is binomial or negative binomial. Whereas such subjective priors are deemed appropriate in the context of problems involving personal decision making, they are often objected to by individuals involved in matters of public discourse pertaining to issues of government policy and safety. Subjective priors are also of concern in adversarial scenarios involving litigation and the judiciary and when dealing with purely scientific investigations, like the assessment of universal constants; e.g. the speed of light. It is because of such concerns that the so-called 'objective priors', like the noninformative and the reference, have come into being. The purpose of this section is twofold; the first is to discuss some systematic approaches for eliciting and coding expert testimonies on survival times and number of events, and the second is to articulate and to overview some key ideas and notions about objective (or default) priors and the controversies that surround them.

5.3.1 Eliciting Prior Distributions via Expert Testimonies

In section 5.2, I have considered four chance distributions as points of discussion: the exponential, the Weibull, the Bernoulli and the Poisson. Since the exponential is a special case of the Weibull, it suffices to focus on the latter, and this is what we will do. The Bernoulli case has been amply covered in section 5.2.3 so that need for discussing methods of eliciting a beta prior for its parameter θ is not too pressing. However, those interested in pursuing this topic further may consult Chaloner and Duncan (1983), or Chen and Singpurwalla (1996); the latter consider a context-dependent scenario involving higher reliability items. What therefore remains to be discussed is the case of the Weibull and the Poisson. In the case of the Weibull, I consider an elicitation of $M(T)$, the median life (section 5.2.2). In the case of the Poisson, I consider an elicitation of $\Lambda(T)$, the mean value function.

Elicitation for the Weibull Distribution

Recall that $M(T) = \gamma \exp(c/\beta)$ where $c = \ln \ln 2$, and γ and β are parameters of the Weibull distribution. The approach given below is based on Singpurwalla (1988b), who adopted, for problems in reliability, Lindley's (1983) general plan for eliciting, modulating and coding expert testimony.

Suppose that an expert $\mathcal{E}$ conceptualizes the uncertainty about $M(T)$ via some distribution with mean m and standard deviation s, independent of what $\mathcal{E}$ thinks about β. That is, in $\mathcal{E}$'s view 50% of the units similar to the one of interest will fail by m; s is a measure of $\mathcal{E}$'s conviction in specifying m. Suppose that $\mathcal{E}$ declares the m and the s to an analyst $\mathcal{A}$ as $\mathcal{E}$'s expert testimony about $M(T)$.

$\mathcal{A}$ uses m and s to construct a likelihood function; the likelihood function incorporates the expertise of $\mathcal{E}$ as perceived by $\mathcal{A}$. Specifically, $\mathcal{A}$'s likelihood for $M(T)$, with m and s specified by $\mathcal{E}$, is of a Gaussian shape with a mode at $(m - a)/b$ and a spread reflected by $\delta s/b$. The constants a, b, and γ are specified by $\mathcal{A}$; they reflect $\mathcal{A}$'s view of $\mathcal{E}$'s bias and precision in declaring m and s. Specifically,

(i) If $b = 1$, then a is a bias term expressing $\mathcal{A}$'s view that $\mathcal{E}$ overestimates $M(T)$ by an amount a.
(ii) If $\mathcal{A}$ feels that $\mathcal{E}$ overestimates (underestimates) $M(T)$ by 10% then $a = 0$ and $b = 1.1(0.9)$.
(iii) If $\mathcal{A}$ feels that $\mathcal{E}$ exaggerates (is overcautious about) the precision s, then $\delta > (<)1$.

If $\mathcal{A}$ has full faith in the expertise of the expert, then $a = 0$ and $b = \delta = 1$. Figure 5.2 illustrates this likelihood.

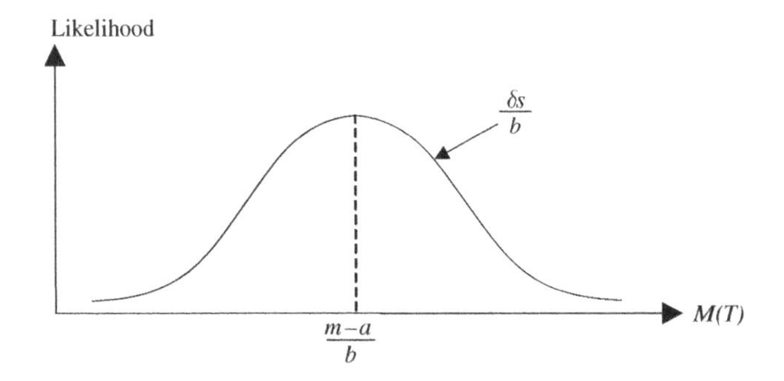

Figure 5.2 $\mathcal{A}$'s likelihood for $M(T)$ for fixed m and s.

Suppose further that s alone provides no information to $\mathcal{A}$ about $M(T)$, and that $\mathcal{A}$'s knowledge about $M(T)$ is weak in comparison with $\mathcal{E}$'s input. The first assumption implies that $\mathcal{A}$'s likelihood of $M(T)$ for a fixed s is a constant. The second assumption implies that $P(M(T))$, $\mathcal{A}$'s prior for $M(T)$, is flat in the region where the likelihood is appreciable. With the above in place, $\mathcal{A}$ invokes Bayes' law to write:

$$
\begin{aligned}
P\,(M(T);\, m,\, s,\, p) &\propto \mathcal{L}(M(T);\, m,\, s,\, p)P(M(T)) \\
&\propto \mathcal{L}(M(T);\, m,\, s,\, p),
\end{aligned}
$$

since $P(M(T))$ is flat where the likelihood is appreciable; the constant p denotes the vector of $\mathcal{A}$'s inputs (a, b, δ). The genesis of $\mathcal{L}(M(T); m, s, p)$ is the probability

$$
P(\mathcal{M}, \mathcal{S} \mid M(T);\, p) = P(\mathcal{M} \mid \mathcal{S}, M(T);\, p)P(\mathcal{S} \mid M(T);\, p),
$$

where $\mathcal{M}$ and $\mathcal{S}$ are viewed by $\mathcal{A}$ as random quantities whose realizations are m and s respectively (section 2.4.3). Since $\mathcal{M}$ and $\mathcal{S}$ have been revealed as m and s, respectively, $P(\mathcal{M} \mid \mathcal{S}, M(T); p)$ becomes $\mathcal{L}(M(T); m, s, p)$ and $P(\mathcal{S} \mid M(T); p)$ becomes $\mathcal{L}(M(T); s, p)$. But s by itself provides no information to $\mathcal{A}$ about $M(T)$; consequently, $\mathcal{L}(M(T); s, p)$ is a constant, and $\mathcal{L}(M(T); m, s, p)$ has the bell-shaped Gaussian form of Figure 5.2. Thus

$$
P\,(M(T);\, m,\, s,\, p) \propto \mathcal{L}(M(T);\, m,\, s,\, p);
$$

this results in the claim that:

$\mathcal{A}$ describes the uncertainty about $M(T)$ via a Gaussian distribution with mean $\mu = (m - a)/b$ and standard deviation $\sigma = \delta s/b$.

The above claim would be satisfactory if $M(T)$ was not restricted to be non negative. Thus, we need to truncate the above Gaussian distribution at 0, making $\mathcal{A}$'s probability density of the median at $M(T)$ of the form

$$
\pi(M(T);\, \mu,\, \sigma) = \frac{K}{\sqrt{2\pi}\sigma} \exp\left(-\frac{1}{2}\left(\frac{M(T) - \mu}{\sigma}\right)^2\right), \quad 0 < M(T) < \infty
$$

$$
\text{where } K = \left(\int_{-\mu/\sigma}^{\infty} \frac{1}{\sqrt{2\pi}} \exp\left(-\frac{u^2}{2}\right) \mathrm{d}u\right)^{-1}; \tag{5.22}
$$

$K \approx 1$ when $\mu \geq 3\sigma$.

Once $\mathcal{A}$ arrives upon $\pi(M(T); \mu, \sigma)$ and a gamma prior for β, namely $\pi(\beta; \alpha, \psi)$ (section 5.2.2) $\mathcal{A}$ can induce a joint distribution for γ and β by first inducing a distribution for γ conditioned on β, say $\pi(\gamma \mid \beta; \mu, \sigma)$ and then multiplying the latter by $\pi(\beta; \alpha, \psi)$. To obtain $\pi(\gamma \mid \beta; \mu, \sigma)$, $\mathcal{A}$ invokes the relationship $M(T) = \gamma \exp(c/\beta)$ in (5.9), and using standard techniques arrives on the form:

$$
\pi(\gamma \mid \beta;\, \mu,\, \sigma) = \frac{K e^{c/\beta}}{\sqrt{2\pi}\sigma} \exp\left(-\frac{1}{2}\left(\frac{\gamma e^{c/\beta} - \mu}{\sigma}\right)^2\right),
$$

for $0 < \gamma < \infty$. Multiplying this equation by $\pi(\beta; \alpha, \psi)$ gives us an elicited joint prior for the parameters γ and β of the Weibull distribution $P\,(T > t \mid \gamma, \beta) = \exp\left(-(t/\gamma)^{\beta}\right)$. This turns out to be (5.9) of section 5.2.2.

The development so far has considered the case of a single expert $\mathcal{E}$. Extensions for the case of multiple experts is relatively straightforward, save for the fact that now $\mathcal{A}$ needs to specify, in addition to the a, b and δ for each expert, the correlations between the declared values of m and s, as perceived by $\mathcal{A}$. The details tend to get messy (cf. Lindley and Singpurwalla, 1986). For example, in the case of two experts who provide (m_1, s_1) and (m_2, s_2) as inputs, and with $\mathcal{A}$ introducing the constants $(a_i, b_i, \delta_i), i = 1, 2$, and ρ_{12}, $\mathcal{A}$'s perceived correlation between the expert testimonies m_1 and m_2, $\mathcal{A}$'s final assessment of $M(T)$ is that $M(T)$ has a Gaussian distribution with mean μ and variance σ^2, where

$$\mu = \frac{(m_1 - a_1)(\beta_1\sigma^{11} + \beta_2\sigma^{12}) + (m_2 - a_2)(\beta_1\sigma^{12} + \beta_2\sigma^{22})}{\beta_1^2\sigma^{11} + \beta_2^2\sigma^{22} + \beta_1\beta_2(\sigma^{12} + \sigma^{21})}, \quad \text{and}$$

$$\sigma = \left(\beta_1^2\sigma^{11} + \beta_2^2\sigma^{22} + \beta_1\beta_2\left(\sigma^{12} + \sigma^{21}\right)\right)^{-1/2};$$

σ^{ij} are the elements of a matrix that is the inverse of a matrix with elements $\sigma_{11} = \delta_1^2 s_1^2$, $\sigma_{22} = \delta_2^2 s_2^2$, and $\sigma_{12} = \sigma_{21} = \rho_{12}(\sigma_{11}\sigma_{22})^{1/2}$. The case of multiple experts follows along similar lines; the details are in Lindley and Singpurwalla (1986).

Whereas the discussion so far has centered around the scenario of assessing a median $M(T)$, we could, in principle, have chosen any other unknown quantity, such as the mode, the mean, or some percentile, about which the expert is able to provide testimony. The procedure for modulating and coding m and s will be the same. Indeed we could also have elicited expert testimony on the lifelength T itself. The procedure outlined above would still be applicable, since its basic premise is to use the expert testimony as data to construct a likelihood based on the expertise of the expert. The assumptions used by us, namely, the flat likelihood of $M(T)$ for a specified s, $\mathcal{A}$'s knowledge about $M(T)$ being flat where the likelihood is appreciable, and the Gaussian likelihood for $M(T)$ given m and s, are purely for technical convenience. Other forms of the likelihood could have been used. For example, Singpurwalla and Song (1986) use a likelihood based on the chi-squared distribution. The generality of the procedure, vis-à-vis its applicability to scenarios involving other chance distributions like the Poisson, can be appreciated via the work of Campodónico and Singpurwalla (1994a, 1995); an overview of which is given below.

Elicitation for the Non-homogenous Poisson Process

We follow here the general scheme used in the case of $M(T)$ for the Weibull distribution involving an analyst $\mathcal{A}$ and an expert $\mathcal{E}$. Following the notation of section 5.2.4 we consider $\mu_i = \Lambda(t_i)$, $i = 1, 2$, for $t_1 < t_2$; we are supposing that the functional form of $\Lambda(t_i)$ entails two parameters. Suppose that $\mathcal{E}$ provides $\mathcal{A}$ testimony about $\mu_1 \leq \mu_2$ via the constants m_i and s_i, $i = 1, 2$. According to $\mathcal{E}$, m_i is a measure of location and s_i a measure of scale about $\mathcal{E}$'s uncertainty about μ_i, $i = 1, 2$.

Suppose that $\mathcal{A}$ views m_i and s_i as realizations of the random quantities $\mathcal{M}_i$ and $\mathcal{S}_i$, and models their distributions based on $\mathcal{A}$'s opinions about $\mathcal{E}$'s expertise and $\mathcal{A}$'s perceived dependence between $\mathcal{M}_1$ and $\mathcal{M}_2$, and $\mathcal{S}_1$ and $\mathcal{S}_2$. These distributions will enable $\mathcal{A}$ to specify the needed likelihood functions. Since $\mu_1 \leq \mu_2$, it is reasonable to suppose that $\mathcal{M}_1$ and $\mathcal{M}_2$ will be positively dependent. Thus with m_1, m_2, s_1 and s_2 given, $\mathcal{A}$'s task is to specify $\pi(\mu_1, \mu_2; m_1, m_2, s_1, s_2)$; this, by Bayes' Law is of the form

$$\pi(\mu_1, \mu_2; \bullet) \propto \mathcal{L}(\mu_1, \mu_2; m_1, m_2, s_1, s_2)\, P(\mu_1, \mu_2),$$

where $P(\mu_1, \mu_2)$ is $\mathcal{A}$'s prior for μ_1 and μ_2.

As was true with the case of $M(T)$, $\mathcal{A}$'s main task is to specify the likelihood $\mathcal{L}(\mu_1, \mu_2; m_1, m_2, s_1, s_2)$. For this we note that the genesis of this likelihood would be the joint probability $P(\mathcal{M}_1, \mathcal{M}_2, \mathcal{S}_1, \mathcal{S}_2 \mid \mu_1, \mu_2)$ which by the multiplication rule can be factored as:

$$P(\mathcal{M}_1, \mathcal{M}_2, \mathcal{S}_1, \mathcal{S}_2 \mid \mu_1, \mu_2) = P(\mathcal{M}_2 | \mathcal{M}_1, \mathcal{S}_1, \mathcal{S}_2, \mu_1, \mu_2) \cdot P(\mathcal{S}_2 | \mathcal{M}_1, \mathcal{S}_1, \mu_1, \mu_2) \cdot P(\mathcal{M}_1 | \mathcal{S}_1, \mu_1, \mu_2) \cdot P(\mathcal{S}_1 | \mu_1, \mu_2). \tag{5.23}$$

$\mathcal{A}$ now makes a series of assumptions grouped in four classes, each class corresponding to a term in (5.23). Specifically, suppose that in $\mathcal{A}$'s view:

- Given μ_1 and $\mathcal{S}_1 = s_1$, $\mathcal{M}_1$ is independent of μ_2; thus

$$P(\mathcal{M}_1 \mid \mathcal{S}_1, \mu_1, \mu_2) = P(\mathcal{M}_1 | \mathcal{S}_1, \mu_1).$$

  Furthermore,
- $\mathcal{M}_1$ has a truncated (at zero) Gaussian distribution with mean $a + b\mu_1$ and standard deviation δs_1.

As before a, b and δ reflect $\mathcal{A}$'s view of the biases and attitudes of $\mathcal{E}$ in declaring m_1 and s_1.

- $\mathcal{M}_2$ is independent of μ_1 given μ_2, $\mathcal{M}_1 = m_1$, $\mathcal{S}_1 = s_1$, and $\mathcal{S}_2 = s_2$; thus

$$P(\mathcal{M}_2 | \mathcal{M}_1, \mathcal{S}_1, \mathcal{S}_2, \mu_1, \mu_2) = P(\mathcal{M}_2 \mid \mathcal{M}_1, \mathcal{S}_1, \mathcal{S}_2, \mu_2).$$

- $\mathcal{M}_2$ has a truncated Gaussian distribution with mean $a + b\mu_2$ and standard deviation δs_2; truncation is on the left and is at the point $m_1 + ks_1$, with k specified by $\mathcal{A}$.

The parameter k reflects $\mathcal{A}$'s view as to how cautious $\mathcal{E}$ is about discriminating between μ_1 and μ_2.

- The distribution of $\mathcal{S}_1$ is exponential with mean $(\mu_2 - \mu_1)$.

This implies that as the disparity between μ_1 and μ_2 increases, the measure of scale $\mathcal{S}_1$ also increases.

- Given $\mathcal{S}_1 = s_1, \mu_1$ and μ_2, $\mathcal{S}_2$ is independent of $\mathcal{M}_1$; thus

$$P(\mathcal{S}_2 | \mathcal{M}_1, \mathcal{S}_1, \mu_1, \mu_2) = P(\mathcal{S}_2 | \mathcal{S}_1, \mu_1, \mu_2).$$

- $\mathcal{S}_2$ has a truncated normal distribution with mean s_1 and variance $(\mu_2 - \mu_1)$; truncation is on the left and is at zero.

The last assumption above implies that $\mathcal{S}_2$ has the same mean as $\mathcal{S}_1$, but that its variance depends on the disparity between μ_1 and μ_2.

Finally, as was so in the case of $M(T)$, $\mathcal{A}$ also assumes that $P(\mu_1, \mu_2)$ is relatively constant over the range of μ_1 and μ_2 on which the likelihood of μ_1 and μ_2 is appreciable. That is, $\mathcal{A}$'s uncertainty about μ_1 and μ_2 is vague in comparison with $\mathcal{E}$'s testimony.

With $\mathcal{M}_1, \mathcal{M}_2, \mathcal{S}_1$ and $\mathcal{S}_2$ revealed as m_1, m_2, s_1 and s_2, respectively, the above four probabilities yield four likelihoods as factorizations of $\mathcal{L}(\mu_1, \mu_2; m_1, m_2, s_1, s_2)$. Specifically, $P(\mathcal{M}_1 | \mathcal{S}_1, \mu_1)$

yields $\mathcal{L}(\mu_1; m_1, s_1)$, $P(\mathcal{M}_2|\mathcal{M}_1, \mathcal{S}_1, \mathcal{S}_2, \mu_2)$ yields $\mathcal{L}(\mu_2; m_1, m_2, s_1, s_2)$, $P(\mathcal{S}_1|\mu_1, \mu_2)$ yields $\mathcal{L}(\mu_2 - \mu_1; s_1)$, and $P(\mathcal{S}_2|\mathcal{S}_1, \mu_1, \mu_2)$ yields $\mathcal{L}(\mu_2 - \mu_1; s_1, s_2)$. These likelihoods when multiplied together yield $\mathcal{L}(\mu_1, \mu_2; m_1, m_2, s_1, s_2)$, which with the assumption that $P(\mu_1, \mu_2)$ is relatively constant results in (5.21) as the joint prior of μ_1 and μ_2. The constants of this prior are: $l_i = (m_i - a)/b$, $r_i = \delta s_i/b$, $i = 1, 2$, $n_1 = -(a/b)$, and $n_2 = (m_1 + ks_1 - a)/b$. The prior incorporates the dependence between $\mathcal{M}_1$ and $\mathcal{M}_2$, but it assumes that $\mathcal{E}$'s biases and precision in declaring $\mathcal{M}_1$ and $\mathcal{S}_1$ are the same as that for $\mathcal{M}_2$ and $\mathcal{S}_2$.

Campódonico and Singpurwalla (1994a) illustrate a use of this prior for predicting the number of software failures using a commonly used Poisson process for describing software failures. In Campodónico and Singpurwalla (1995) this prior is used for predicting the number of defects in railroad tracks. In both cases, numerical methods were used and a computer code developed.

5.3.2 Using Objective (or Default) Priors

When the prior distribution is personal to an investigator, that is when the entire distribution is elicited – as was done in section 5.3.1 – or when the parameters of a suitably chosen distribution are elicited – as was done in section 5.2.1 and 5.2.2, the priors are said to be subjective. By contrast non-subjective or objective priors are those in which the personal views of any and all investigators have no role to play. Objective priors are chosen by convention or through structural rules and are to be viewed as a standard of reference. According to Dawid (1997), 'no theory which incorporates non-subjective priors can truly be called Bayesian'.

The class of objective priors includes those that claim to be non-informative priors and those that are labeled 'reference priors', though some like Bernardo (1997) maintain that there is no such thing as a non-informative prior; he claims that 'any prior reflects some form of knowledge'. Objective priors are also referred to as 'default priors' or 'neutral priors'. Such priors have been introduced by us in section 2.7.3 in the context of testing a sharp null hypothesis about the mean of an exponential distribution, and in section 5.2.3 in the context of Bernoulli trials. It is often the case that objective priors suffer from impropriety, are dependent on the probability model that is assumed and could result in posteriors that are also improper (cf. Casella, 1996). The purpose of this section is to overview the nature of such priors and to discuss arguments used to rationalize some of their unattractive features. In the sequel, I also present the kind of improper priors that have been proposed for some of the chance distributions of section 5.2.

Historical Background

Non-subjective priors were used by default in the works of Bayes (1763) and Laplace (1812). Bayes used a uniform prior for the Bernoulli θ in a binomial setting, and Laplace used an improper uniform prior for the mean of a Gaussian distribution. During those times no one considered using priors that were different from the uniform. The rationale was that any prior other than the uniform would reflect specific knowledge, and in doing so would violate the principle of insufficient reason. However, in the early 1920s it was felt that the universal use of uniform priors did not make much sense. For one, a uniform prior on the Bernoulli θ does not imply a uniform prior on θ^2, and this is tantamount to claiming indifference between all values of θ but a specific knowledge about values of θ^2. Thus the uniform distribution on θ is not invariant vis-à-vis the principle of insufficient reason. Since most scientists of that day were reluctant to use personal priors in their scientific work, methods alternate to Bayes' and Laplace's were developed, and were labeled as 'objective methods of scientific indifference'; frequentist statistics is one such method. A revival of the Bayes–Laplace paradigm came about in the 1940s with the publication of Jeffreys (1946), who produced an alternative to the uniform as a non-subjective prior; this prior is now known as the **Jeffreys' prior**.

Jeffreys' General Rule and Jeffreys' Priors

Jeffreys work started with a collection of rules for several scenarios, each scenario treated separately. The simplest is the case of a finite parameter space; i.e. θ takes values $\theta_1, \theta_2, \ldots, \theta_n$, for $n < \infty$. He adhered to the principle of insufficient reason and assigned a probability $1/n$ to each value of θ. Jeffreys then considered the case of bounded intervals so that $\theta \in [a, b]$ for constants $a < b$, $a \geq 0$, $b < \infty$. The prior density in this case was taken to be the uniform over $[a, b]$. When the parameter space was the interval $(-\infty, +\infty)$, as is the case for a Gaussian mean, Jeffreys advocated that the prior density be constant over $(-\infty, +\infty)$. The above priors were in keeping with Bayes–Laplace tradition of insufficient reason. Note that the second prior results in impropriety, a matter that did not appear to have been of concern to Jeffreys. When the parameter space is the interval $(0, \infty)$, as is the case with the scale parameter of some well-known chance distributions like the exponential, the Gaussian and the Weibull, Jeffreys proposed the prior $\pi(\theta) = 1/\theta$. His justification for this choice was invariance under power transformation of the scale parameter θ. That is, if $\lambda = \theta^2$, then by a change of variables technique, it is verified that $\pi(\lambda)$ is of a similar form, namely $\pi(\lambda) \propto 1/\lambda$; see Kass and Wasserman's (1996) comprehensive review of objective priors. Motivated by considerations of invariance, Jeffreys' 1946 paper proposes a 'general rule' for obtaining priors. For a probability model entailing a single parameter θ, the prior on θ is of the form

$$\pi(\theta) \propto \left(E_{X|\theta}\left[-\frac{\partial^2}{\partial \theta^2} \log p(X|\theta) \right] \right)^{1/2}, \tag{5.24}$$

where the expectation of X is with respect to the probability model $p(X|\theta)$.

Jeffreys' general rule, which yields Jeffreys' priors, is not based on intervals in which the parameter lies, and could conflict with the rules based on intervals. For example, when $p(X|\theta)$ is a binomial, the general rule results in $p(\theta) \propto \theta^{-1/2}(1-\theta)^{-1/2}$, whereas the rule based on intervals has $p(\theta) = 1$. Interestingly, Jeffreys adhered to the principle of insufficient reason for priors on bounded intervals and used $p(\theta) = 1$ for the Bernoulli θ (cf. Geisser, 1984). When $p(X|\theta)$ is the **negative binomial**, i.e. when X is the number of trials to the first success in Bernoulli outcomes, so that $p(X|\theta) = (1-\theta)^{x-1}\theta$, Jeffreys' general rule yields $\pi(\theta) = \theta^{-1/2}(1-\theta)^{-1}$ as a prior for θ; this prior is improper. The Bernoulli scenario illustrates a feature of Jeffreys' priors, namely the dependence of the prior on the model: $\theta^{-1/2}(1-\theta)^{-1/2}$ for the binomial, and $\theta^{-1/2}(1-\theta)^{-1}$ for the negative binomial. Since θ as the limit of observable zero-one sequences has a physical connotation, namely that θ is a chance (section 3.1.3), dependence of the prior for θ on the model for X makes little sense.

An extension of Jeffreys' general rule – equation (5.24), when $\theta = (\theta_1, \theta_2)$ is a vector, takes the form

$$\pi(\theta) \propto \det(I(\theta))^{1/2},$$

where $\det(\bullet)$ denotes a determinant, and $I(\theta)$ is the matrix

$$I(\theta) = E_{X|\theta}\left(-\frac{\partial^2}{\partial \theta_1 \partial \theta_2} \log p(X|\theta) \right). \tag{5.25}$$

Jeffreys observed that the results of this rule could also conflict with the results provided by the rule based on intervals. For example, with $p(X|\theta_1, \theta_2)$ a Gaussian with mean θ_1 and variance θ_2^2, the general rule gives $\pi(\theta_1, \theta_2) \propto 1/\theta_2^2$, whereas the rule based on intervals gives $\pi(\theta_1, \theta_2) \propto 1/\theta_2$; similarly for the lognormal with parameters θ_1 and θ_2^2. Jeffreys addressed this problem by stating that θ_1 and θ_2 should be judged independent a priori and the general rule applied for θ_1 given θ_2 fixed and for θ_2 given θ_1 fixed. Thus with θ_2 fixed, the general rule gives a uniform prior for θ_1 and with θ_1 fixed the general rule gives $\pi(\theta_2) = 1/\theta_2$. It is because of the above that many view Jeffreys' priors as a collection of ad hoc rules.

Reference Priors: The Berger–Bernardo Method

Prompted by the above concern, and motivated by some work of Good (1966) and Lindley (1956), Bernardo (1979a) proposed a method for constructing objective priors that he labeled 'reference priors'. Like Jeffreys' priors, the reference priors also depend on the underlying probability model. However, the rationale for this dependence is more transparent in the case of reference priors than in the case of Jeffreys' priors. To see why, we first note that Bernardo's method is based on the notion of 'missing information', a notion that is characterized by the **Kullback–Leibler distance** between the posterior and the prior densities. Specifically, for a prior $\pi(\theta)$ and its posterior $\pi(\theta; x^{(n)})$, the Kullback–Leibler distance is defined as

$$K_n\left[\pi(\theta; x^{(n)}), \pi(\theta)\right] = \int_\theta \pi(\theta; x^{(n)}) \log\left[\frac{\pi(\theta; x^{(n)})}{\pi(\theta)}\right] \mathrm{d}\theta,$$

where $x^{(n)}$ is a realization of $X^{(n)} = (X_1, \ldots, X_n)$, and the X_is are independent and identically distributed with $p(X|\theta)$ as a probability model. The quantity $K_n[\bullet]$ is to be interpreted as the gain in information about θ provided by the experiment that yields $x^{(n)}$, and $K_n^\pi[\bullet] = E(K_n[\bullet])$ is the **expected gain in information**. The expectation is with respect to the predictive distribution of $X^{(n)}$; i.e.

$$p(X^{(n)}) = \int p(X^{(n)}|\theta)\pi(\theta)\mathrm{d}\theta;$$

more about this is said later, in section 5.5.1. Bernardo's idea was to find that $\pi(\theta)$ for which $K_\infty^\pi[\bullet] = \lim_{n\to\infty} K_n^\pi[\bullet]$ is a maximum; the rational here being that $K_\infty^\pi[\bullet]$ is a measure of the missing information in the experiment. However, there is a problem with this scheme, in the sense that K_∞^π could be infinite. To circumvent this difficulty, one finds the prior π_n which maximizes $K_n^\pi[\bullet]$, and computes a posterior based on this π_n. One then finds the limit of the sequence of posteriors based on π_n; denote this limit by $\pi_\infty(\theta; x^{(n)})$. Then the reference prior is that prior which by a direct application of Bayes' law would yield $\pi_\infty(\theta; x^{(n)})$ (Berger and Bernardo, 1992). Under certain regularity conditions, the reference prior turns out to be that given by (5.24) and (5.25) for continuous parameter spaces, and a uniform prior for finite parameter spaces. Thus the aforementioned approach, now known as the Berger–Bernardo method (cf. Kass and Wasserman, 1996), provides, at least under some regularity conditions, a rationale for the construction of Jeffreys' priors. Consequently, the reference prior for the Bernoulli θ under binomial sampling is $\theta^{-1/2}(1-\theta)^{-1/2}$, and it is $\theta^{-1/2}(1-\theta)^{-1}$ under negative binomial sampling. For θ uniform on $[0, \theta]$, the reference prior turns out to be $\pi(\theta) \propto \theta^{-1}$, and the **reference posterior**, i.e. the posterior distribution resulting from a reference prior, is a Pareto distribution (Bernardo and Smith, 1994, p. 438).

When the chance distribution is a Weibull with scale parameter γ and shape parameter β, the reference prior turns out to be $\pi(\gamma, \beta) = (\gamma\beta)^{-1}$, whereas Jeffreys' rules would lead to the prior $1/\gamma$ (cf. Berger and Pericchi, 1996). With the exponential as a chance distribution, the parameter β is one, so that $\pi(\gamma) = 1/\gamma$ is both the reference and the Jeffreys' priors.

To complete our discussion of the Weibull distribution, I cite here the work of Sun (1997) and that of Green, *et al.* (1994), who considered the case of a three-parameter Weibull chance distribution with a threshold parameter μ. Specifically,

$$P(T > t|\mu, \gamma, \beta) = \exp\left(-\frac{(t-\mu)}{\gamma}\right)^\beta, \quad \text{for } \mu, \gamma, \beta > 0.$$

This distribution is deemed appropriate where T has to be logically strictly greater than zero, as is the case with diameters of tree trunks. In the context of reliability and survival analysis, a use of the three-parameter Weibull precludes the possibility of an item's failure at conception. The authors cited above, and also Sinha and Sloan (1988), assume that the three parameters are a priori independent, and assign $1/\gamma$ and $1/\beta$ as priors for γ and β, respectively. The prior for μ is assumed to be a constant over $[0, \infty)$. Whereas the choice $1/\gamma$ can be justified on the grounds that it is Jeffreys' prior, the choice $1/\beta$ for the prior on β is unconventional. All the same, the novelty of Green *et al.* is the consideration of a Bayesian analysis of threshold parameters, μ in this particular case.

When the chance distribution is Poisson, then reference priors for the mean value function of the underlying Poisson process can be induced from the reference priors for the parameters of the distribution of the time to occurrence of the first event in the Poisson process. Thus, for example, if the underlying Poisson process is homogenous with a mean value function $\Lambda(t) = t/\theta$, then $P(T > t \mid \theta) = \exp(-t/\theta)$, and the reference prior on $\Lambda(t)$ induced from the reference prior on θ, namely, $1/\theta$, is $\pi(\Lambda(t)) = 1/\Lambda(t)$. Similarly, with $\Lambda(t) = \exp(-(t/\gamma)^{\beta})$, though the calculation of $\pi(\Lambda(t))$ in this case would be more cumbersome.

Reference priors have several attractive features, namely, invariance under one-to-one transformations, enable the data to play a dominant role in the construction of the posterior distribution and generally produce posterior distributions that are proper. More important, under some regularity conditions, the reference priors coincide with Jeffreys' priors, the latter having been justified from many different viewpoints (question 20 of Bernardo, 1997). Their main disadvantage is impropriety and dependence on the underlying probability model. The latter tantamounts to a violation of the likelihood principle. On the matter of impropriety, Bernardo's claim is that non-subjective priors should not be viewed as probability distributions. Rather, they should be viewed as positive fractions which serve to produce non-subjective posteriors via a formal use of Bayes' law. The non-subjective posteriors, namely the reference posteriors, are to be used as a benchmark or a standard for performing a sensitivity analysis against other, possibly subjective priors; see answers to questions 7 and 23 in Bernardo (1997). It is crucial that reference posteriors are proper and their main purpose is to describe the inferential content of the data in scientific communication. Reference posteriors need not be used in betting or other scenarios of personal decision making because such posteriors may not be an honest reflection of 'personal beliefs'.

Other Methods for Constructing Objective Priors

Kass and Wasserman (1996) describe several other methods for constructing non-subjective priors. Noteworthy among these are the maximum entropy priors of Jaynes reviewed by Zellner (1991) and by Press (1996), Zellner and Min's (1993) maximal data information prior, priors of Chernoff (1954) and Good (1969) based on decision-theoretic arguments, priors based on game theoretic arguments of Kashyap (1971), the prior by Rissanen (1983) based on coding theory arguments, and a class of improper priors by Novick and Hall (1965) called 'indifference priors'. For the case of the Bernoulli θ, Novick and Hall's indifference prior turns out to be $\{\theta(1-\theta)\}^{-1}$, which is improper but which induces a proper posterior as soon as one trial is performed. Several of the above methods boil down to being the methods of Jeffreys.

5.4 PREDICTIVE DISTRIBUTIONS INCORPORATING FAILURE DATA

I consider here the general set-up of (5.4) and (5.7), where I recall that $\tau^{(n)} = (\tau_1, \ldots, \tau_n)$, and that τ_i is the realization of T_i, the lifelength if the i-th item. The vector $\tau^{(n)}$ constitutes one form of the observed failure data. But failure data can also be had in other forms, for example survival times, wherein all that the analyst has is knowledge of the fact that T_i was greater than some τ_i^*. A knowledge of survival times is common in the bio-medical set-up wherein subjects have a tendency to drop-off an observational study, or when the study itself is terminated at some time say τ^*. The same can also be true in scenarios involving industrial life-testing for quality control purposes. Conversely, failure data can also be had in the form $T_i < \tau_i^{**}$; that is, the analyst has knowledge of the fact that the i-th item had failed by time τ_i^{**}, but is not aware of the exact time of failure. This scenario is prevalent in industrial settings involving inspection and maintenance at times τ_i^{**}, $i = 1, \ldots, n$. Thus, by the generic term 'failure data', I mean to include observations of the kind $T_i = \tau_i$, or $\tau_i > T_i^*$, or $T_i < \tau_i^{**}$, $i = 1, \ldots, n$. Often failure data is also had in terms of outcomes of Bernoulli trials so that $X_i = x_i$, with $x_i = 1$ if $T_i > t_i$ and $x_i = 0$ otherwise, for some fixed t_i, $i = 1, \ldots, n$.

How is failure data obtained? Most often it is reported from field experience. For example, data on the failure of automobile parts is reported to the manufacturer by the automobile dealers, who in turn receive this information from the user or from maintenance and repair shops. Similarly, in the bio-medical scenarios wherein doctors and hospitals report survival and failure histories to drug manufacturers, such data are called **retrospective failure data**. However, failure data can also be obtained from carefully controlled and well-designed experiments called **clinical trials** in the bio-medical set-up, and **life-testing** in the industrial set-up. Clinical trials entail tests on human subjects and are therefore complicated to discuss because they involve ethical and moral issues (cf. Kadane and Sedransk, 1979). By contrast industrial life-testing involves tests on physical units and is thus devoid of moral issues; it is therefore easier to discuss. The same is also true of drug testing on non-human subjects such as mice and monkeys – though this too could be a moral issue! In what follows I overview some of the most commonly used design strategies for industrial life-testing and their relevance for assessing predictive distributions based on the data they produce.

5.4.1 Design Strategies for Industrial Life-testing

The best way to learn about the credible performance of an item is via retrospective data. However, for newly designed items such data are not available and so we need life-testing experiments. Such experiments provide information via failure data and this enables us to make more realistic assessments of an item's reliability. Such assessments are used for designing systems comprising the units in question, for satisfying safety requirements, or for meeting contractual obligations via acceptance sampling.

Life tests are generally conducted under tightly controlled conditions involving environmental stresses that are constant during the test or which change over time, either according to a specified pattern or randomly. The former are called **constant stress tests** and the latter, **dynamic stress tests**. Tests involving environmental conditions that change with time according to a specified pattern could be of several types: continuously increasing stress, step-stress or cyclical stress; Figure 5.2 illustrates the nature of these patterns.

With a continuously increasing stress test, we monitor the lifelength of an item that is subjected to a stress whose intensity increases with time in a manner illustrated by Figure 5.3(a). With step-stress test, the stress on an item is increased in steps at several pre-specified times $s_1, s_2, \ldots$, until the item fails or the test is stopped; this is illustrated in Figure 5.3(b). Figure 5.3(c) illustrates a cyclical stress that is sinusoidal. Such stress patterns are suitable for items experiencing vibrations due to rotation or other periodic phenomena. When the change in stress over time is random, the stress–time plots of Figure 5.3 will be the sample path of a stochastic process.

It is often the case that continuously increasing and step-stress tests are conducted to produce early failures of items that are highly reliable. For such items, obtaining failure times under conditions in which they are designed to operate would entail long waits. Therefore such tests are a form of accelerated life tests, though most often accelerated tests entail the testing of items

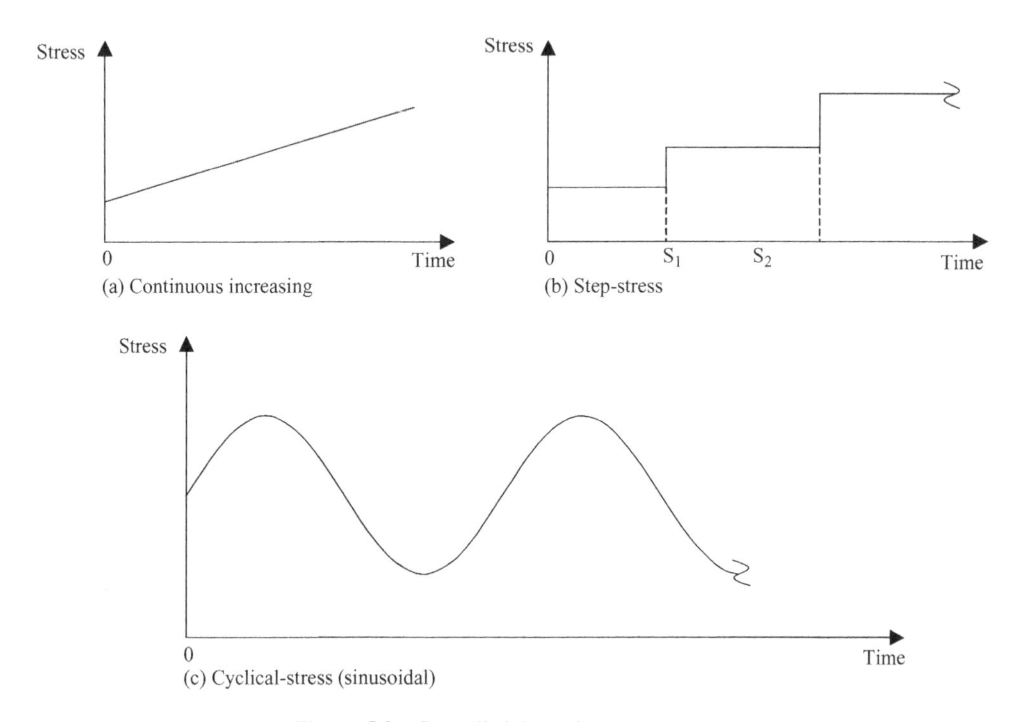

Figure 5.3 Controlled dynamic stress patterns.

under a constant but elevated (from the design) stress. The opposite of accelerated testing is decelerated testing. This form of testing entails collecting lifelength information (be it failures or survival) at low stresses to assess lifelength behavior at higher stresses. Little has been written on this topic and the assessment methodology here entails the use of some physical models of degradation together with expert testimonies (Singpurwalla, 2005).

With the above appreciation of the environmental conditions under which life tests can be conducted, the next matter pertains to the statistical aspects of the testing strategies. The simplest testing scenario is the one involving **item censoring**, also known as **Type II censoring**. Here n items are put on a test, and the test is stopped when the r-th item fails. Both n and r are specified in advance, and the r times of failure $\tau_1, \ldots, \tau_r$ are noted; when $r = n$, the test is terminated when all the n items on test fail. The choice of n and r is an important matter and involves a trade-off between the cost of testing and the utility of the added knowledge obtained via the test. With **Type I censoring**, also known as a **time truncated life test**, n items are put on test, and the test is stopped at some pre-specified time t. The number of observed failures, k, $k = 0, 1, \ldots, n$, and the times of these k failures $\tau_1, \tau_2, \ldots, \tau_k$ are noted; here k is random. Once again the choice of n and t entails a trade-off between the costs of testing and the added knowledge provided by the test. It is important to bear in mind that the added knowledge need not necessarily result in the reduction of uncertainties about lifetimes. It could happen that the results of a life-test could increase our uncertainties. There are other types of life-tests, such as a combination of Type I and Type II censoring, wherein n items are put on test and the test is stopped at time t or at time τ_r, the time of the r-th failure, whichever comes first.

In the context of life-testing, there is an important issue that arises, namely the relevance of the design of the life-test to an assessment of the predictive distribution. Specifically, given a collection of failure times, say $\tau_1, \ldots, \tau_u$, does it matter whether these data were obtained under

Type I censoring, or Type II censoring, or a combination of Type I and Type II censoring? In other words, does the 'stopping rule' – which is a prescription of when to stop the test – matter? If it does matter, then how can one proceed with the analyses of retrospective data wherein it is rare, if not impossible, to know what the stopping rule was? My claim is that, in general, the stopping rule does not matter. However, there are certain circumstances wherein it does matter; these are encapsulated under the label 'informative stopping rules'. By contrast, under the frequentist or sample theoretic paradigm, the stopping rule almost always has a role to play. In what follows I introduce the notion of a stopping rule and describe the conditions under which it is informative and deal with what it means to say that a stopping rule is noninformative.

5.4.2 Stopping Rules: Non-informative and Informative

The rule by which the sample size is determined is called a **stopping rule**. It is expressed as the probability that sampling continued as long as it did and then stopped based on the stopping rule. The said probability can be conditional on the parameters of the underlying probability model and/or the values of the observations already made. The stopping rule plays an important role when inference is based on the sample-theory approach. Specifically, point and interval estimation and tests of hypotheses depend on the stopping rule, because it is the latter which determines the sampling distribution. By contrast, under Bayesian inference, the role of the stopping rule needs to be carefully explored to ensure whether it is relevant or not. The above difference is best appreciated via the case of Bernoulli trials.

Suppose that r successes have been observed in n Bernoulli trials, with θ as the probability of success of each trial. Then the likelihood for θ with n fixed and $R=r$ as the observable is binomial, whereas it is negatively binomial if r is fixed and $N=n$ is observed. Specifically, the likelihoods for the binomial and the negative binomial cases are

$$\binom{n}{r}\theta^r(1-\theta)^{n-r} \text{ and } \binom{n-1}{r-1}\theta^r(1-\theta)^{n-r},$$

respectively. The probability models that generate these likelihoods are different, and consequently the sampling distributions for the two scenarios are different. The result is that sample-theory-based inferences for θ under the two models will not be alike. For example, the minimum variance unbiased estimate of θ, under binomial sampling, is r/n whereas that under negative binomial sampling is $(r-1)/(n-1)$. By contrast, under the Bayesian approach, inference for θ will be the same, no matter how the r and n were obtained. This is because the stopping rules for the two protocols which yield the observable data are non-informative.

To appreciate this notion I start by considering the probability model for a sequence of exactly n Bernoulli trials, $X_1, \ldots, X_n$, conditional on the parameter θ and a nuisance parameter ψ, where $P(X_i=1 \mid \theta)=\theta$, $i=1,\ldots,n$; I denote this probability by $P(X_1,\ldots,X_n \mid \theta,\psi)$. The model has two components: one for the process that generates the realizations of each X_i and the other for the process which enables us to conduct the actual trials. Specifically, let $P(1 \mid \theta,\psi)$ denote the probability that a first trial (leading to an outcome X_1) will be performed; the nuisance parameter ψ is associated with the probability model for the decision to perform a trial. In general, let $P((k+1) \mid X_1,\ldots,X_k,\theta,\psi)$ denote the probability that one more trial will be performed after the k-th trial. Then,

$$\begin{aligned}P(X_1,\ldots,X_n \mid \theta,\psi) = {} & P(1 \mid \theta,\psi)P(X_1 \mid \theta,\psi)\cdot P(2 \mid X_1,\theta,\psi)P(X_2 \mid X_1,\theta,\psi)\ldots \\ & \ldots P(n \mid X_1,\ldots,X_{n-1},\theta,\psi)\cdot P(X_n \mid X_1,\ldots,X_{n-1},\theta,\psi) \\ & \cdot\{1-P((n+1) \mid X_1,\ldots,X_n,\theta,\psi)\}.\end{aligned}$$

Since X_i is independent of $X_i, \ldots, X_{i-1}$ and ψ, given θ,

$$P(X_1, \ldots, X_n \mid \theta, \psi) = \prod_{i=1}^{n} P(X_i \mid \theta) Q(n | X^{(n)}, \theta, \psi),$$

where

$$Q\left(n \mid X^{(n)}, \theta, \psi\right) = P(1 \mid \theta, \psi) \cdot P(2 \mid X_1, \theta, \psi) \ldots P(n \mid X_1, \ldots, X_{n-1}, \theta, \psi) \cdot \{1 - P((n+1) \mid X_1, \ldots, X_n, \theta, \psi)\}.$$

Consequently, once the $X_1, \ldots, X_n$ have been observed as $x_1, \ldots, x_n$, respectively, then $\mathcal{L}(\theta, \psi; x^{(n)})$, the likelihood generated by $P(X_1, \ldots, X_n | \theta, \psi)$ for a fixed $x^{(n)} = (x_1, \ldots, x_n)$ is

$$\begin{aligned} \mathcal{L}\left(\theta, \psi; x^{(n)}\right) &= \prod_{i=1}^{n} \mathcal{L}(\theta; x_i) \cdot \mathcal{L}(\theta, \psi; n, x^{(n)}) \\ &= \theta^r (1-\theta)^{n-r} \cdot \mathcal{L}(\theta, \psi; n, x^{(n)}), \end{aligned} \tag{5.26}$$

where $r = \sum_1^n x_i$. Thus the likelihood for θ (and ψ) given by the sample $x_1, \ldots, x_n$ factors into two parts, one part pertaining to the process that generates each x_i via the term $\theta^r(1-\theta)^{n-r}$ and the other pertaining to the process that causes us to stop at the n-th trial, namely the stopping rule, via the term $\mathcal{L}(\theta, \psi; n, x^{(n)})$. The quantity $Q(n | X^{(n)}, \theta, \psi)$ with $X^{(n)} = (X_1, \ldots, X_n)$ is called the **stopping probability**.

Suppose now that the stopping probability is independent of θ, given $X^{(n)}$ and ψ. Then $Q(n \mid X^{(n)}, \theta, \psi) = Q(n \mid X^{(n)}, \psi)$, and consequently $\mathcal{L}(\theta, \psi; n, x^{(n)}) = \mathcal{L}(\psi; n, x^{(n)})$. This means that the factor $\mathcal{L}(\theta, \psi; n, x^{(n)})$ of the likelihood (5.26) will be constant in θ and will have no effect on it. Consequently, the likelihood for θ will depend only on $\theta^r(1-\theta)^{n-r}$. In such cases, the stopping rule is said to be *non-informative for* θ. The same is also true if θ and ψ are a priori independent. Otherwise the stopping rule is said to be informative.

In the context of Bernoulli trials, were n is fixed by an act of free will, then exactly n trials will be conducted so that

$$P((k+1) \mid X_1, \ldots, X_k, \theta, \psi) = \begin{cases} 1, \text{ if } k < n, \text{ and} \\ 0, \text{ if } k \geq n; \end{cases}$$

consequently, $\mathcal{L}(\theta, \psi; n, x^{(n)}) = 1$, a constant. Similarly, were r fixed by an act of free will

$$P((k+1) \mid X_1, \ldots, X_k, \theta, \psi) = \begin{cases} 1, \text{ if } \sum^k X_i < r, \text{ and} \\ 0, \text{ if } \sum^k X_i \geq r; \end{cases}$$

and here again $\mathcal{L}(\theta, \psi; n, x^{(n)}) = 1$.

Thus under either scenario, n fixed or r fixed,

$$\mathcal{L}(\theta, \psi; x^{(n)}) = \theta^r (1-\theta)^{n-r} \cdot 1$$

and Bayesian inference for θ is independent of the stopping rules, these being non-informative about θ. The essence of the above development is the feature that with Bayesian inference for the Bernoulli θ, a knowledge of the stopping process is not essential. Whereas non-informative stopping rules appear to be the norm, there are scenarios involving informative stopping rules; examples are in Raiffa and Schaifer (1961) and in Roberts (1967).

Suppose that instead of being given the actual sequence $x_1, \ldots, x_n$, we were only told that n trials resulted in r successes. Then the development proceeds along lines parallel to those leading to (5.26), with the resulting likelihood still being $\theta^r(1-\theta)^{n-r} \cdot 1$. However, if in addition to n and r, we were also told that n was fixed in advance, the resulting likelihood will be $\binom{n}{r}\theta^r(1-\theta)^{n-r} \cdot 1$. But Bayesian inference about θ is not affected by the constant $\binom{n}{r}$, which cancels out with an application of Bayes' law. Specifically,

$$\pi(\theta; n, R=r) = \frac{\binom{n}{r}\theta^r(1-\theta)^{n-r}\pi(\theta)}{\int \binom{n}{r}\theta^r(1-\theta)^{n-r}\pi(\theta)\mathrm{d}\theta} = \frac{\theta^r(1-\theta)^{n-r}\pi(\theta)}{\int \theta^r(1-\theta)^{n-r}\pi(\theta)\mathrm{d}\theta}.$$

Similarly, for the case of r fixed in advance wherein the constant preceeding $\theta^r(1-\theta)^{n-r}$ is $\binom{n-1}{r-1}$.

In general, with non-informative stopping rules, the constants associated with the stopping rule – such as 1 in the case of Bernoulli trials – and those associated with the data-generating process – such as $\binom{n}{r}$ or $\binom{n-1}{r-1}$ – combine and then cancel out with an application of Bayes' law. The resulting inference is therefore not affected by such constants. In what follows, I illustrate the workings of the aforementioned by considering the scenarios of life-testing for reliability assessment. However, to do so, it is helpful to introduce an important statistic, namely the 'total time on test statistic.'

5.4.3 The Total Time on Test

With life-testing involving the continuous monitoring of lifetimes, the notion of the total time on test plays a key role, in the case of both Bayesian and non-Bayesian inference. To appreciate this, consider the case of Type II censoring, wherein n items are put on some carefully controlled and well-designed life-test that is terminated at the time of the r-th failure. Both n and r are specified in advance and the r times of failure $\tau_{(1)} \leq \tau_{(2)} \leq \cdots \leq \tau_{(r)}$ are noted. Then, the **total time on test**, say $T(n, r)$, is defined as

$$T(n, r) = n\tau_{(1)} + (n-1)(\tau_{(2)} - \tau_{(1)}) + \cdots + (n-r+1)(\tau_{(r)} - \tau_{(r-1)}) = \sum_{1}^{r} \tau_{(i)} + (n-r)\tau_{(r)};$$

it is the total lifetime of all the n units on test, starting from time zero, until time τ_r, the time to failure of the r-th item. Observe that when $r = n$, $T(n, n) = \sum_{1}^{n} \tau_{(i)}$, the sum of the lifetimes of all the n units.

Similarly, with Type I censoring wherein n items are put on a test that is terminated at some time t, with n and t pre-specified, and k failures observed at $\tau_{(1)} \leq \tau_{(2)} \leq \cdots \leq \tau_{(k)}$, the total time on test, $T(n, t)$, is

$$T(n, t) = \sum_{1}^{n} \tau_{(i)} + (n-k)(t - \tau_{(k)});$$

here k is random. When $k = n$, $T(n, t)$ is simply the sum of lifetimes $\sum_1^n \tau_{(i)}$. When each item has its own truncation time τ_i^*, $i = 1, \ldots, n$, then $T(n, \tau_1^*, \ldots, \tau_n^*)$ the total time on test is $\sum_1^n T(1, \tau_i^*)$, where

$$T(1, \tau_i^*) = \begin{cases} \tau_i, & \text{if } T_i \leq \tau_i^* \\ \tau_i^*, & \text{if } T_i > \tau_i^*; \end{cases}$$

τ_i is the realization of T_i, the lifelength of the i-th item. When the test truncation times τ_i^* are ordered, so that $\tau_1^* \leq \tau_2^* \leq \cdots \leq \tau_n^*$, the Type I censored life-test is said to be **progressively censored**, and if the τ_i^*s are themselves random we have the case of random censoring. Inference under random censoring would entail the specification of a probability model for censoring.

Finally, if the life-test is a combination of Type I and Type II tests, also known as **hybrid testing**, the total time on test $T(n, r, t)$ is given by

$$T(n, r, t) = \begin{cases} T(n, r), & \text{if } T_{(r)} \leq t \\ T(n, t), & \text{if } T_{(r)} > t; \end{cases}$$

the quantities n, r and t are pre-specified.

Thus, no matter what testing protocol and censoring schemes are specified, a total time on test for the life-testing experiment can be easily specified. The total time on test statistic also plays a key role in the mathematical theory of reliability. For example, Barlow and Campo (1975) use it to assess the failure rates of chance distributions used in reliability and Chandra and Singpurwalla (1981) relate it to notions of income inequalities in economics. I now show that for Bayesian inference about parameters of chance distributions used in reliability all that we need to know is the total time on test; we need not know the actual failure and survival times, nor do we need to know the censoring protocols, unless of course they are informative.

5.4.4 Exponential Life-testing Procedures

Suppose that the underlying chance distribution is assumed to be an exponential with $P(T > t \mid \lambda) = \exp(-\lambda t)$ (section 5.2.1) so that (5.4) takes the form

$$P(T_{n+1} \geq t_{n+1}; \bullet) = \int_\lambda \exp(-\lambda t) F(\mathrm{d}\lambda; \bullet),$$

with $\bullet$ denoting data from a life-test, and

$$F(\mathrm{d}\lambda; \bullet) \propto \mathcal{L}(\lambda; \bullet) F(\mathrm{d}\lambda); \tag{5.27}$$

see equation (5.5). Depending on the testing protocol and censoring scheme used, the data from the life-test could arrive in one of several forms discussed in section 5.4.3. For example, under Type II censoring with n and r fixed, we would observe the r ordered failure times $\tau_{(1)} \leq \tau_{(2)} \leq \cdots \leq \tau_{(r)}$, and the likelihood of λ, $\mathcal{L}(\lambda; \bullet)$, would be of the form

$$\begin{aligned} \mathcal{L}(\lambda; \bullet) &= \frac{n!}{(n-r)!} \prod_{i=1}^{r} \lambda \mathrm{e}^{-\lambda \tau_{(i)}} \mathrm{e}^{-\lambda(n-r)\tau(r)} \\ &= \frac{n!}{(n-r)!} \lambda^r \mathrm{e}^{-\lambda T(n,r)}. \end{aligned} \tag{5.28}$$

where $T(n, r)$ is the total time on test. This likelihood is a consequence of the fact the stopping rule for the test is non-informative – an issue than can be verified using the same line of argument used for the case of Bernoulli testing with n and r specified (section 5.4.2) and the fact that a probability model for the first r order statistics $T_{(1)} \leq T_{(2)} \leq \cdots \leq T_{(r)}$ of $T_1, \ldots, T_n$ has density at $\tau_{(1)} \leq \tau_{(2)} \leq \cdots \leq \tau_{(r)}$ of the form

$$\frac{n!}{(n-r)!} \prod_{i=1}^{r} f(\tau_{(i)} \mid \lambda)(P(T > \tau_{(i)} \mid \lambda))^{n-r}, \tag{5.29}$$

where the T_is are independent (given λ) and identically distributed with distribution $P(T \geq t \mid \lambda)$ whose density at t is $f(t \mid \lambda)$. The **order statistics** of a collection of random variables $T_1, \ldots, T_n$ is an ordering of these random variables from the smallest to the largest, so that $T_{(1)} \leq T_{(2)} \leq \cdots \leq T_{(n)}$; thus $T_{(1)} = \min_i T_i$ and $T_{(n)} = \max_i T_i$. When $r = n$, (5.29) simplifies as $n! \prod_1^n f(\tau_{(i)} \mid \lambda)$, implying that the ordering of (conditionally) independent random variables renders them dependent.

Under Type I censoring, with t specified by free will, the stopping rule is again non-informative, and the likelihood of λ for an observed k (the number of failures), and $\tau_{(1)} \leq \tau_{(2)} \leq \cdots \leq \tau_{(k)} \leq t$ is of the form

$$\begin{aligned}\mathcal{L}(\lambda; \bullet) &= \frac{n!}{(n-k)!} \prod_{i=1}^{k} \lambda e^{-\lambda \tau_{(i)}} e^{-\lambda(n-k)(t-\tau_{(k)})} \\ &= \frac{n!}{(n-k)!} \lambda^k e^{-\lambda T(n,t)}.\end{aligned} \tag{5.30}$$

where $T(n, k)$ is the total time on test. The line of reasoning used to obtain (5.28) is also used here, except that now we need to consider the joint distribution of the first k order statistics out of n.

When censoring is a combination of Type I and Type II, the likelihood will depend on whether $\tau_{(r)} \leq t$ or $\tau_{(r)} > t$. In the first case it will be the product of the right-hand side of (5.28) and the probability that $T_{(r)} \leq t$, which is the distribution function of the r-th order statistic. In the second case, it will be the product of the right-hand side of (5.30) and the probability that $T_{(r)} > t$. Specifically, with hybrid testing, $\mathcal{L}(\lambda; \bullet)$ will take the form:

$$\left(\frac{n!}{(n-r)!} \lambda^r e^{-\lambda T(n,r)}\right) \left(\sum_{i=r}^{n} \binom{n}{i} \left(1 - e^{-\lambda t}\right)^i \left(e^{-\lambda(n-i)t}\right)\right), \quad \text{if } \tau_{(t)} \leq t,$$

and for $\tau_{(r)} > t$, with $k < r$ failures observed by t, it will be

$$\left(\frac{n!}{(n-k)!} \lambda^k e^{-\lambda T(n,t)}\right) \left(\sum_{i=0}^{r-1} \binom{n}{i} \left(1 - e^{-\lambda t}\right)^i \left(e^{-\lambda(n-i)t}\right)\right). \tag{5.31}$$

With retrospective data it is often the case that the censoring rule is not known, and all that we have at our disposal is n,the number of items under observation, the ordered times to failure $\tau_{(1)} \leq \tau_{(2)} \leq \cdots \leq \tau_{(k)}$, and now the time, say t; i.e. t is the time at which observation on the failure/survival process is terminated. The stopping rule will be non-informative as long as t was chosen at free will, or it was determined by either k or by $\tau_{(1)} \leq \tau_{(2)} \leq \cdots \leq \tau_{(k)}$. That is, t should not depend on λ. When such is the case, the likelihood for λ is that given by (5.30), so that with a Bayesian analysis all that matters is $T(n, t)$, the total time on test; similarly with the case of progressive censoring.

In the context of life-testing the role of informative stopping rules becomes more transparent via the scenario of random censoring. To see how, consider the case of a single item whose lifelength T has the exponential chance distribution considered. Suppose that the choice of testing a single item (to failure or until it gets censored) is done by free will. Now suppose that the censoring time Y is random with $P(Y > t \mid \mu)$ having a density $g(t \mid \mu)$ at t. Observation on the item stops when the item fails or gets censored. The parameter of interest is λ, the one associated with lifelength T; μ is a nuisance parameter. Then, the likelihood for λ, were the item to fail at t, will be governed by the probability $(\lambda e^{-\lambda t}) \cdot P(Y > t \mid \mu)$, whereas it is governed by the probability $(\mu e^{-\mu t}) \cdot P(T > t \mid \lambda)$ if the item were censored at t. The first term of the above two likelihoods define the stopping rule, which in the second case will be non-informative for λ, if μ and λ are independent; it will be informative if $\mu = \lambda$ or if μ depends on λ. As a special case of the above, suppose that Y is also exponential with a scale parameter μ; i.e. $P(Y > t \mid \mu) = e^{-\mu t}$. Then the said likelihoods are $\lambda e^{-t(\lambda+\mu)}$ and $\mu e^{-t(\lambda+\mu)}$, respectively; t is the total time on test.

With n items under observation, where n is chosen by free will, the likelihood will be the product of the above two likelihoods, the first for every item that fails and the second for every item that gets censored. Thus, for example, suppose that of the n items under test, k experience a failure at times $\tau_1, \tau_2, \ldots, \tau_k$, and the remaining $(n-k)$ get censored at times $\tau^*_{k+1}, \ldots, \tau^*_n$. Then, the likelihood of λ and μ is

$$\lambda^k \mu^{n-k} \exp\left\{-(\lambda+\mu)\left(\sum_{i=1}^{k} \tau_i + \sum_{j=k+1}^{n} \tau^*_j\right)\right\},$$

where $\left(\sum_{i=1}^{k} \tau_i + \sum_{j=k+1}^{n} \tau^*_j\right)$ is the total time on test.

If μ and λ are independent then the stopping rule given by the censoring times is non-informative for λ, and the term μ (total time on test) cancels out as a constant in an application of Bayes' law. Thus the part of the likelihood that is germane to inference about λ is

$$\lambda^k \exp\left\{-\lambda\left(\sum_{i=1}^{k} \tau_i + \sum_{j=k+1}^{n} \tau^*_j\right)\right\}. \tag{5.32a}$$

If $\mu = \lambda$, the likelihood is

$$\lambda^n \exp\left\{-2\lambda\left(\sum_{i=1}^{k} \tau_i + \sum_{j=k+1}^{n} \tau^*_j\right)\right\}, \tag{5.32b}$$

and the stopping rule contributes to inference about λ; it is therefore informative.

In any case, the role of the total time on test for inference about λ is central with regard to all the censoring schemes discussed by us.

Going back to where we left off with (5.27), suppose that $F(d\lambda)$ is the gamma density with a scale parameter ψ and a shape parameter α, where ψ and α are elicited via the approach prescribed in section 5.2; thus

$$F(d\lambda) = \frac{\psi^\alpha \lambda^{\alpha-1} e^{-\psi\lambda}}{\Gamma(\alpha)}. \tag{5.33}$$

Then for $\mathcal{L}(\lambda; \cdot)$ given by (5.28) – the case of Type II censoring – it is straightforward to verify that the posterior for λ is also a gamma density with a scale parameter $(\psi + T(n, r))$ and a shape parameter $(\alpha + n)$; i.e.

$$F(d\lambda; \bullet) = \frac{(\psi + T(n, r))^{\alpha+n} \lambda^{\alpha+n-1} e^{-\lambda(\psi+T(n,r))}}{\Gamma(\alpha+n)}. \tag{5.34}$$

Consequently, $E(\lambda; \bullet) = (\alpha + n)/(\psi + T(n, r))$, the role of n and $T(n, r)$ being to change the mean of λ from α/ψ to that given above.

Similarly, with Type I censoring, when $\mathcal{L}(\lambda; \bullet)$ is given by (5.30), the gamma prior on λ results in a gamma posterior for λ of the same form as (5.34), but with $T(n, r)$ replaced by $T(n, t)$. In general, the above result will also hold when the observed failure data is retrospective, or the testing is progressive, or the censoring is random and non-informative; all that matters is a use of the appropriate total time on test statistic. The only exception is the case of informative censoring wherein the likelihood will entail additional terms involving λ. In the case of hybrid censoring, it can be verified (cf. Gupta and Kundu, 1998) that the likelihood (equation (5.31)) can also be written as

$$\frac{n!}{(n - r^*)!}\lambda^{r^*} \exp(-\lambda S), \tag{5.35}$$

where r^* is the number of units that fail in $[0, t^*]$, with $t^* = \min(\tau_{(r)}, t)$ and S the total time on test is

$$\begin{aligned} S &= \sum_{i=1}^{r^*} \tau_i + (n - r^*)t^*, \text{ if } r^* \geq 1 \\ &= nt, \qquad\qquad\qquad\quad \text{if } r^* = 0. \end{aligned}$$

This form of the likelihood parallels that of (5.28) and (5.30), and thus with hybrid testing a result of the form given by (5.34) continues to hold with $T(n, r)$ replaced by S.

Once $F(\mathrm{d}\lambda; \bullet)$ is assessed, it can be used in (5.4) and (5.7) to provide predictions of future lifetimes. Furthermore, since λ is the model failure rate of $P(T > t \mid \lambda)$, (5.34) provides inference about λ, should this be of interest. Specifically, from (5.4), we can derive the result that

$$P(T_{n+1} \geq t_{n+1}; \bullet) = \left(\frac{\psi + T(n, r)}{\psi + T(n, r) + t_{n+1}}\right)^{\alpha+n}$$

and from (5.7) the result that

$$P(T_{n+1} \geq t_{n+1}, \ldots, T_{n+m} \geq t_{n+m}; \bullet) = \left(\frac{\psi + T(n, r)}{\psi + T(n, r) + t^*}\right)^{\alpha+n},$$

where $t^* = \sum_{i=1}^{m} t_{n+i}$. Both the predictive distributions are Pareto, the former indexed by t_{n+1}, the latter by t^*. Indexing by t^* suggests that in the scenario considered here, what matters is t^*, the total horizon of prediction. The individual times $t_{n+1}, \ldots, t_{n+m}$ do not matter; only their sum does. This feature can be seen as a manifestation of the lack of memory property of the exponential distribution and will always be true, irrespective of the prior distribution λ that is used.

Instead of assigning the gamma prior distribution $F(\mathrm{d}\lambda; \psi, \alpha)$ on λ (equation (5.33)) one may prefer to assign a prior distribution on $1/\lambda \stackrel{\text{def}}{=} \mu$, the mean time to failure. A possible choice is the inverted gamma distribution with parameters ψ and α; here for $\psi, \alpha, \theta > 0$,

$$F(\mathrm{d}\mu; \psi, \alpha) = \frac{\psi^\alpha \mu^{-(\alpha+1)} \exp(-\psi/\mu)}{\Gamma(\alpha)}.$$

The mean and variance of this distribution are $\psi/(\alpha - 1)$ and $\psi^2/(\alpha - 1)^2(\alpha - 2)$; these exist only if $\alpha > 2$. With this choice for a prior, it can be seen that (5.12) will continue to hold and that the predictive distribution of T_{n+1} is precisely the same Pareto distribution obtained when the prior on λ was a gamma distribution; i.e.

$$P(T_{n+1} \geq t; \bullet) = \left(\frac{\psi + T(n,r)}{\psi + T(n,r) + t} \right)^{\alpha+n}.$$

This is because a gamma prior on λ induces an inverted gamma prior on μ, and coherence of the probability calculus will ensure that the predictive distribution remains the same.

There are some other interesting features about the choice of the inverted gamma prior on μ. The first thing to note is that $E(T_{n+1}; \bullet)$, the mean of the predictive distribution is $(\psi + T(n,r))/(\alpha + n + 1)$ – the mean of a Pareto with parameters $(\psi + T(n,r))$ and $(\alpha + n)$. Consequently, were there to be no failure data under this set-up, both $T(n,r)$ and n will not be there so that $E(T_{n+1}; \psi, \alpha)$, the mean of the predictive distribution will be $\psi/(\alpha - 1)$. But $\psi/(\alpha - 1)$ is the mean of the inverted gamma prior on μ. Thus here we have the curious result that in the absence of any life-testing data, the mean of the prior and the predictive distributions are identical, provided that they exist; i.e. provided that $\alpha > 1$.

This completes our discussion on inference and predictions using an exponential chance distribution assuming various informative and non-informative censoring schemes. In principle, the case of the gamma and the Weibull distribution, or for that matter any other chance distribution, will follow along similar lines, the main difficulty being an assessment of multi-dimensional parameters and the computational issues that such matters will entail. Here the total time on test statistic will no more play the central role it plays in the case of the exponential distribution, and expressions having the closed form nature of (5.34) are hard to come by. In section 5.4.5 the nature of these difficulties via the case of the Weibull distribution is illustrated. However, before doing so I need to draw attention to the following important point about assessing chance distributions.

Estimated Reliability – An Approximation to the Predictive Distribution

I start by drawing attention to the feature that what has been discussed thus far pertains to a coherent assessment of predictive distributions, and the parameter λ of the underlying exponential chance distribution $\exp(-\lambda t)$. The coherent assessment of this chance distribution itself, namely the reliability function, has not been mentioned. This may seem strange because much of the non-Bayesian literature in reliability and life-testing focuses on estimating the reliability function. Why this omission?

Our response to the above question is that in practice, what matters is assessing predictive distributions of observable quantities like T, and not the predictive distribution of T conditioned on unobservable parameters like λ, which is what the reliability function is. However, if for some reason – for instance section 6.2.3 – an assessment of a chance distribution is still desired, then the best that one can hope to do is to replace λ by $E(\lambda; \bullet)$, the mean of its posterior distribution. It is important to note that there is no rule of probability that justifies this step. The closest that comes to a rationale is that $\exp(-E(\lambda; \bullet)t)$ is an approximation to the predictive distribution of T. Specifically, by an infinite series expansion of $\exp(-E(\lambda; \bullet)t)$ and of $\exp(-\lambda t)$, we can verify that the predictive distribution of T

$$\int_0^\infty \exp(-\lambda t) F(d\lambda; \bullet) \approx \exp(-E(\lambda; \bullet)t).$$

Thus our proposed estimate of the reliability function can be seen as a surrogate for the predictive distribution of T.

5.4.5 Weibull Life-testing Procedures

Suppose that $P(T > t|\lambda, \beta) = \exp(-\lambda t^{\beta})$, for $\lambda, \beta > 0$ and $t \geq 0$, and to keep matters simple we focus attention on the case of Type II censoring. Thus given the r ordered failure times $\tau_{(1)} \leq \tau_{(2)} \leq \cdots \leq \tau_{(r)}$, the likelihood of λ and β is

$$\mathcal{L}(\lambda, \beta; \bullet) = \frac{n!}{(n-r)!} (\lambda\beta)^r \exp\left[-\lambda\left(\sum_{i=1}^{r} \tau_{(i)}^{\beta} + (n-r)\,\tau_{(r)}^{\beta}\right)\right] \prod_{i=1}^{r} \tau_{(i)}^{\beta}, \tag{5.36}$$

which when re-parameterized as $\mathcal{L}(\lambda, \gamma; \bullet)$ with $\lambda = \gamma^{-\beta}$, and when multiplied by a prior such as that given by (5.9) yields a posterior $F(\mathrm{d}\lambda, \mathrm{d}\gamma; \bullet)$. This posterior when used in (5.4) and (5.7) provides predictions of future lifetimes. The computational challenge that such a scenario involving the integral of a non-linear function of the parameters is daunting. The situation is no better if other forms of censoring were to be considered or even with the absence of any censoring. The material in Singpurwalla (1988b), which is based on a conventional numerical perspective, provides a feel of the ensuing difficulties. It is because of such obstacles that some Bayesians have abandoned a use of proper, subjectively induced, priors and have then appealed to computer-intensive approaches such as simulation by the Markov Chain Monte Carlo (MCMC) (cf. Upadhyay, Vasishta and Smith, 2001). The practical importance of such methods mandates that some of these be overviewed and this is done in the appendix; the remainder of this section is devoted to an application of the MCMC method to the scenario at hand. Readers interested in a better appreciation of the material that follows are urged to first consult Appendix A.

Inference for the Weibull via MCMC

In the interest of generality, suppose that the Weibull chance distribution considered here has three parameters, with μ as a threshold; recall (section 5.3.2) that this was the model considered by Green *et al.* (1994). Suppose further, that in (5.36), $n = r$ so that there is no censoring; from an MCMC point of view this is no more a limitation (Appendix A). Then, with the re-parameterization $\lambda = \gamma^{-\beta}$, the likelihood of (5.36) becomes

$$\frac{\beta^n}{\gamma^{n\beta}} \prod_{i=1}^{n} \left(\tau_{(i)} - \mu\right)^{\beta} \exp\left(\sum_{i=1}^{n} \left(\frac{\tau_{(i)} - \mu}{\gamma}\right)^{\beta}\right), \tag{5.37}$$

so that with the joint prior $F(\mathrm{d}\gamma, \mathrm{d}\beta, \mathrm{d}\mu) \propto (\gamma\beta)^{-1}$ of Green *et al.* (1994) (also Sinha and Sloan, 1988), the posterior is proportional to

$$\frac{\beta^{n-1}}{\gamma^{n\beta+1}} \prod_{i=1}^{n} \left(\tau_{(i)} - \mu\right)^{\beta-1} \exp\left(-\sum_{i=1}^{n} \left(\frac{\tau_{(i)} - \mu}{\gamma}\right)^{\beta}\right). \tag{5.38}$$

We need to emphasize that the prior mentioned above is improper, and from a purely subjective Bayesian point of view, it should not be entertained. Besides convenience, its virtue is that it facilitates a straightforward application of the Gibbs sampler; my aim in introducing this development is to ensure completeness of coverage.

The full conditionals generated by the posterior of (5.38), with $\boldsymbol{\tau} = (\tau_{(1)}, \ldots, \tau_{(n)})$, can be written as:

$$p(\gamma|\beta, \mu; \boldsymbol{\tau}) \propto \frac{1}{\gamma^{n\beta+1}} \exp\left(-\sum_{i=1}^{n} \left(\frac{\tau_{(i)} - \mu}{\gamma}\right)^{\beta}\right),$$

$$p(\beta \mid \gamma, \mu; \boldsymbol{\tau}) \propto \frac{\beta^{n-1}}{\gamma^{n\beta}} \prod_{i=1}^{n} \left(\tau_{(i)} - \mu\right)^{\beta} \exp\left(-\sum_{i=1}^{n} \left(\frac{\tau_{(i)} - \mu}{\gamma}\right)^{\beta}\right), \quad \text{and}$$

$$p(\mu \mid \gamma, \beta; \boldsymbol{\tau}) \propto \prod_{i=1}^{n} \left(\tau_{(i)} - \mu\right)^{\beta-1} \exp\left(-\sum_{i=1}^{n} \left(\frac{\tau_{(i)} - \mu}{\gamma}\right)^{\beta}\right).$$

Generating random samples from the first of the above three conditionals is straightforward, since $p(\gamma \mid \beta, \mu; \tau)$ is, de facto, a gamma density. Generating random samples from $p(\beta \mid \gamma, \mu; \tau)$ is via a procedure due to Gilks and Wild (1992), whereas those from $p(\mu \mid \gamma, \beta; \tau)$ is via a scheme proposed by Upadhyay, Vasishta and Smith (2001). This latter paper is noteworthy because it illustrates the above approach by considering some data on the fatigue life of bearings.

To conclude, the success of the Gibbs sampling approach for analyzing failure-time data assumed to be meaningfully described by a Weibull chance distribution depends on a key feature, namely the use of a prior that facilitates a generation of random samples from the conditionals of the posterior.

5.4.6 Life-testing Under the Generalized Gamma and the Inverse Gaussian

The generalized gamma, as a chance distribution, was introduced in section 5.2.5. This distribution is attractive because it encompasses the exponential, the gamma, the Weibull and the lognormal as special cases. Thus, in principle, a Bayesian analysis of life-test data under the aegis of the generalized gamma would be very encompassing. However, the subjective assessment of a joint prior on the four parameters of this distribution is a demanding task. Furthermore, even if the above were to be meaningfully done, the computational challenge would pose a serious obstacle. It is because of the above reasons that Upadhyay, Vasishta and Smith (2001) consider the following three-parameter version of this distribution with density at t of the form

$$\frac{\beta}{\Gamma(\nu)} \frac{t^{\beta\nu-1}}{\gamma^{\beta\nu}} \exp\left(-\left(\frac{t}{\gamma}\right)^{\beta}\right), \quad \text{for } t > 0,\ \gamma, \beta, \nu > 0,$$

and assign the following independent priors:

$$p(\gamma) \propto \frac{1}{\gamma}$$

$$p(\beta) \propto \mathcal{G}(a_1, b_1), \quad \text{and}$$

$$p(\nu) \propto \mathcal{G}(a_2, b_2), \quad \text{where}$$

$\mathcal{G}(a, b)$ denotes a gamma density with shape parameter a and scale parameter b. Assuming a complete (uncensored) sample of size n from this distribution, the likelihood can be written out in a straightforward manner, and this when multiplied by the above prior yields a posterior whose full conditionals, for $\boldsymbol{\tau} = (\tau_{(1)}, \ldots, \tau_{(n)})$, are:

$$p(\gamma \mid \beta, \nu; \boldsymbol{\tau}) = \frac{1}{\gamma^{n\beta\upsilon+1}} \exp\left(-\sum_{i=1}^{n} \left(\frac{\tau_{(i)}}{\gamma}\right)^{\beta}\right),$$

$$p(\beta \mid \gamma, \nu; \boldsymbol{\tau}) = \frac{\beta^{n+a_1-1}}{\gamma^{n\beta\upsilon}} \prod_{i=1}^{n} \tau_{(i)}^{\beta\upsilon} \exp\left(-\left(\frac{\beta}{b_1} + \sum_{i=1}^{n} \left(\frac{\tau_{(i)}}{\gamma}\right)^{\beta}\right)\right), \quad \text{and}$$

$$p(\nu \mid \gamma, \beta; \boldsymbol{\tau}) = \left(\frac{1}{\Gamma(\nu)}\right)^{n} \frac{\nu^{a_2-1}}{\gamma^{n\beta\upsilon}} \prod_{i=1}^{n} \tau_{(i)}^{\beta\upsilon} \exp\left(-\left(\frac{\nu}{b_2}\right)\right).$$

Generating random samples from the first of the above three conditionals is straightforward because the distribution is de facto a gamma. For the other two conditionals, the procedure of Gilks and Wild (1992) can be invoked to generate the required samples. Upadhyay, Vasishta and Smith (2001) illustrate the workings of this procedure via some failure data on ball-bearings. Furthermore, they also discuss the matter of model comparisons and model validity via simulated predictive distributions. Whereas aspects of this work cannot be said as being subjective Bayesian, the paper of Upadhyay, Vasishta and Smith (2001) represents a significant forward leap in approaches for analyzing failure time data.

When $\boldsymbol{\tau} = (\tau_{(1)}, \ldots, \tau_{(n)})$ is a complete sample of size n from an Inverse Gaussian distribution with a joint prior on μ and ϕ of the form given in section 5.2.6, the joint posterior at μ and ϕ is:

$$C\phi^{(\nu-1)/2}\mu^{(n-3)/2}\exp\left\{-\phi\left(\frac{\nu_1}{2\mu} - \nu_2 + \frac{\nu_3\mu}{2}\right)\right\},$$

where $\nu = n - 2\gamma$, $\nu_1 = n\overline{\tau} + \omega\eta$, $\nu_2 = n + \omega - \alpha$, $\nu_3 = n\overline{\overline{\tau}} + \omega/\eta$, with

$$\overline{\tau} = \frac{1}{n}\sum_{i=1}^{n}\tau_{(i)}, \quad \overline{\overline{\tau}} = \frac{1}{n}\sum_{i=1}^{n}\left(\tau_{(i)}\right)^{-1}, \quad \text{and}$$

C is a normalizing constant that can be iteratively evaluated in a manner that parallels an evaluation of the percentiles of a 'Student-t' distribution. The details are in Betro and Rotondi (1991). The parameter μ being the mean of the IG distribution conveys a practical import. Its posterior distribution – obtained by integrating out the parameter ϕ in the joint posterior of μ and ϕ – turns out to be

$$C\Gamma\left(\frac{1}{2}(\nu+1)\right)\frac{\mu^{(n-3)/2}}{\left\{\frac{\nu_1}{2\mu} - \nu_2 + \frac{\nu_3\mu}{2}\right\}^{(\nu+1)/2}},$$

where $\Gamma(\bullet)$ is the gamma function.

5.4.7 Bernoulli Life-testing Procedures

The Bernoulli, as a chance distribution for reliability studies was motivated in section 5.2.3, and priors for the Bernoulli parameters were also discussed there. In section 5.4.2 the notion of informative and non-informative stopping rules was illustrated with Bernoulli trials as a motivating scenario. Proper priors for the Bernoulli parameter θ are given by (5.11) and (5.12), and in what follows attention is focused on the former. Thus if r successes are observed in n Bernoulli trials, each having θ as the probability of success, (5.4) takes the form

$$P(X_{n+1} = x; (n, r)) = \int_0^1 \theta^x(1-\theta)^{1-x}F(\mathrm{d}\theta; (n, r)), \tag{5.39}$$

$$\text{with} \quad F(\mathrm{d}\theta; (n, r)) \propto \mathcal{L}(\theta; (n, r))F(\mathrm{d}\theta) \tag{5.40}$$

where $F(\mathrm{d}\theta)$ is given by (5.11) and $\mathcal{L}(\theta; (n, r))$ is proportional to $\theta^r(1-\theta)^{n-r}$; $x = 1(0)$ for success (failure). It can now be seen (details left as an exercise) that (with α and β suppressed) (5.39) simplifies as:

$$P(X_{n+1} = x; (n, r)) = \frac{\Gamma(\alpha+\beta+n)}{\Gamma(\alpha+r)\Gamma(\beta+n-r)}\frac{\Gamma(x+\alpha+r)\Gamma(\beta+n-r+1-x)}{\Gamma(\alpha+\beta+n+1)}, \tag{5.41}$$

and that

$$F(\mathrm{d}\theta;\,(n,r)) = \frac{\Gamma(\alpha+\beta+n)}{\Gamma(\alpha+r)\Gamma(\beta+n-r)}\theta^{\alpha+r-1}\theta^{\beta+n-r-1}. \tag{5.42}$$

The predictive distribution (equation (5.41)) is a 'beta binomial distribution', and the posterior distribution of θ (equation (5.42)) is again a beta distribution.

The family of beta distributions is said to be a **natural conjugate family** under Bernoulli sampling, because both the prior and the posterior distributions are members of this (same) family. Similarly, the family of gamma distributions is a natural conjugate family for the parameter λ encountered in section 5.4.4 on exponential life-testing procedures (equations (5.33) and (5.34)).

Since $E(\theta;\,\alpha,\beta) = \alpha/(\alpha+\beta)$, it follows from (5.42) that $E(\theta;\,(n,r),\alpha,\beta) = (\alpha+r)/(\alpha+\beta+n)$; thus the role of the data is to change the mean of θ from $\alpha/(\alpha+\beta)$ to that given above. If the prior on θ is a uniform, that is, if $\alpha=\beta=1$, then $E(\theta;\,\alpha,\beta)=1/2$ and $E(\theta;\,(n,r),\alpha,\beta)=(r+1)/(n+2)$, instead of the usual r/n that one normally tends to use.

The above development generalizes to the case of predicting the number of successes in m future trials, given r successes in n trials; i.e. the scenario of (5.7). It can be seen – details left as an exercise for the reader – that for $x=0,1,\ldots,m$

$P(x$ successes in m future trials; $(n,r),\alpha,\beta)$

$$= \binom{m}{x}(B(a,b))\,(B(a+x,b+m-x))^{-1}, \tag{5.43}$$

where $B(a,b) = \frac{\Gamma(a+b)}{\Gamma(a)\Gamma(b)}$, $a=\alpha+r$ and $b=\beta+n-r$.

Equation (5.43) is called the **beta binomial distribution**. It has played a key role in philosophical arguments involving the assertion of a scientific law based on empirical evidence alone. A feel for these can be best described by the scenario given below.

Infinite Number of Successes Cannot Assert the Certainty of Reliability

Suppose that n identical copies of a unit are tested for some pre-specified time and all of these result in success; i.e. $r=n$. If a Bernoulli chance distribution with a parameter θ is used to describe the outcome of each trial, and if the prior on θ is taken to be a uniform on $(0,1)$, then the probability that the next m trials will lead to m successes is, by (5.43), given as $(n+1)/(m+n+1)$. Suppose now, that n is large and $m << n$, say $m=1$. Then the above probability tends to (but is not equal to) one. However, if $m >> n$, and if n is large, conceptually infinite, then the above probability decreases to zero! This means that under a Bayes–Laplace type uniform prior, an infinite number of successful tests of a unit cannot guarantee that a large number of future tests will all result in successes. Jeffreys (1980) used this type of an argument to claim that empirical observations alone cannot prove a scientific law. He then went on to construct a prior under which the predictive distribution provides results that are more in accord with the disposition of experimental scientists; also see Singpurwalla and Wilson (2004).

Bernoulli Testing Under an Exponential-gamma Induced Prior

We consider here the case of Bernoulli testing when the prior $F(\mathrm{d}\theta;\,\psi,\alpha,t)$ is of the form given by (5.12) of section 5.2.3. Recall that under this prior for θ, the mission time t is a parameter, and that $(\psi/(\psi+t))^{\alpha}$ is the mean value function of this prior. Whereas the merits of such a prior need to be explored, the results given below could be useful, especially when the lifetimes are judged to be exponential and the actual times to failure are not available. For example, when observations can be taken only at time t.

Suppose, then, that r successes are observed in n trials, where each trial entails the life-testing of an item until time t, where both n and t are chosen at will; i.e., stopping rule is non-informative. Then, it can be shown, albeit after some cumbersome algebra, that the posterior distribution of θ, $F(\mathrm{d}\theta; (n, r), \psi, \alpha, t)$ is of the form

$$\left[\sum_{j=0}^{n-r}\binom{n-r}{j}(-1)^j\frac{\Gamma(\alpha)}{\omega(j)}\right]^{-1}\theta^{\left(\frac{\psi}{t}+r-1\right)}(1-\theta)^{n-r}\left(\ln\left(\frac{1}{\theta}\right)\right)^{\alpha-1}, \tag{5.44}$$

where $\omega(j)=\left(\frac{\psi}{t}+r+j\right)^{\alpha}$. The details are left as an exercise.

The mean of the above distribution gives us the predictive distribution $P(X_{n+1} = 1; (n, r), \psi, \alpha, t)$. This can be seen (details left as an exercise) as being

$$\begin{aligned} E(\theta; (n, r), \psi, \alpha, t) = &\left[\sum_{j=0}^{n-r}\binom{n-r}{j}(-1)^j\frac{\Gamma(\alpha)}{\left(\frac{\psi}{t}+r+j+1\right)^{\alpha}}\right] \\ &\cdot\left[\sum_{j=0}^{n-r}\binom{n-r}{j}(-1)^j\frac{\Gamma(\alpha)}{(\omega(j))^{\alpha}}\right]^{-1}. \end{aligned} \tag{5.45}$$

Thus the effect of the data is to change the mean of θ from $(\psi/(\psi+t))^{\alpha}$ to the elaborate form given above. Contrast this with the case of a beta prior on θ, wherein the effect of the data is to change the mean from $\alpha/(\alpha+\beta)$ to $(\alpha+r)/(\alpha+\beta+n)$.

The posterior probability that θ exceeds θ_0 is obtained using the incomplete gamma function. Namely

$$\begin{aligned} P(\theta \geq \theta_0; (n, r), \psi, \alpha, t) = &\left[\sum_{j=0}^{n-r}\binom{n-r}{j}(-1)^j\frac{\Gamma(\alpha)}{(\omega(j))^{\alpha}}\right]^{-1} \\ &\sum_{j=0}^{n-r}\binom{n-r}{j}(-1)^j\frac{\Gamma(\alpha)}{(\omega(j))^{\alpha}}\int_0^{\ln(1/\theta_0)}\frac{\mathrm{e}^{-s\omega(j)}(\omega(j))^{\alpha}s^{\alpha-1}}{\Gamma(\alpha)}\mathrm{d}s. \end{aligned}$$

The development of the above material assumes that t is both the mission time for future trials (namely the X_{n+1} case) and also the test time for n trials constituting the observed data. This need not be so. If the test time is t and the mission time is τ, $\tau \neq t$, and if $\Delta = t/\tau$, then all of the above results go through with the following changes:

(c1) The $\omega(j)$'s get replaced by $\left(\frac{\psi}{t}+\Delta(r+j)\right)^{\alpha}$;
(c2) In (5.45), $\left(\frac{\psi}{t}+r+j+1\right)$ gets replaced by $\left(\frac{\psi}{t}+\Delta(r+j)+1\right)$;
(c3) In (5.44), $\left(\frac{\psi}{t}+r-1\right)$ gets replaced by $\left(\frac{\psi}{t}+\Delta r-1\right)$ and $(1-\theta)^{n-r}$ gets replaced by $(1-\theta^{\Delta})^{n-r}$.

A verification of the above is left as an exercise for the reader.

5.4.8 Life-testing and Inference Under the BVE

Thus far, our discussion on life-testing has centered around univariate chance distributions like the exponential, the Weibull, the generalized gamma, and the Bernoulli variables that these distributions could induce. We have noticed that a key ingredient for producing the predictive distributions of interest is the posterior distribution of the parameters of the underlying chance distribution. The same will also be true when dealing with multivariate chance distributions like those of Gumbel, Freund of Marshall and Olkin's BVE. The posterior distribution of the underlying parameters is obtained via the product of the likelihood and the prior, where it has been traditionally so that the likelihood is based on the probability density function of the assumed chance distribution. Equations (5.28) and (5.46) are examples. Since all the bivariate distributions discussed in section 4.7, save the BVE, are absolutely continuous (Table 4.1) they generate probability densities (with respect to the two-dimensional Lebesgue measure). These densities provide a basis for writing out the appropriate likelihoods. The difficulty arises with the BVE since this distribution does not have a density with respect to the two-dimensional Lebesgue measure – Claim 4.7 of section 4.7.4. However, Bemis, Bain and Higgins (1972) have proposed a way of overcoming this difficulty by introducing

$$f(t_1, t_2) = \begin{cases} \lambda_1(\lambda_2 + \lambda_{12})\overline{F}(t_1, t_2), & t_2 > t_1 > 0, \\ \lambda_2(\lambda_1 + \lambda_{12})\overline{F}(t_1, t_2), & t_1 > t_2 > 0, \text{ and} \\ \lambda_{12}\overline{F}(t_1, t_2), & t_1 = t_2 > 0, \end{cases} \tag{5.46}$$

and claiming that $f(t_1, t_2)$ is a density with the understanding that the first two equalities of (5.46) are densities with respect to the two-dimensional Lebesgue measure, and the third equality a density with respect to the one-dimensional Lebesgue measure. This result has enabled the application of any theory of inference based on densities, as was done by Bhattacharyya and Johnson (1973), Shamseldin and Press (1984) and Peña and Gupta (1990). However, there is another difficulty that the BVE poses which the above did not attempt to cope with, namely the problem of identifiability mentioned in section 4.7.4. Specifically, the matter of identifiability arises in the context of observing failure data on parallel redundant systems when the first failure is an individual failure. Under such circumstances, the cause of the second failure cannot be exactly determined. A consequence is that data on the second unit to fail cannot be used for learning about the dependency parameter λ_{12} and the parameter λ_1 or λ_2, depending on which unit is the second to fail. A way of overcoming this obstacle is through the technique of data augmentation (mentioned in Appendix A) and using the Gibbs sampling algorithm on the augmented data. A use of this strategy for the problem at hand is outlined below.

Data Augmentation and Gibbs Sampling for the BVE

A Bayesian approach for inference about the parameters λ_1, λ_2 and λ_{12} of the BVE has been proposed by Shamseldin and Press (1984) and by Pena and Gupta (1990) but without considering the scenario of life-testing for parallel systems; thus the matter of identifiability has not been considered by them. Shamseldin and Press (1984) assume that $\lambda_1 = \lambda_2$ and assign independent priors for the λ_is and λ_{12}; Pena and Gupta (1990) allow dependence. We shall assume independent gamma priors for λ_1, λ_2 and λ_{12}, our main objective being an illustration of the use of the technique of data augmentation. Accordingly, we assume that the joint prior on λ_1, λ_2 and λ_{12} has a density

$$\pi_1(\lambda_1) \cdot \pi_2(\lambda_2) \cdot \pi_3(\lambda_{12})$$

where $\pi_i(\lambda) \propto \exp(-\beta_i \lambda)\lambda^{\alpha_i - 1}$, for $\alpha_i, \beta_i > 0,\ i = 1, 2$.

For the parallel sampling perspective adopted here, let $\{\mathbf{t} = (t_{1j}, t_{2j})\},\ j = 1, \ldots, n$ be the data, with t_{1j} denoting time to failure of unit 1, and t_{2j} the time to failure of unit 2 in a two-component parallel system; the number of observations is n. Furthermore suppose that of these n observations, n_1 is the number of times $t_{1j} < t_{2j}$, n_2 the number of times $t_{1j} > t_{2j}$, and n_{12} the number of times $t_{1j} = t_{2j}$; thus $n = n_1 + n_2 + n_{12}$. Thus n_i is the number of times the failure is caused by the shock-generating process $N_i(t),\ t \geq 0,\ i = 1, 2, 3$.

Since information about the cause of the second failure is not contained in the data **t**, the data augmentation process requires that we introduce this information via suitable random variables; this is the essence of the technique. We are therefore motivated to introduce two new random variables M_1 and M_2 where M_i denotes the number of times the failure of component i is caused by the process $N_i(t)$, when component i is the second to fail. Since component 1 is the first to fail n_2 times, $0 \le M_1 \le n_2$; similarly for M_2. Thus conditional on M_1 and M_2, the likelihood of λ_1, λ_2 and λ_{12}, in the light of the data **t** is

$$L(\lambda_1, \lambda_2, \lambda_{12} \mid M_1 = m_1, M_2 = m_2; \mathbf{t}) = \left((\lambda_1)^{n_1+m_1} e^{-\lambda_1 \tau_1}\right)\left((\lambda_2)^{n_2+m_2} e^{-\lambda_2 \tau_2}\right) \cdot \left((\lambda_1)^{n-m_1-m_2} e^{-\lambda_{12}\tau_{12}}\right),$$

where $\tau_1 = \sum_{j=1}^{n_1} t_{1j}$, $\tau_2 = \sum_{j=1}^{n_2} t_{2j}$ and $\tau_{12} = \sum_{j=1}^{n} \max(t_{1j}, t_{2j})$.

The effect of data augmentation has been to factorize the likelihood, because in the absence of M_1 and M_2 the likelihood would have been

$$(\lambda_1 (\lambda_2 \lambda_{12}))^{n_1} (\lambda_2 (\lambda_1 \lambda_{12}))^{n_2} \lambda_{12}^{n_{12}} \exp(-\lambda_1 \tau_1 - \lambda_2 \tau_2 - \lambda_{12} \tau_{12}).$$

As a consequence of the independent priors on λ_1, λ_2 and λ_{12} and a factorization of the likelihood, λ_1, λ_2 and λ_{12} are a posteriori independent, with λ_i having a gamma distribution with scale parameter $(\beta_i + \tau_i)$ and a shape parameter $(\alpha_i + n_i + m_i)$, $i = 1, 2$, and λ_{12} also having a gamma distribution with scale parameter $(\beta_{12} + \tau_{12})$ and shape parameter $(\alpha_{12} + n - m_1 - m_2)$. Sampling from these distributions to generate a Gibbs sample is therefore straightforward.

The other conditional distributions required are those of M_1 and M_2. Since M_1 is the number of times a unit fails due to the process $N_1(t)$ as opposed to $N_3(t)$, out of n_2 items on test, M_1 has a binomial distribution with parameters n_2 and $\lambda_1/(\lambda_1 + \lambda_{12})$. The probability that an item fails due to the process $N_1(t)$ rather than $N_3(t)$ is $\lambda_1/(\lambda_1 + \lambda_{12})$. Similarly, M_2 has a binomial distribution with parameters n_1 and $\lambda_2/(\lambda_2 + \lambda_{12})$. All the ingredients needed to invoke the Gibbs sampling algorithm are at hand. The posterior distributions of λ_1, λ_2 and λ_{12} can now be obtained.

5.5 INFORMATION FROM LIFE-TESTS: LEARNING FROM DATA

I have said that life-testing experiments yield 'information' via failure and survival times, and that such information helps provide realistic assessments of an item's survivability. I have also said that data from life-tests enables us to 'learn more' about the underlying parameters of chance distributions, such knowledge being of value when taking actions about the acceptance or not of an item, or for obtaining an enhanced appreciation (i.e. inference) of the nature of the parameters. But what do such commonly used terms like 'information', 'learning' and 'knowledge' mean? The word 'information' has already appeared in section 5.3.2 in the context of reference priors, but little has been said about its basic import. The purpose of this section is to make explicit notions such as entropy, information, and learning, and to articulate their role in reliability and survival analysis.

5.5.1 Preliminaries: Entropy and Information

By the term 'information', we mean anything that changes our probability distribution about an unknown quantity, say θ, which for convenience is assumed to be a parameter of some chance distribution. A consequence of a change in the probability distribution of θ could be a change in some action that is taken based on an appreciation of θ. Thus one way to measure the change in the probability distribution of θ is via a utility function $\mathcal{U}(\theta, d(T^*, n))$, where $d(T^*, n)$ is some action or decision that we may take based on data T^* obtained via n observations. In the context of life-testing, the total time on test (section 5.4.3) – best encapsulates T^*. Recall that the notion of a utility was introduced in section 1.4, and utilities were discussed in section 2.8.1 where the notation $u(C_{ij})$ was used to denote the utility of an action j when the state of nature was i. In $\mathcal{U}(\theta, d(T^*, n))$, θ is to be identified with i and $d(T^*, n)$ with j. In what follows $d(T^*, n)$ will be denoted by just d.

Suppose now that $F(\theta)$ is our prior distribution of θ, and $F(\theta|(T^*, n))$ our posterior for θ were we to take n observations and observe T^* as data. Then, by the principle of maximization of expected utility (section 2.8.2) the **expected gain in utility** due to (T^*, n) is measured as

$$g(n) = E_{T^*}\left\{\max_d \int_\theta \mathcal{U}(\theta, d)F(\mathrm{d}\theta|(T^*, n))\right\} - \left\{\max_d \int_\theta \mathcal{U}(\theta, d)F(\mathrm{d}\theta)\right\}; \qquad (5.47)$$

it can be seen that $g(n) \geq 0$. The expectation above is with respect to the marginal distribution of T^*, and n, a decision variable, is assumed to be specified. Determining an optimum n is discussed later, in section 5.6. The quantity $g(n)$ is also known as the **expected information** about θ that would be provided by the data T^* were n observations to be taken. It is important to bear in mind that $g(n)$ is in the subjunctive mood; i.e. $g(n)$ is based on the contemplative assumption of taking n observations and obtaining T^* as data. It is for this reason that expectation with respect to the marginal T^* is taken. In Bayesian statistics, such contemplative analyses are known as **preposterior analyses** because they precede the obtaining of any actual posterior.

What are the possible choices for $d(T^*, n)$ and what possible forms can $\mathcal{U}(\theta, d)$ take? The answer depends on the practicalities of the situation at hand. The simplest scenario is one wherein the decision is a tangible course of action, such as accepting or rejecting a batch of items. In this case, with d as the decision to accept and θ a measure of quality, it is reasonable to let $\mathcal{U}(\theta, d)$ be an increasing function of θ. More about scenarios involving tangible courses of action is said later in sections 5.5.3 and 5.6. For now we shall consider a more subtle case wherein the decision is an enhanced appreciation of θ.

Choice of Utility Functions for Inference (or Enhanced Appreciation)

Since a probability density best encapsulates our appreciation of uncertainty about θ, our choice of d should be some density, say $p(\bullet)$. Thus $\mathcal{U}(\theta, d)$ will be of the form $\mathcal{U}(\theta, p(\bullet))$, which is the utility of declaring the density $p(\bullet)$, when the parameter takes the value θ. But it is the posterior distribution of θ that accords well with our belief about θ. Thus the question arises as to what functional form of $\mathcal{U}(\theta, p(\bullet))$ will result in the choice that an optimum $p(\bullet)$ is the posterior distribution of θ? Bernardo (1979b), in perhaps one of the most striking results in Bayesian decision theory, has shown that the utility function has to be of the logarithmic form; that is $\mathcal{U}(\theta, p(\bullet)) = \log p(\theta)$. Consequently, if this form of the utility function is invoked in (5.47), then

$$g(n) = E_{T^*}\left\{\int_\theta \log(F(\mathrm{d}\theta|(T^*, n)))F(\mathrm{d}\theta|(T^*, n))\right\} - \left\{\int_\theta \log(F(\mathrm{d}\theta))F(\mathrm{d}\theta)\right\}. \qquad (5.48)$$

In order to connect the above notions from Bayesian decision theory to those used in communication theory, we remark that the **Shannon information** about θ, whose uncertainty is given by $F(\mathrm{d}\theta)$, is

$$\mathbf{I}(\theta) = \int_\theta \log F(\mathrm{d}\theta)F(\mathrm{d}\theta); \qquad (5.49)$$

$-\mathbf{I}(\theta)$ is known as the **entropy** of $F(\mathrm{d}\theta)$. Similarly with $F(\mathrm{d}\theta|(T^*, n))$, we have the entities $\mathbf{I}(\theta|(T^*, n))$ and $-\mathbf{I}(\theta|(T^*, n))$. Consequently, $g(n)$ is $E_{T^*}[\mathbf{I}(\theta|T^*, n)] - \mathbf{I}(\theta)$.

If the distribution function of T^* is denoted by $F(T^*)$ – recall that T^* has not as yet been observed – then another way of writing $g(n)$ is

$$g(n) = \int_{T^*} \int_{\theta} \log \left(\frac{F(\mathrm{d}\theta, \mathrm{d}T^*)}{F(\mathrm{d}\theta)F(\mathrm{d}T^*)} \right) F(\mathrm{d}\theta, \mathrm{d}T^*); \tag{5.50}$$

here $F(\theta, T^*)$ is the joint distribution function of θ ant T^*. To see why, we note that the first term of (5.48) is

$$\int_{T^*} \int_{\theta} \log (F(\mathrm{d}\theta)|T^*, n)\, F(\mathrm{d}\theta|T^*, n) F(\mathrm{d}T^*) = \int_{T^*} \int_{\theta} \log (F(\mathrm{d}\theta)|T^*, n)\, F(\mathrm{d}\theta, \mathrm{d}T^*),$$

and that the second term is

$$\int_{T^*} \int_{\theta} \log F(\mathrm{d}\theta) F(\mathrm{d}\theta, \mathrm{d}T^*).$$

Consequently,

$$g(n) = E_{(T^*, \theta)} \left\{ \log \left(\frac{F(\mathrm{d}\theta, \mathrm{d}T^*)}{F(\mathrm{d}\theta)F(\mathrm{d}T^*)} \right) \right\}, \tag{5.51}$$

where the expectation is with respect to $F(\theta, T^*)$.

When $g(n)$ is written as above, it is interpreted as the mutual information between θ and T^*, or as the Kullback–Leibler distance for discrimination between $F(\mathrm{d}\theta|(T^*, n))$ and $F(\mathrm{d}\theta)$. Soofi (2000) provides a recent account on these and related matters. Mutual information has the property that $g(n) \geq 0$, with equality, if and only if θ and T^* are independent. In the latter case, knowledge of T^* gives us no information about θ. In life-testing, one endeavors to collect that data which results in a T^* for which $g(n)$ is a maximum. In communication theory (Shannon (1948) and Gallager (1968)), the maximum value of $g(n)$ is known as **channel capacity**; here θ is interpreted as the message to be sent, and T^* interpreted as the message received. The connection between information theory and the prior to posterior transformation in Bayesian inference has been articulated by Lindley (1956) in perhaps one of his several landmark papers. Other references relating the above ideas to the design of (life-testing) experiments are in Verdinelli, Polson and Singpurwalla (1993), and in Polson (1992).

Our discussion thus far has been conducted on the premise that the n observations have yet to be taken and the T^* remains to be observed; i.e. what we have been doing is a pre-posterior analysis. Thus $g(n)$ represents the expected gain in utility or the expected information. Suppose now that the n observations are actually taken and that the data T^* has indeed been obtained. Then, by analogy with (5.47), the **observed information** or the **observed change in utility** is

$$g(T^*, n) = \max_d \int_{\theta} \mathcal{U}(\theta, d) F(\mathrm{d}\theta; T^*, n) - \max_d \int_{\theta} \mathcal{U}(\theta, d) F(\mathrm{d}\theta).$$

An unpalatable feature of $g(T^*, n)$ is that unlike $g(n)$, which is always non negative, $g(T^*, n)$ could be negative. A negative value of $g(T^*, n)$ signals the fact that data could result in negative information. This of course, though contrary to popular belief, is still reasonable, because in actuality one could be surprised by the data one sees, and as a consequence be less sure about θ than one was prior to observing the data.

One attempt at avoiding the possible negativity of $g(T^*, n)$ is to introduce $\tilde{g}(T^*, n)$, the conditional value of sample information (cf. De Groot, 1984). Here, let d_0 denote that value of d which maximizes

$$U(F, d) \stackrel{\text{def}}{=} \int_{\theta} \mathcal{U}(\theta, d) F(d\theta);$$

d_0 is known as the Bayes' decision with respect to the prior $F(d\theta)$. We assume that d_0 exists and is unique. Then

$$\tilde{g}(T^*, n) \stackrel{\text{def}}{=} \max_{d} U\left[F(\theta|T^*, n), d\right] - U\left[F(\theta|T^*, n), d_0\right],$$

the difference between the expected utility from the Bayes' decision using the data (T^*, n) and the expected utility using d_0, the decision that would have been chosen had the data not been available. By definition, it is true that $\tilde{g}(T^*, n) \geq 0$ for every T^*, and that its expected value with respect to $F(T^*)$, the marginal of T^*, is indeed $g(n)$.

5.5.2 Learning for Inference from Life-test Data: Testing for Confidence

For the purpose of this section, we shall suppose that the intent of learning is an enhanced appreciation about an unknown parameter of a chance distribution. The matter of learning for taking a tangible action such as acceptance or rejection is considered the in section 5.5.3. To simplify matters, suppose that the chance distribution is an exponential with a parameter λ, so that for $\lambda > 0$, $P(T > t|\lambda) = \exp(-\lambda t)$, for $t \geq 0$; λ is the failure rate. Let $\xi = \lambda^{-1}$ be the mean time to failure of the exponential distribution of T, and as before let T^* denote data from a life-test.

In practice, life-testing is commonly done to learn more about the mean time to failure of an item; i.e. to gain an enhanced appreciation of the parameter ξ. Whereas learning more about ξ may have limited tangible merit, it is often the case that contractual requirements are specified in terms of ξ. Thus it behoves us to consider the information about ξ provided by the data T^*, and with that in mind we wish to address the question of how much testing should be done to obtain sufficient knowledge about ξ? For this, a strategy has been proposed by Lindley (1957b), namely, to continue testing until $I(\xi; T^*) = \int \log F(d\xi; T^*) F(d\xi; T^*)$, the Shannon information about ξ (equation (5.49) reaches a prescribed value. Since prescribing a value for $I(\xi; T^*)$ seems elusive, one possibility is to continue testing until $I(\xi; T^*)$ fails to reflect a significant increase. Lindley's proposed strategy was for the Bernoulli chance distribution, but the idea is general enough to be invoked for other distributions as well.

To facilitate calculations, suppose that $F(\lambda)$, the prior distribution of λ is a gamma with shape (scale) $\alpha(\psi)$ (section 5.2.1). Then, it is easy to see that upon observing a single failure at time t the posterior distribution of λ is also a gamma with shape $\alpha + 1$ and scale $\psi + t$. This fact allows us to display the testing scheme on a diagram with axes giving the values of the gamma scale and shape parameters. Specifically, as has been shown by El-Sayyad (1969), when $F(\lambda)$ is a gamma distribution with shape (scale) parameter $a(b)$, the Shannon information is to a good approximation, and for large values of a, of the form $I(\lambda; a, b) \approx \frac{1}{2}\log(b^2/2\pi a) - \frac{1}{2} + \frac{1}{2}a$. Thus testing should be continued until the values of a and b that are obtained are such that a prescribed value for $I(\lambda; \bullet)$ is achieved. For values of $a > 5$, the relationship between a, b and $I(\lambda; \bullet)$ is parabolic, and of the form $b^2 = Ca$ where C depends on $I(\lambda; \bullet)$, the amount of information that is desired. This relationship is graphically portrayed in Figure 5.4 with the prior parameters α and ψ as the starting values of a and b, respectively, and $t_1, t_2, \ldots,$ as failure times, assuming the testing of one item at a time.

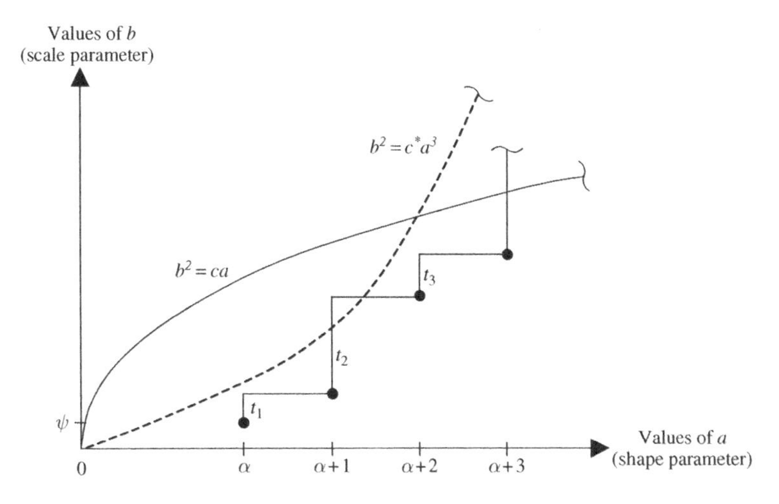

Figure 5.4 Values of the gamma scale and shape parameters for desired shannon information.

In Figure 5.4 testing will stop after Three failures for a specified value of C, where C is determined by a desired value of $\mathbf{I}(\lambda; \bullet)$. In the schemata discussed above, stopping between failures is not considered.

It is well known (cf. Lindley, 1956) that $\mathbf{I}(\lambda; \bullet)$ is not invariant under transformations of λ, so that when interest centers around ξ the mean time to failure, $I(\xi; \bullet) = I(\lambda; \bullet) + 2E_\lambda[\log \lambda]$; (El-Sayyad, 1969). In this case the boundary in the (a, b) diagram is $b^2 \approx C^* a^3$ where the constant C^* depends on $I(\xi; \bullet)$, the amount of Shannon information that is desired. This boundary is illustrated by the convex dotted curve of Figure 5.4. A consequence is that for any fixed value of α and ψ, the amount of testing needed to get the same amount of information about λ and ξ will be different. Thus given a fixed amount of test time and a fixed number of test items, it matters whether a desired amount of information is needed about the failure rate λ, or about the mean ξ. More details pertaining to the implications of the concave and the convex boundaries of Figure 5.4 are in Singpurwalla (1996). Also discussed therein is an observation by Abel and Singpurwalla (1993), which shows that at any given point in time t, a failure is more informative than a survival, if interest centers around ξ the mean time to failure, but that the opposite is true when interest centers around the failure rate $\lambda = \xi^{-1}$. Specifically, $I(\xi;$ failure at $t) > I(\xi;$ survival at $t)$ for all t, but that $I(\lambda;$ failure at $t) < I(\lambda;$ survival at $t)$ for all t; (Figure 5.5).

Information Loss Due to Censoring

Recall, (section 5.4.1, that in practice life-tests are often censored to save test time and costs. However, there is a price to be paid for this saving, namely, the loss of information. The purpose of this sub-section is to describe ways of assessing the said loss. I start with the case of Type I (or time truncated) censoring wherein n items are put on a life test that is terminated at some pre-specified time t. Suppose that $k \geq 0$ failures at times $\tau_1 \leq \tau_2 \leq \cdots \leq \tau_k$ are observed so that the available data is $D_t = \{k, \tau_1, \tau_2, \ldots, \tau_k\}$. Following Brooks (1982), I focus attention on the exponential case $P(T > t \mid \lambda) = e^{-\lambda t}$ for $\lambda > 0$ and $t \geq 0$, and compute the expected information in D_t about λ – via (5.48) – as

$$I_C = E_{D_t}\left[I(\lambda | D_t, n) - I(\lambda)\right],$$

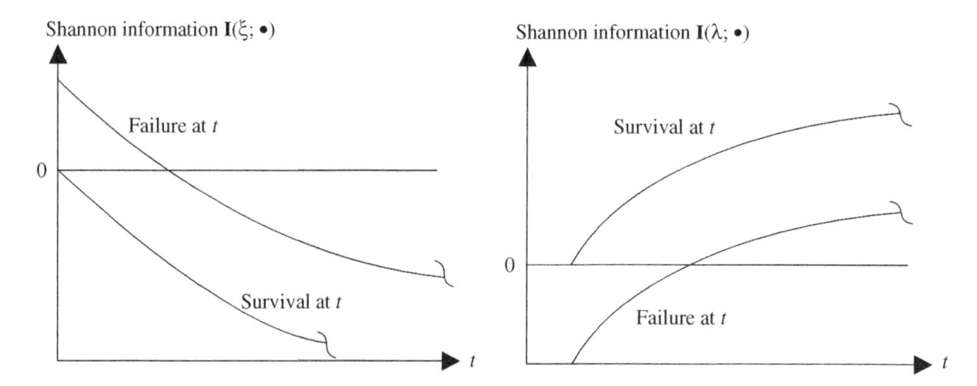

Figure 5.5 A comparison of shannon information about failure rate and mean given failure or survival at t.

where the expectation is taken with respect to the marginal distribution of D_t. Similarly, we compute

$$I_{\overline{C}} = E_{D_\infty}\left[I(\lambda|D_\infty, n) - I(\lambda)\right],$$

where D_∞ denotes the case of no censoring; i.e. by setting $t = \infty$. The difference between $I_{\overline{C}}$ and I_C represents the loss of information due to censoring; we would expect that $I_{\overline{C}} \geq I_C$. The detailed computations tend to be cumbersome, but if a gamma prior with shape (scale) $\alpha(\psi)$ is assumed, then according to Brooks (1982), $I_C - I_{\overline{C}} \approx -\frac{1}{2}E_\lambda[\log(1 - e^{-\lambda t})]$. An interesting follow on to the above is the work of Barlow and Hsiung (1983) who show that for a 'time transformed exponential', namely, the one wherein $P(T > s \mid \lambda) = e^{-\lambda R(s)}$, where $R(s)$ is a known function of s, I_C is concave and increasing in both $R(s)$ and n. The prior on λ is a gamma with shape (scale) $\alpha(\psi)$. Note that with $R(s) = s$, we have the exponential as a chance distribution, and that with $R(s; \beta) = s^\beta$, the Weibull distribution with a known shape parameter $\beta > 0$. The concavity of I_C enables us to assess the effects of increasing the test truncation time t and n, the number of items to be tested; it implies that there is a limit beyond which increasing n and/or t gives little added information. Figure 5.6 taken from Barlow and Hsiung (1983) illustrates this point; here $\alpha = 2.79$ and $\psi = 4.78$.

The concavity of expected information has also been noted by Ebrahimi and Soofi (1990), who consider the case of Type II censoring, namely subjecting n items to a test that is stopped at the r-th failure so that the observed data is $D_r = \{k, \tau_1, \tau_2, \ldots, \tau_r\}$. Furthermore, the authors remark that

$$I_C^* = E_{D_r}\left[I(\lambda|D_r, n) - I(\lambda)\right]$$

increases in r but decreases in α, the shape parameter of the gamma prior distribution of λ.

5.5.3 Life-testing for Decision Making: Acceptance Sampling

Whereas life-testing for enhanced appreciation of unknown parameters has merit in scientific contexts, in the industrial set-up life-testing is most likely done with the aim of making tangible decisions such as the acceptance of a batch of items. The best example of this application is the U.S. Department of Defense's Military Standard 781 C plan of 1977. Here a sample of n items from a batch of N items is subjected to a life-test, and based on the results of the test, the untested items are accepted or rejected. We shall suppose that items which are surviving at the termination of the test are discarded and that n is specified. The determination of an optimum n

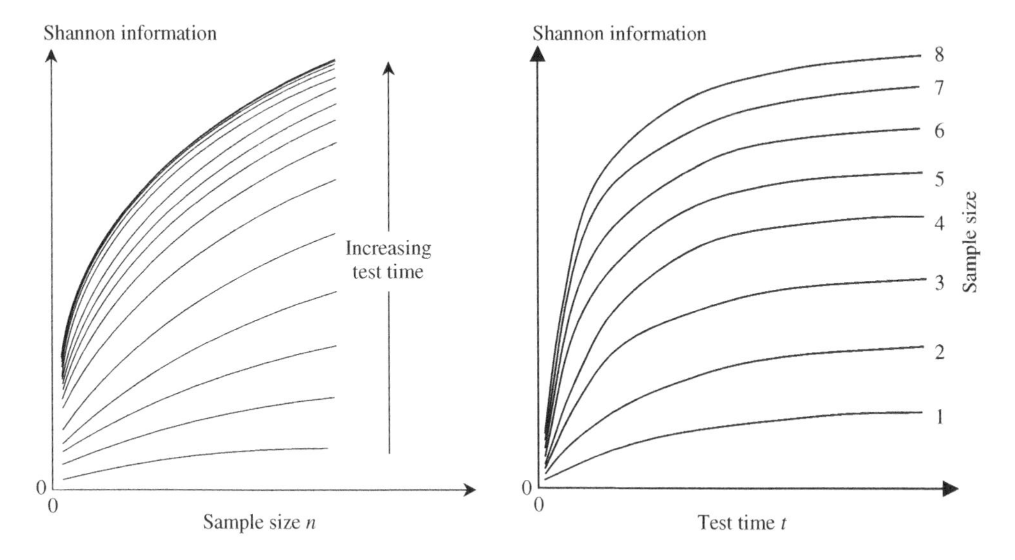

Figure 5.6 The effect of sample size n and test time t on Shannon information (from Barlow and Hsiung, 1983).

is discussed later in section 5.6. In what follows, we describe a strategy to be used by a consumer $\mathcal{C}$ who tests the $n \geq 0$ items and then makes an accept/reject decision.

Since n is assumes known, $\mathcal{C}$'s decision tree will have one decision node and three random nodes (section 1.4). This decision tree is shown in Figure 5.7.

With n given, $\mathcal{C}$ tests the n items and obtains as data – at the random node $\mathcal{R}_1$ – the quantity T^*. Once T^* is obtained $\mathcal{C}$ takes an action at the decision node $\mathcal{D}_1$ to either accept (A) or to reject (R) the remaining ($N - n$) items. Suppose that $\mathcal{C}$ chooses to accept; then at the random node $\mathcal{R}_2$ nature takes its course of action in that the value of the unknown parameter happens to be θ. For this sequence of events, working backwards from θ, $\mathcal{C}$ experiences a utility $\mathcal{U}(\theta, A, T^*, n)$. Similarly when $\mathcal{C}$ chooses to reject and nature takes the action θ at $\mathcal{R}_3$, the utility to $\mathcal{C}$ is $\mathcal{U}(\theta, R, T^*, n)$.

Suppose that $\mathcal{C}$'s prior on θ is $F(\mathrm{d}\theta)$; then having obtained T^*, $\mathcal{C}$ computes $F(\mathrm{d}\theta; T^*)$, the posterior for θ. Furthermore, it is reasonable to suppose that

$$\mathcal{U}(\theta, A, T^*, n) = \mathcal{U}(\theta, A), \text{ and that}$$

$$\mathcal{U}(\theta, R, T^*, n) = \mathcal{U}(\theta, R);$$

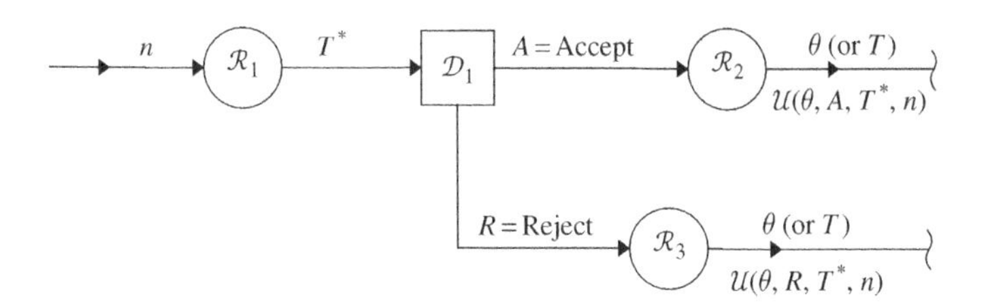

Figure 5.7 Decision tree for acceptance or rejection by $\mathcal{C}$.

this is because n is chosen and the observed data T^* will not have any effect on $\mathcal{C}$'s utility of accepting or rejecting, once θ is treated as being known. $\mathcal{C}$ next invokes the principle of maximization of expected utility and accepts the $(N-n)$ untested items if:

$$\int_{\theta} \mathcal{U}(\theta, A)F(\mathrm{d}\theta; T^*) \geq \int_{\theta} \mathcal{U}(\theta, R)F(\mathrm{d}\theta; T^*); \tag{5.52}$$

otherwise $\mathcal{C}$ will reject. As stated before, in section 5.5.1, if θ is a measure of the quality of the item, say the mean time to failure, then $\mathcal{U}(\theta, A)$ will be an increasing function of θ. For example, $\mathcal{U}(\theta, A) = \log\theta$, suggesting that the utiles to $\mathcal{C}$ for very large values of θ get de facto saturated at some constant value.

There is a reinforcement to the argument which leads to (5.52) above. This is based on the premise that utilities are better understood and appreciated if they are defined in terms of observable entities, like lifelengths, instead of abstract parameters like θ. With that in mind, suppose then that $T_1, T_2, \ldots$, the lifelengths of all the N items in the batch are judged exchangeable, and suppose that T denotes the generic lifelength of any one of the $(N-n)$ untested items. Then given T^*, the predictive distribution of T is

$$P(T > t; T^*) = \int_{\theta} P(T \geq t|\theta)F(\mathrm{d}\theta; T^*). \tag{5.53}$$

With the above in place, I now suppose that at nodes $\mathcal{R}_2$ and $\mathcal{R}_3$, nature will yield a life-time T (instead of θ) so that $\mathcal{C}$'s terminal utilities are $\mathcal{U}(T, A, T^*, n)$ and $\mathcal{U}(T, R, T^*, n)$. As before, I suppose that the above utilities do not depend on T^* and n, so that $\mathcal{C}$'s criterion for accepting the untested items will be

$$\int_{\theta} \mathcal{U}(t, A)F(\mathrm{d}t; T^*) \geq \int_{\theta} \mathcal{U}(t, R)F(\mathrm{d}t; T^*), \tag{5.54}$$

where $\mathcal{U}(t, A)$ is $\mathcal{C}$'s utility for acceptance when $T = t$, and $F(\mathrm{d}t; T^*)$ is the probability density at t generated by (5.53), the predictive distribution of T; similarly $\mathcal{U}(t, R)$. Here again it is reasonable to suppose that $\mathcal{U}(t, A)$ is an increasing function of t. What about $\mathcal{U}(t, R)$, the utility to $\mathcal{C}$ of rejecting an item whose lifelength T is t? One possibility is to let $\mathcal{U}(t, R) = 0$, assuming that there is no tangible regret in having rejected a very good item. Another possibility is to let $\mathcal{U}(t, R) = a_3$, where the constant a_3 encapsulates the utility of a lost opportunity.

Besides ease of interpreting utilities, there is another advantage to the formulation which leads us to (5.54). This is because the acceptance (or the rejection) criterion of (5.54) can come into play at any time during the life-test, not just at the time of a failure. The practical merit of this feature is timely decision making. To see how, suppose that the underlying chance distribution of T is an exponential with ξ as the mean time to failure; i.e. $P(T > t \mid \xi) = \exp(-(t/\xi))$, $t \geq 0$, $\xi > 0$. Suppose that the prior on ξ is an **inverted gamma** distribution with shape (scale) parameter $\alpha(\psi)$; i.e. the prior distribution of $\lambda = 1/\xi$ is a gamma with shape (scale) parameter $\alpha(\psi)$. Then

$$F(\mathrm{d}\xi; \alpha, \psi) = \frac{\psi^{\alpha}}{\Gamma(\alpha)} \exp\left(-\frac{\psi}{\xi}\right) \xi^{-(\alpha+1)}. \tag{5.55}$$

If n items are tested and k failures observed at time t, then the total time on test $\tau = \sum_1^k \tau_i + (n-k)t$, where τ_i is the time of the i-th failure; recall that $\tau_1 \leq \tau_2 \leq \cdots \leq \tau_k$. It is evident (details left as an exercise for the reader) that the posterior distribution of ξ is also an inverted gamma with shape $(\alpha+k)$ and scale $(\psi+\tau)$. That is

$$F(\mathrm{d}\xi;\, \alpha, \psi, \tau, k) = \frac{(\psi+\tau)^{\alpha+k}}{\Gamma(\alpha+k)} \exp\left(-\frac{\psi+\tau}{\xi}\right) \xi^{-(\alpha+k+1)}. \tag{5.56}$$

We now suppose that $\mathcal{U}(t, A) = a_1 t - a_2$ and $\mathcal{U}(t, R) = a_3$, where a_1 is a utile to $\mathcal{C}$ for every unit of time that an item functions, and $-a_2$ the utile to $\mathcal{C}$ in accepting an item that does not function at all; a_2 encompasses the cost to $\mathcal{C}$ of purchasing and installing an item. With the above in place, an application of (5.54) would result in accepting the untested items as soon as

$$\frac{\psi+\tau}{\alpha+k-1} > \frac{a_2+a_3}{a_1}; \tag{5.57}$$

the details which entail routine technical manipulations are left as an exercise for the interested reader. The essence of the intent of (5.57) can be graphically portrayed with the shape parameter on the horizontal axis and the scale parameter on the vertical axis. The boundary between $A(n)$, the region of acceptance, and $R(n)$, the region rejection, is a line with slope $(a_2+a_3)/a_1$ (Figure 5.8).

It is instructive to note the parallels between Figure 5.8 and Figure 5.4. In Figure 5.8, the sample path of the $(\alpha+k, \psi+\tau)$ curve takes jumps of size $nt_1, (n-1)(\tau_2-\tau_1), \ldots, (n-k)(\tau_{k+1}-\tau_k), \ldots, 2(\tau_{n-1}-\tau_{n-2}), (\tau_n-\tau_{n-1})$. Acceptance occurs at the point marked as (a) on the boundary, after the third failure, but prior to the fourth.

A more realistic version of $\mathcal{U}(t, A)$ would be $\mathcal{U}(t, A) = a_1 t^p - a_2$ for $p \leq 1$; this would make the utility function increasing but concave. When such is the case, $\mathcal{C}$'s criterion for acceptance will be

$$\frac{(\psi+\tau)^p\,\Gamma(\alpha+k-p)}{\Gamma(\alpha+k)} > \frac{a_2+a_3}{a_1}; \tag{5.58}$$

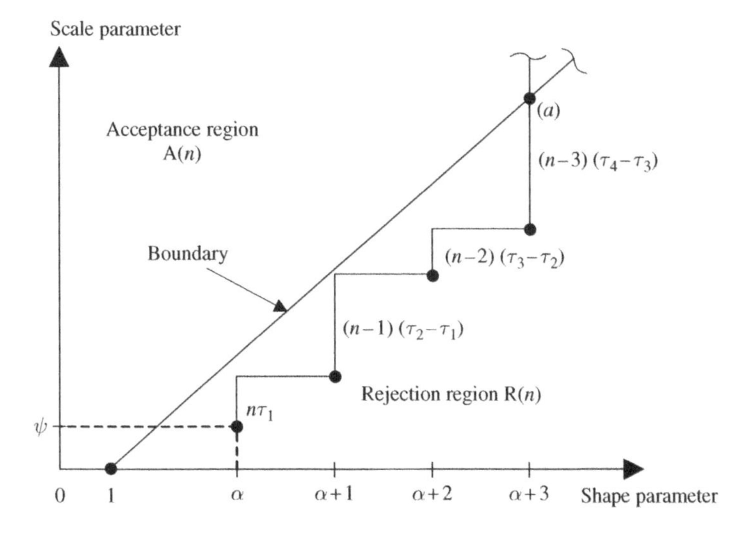

Figure 5.8 Parameters of the inverted gamma and $\mathcal{C}$'s acceptance–rejection region.

this suggests that the boundary between acceptance and rejection is a curve instead of the straight line of Figure 5.8.

The starting point (α, ψ) in Figure 5.8 is shown to lie in the region of rejection. This means that based on prior knowledge alone, $\mathcal{C}$ is unable to accept the lot of untested items. Had the point (α, ψ) been in the region of acceptance, $\mathcal{C}$ would accept the lot without any testing. There is an additional feature to the scheme illustrated by Figure 5.8. This pertains to the fact that had testing been continued after the point (a) had been traversed by the sample path of $(\alpha + k, \psi + \tau)$ – say until all the n items on test fail – then it is possible that the sample path would cross the boundary line again and re-enter the region of rejection. That is, acceptance could be premature. Assuming that n is optimally chosen (section 5.6), the implication of such a switch from acceptance to rejection would be that added evidence from the life-test suggests rejection as the prudent course of action. With this caveat in mind, why then should one stop testing as soon as the sample path of $(\alpha + k, \psi + \tau)$ reaches the point (a)? One answer to this question is that since the Shannon information is concave in the test time (Figure 5.6) I would have gained a sufficient amount of knowledge about T by the time we accept so that a second crossing is unlikely. That is, our decision to accept is fairly robust. Furthermore, assuming that acceptance does not occur very early on in the test, a second crossing can happen if the sample path takes a small jump. But small jumps correspond to very small inter-failure times, and such inter-failure times could be seen as outliers, whose effects on the decision process should be tempered. To account for such outliers, it is important that the predictive distribution of T have heavy tails. Finally, but more importantly, our set-up had not incorporated a disutility (i.e. a negative utility) associated with the time needed to conduct the test and the utility gained by making early decisions. Were such utilities to be incorporated for deigning $\mathcal{C}$'s actions, then the cost of testing would offset the added information obtained by additional testing. Multiple crossings are the consequence of the sample path being close to the accept/reject boundary.

5.6 OPTIMAL TESTING: DESIGN OF LIFE-TESTING EXPERIMENTS

The material of section 5.5 assumed that n, the number of items tested, is specified. In actuality, n is also a decision variable – like accept or reject – and therefore, it should also be chosen by the principle of maximization of expected utility. Since the testing of each item entails costs, namely the cost of the item tested plus the cost of the actual conduct of the test, there is a disutility associated with the testing of an item. In what follows we assume that any tested item that does not fail during the test is worthless; i.e. it is discarded. As a start, we shall assume that the cost, in utiles, of procuring and testing an item to failure is a fixed constant s, irrespective of how long it takes for the item to fail. Furthermore, the set-up costs associated with the conduct of the life-test are assumed to be negligible.

Figure 5.9 shows a decision tree that illustrates the problem of choosing an optimal sample size n for a life-testing experiment. We assume that the cost of sampling and testing is borne by a consumer $\mathcal{C}$, who chooses n on the premise that all the n items will be tested to failure. The adversarial scenario wherein the sampling and testing costs are borne by a manufacturer $\mathcal{M}$

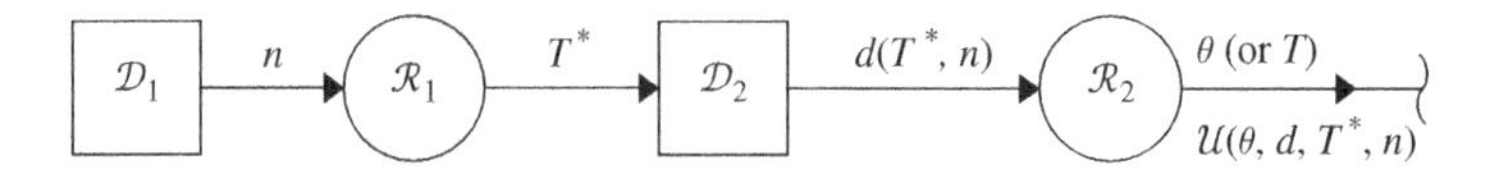

Figure 5.9 $\mathcal{C}$'s Decision tree for sample size selection in a life-test.

will be considered in section 5.7. In Figure 5.9, the convention of Figures 1.1 and 5.7, and the notation of section 5.5 are used.

Figure 5.9 is an elaboration of the decision tree of Figures 5.7 in the sense that it has two decision nodes, $\mathcal{D}_1$ at which $\mathcal{C}$ selects an n, and $\mathcal{D}_2$ at which $\mathcal{C}$ takes an action $d(T^*, n)$ based on the data T^* revealed at the random node $\mathcal{R}_1$. The decision $d(T^*, n)$ could be a tangible action, like the accept/reject choices of section 5.5.3, or it could be a probability distribution for an enhanced appreciation of θ (sections 5.5.1 and 5.5.2). At the random node $\mathcal{R}_2$ nature takes the action θ (or equivalently reveals a T) so that at the terminus of the tree $\mathcal{C}$ experiences a utility $\mathcal{U}(\theta, d, T^*, n)$, where $d = d(T^*, n)$. Following Lindley (1978) (and also the references therein), we assume that $\mathcal{U}(\theta, d, T^*, n)$ does not depend on T^* and that it is additive and linear in n. Thus

$$\mathcal{U}(\theta, d, T^*, n) = \mathcal{U}(\theta, d) - ns. \tag{5.59}$$

Working the tree backwards, $\mathcal{C}$ will choose at the decision node $\mathcal{D}_2$ that d for which

$$\int_\theta \mathcal{U}(\theta, d) F(\mathrm{d}\theta | T^*, n)$$

is a maximum. At the decision node $\mathcal{D}_1$, $\mathcal{C}$ will choose that n for which

$$\left[E_{T^*} \left\{ \max_d \int_\theta \mathcal{U}(\theta, d) F(\mathrm{d}\theta | T^*, n) \right\} - ns \right]$$

is a maximum. Were $\mathcal{C}$ to choose d and n as indicated above, then $\mathcal{C}$'s expected utility will be

$$\max_n \left[E_{T^*} \left\{ \max_d \int_\theta \mathcal{U}(\theta, d) F(\mathrm{d}\theta | T^*, n) \right\} - ns \right]. \tag{5.60}$$

Observe that (5.60) is an extension of the first part of (5.47), the expected utility due to T^* and n. In principle, the above pre-posterior analysis of $\mathcal{C}$'s actions provides a normative solution to the problem of $\mathcal{C}$ choosing an optimal sample size.

If the goal of life-testing is to obtain an enhanced appreciation of θ, then $\mathcal{U}(\theta, d) = \mathcal{U}(\theta, p(\bullet)) = \log p(\theta)$ (section 5.5.1). In this case $\mathcal{C}$ will choose that n which yields

$$\max_n \left[E_{T^*} \left\{ \int_\theta \log(F(\mathrm{d}\theta | T^*, n)) \right\} - ns \right]. \tag{5.61}$$

If the goal of life-testing is for $\mathcal{C}$ to make a rational accept/reject decision, then it is best to formulate the problem in terms of the observable T (instead of the θ) as was done in section 5.5.3. Once this is done, $\mathcal{C}$ will choose that n which results in

$$\max_n \left[E_{T^*} \left\{ \max_{A,R} \mathcal{U}(t, \bullet) F(\mathrm{d}t | T^*, n) \right\} - ns \right], \tag{5.62}$$

where $\mathcal{U}(t, \bullet)$ is $\mathcal{U}(t, A)$ or $\mathcal{U}(t, R)$, and $F(\mathrm{d}t | T^*, n)$ is the posterior predictive distribution of T, were $\mathcal{C}$ to take a sample of size n and observe T^* as data. The details leading to (5.62) are relatively straightforward; they are left as an exercise for the reader. It is important to note that even though $\mathcal{C}$ can make a decision to accept prior to the failure of all the n items on test, $\mathcal{C}$'s pre-posterior analysis for choosing an optimal should be based on the premise that testing will continue until all n items fail.

An obstacle to using (5.61) and (5.62) is the difficulty of performing the underlying computations. Even the simplest assumptions regarding the chance distributions and the priors on their parameters, lead to complex calculations. Computing expectations with respect to the distribution of T^* is certainly a hurdle. Thus efficient simulations and approximation schemes, suitably codified for computer use, are highly desirable. The situation described here gets more complicated if we wish to incorporate the scenario in which the cost of life-testing is a function of both s, the cost of procuring an item, and also the duration of the life-test. If $\mathcal{C}$ chooses an n, then $\mathcal{C}$ expects the life-test to last until τ_n, the failure time of the last item to fail. Thus if $\mathcal{C}$ incurs a disutility of s^* utiles for each unit of time the test is in progress, then the expected disutility incurred by $\mathcal{C}$ due to the conduct of the test is $ns + s^* E(\tau_n)$; $E(\tau_n)$ is the expected value of τ_n, the largest order statistic (out of n) of the predictive distribution of the n times to failure in the test. When this disutility is accounted for, (5.60) becomes

$$\max_n \left[E_{T^*} \left\{ \max_d \int_\theta \mathcal{U}(\theta, d) F(\mathrm{d}\theta | T^*, n) \right\} - ns - s^* E(\tau_n) \right]. \tag{5.63}$$

Since $E(\tau_n)$ will entail n, the inclusion of $E(\tau_n)$ in the term to be maximized over n is meaningful. However, a question may arise as to why $E(\tau_n)$ is not included in the term within the braces over which an expectation with respect to the distribution of T^* is taken. This is a subtle matter that deserves mention. It has to do with the fact that when $\mathcal{C}$ is contemplating n at the decision node $\mathcal{D}_1$, all that is available to $\mathcal{C}$ is the predictive distribution of the n times to failure based on $\mathcal{C}$'s prior alone. However, at $\mathcal{D}_2$ when $\mathcal{C}$ is contemplating an action to take, $\mathcal{C}$ has, in addition to the prior, the T^* that $\mathcal{C}$ hopes to observe. Thus the expectation with respect to T^* of the quantity in braces of (5.63). The above points are best illustrated when we consider an evaluation of $E(\tau_n)$. The predictive distribution of the times to failure of the items on test is based on the prior alone. Thus

$$P(T > t) = \int_\theta P(T > t | \theta) F(\mathrm{d}\theta).$$

Since τ_n is $\max(\tau_1, \tau_2, \ldots, \tau_n)$, its distribution function is

$$P(\tau_n \leq t) = (1 - P(T \leq t))^n = \left[1 - \int_\theta P(T > t | \theta) F(\mathrm{d}\theta) \right]^n.$$

Let $F(\mathrm{d}\tau_n)$ denote the probability density generated by the above distribution at the point τ_n. Then

$$E(\tau_n) = \int_0^\infty \tau_n F(\mathrm{d}\tau_n);$$

$\mathbf{E}(\tau_n)$ depends on n since $F(\mathrm{d}\tau_n)$ depends on n – thus its inclusion in the term within brackets over which n is maximized.

Thus to summarize, should $\mathcal{C}$ wish to incorporate the cost of testing into a pre-posterior analysis, then $\mathcal{C}$ will choose that n which yields:

$$\max_n \left[E_{T^*} \left\{ \max_d \int_\theta \mathcal{U}(\theta, d) F(\mathrm{d}\theta | T^*, n) \right\} - ns - s^* \int_0^\infty \tau_n F(\mathrm{d}\tau_n) \right]. \tag{5.64}$$

In order to compute an expectation with respect to T^*, $\mathcal{C}$ needs to obtain the marginal distribution of T^* when n items are tested. Let $F(T^*; n)$ denote the marginal distribution of T^*, and $F_{\mathcal{C}}(\theta)$, $\mathcal{C}$'s prior distribution forθ. Then for $\mathcal{C}$

$$F(T^*; n) = \int_{\theta} F(T^* | \theta; n) F_{\mathcal{C}}(\mathrm{d}\theta); \tag{5.65}$$

let $F(\mathrm{d}T^*; n)$ be the probability density at T^* generated by $F(T^*; n)$. Then (5.64) becomes

$$\max_n \left[\int_{T^*} \left\{ \max_d \int_{\theta} \mathcal{U}(\theta, d) F(\mathrm{d}\theta | T^*, n) \right\} F(\mathrm{d}T^*; n) - ns - s^* \int_0^{\infty} \tau_n F(\mathrm{d}\tau_n) \right]. \tag{5.66}$$

Evaluating the above, either via numerical approximations or via simulations, is an open problem – perhaps a challenge – that needs to be addressed. Since $\mathcal{C}$ has the option of conducting a time truncated or an item censored test, T^* is best encapsulated by the total time on test statistic (section 5.4.3).

5.7 ADVERSARIAL LIFE-TESTING AND ACCEPTANCE SAMPLING

Consider the scenario of section 5.5.3 wherein a consumer $\mathcal{C}$, who needs a batch of items from a manufacturer $\mathcal{M}$, either accepts or rejects the items offered by $\mathcal{M}$. With reference to Figure 5.8, suppose that for $\mathcal{C}$, the parameters α and ψ of the inverted gamma prior on ξ – the mean time to failure – lie in the region of rejection. Then $\mathcal{C}$'s decision is to reject the lot and consider alternate manufacturers. We are supposing here that $\mathcal{C}$ is unwilling to incur any disutilities by conducting a life-test on a sample of items from the batch to either refute or to re-affirm the decision to reject. $\mathcal{C}$'s conduct is therefore adversarial. The problem that we wish to consider here is the case wherein upon rejection by $\mathcal{C}$, $\mathcal{M}$ considers the option of offering a free sample to $\mathcal{C}$ for the purpose of testing, in the hope that subsequent to testing $\mathcal{C}$ will change his (her) mind and accept the lot; the cost of testing is also to be borne by $\mathcal{M}$. Should $\mathcal{M}$ offer $\mathcal{C}$ a sample, and if so, how large a sample should it be?

The above problem has been discussed by Lindley and Singpurwalla (1991 and 1993); the former in the context of acceptance sampling for quality control involving Bernoulli, Poisson, and Gaussian sampling, and the latter in the context of reliability involving exponential life-testing. In what follows I give an overview of the life-testing scenario by focusing attention on $\mathcal{M}$'s decision tree (Figure 5.10). Once $\mathcal{M}$ offers $\mathcal{C}$ a sample for life-testing, and if $\mathcal{C}$ agrees to accept this offer, then $\mathcal{C}$'s subsequent actions will be determined by the material of section 5.5.3, and in particular by the dictates of Figure 5.7.

$\mathcal{M}$'s decision of how large a sample $n \geq 0$ to offer $\mathcal{C}$ will be based on $\mathcal{M}$'s prior about ξ and a knowledge of $\mathcal{C}$'s accept/reject criteria. $\mathcal{M}$ and $\mathcal{C}$ need not agree on a prior neither should they be required to pool their priors; furthermore, $\mathcal{M}$ need to know $\mathcal{C}$'s prior. However, $\mathcal{M}$ needs to be informed of $\mathcal{C}$'s accept/reject criteria which are determined by $\mathcal{C}$'s prior and $\mathcal{C}$'s utilities. For example, in Figure 5.8, $\mathcal{C}$'s accept/reject boundary is a straight line, and $\mathcal{C}$'s criterion for acceptance is

$$\frac{\psi + \tau}{\alpha + k - 1} > \frac{a_2 + a_3}{a_1}.$$

If the optimal value of $\mathcal{M}$'s n turns out to be zero, then $\mathcal{M}$ does not offer $\mathcal{C}$ a sample and the adversarial game is concluded. With $n = 0$, the implication is that $\mathcal{C}$'s prior probability

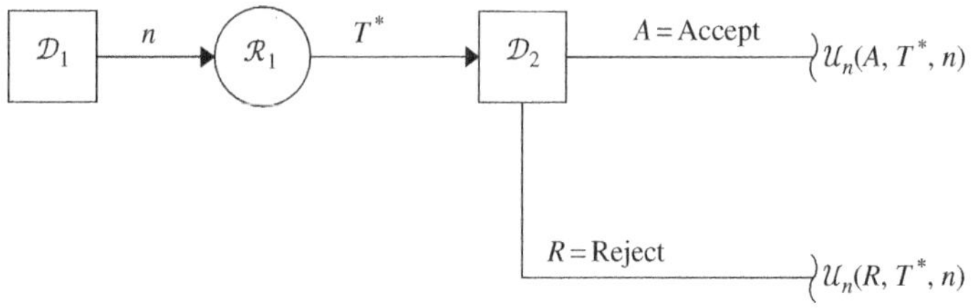

Figure 5.10 $\mathcal{M}$'s Decision tree with $\mathcal{C}$ taking an action at $\mathcal{D}_2$.

and utilities are such that, in $\mathcal{M}$'s judgment (based on $\mathcal{M}$'s prior and utilities), no amount of testing can convince $\mathcal{C}$ to change his/her mind. Thus as far as $\mathcal{M}$ is concerned, offering $\mathcal{C}$ a free sample is a losing proposition.

Figure 5.10 shows $\mathcal{M}$'s decision tree for the set-up described above. What is novel here is that at the decision node labeled $\mathcal{D}_2$, it is $\mathcal{C}$ (not $\mathcal{M}$) who takes an action. In the decision trees of Figures 5.7 and 5.9, it was $\mathcal{C}$ who was taking actions at all the decision nodes.

At the decision node $\mathcal{D}_1$, $\mathcal{M}$ takes the action of choosing an n (much like $\mathcal{C}$ does in Figure 5.9), and at the random node $\mathcal{R}_1$, nature reveals T^* as data. Subsequent to this, $\mathcal{C}$ makes the decision to either accept or to reject the lot based on $\mathcal{C}$'s accept/reject criteria. That is $\mathcal{C}$ takes those actions which maximize $\mathcal{C}$'s expected utility. $\mathcal{C}$'s accept/reject criteria boil down to T^* belonging to an acceptance region, say $\mathcal{A}(n)$, for acceptance and a rejection region, say $\mathcal{R}(n)$, for rejection (Figure 5.8). Once $\mathcal{C}$ takes an action, $\mathcal{M}$ experiences a terminal utility $\mathcal{U}_M(A, T^*, n)$ or $\mathcal{U}_M(R, T^*, n)$. Note that unlike the decision trees of Figures 5.7 and 5.9, the terminal utility to $\mathcal{M}$ need not be determined by θ (or by T) which encapsulate nature's course of action. All that could matter to $\mathcal{M}$ is acceptance or rejection by $\mathcal{C}$. The decision tree of Figure 5.10 pertains to $\mathcal{M}$ contemplating what n to offer $\mathcal{C}$. Thus the analyses that lead to an optimal course of action by $\mathcal{M}$ is a pre-posterior analysis. Consequently, to $\mathcal{M}$, the decision node $\mathcal{D}_2$ is no different from a random node in the sense that instead of nature taking an action it is $\mathcal{C}$ who takes the action. Furthermore $\mathcal{C}$'s action at $\mathcal{D}_2$ will depend on the data T^* that will be observed by $\mathcal{C}$, and whether T^* lies in the acceptance region $\mathcal{A}(n)$, or the rejection region $\mathcal{R}(n)$.

Following the principle of maximization of expected utility, $\mathcal{M}$ will fold the decision tree of Figure 5.10 backwards, and will offer $\mathcal{C}$ that sample size n which results in

$$\max_n \left\{ \int_{T^* \in \mathcal{A}(n)} \mathcal{U}_M(A, T^*, n) F(\mathrm{d}T^*; n) + \int_{T^* \in \mathcal{R}(n)} \mathcal{U}_M(R, T^*, n) F(\mathrm{d}T^*; n) \right\},$$

where $F(\mathrm{d}T^*; n)$ is the probability density at T^* generated by $F(T^*; n)$, the marginal distribution of T^* as perceived by $\mathcal{M}$, were $\mathcal{M}$ to offer $\mathcal{C}$ a sample of size n. If $F_{\mathcal{M}}(\mathrm{d}\theta)$ denotes $\mathcal{M}$'s prior distribution for θ, then

$$F(T^*; n) = \int_{\theta} F(T^* | \theta; n) F_{\mathcal{M}}(\mathrm{d}\theta). \tag{5.67}$$

Contrast (5.67) with (5.65); the former is based on $\mathcal{M}$'s prior for θ, whereas the latter is on $\mathcal{C}$'s prior for θ. Since $\mathcal{M}$ is to bear the cost of testing,

$$\mathcal{U}_M(A, T^*, n) = \mathcal{U}_M(A, n) = b_1 - b_2 n,$$

assuming that $\mathcal{M}$'s utility for acceptance or rejection by $\mathcal{C}$ does not depend on T^*; $b_1 > 0$ is the utility to $\mathcal{M}$ (in utiles) for every item accepted by $\mathcal{C}$ and $b_2 > 0$, the disutility incurred by $\mathcal{M}$ for every item tested. Similarly, $\mathcal{U}_M(R, T^*, n) = b_3 + b_2 n$ is the disutility to $\mathcal{M}$ for an item rejected by $\mathcal{C}$ subsequent to the testing of n items; $b_3 > 0$. With the above in place, $\mathcal{M}$ will offer $\mathcal{C}$ that sample size n which results in

$$\max_n \left\{ \int_{T^* \in \mathcal{A}(n)} (b_1 - b_2 n)\, F(\mathrm{d}T^*; n) - \int_{T^* \in \mathcal{R}(n)} (b_3 + b_2 n)\, F(\mathrm{d}T^*; n) \right\}.$$

More details and specific cases can be found in Lindley and Singpurwalla (1991, 1993).

5.8 ACCELERATED LIFE-TESTING AND DOSE–RESPONSE EXPERIMENTS

Accelerated life-testing was alluded to in section 5.4.1 as a way of economizing on test time by elevating the environmental conditions under which a life test is performed. **Nominal** or **use conditions stress** means the environmental conditions under which an item is designed to be used. The elevated conditions under which identical copies of the item are tested are known as **accelerated stresses**. An **accelerated life-test** is a life-test in which items are tested under accelerated stresses to gain information about its lifelength under nominal stress. This type of testing is done when it is not physically possible to conduct life-tests under use conditions stress, or when testing under use conditions entails an excessive amount of test time to yield failures. Accelerated life-testing came into being in the space age; however, to the best of our knowledge, it was pioneered at the Bell Telephone Laboratories in the context of assessing the reliability of underwater telephone cables.

The viability of accelerated life-testing as a way to learn about lifetimes under use conditions stress depends on an ability to relate the information gained at one stress level to that which would be gained at lower stress levels. Thus a knowledge of the underlying physics of failure plays a key role here. Such knowledge is encapsulated in what is known as the **time transformation function**. If the time transformation function is ill chosen, then the resulting inferences can be misleading. In principle, there could be several time transformation functions, each appropriate for a particular stress regime. Subsequent to the initial papers that laid out a statistical framework to model this problem (Singpurwalla 1971a, b) much has been written on the topic of accelerated life testing. A recent treatise on this topic is by Bagdonavicius and Nikulin (2001). In section 5.8.1, I start with a formal introduction to the accelerated life-testing problem by presenting two commonly occurring versions; one based on time to failure data and the other based on the proportion of observed failures. Section 5.8.2 is an overview of the technology of Kalman filtering, my proposed method for dealing with inferential issues underlying the accelerated life-testing problem. Section 5.8.3 pertains to an illustration of the application of the Kalman filter algorithm for inference and extrapolation in accelerated tests; the material here is expository with details kept to a minimum. Section 5.8 ends with the important issue of the design of experiments for accelerated testing; the material of section 5.8.4 draws on the notions of entropy and information discussed in section 5.5.1.

5.8.1 Formulating Accelerated Life-testing Problems

Suppose that T_j denotes the time to failure of an item under stress S_j, $j = 1, \ldots, m$, where $S_1 < S_2 < \cdots < S_m$; '$S_j < S_k$' denotes the fact that S_j is less severe than S_k. Severity of stress manifests itself in the fact that $S_j < S_k \Rightarrow T_j \overset{st}{\geq} T_k$; i.e. for all $t \geq 0$, $P(T_j \geq t) \geq P(T_k \geq t)$. Let S_u be the use conditions stress and suppose that $S_u < S_1$; then $S_1, S_2, \ldots, S_m$ are accelerated stresses. We assume that it is possible to conduct life tests at all the m accelerated stresses, but that it may or may not be possible to do any testing at use conditions stress S_u and any stress $S_{\overline{u}} < S_u$.

To keep our discussion simple, suppose that n items are tested under S_j and that the test is stopped at the time of occurrence of the k-th failure; for $j = 1, \ldots, m$; i.e. we have a Type II censored test at all the accelerated stress levels. In actuality, it is most likely that a large number of items will be tested at low stress levels so that $n_1 \geq n_2 \geq \cdots \geq n_m$, where n_j denotes the number of items tested under stress S_j; our assumption that $n_j = n$ for all j simplifies the notation without a compromise in generality. Similarly, with the assumption of a common k over all stress levels. Let T_{jl}, $l = 1, 2, \ldots, k$ denote the time to failure of the l-th item when it experiences a stress S_j, $j = 1, \ldots, m$. Let $T_{(jl)}$ denote the l-th order statistic of the T_{jl}s; i.e. $T_{(j1)} \leq T_{(j2)} \leq \cdots \leq T_{(jk)}$. Given that the $t_{(jl)}$'s, $j = 1, \ldots, m$, $l = 1, \ldots, k$, our aim is to make statements of uncertainty about T_u; $t_{(jl)}$ is a realization of $T_{(jl)}$.

In order to do the above, we need to specify a relationship between T_j and T_u for each j, $j = 1, \ldots, m$. It is important that there be some physical or practical justification for such relationships. These relationships are the time transformation functions mentioned before, and in principle are under the preview of subject matter specialists. Since specifying several relationships, one for each T_j, can be cumbersome, a simplifying strategy is to assume a common general form over all the T_js with the only variable being the S_js. An example is the famous **Power Law Model** in which for $j = 1, \ldots, m$, the random quantities $\overline{F}_j(T_{(jl)})$, $l = 1, \ldots, k$, are related to each other via the unknown parameters $\alpha_1, \alpha_2 > 0$ as:

$$\overline{F}_j(T_{(jl)}) = \exp(-\alpha_1 S_j^{\alpha_2} T_{(jl)}); \tag{5.68}$$

here, $\overline{F}_j(t) = P(T_j \geq t)$. Implicit to the above relationship is the assumption that the underlying chance distribution is exponential.

Traditionally, the Power Law and other such time transformation functions such as the Arrhenius and Eyring laws (Mann, Schafer and Singpurwalla, 1974), have been stated in terms of the unknown parameters of the underlying chance distributions. The specification of (5.68) is due to Blackwell and Singpurwalla (1988). It is a generalization of a traditional power law model for accelerated life-testing, namely, $P(T_j > t|\theta_j) = \exp(-t/\theta_j)$, with $\theta_j = \alpha_1 S_j^{\alpha_2}$. Thus in $\overline{F}_j(T_{(jl)}) = \exp(-\alpha_1 S_j^{\alpha_2} t)$, we replace t by $T_{(jl)}$ to produce the model of (5.68). There are two advantages to our proposed model. One is that it is a relationship between the observables like the $T_{(jl)}$s instead of parameters like the θ_j's. The second is that the model provides a framework for invoking a computationally efficient filtering algorithm. A similar strategy may be adopted when the underlying chance distribution is a Weibull. However, the filtering mechanism will entail computational difficulties due to non-linearities.

The data from an accelerated life-test $t_{(jl)}$, $j = 1, \ldots, m$, $l = 1, \ldots, k$, is used to make inferences about α_1 and α_2, and these in turn are used to a obtain predictive distribution for T_u, namely $P(T_u \geq t_u)$. The Kalman filter model, also known as the Dynamic Linear Model or a State Space Model, provides a nice mechanism for accomplishing the said inference. The Kalman filter model is a general methodology with applications to other problems in reliability. It is overviewed in section 5.8.2, following which I will continue with the problem posed above. But first I give below another version of the accelerated life-testing problem, a version that is prompted by scenarios in dose–response studies in the biomedical context, and in damage assessment experiments in the engineering reliability context. This second version of the accelerated life-testing problem is also amenable to analysis via a Kalman filter model.

Dose–Response and Damage Assessment Studies

In dose–response studies, several subjects are treated to a dose, and the proportion of subjects that respond to the dose are noted. The dose is a proxy for the stress in accelerated life-testing, and the response could be a subject's failure or survival. Thus in a dose–response experiment the response is a binary variable, whereas in conventional accelerated life-testing the response is the time to failure. The testing is done at several levels of the dose, and generally, these levels are higher than the nominal dose. The aim is to make an inference about the subject's response at the nominal dose. In damage assessment studies, the situation is slightly different in the sense that often it is not possible to test an item at any desired stress that is pre-specified; the consequence that only one item can be tested at any level of the stress. The response in damage assessment studies is a number between zero and one (both inclusive) indicating the extent of damage done; often, the response is a subjective judgment. Here again, items are tested at the several levels of the accelerated stresses, the aim being to assess damage at a nominal stress. Both the dose–response and the damage assessment scenario can be formulated in a unified way as described below. In the interest of clarity, the notation used below is different from that used in the context of conventional accelerated testing.

Quantified values of the dose (or the damage inflicting stress) will be denoted by X, where X takes values x, $0 \leq x < \infty$. The response of a subject to a dose x will be denoted by $Y(x)$, where $0 \leq Y(x) \leq 1$, for any x. In drug testing, wherein x denotes the dose that is administered to a patient, the relationship between $Y(x)$ and x is known as the **potency curve** or the **dose–response function**. In the context of damage assessment, $Y(x) = 0$ implies the total demolition of an item under study, and $Y(x) = 1$ denotes a total resistance to the damage causing agent. It is reasonable to suppose that $Y(x)$ is non-increasing in x, with $Y(0) = 1$ and $Y(\infty) = 0$. Whereas one can propose several plausible models for describing this behavior of $Y(x)$, the model I consider here is of the form

$$E(Y(x)|\alpha, \beta) = \exp(-\alpha x^{\beta}), \tag{5.69}$$

where $\alpha, \beta > 0$ are unknown parameters. Alternate strategies involve taking a non-parametric approach wherein no parametric function relating $Y(x)$ to x is the sole basis of an analysis. More about this is said later in section 9.3.

Observe that the right-hand side of equation (5.69) is the survival function of a Wiebull distribution. Arguments that support this choice for a model are given by Meinhold and Singpurwalla (1987), who illustrate its generality for describing the several possible shapes that $E(Y(x)|\alpha, \beta)$, the expected dose–response function, can take. Figure 5.11, taken from the above reference, illustrates this feature.

The data from a dose–response (or a damage assessment) experiment involving testing at doses $x_1, \ldots, x_m$ will consist of the responses $Y(x_1), \ldots, Y(x_m)$. The pair $(x_j, Y(x_j))$, $j = 1, \ldots, m$, will be used in a Kalman filter model for making inferences about α and β, and these in turn will enable us to assess $Y(x_u)$, the response at nominal dose x_u. The specifics on how to proceed with the above can be best appreciated once the Kalman filter model and the filtering algorithm are introduced. I therefore digress from the problems at hand and return to them in section 5.8.3, following an overview of Kalman filtering.

5.8.2 The Kalman Filter Model for Prediction and Smoothing

The Kalman filter model (KF) is based on two equations, an 'observation equation', and a 'system or state equation', with each equation containing an error term having a Gaussian distribution with parameters assumed known. The observation equation, given below, says that

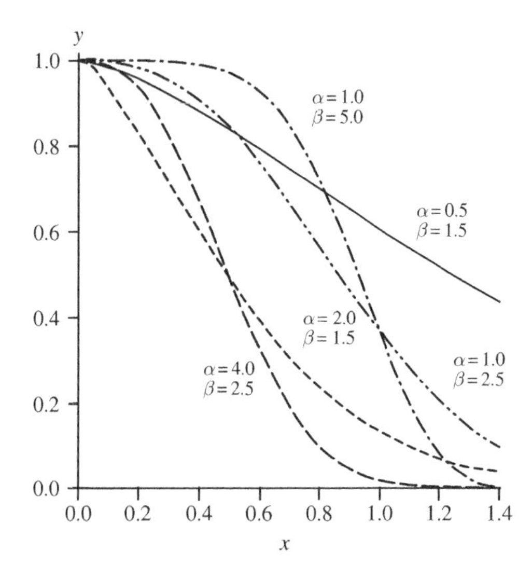

Figure 5.11 A Plot of $E(Y(x)) = \exp(-\alpha X^{\beta})$ for different choices of values of α and β.

for some index, say x_j, the response $Y(x_j)$ is related to an unknown state of nature $\theta(x_j)$, via the relationship:

$$Y(x_j) = F(x_j)\theta(x_j) + v(x_j), \tag{5.70}$$

where $F(x_j)$ is a known scalar. The quantity $v(x_j)$ is an error term, assumed Gaussian with mean 0 and variance $V(x_j)$, also assumed known; this is denoted as '$v(x_j) \sim N(0, V(x_j))$'. The state of nature $\theta(x_j)$ is assumed to be related to its predecessor $\theta(x_{j-1})$ via the **system equation**:

$$\theta(x_j) = G(x_j)\theta(x_{j-1}) + w(x_j), \tag{5.71}$$

where $w(x_j) \sim N(0, W(x_j))$ with $W(x_j)$ assumed known. Also, the $v(x_j)$ and the $w(x_j)$, $j = 1, 2, \ldots,$ are serially and contemporaneously uncorrelated; furthermore, the $G(x_j)$ is also assumed known. For equation (5.71) to be meaningful, the x_j's need to be ordered, as say $x_1 \geq x_2 \geq \cdots \geq x_j \geq \cdots$; this will ensure that the $\theta(x_j)$'s will also be ordered. Equations (5.70) and (5.71) are given in scalar form; in principle, they could also be in vector form. Furthermore, some, but not all, of the error theory assumptions can be relaxed to encompass inter-dependence and non-Gaussianity; see, for example, Meinhold and Singpurwalla, (1989).

Traditionally, the KF model, which is comprised of the observation and the system equations, is introduced and discussed with time as an index. My choice of the x_j's as indices is to facilitate linkage with the accelerated life-testing scenario of section 5.8.1, in particular the dose–response and the damage assessment models. Thus in what follows, it makes sense to replace x_j by j, so that $Y(x_j)$ is simply Y_j and $\theta(x_j)$ is θ_j; similarly the other quantities. With the above notation in place, suppose now that $\theta_{j-1} \sim N(\widehat{\theta}_{j-1}, \Sigma_{j-1})$, for $j = 1, 2, \ldots$; for $j = 1$, we assume that $\widehat{\theta}_0$ and Σ_0 are user specified. Then, the KF mechanism yields the recursive result that on observing Y_j,

$$\theta_j \sim N(\widehat{\theta}_j, \Sigma_j) \tag{5.72}$$

where $\widehat{\theta}_j$ and Σ_j are given by the **forward recursive equations** as:

$$\begin{aligned}
\widehat{\theta}_j &= G_j\widehat{\theta}_{j-1} + K_j(Y_j - F_jG_j\widehat{\theta}_{j-1}),\\
\Sigma_j &= (1 - K_jF_j)R_j,\\
R_j &= G_j^2\Sigma_{j-1} + W_j, \quad \text{and}\\
K_j &= R_jF_j(F_j^2R_j + V_j)^{-1};
\end{aligned} \tag{5.73}$$

the details, which entail a Bayesian prior to posterior updating, are in Meinhold and Singpurwalla (1983).

The recursive nature of (5.72) makes it suitable for the situation wherein given any index j, the focus is on predicting Y_{j+1} via inference about θ_j, the latter being based on $Y_j, Y_{j-1}, \ldots, Y_1$. As an illustration, suppose that $F_j = G_j = 1$ for all j; then from (5.70) and (5.71) we see that $Y_{j+1} = \theta_{j+1} + v_{j+1} = \theta_j + w_{j+1} + v_{j+1}$. But $\theta_j \sim N(\widehat{\theta}_j, \Sigma_j)$, where $\widehat{\theta}_j$ and Σ_j are given by (5.73). Consequently, given $Y_j, Y_{j-1}, \ldots, Y_1$, we note that the predictive distribution of Y_{j+1} is a Gaussian with mean $\widehat{\theta}_j$ and variance $\Sigma_j + W_{j+1} + V_{j+1}$. Thus the KF model provides a recursive mechanism for predicting the future values of an observable given its past values.

The above predictive scheme is appropriate in the context of a **time series**, that is, when the index j represents time. A time series is a collection of observations that are obtained over time; for example, the hourly temperature reading at a certain location. Once Y_{j+1} becomes available, attention will shift to θ_{j+1} and Y_{j+2}; the assessment of θ_j, which has been based on $Y_1, Y_2, \ldots, Y_j$, will not be revised. On the other hand, for those situations wherein all the Y_js, say m in all, are simultaneously presented, it is meaningful to base inference about any θ_j on all the available Y_js – not just $Y_1, Y_2, \ldots, Y_j$. Such an approach is called **Kalman filter smoothing**, and is appropriate in the context of dose–response and damage assessment studies wherein all the Y_js are simultaneously available.

To describe the smoothing formulae, we need to slightly expand our notation. Let $\widehat{\theta}_j(m)$ and $\Sigma_j(m)$ denote the mean and the variance of the distribution of θ_j based on m observations $Y_1, Y_2, \ldots, Y_m$. Then, it can be shown that the **backwards recursion equations** (Meinhold and Singpurwalla, 1987) yield:

$$\begin{aligned}
\widehat{\theta}_j(m) &= \widehat{\theta}_j(j) + J_{j+1}[\widehat{\theta}_{j+1}(m) - G_{j+1}\widehat{\theta}_j(j)], \quad \text{and}\\
\Sigma_j(m) &= \Sigma_j(j) - J_{j+1}[\Sigma_{j+1}(m) - R_{j+1}]J_{j+1},
\end{aligned} \tag{5.74}$$

where $J_j = \Sigma_{j-1}(j-1)G_j(R_j)^{-1}$. The details are in Appendix A of the paper mentioned above. Note that $\widehat{\theta}_j(j)$ and $\Sigma_j(j)$ are the mean and variance, respectively, of the distribution of θ_j based on $Y_1, Y_2, \ldots, Y_j$ alone, and are obtained via (5.73). The predictive distribution of Y_{m+1} will be based on inference about θ_m, namely, that $\theta_m \sim N(\widehat{\theta}_m(m), \Sigma_m(m))$, with $\widehat{\theta}_m(m)$ and $\Sigma_m(m)$ given by (5.74). This completes our review of the KF model.

5.8.3 Inference from Accelerated Tests Using the Kalman Filter

The time transformation functions of section 5.8.1 are natural candidates for prescribing the observation equations of the Kalman filter model. Once this connection is made, and appropriate change of variables introduced to ensure that the error theory assumptions of the underlying KF models are satisfied, inference from accelerated life-tests proceeds, at least in principle, in a straightforward manner; the details tend to get messy and are therefore kept to a minimum. To see this, let us first consider the dose–response (damage assessment) scenario of section 5.8.1.

Filtering and Smoothing Dose–Response (Damage Assessment) Data
Recall the time transformation function of (5.69), namely $E(Y(x)|\alpha, \beta) = \exp(-\alpha x^{\beta})$, where $x \in [0, \infty)$ is the dose. To facilitate an application of the KF algorithm, we need to introduce a transformation of $Y(x)$ that has a Gaussian distribution. For this, we define a random variable $Y^*(x) = \log\{-\log(Y(x))\}$, and require that $Y^*(x)$ have a Gaussian distribution with mean $\mu(x)$ and variance $\sigma^2(x)$. The merits of this transformation will become clear in the sequel, but for now it suffices to say that with $Y^*(x) \sim N(\mu(x), \sigma^2(x))$, $Y(x)$ has what Meinhold and Singpurwalla (1987) call a **double lognormal distribution**; its probability density at $y(x)$ is of the form:

$$\frac{1}{\sqrt{2\pi}\sigma(x)y(x)(-\log y(x))} \exp\left\{-\frac{1}{2\sigma^2(x)}(\log(-\log y(x)) - \mu(x))^2\right\}, \quad \text{for } 0 \leq y(x) \leq 1.$$

Figure 5.12 taken from Meinhold and Singpurwalla (1987) shows plots of this density for different values of $\mu(x) = \mu$, and $\sigma(x) = \sigma$. These plots show the versatility of this distribution to represent one's subjective opinions about the damage phenomena in the [0, 1] interval.

Properties of the double lognormal distribution are given in Meinhold and Singpurwalla (1987). The one of immediate interest to us here is its mean $E(Y(x)|\mu(x), \sigma^2(x))$, which for small values of $\sigma(x)$ is approximately $\exp(-\exp(\mu(x)))$. However, $E(Y(x)|\alpha, \beta)$ is also equal to $\exp(-\alpha x^{\beta})$, and so $E(Y^*(x)|\mu(x)) = \mu(x) \approx \log \alpha + \beta \log x$. This relationship forms the basis

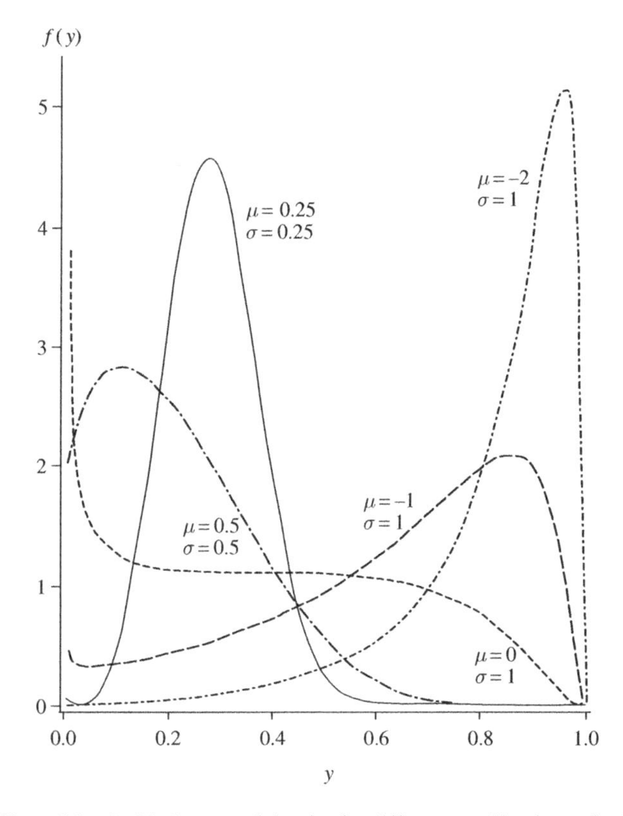

Figure 5.12 Plots of the double lognormal density for different combinations of values of μ and σ.

of our observation equation

$$Y^*(x_j) = \log\alpha + \beta\log(x_j) + v(x_j) \tag{5.75}$$

with $v(x_j) \sim N(0, V(x_j))$.

If we let $\boldsymbol{F}(x_j) = [1, \log(x_j)]$, a row vector, and $\boldsymbol{\theta}(x_j) = [\boldsymbol{\gamma}_j, \boldsymbol{\beta}_j]' = [\boldsymbol{\gamma}, \boldsymbol{\beta}]'_j$, where $\boldsymbol{\gamma}_j = \log\alpha_j$, then we may rewrite (5.75) in vector form as:

$$Y^*(x_j) = \boldsymbol{F}(x_j)\boldsymbol{\theta}(x_j) + v(x_j).$$

This is our observation equation in vector form; it is identical to the form of (5.70) save for the fact that $\boldsymbol{F}(x_j)$ and $\boldsymbol{\theta}(x_j)$ are vectors. To prescribe a system equation for the state of nature $\boldsymbol{\theta}(x_j)$, I use a commonly used strategy in Kalman filtering, namely that of employing the steady model. Specifically, our system equation is written as:

$$\begin{pmatrix}\boldsymbol{\gamma}\\ \boldsymbol{\beta}\end{pmatrix}_j = \begin{pmatrix}\boldsymbol{\gamma}\\ \boldsymbol{\beta}\end{pmatrix}_{j-1} + \boldsymbol{w}(x_j), \tag{5.76}$$

where the vector $\boldsymbol{w}(x_j) \sim N(0, W(x_j))$. The above equation parallels that of (5.71) save for the feature that the state of nature here is a vector quantity and the $\boldsymbol{G}(x_j)$ of (5.71) is a 2×2 identity matrix $\boldsymbol{G}_j \equiv \mathbf{I}$.

It is important to note that the unknown state of nature $[\boldsymbol{\gamma}, \boldsymbol{\beta}]$ is indexed by j, even though our time transformation model did not postulate that the parameters change with dose. The advantage of indexing the parameter vector by j is added flexibility; it makes allowances for departures from the postulated time transformation function. In vector notation, the forward recursion equations of (5.73) change, but only slightly. Specifically, $\boldsymbol{\Sigma}_j$, $\boldsymbol{R}_j$ and $\boldsymbol{K}_j$ become: $\boldsymbol{\Sigma}_j = (\boldsymbol{I} - \boldsymbol{K}_j\boldsymbol{F}_j)\boldsymbol{R}_j$, $\boldsymbol{R}_j = \boldsymbol{G}_j\boldsymbol{\Sigma}_{j-1}\boldsymbol{G}'_j + \boldsymbol{W}_j$, and $\boldsymbol{K}_j = \boldsymbol{R}_j\boldsymbol{F}'_j(\boldsymbol{F}_j\boldsymbol{R}_j\boldsymbol{F}'_j + \boldsymbol{V}_j)^{-1}$; the formula for $\widehat{\boldsymbol{\theta}}_j$ remains the same, except that now it is a vector.

Although the dose levels for the experimental situation considered here may be ordered, there is no reason to suppose that the experiments will be conducted in any particular dose order. Thus inference is more reasonably conducted by smoothing than by filtering. However, in order to invoke the smoothing algorithm, we need to specify, in addition to $\boldsymbol{V}(x_j)$ and $\boldsymbol{W}(x_j)$, the starting values $\widehat{\boldsymbol{\theta}}_0$ and $\boldsymbol{\Sigma}_0$. The latter specification raises the issue of what dose level is to be treated as x_0, the dose from which the forward recursion process is to commence. One strategy is to label x_0 as that dose at which the analyst has the best prior knowledge about $\boldsymbol{Y}(x_0)$. Presumably, this will be a high level of dose, because it is at such dose levels that information is most readily available. As a potential starting value for $\boldsymbol{\beta}_0$, we may use (5.75) to observe that for large x_0, $\boldsymbol{Y}^*(x_0)/\log(x_0) \approx \boldsymbol{\beta}_0$. If x_m is the smallest dose, then using the forward recursion equations we filter from x_0 to x_m, and then use the backward recursion to smooth out the filtered results; this will produce the collection $[\widehat{\boldsymbol{\gamma}}_j, \widehat{\boldsymbol{\beta}}_j]$, $j = 0, 1, \ldots, m$. The data used for filtering and smoothing is the collection $(\boldsymbol{Y}^*(x_j), x_j)$, $j = 1, \ldots, m$. The stability of the $\widehat{\boldsymbol{\gamma}}_j$'s and the $\widehat{\boldsymbol{\beta}}_j$'s reflects the appropriateness of the time transformation function. When such is the case, the smoothed $(\widehat{\boldsymbol{\gamma}}_m, \widehat{\boldsymbol{\beta}}_m)$ pair can be used for extrapolation to x_u, namely to assess $\boldsymbol{Y}(x_u)$ via equations (5.75) and (5.76). Some ad hoc guidelines for choosing $\boldsymbol{V}(x_j)$ and $\boldsymbol{W}(x_j)$ are outlined by Meinhold and Singpurwalla (1987).

Filtering Conventional Accelerated Life-test Data

We start with the time transformation function of (5.68), namely

$$\overline{\boldsymbol{F}}_j\left(\boldsymbol{T}_{(jl)}\right) = \exp\left(-\alpha_1 \boldsymbol{S}_j^{\alpha_2} \boldsymbol{T}_{(jl)}\right),$$

and take logarithms on both sides to obtain

$$\ln\left\{-\ln(\overline{\boldsymbol{F}}_j(\boldsymbol{T}_{(jl)}))\right\} = \ln\alpha_1 + \alpha_2 \ln \boldsymbol{S}_j + \ln(\boldsymbol{T}_{(jl)}).$$

Letting $\boldsymbol{v}^*_{(jl)} = \ln\{-\ln(\overline{\boldsymbol{F}}_j(\boldsymbol{T}_{(jl)}))\}$, $\boldsymbol{Y}^*_{(jl)} = \ln(\boldsymbol{T}_{(jl)})$, $\boldsymbol{F}_j = (-1, -\ln \boldsymbol{S}_j)$, a vector, and $\boldsymbol{\theta}_{(jl)} = (\ln\alpha_1, \alpha_2)$, also a vector, the above equation may be rewritten as:

$$\boldsymbol{Y}_{(jl)} = \boldsymbol{F}_j\boldsymbol{\theta}_{(jl)} + \boldsymbol{v}^*_{(jl)}.$$

This relationship motivates us to consider a 'tentative' observation equation as

$$\boldsymbol{y}^*_{(jl)} = \boldsymbol{F}_j\boldsymbol{\theta}_{(jl)} + \boldsymbol{v}^*_{(jl)}, \tag{5.77}$$

where $\boldsymbol{y}^*_{(jl)}$ is a realization of $\boldsymbol{Y}^*_{(jl)}$.

Since $\overline{\boldsymbol{F}}_j(\boldsymbol{T}_{(jl)})$ has a beta distribution, some straightforward calculations (details left as an exercise to the reader) show that

$$E(\boldsymbol{Y}^*_{(jl)}) = l\binom{n}{l}\sum_{i=0}^{l-1}\binom{l-1}{i}\left[\frac{(-1)^i}{n-l+i+1}\right][-c-\ln(n-l+i+1)],$$

where $c \approx 0.577216$ is Euler's constant. Because $E(\boldsymbol{v}^*_{(jl)}) \neq 0$, we subtract it from both sides of (5.77) to write the observation equation as:

$$\boldsymbol{y}_{(jl)} = \boldsymbol{F}_j\boldsymbol{\theta}_{(jl)} + \boldsymbol{v}_{(jl)}, \tag{5.78}$$

where $\boldsymbol{y}_{(jl)} = \boldsymbol{y}^*_{(jl)} - E(\boldsymbol{v}^*_{(jl)})$ and $\boldsymbol{v}_{(jl)} = \boldsymbol{v}^*_{(jl)} - E(\boldsymbol{v}^*_{(jl)})$; this set-up ensures that $E(\boldsymbol{v}_{(jl)}) = 0$.

For the system equation, I propose the steady model

$$\boldsymbol{\theta}_{(jl)} = \boldsymbol{G}_j\boldsymbol{\theta}_{(j-1)l} + \boldsymbol{w}_{(jl)}, \tag{5.79}$$

where $\boldsymbol{G}_j$ is a 2×2 identity matrix $\mathbf{I}$, and $\boldsymbol{w}_{(jl)}$ is a vector $(\boldsymbol{w}_{1,jl}, \boldsymbol{w}_{2,jl})$ whose elements are Gaussian error terms with means 0 and covariance Σ_w.

Based on an empirical analysis of some numerical results, Blackwell and Singpurwalla (1988) are able to argue that for n as small as 10 and l close to n, $\boldsymbol{v}_{(jl)}$ is approximately Gaussian. The same is also true when n is greater than 10 and l is moderate to large. When n is large, say 25 or more, values of l as small as 6 will suffice to assume that $\boldsymbol{v}_{(jl)}$ is approximately Gaussian. Thus for small values of n – the number of items tested at each stress level – it is only the last few failure times that can be used in the filtering algorithm. It therefore makes sense to test a large number of items at low-stress levels.

Since the $\boldsymbol{v}_{(jl)}$'s are based on order statistics, it is reasonable to suppose that for any j, the covariance between $\boldsymbol{v}_{(jl)}$ and $\boldsymbol{v}_{(jp)}$, $l \neq p$ will be non-zero. However, the covariance between $v_{(jl)}$ and $v_{(pl)}$, $j \neq p$ can be assumed to be zero. That is, the error terms across stress levels are independent but within stress levels they are not. When such is the case the filtering algorithm tends to get involved; engineers refer to such scenarios as filtering in colored noise. Blackwell and Singpurwalla (1988) discuss this type of filtering for the problem at hand. Should we ignore the aforementioned covariance and assume independence of all the $\boldsymbol{v}_{(jl)}$'s, then one can invoke the filtering algorithm of section 5.8.2 starting from the highest stress S_m and filter down to the lowest stress S_1 to obtain the joint posterior distribution of $\ln\alpha_1$ and α_2; this will be a bivariate Gaussian. Once the above is at hand, inference about T_u follows along the lines indicated below.

Let $\ln h_u = \ln \alpha_1 + \alpha_2 \ln S_u$; then $\ln h_u$ has a Gaussian distribution with a mean μ_u and variance σ_u^2. These quantities will depend on the bivariate Gaussian distribution of $\ln \alpha_1$ and α_2. This in turn implies that h_u will have a lognormal distribution with density $p(h_u)$. Consequently

$$P(T_u > t) = \int_0^\infty \exp(-th_u)p(h_u)\mathrm{d}h_u,$$

which is the Laplace transform of a lognormal distribution. Lindley and Singpurwalla (1986) prescribe an approximation to this Laplace transform, which yields the result that

$$P(T_u > t) \approx \left(\frac{\sigma_h t}{\exp(-\mu_h)} + 1\right)^{-1/\sigma_h}.$$

5.8.4 Designing Accelerated Life-testing Experiments

In order to conduct an accelerated life test, an experimenter needs to choose several design variables. With respect to the notation of section 5.8.1, the design variables are: m, the number of accelerated stress levels at which to perform the tests; the values $S_1, ldots, S_m$ at which the tests are to be performed; n, the number of items tested at each stress, and k, the number of items to be tested until failure at each stress level, $k \leq n$. The design variables are under the control of the experimenter, and the collection of all possible designs will be donoted by $\mathcal{A}$.

The purpose of this section is to prescribe a procedure for designing an accelerated life test, where by 'design' I mean a specification of the elements of $\mathcal{A}$. However, in order to do so, we must first describe a general principle for designing experiments. The essence of this principle is grounded in the notions of mutual information and channel capacity; these were introduced in section 5.5.1, but in the context of learning about an unknown parameter. From a more general point of view, consider two random variables X and Y, whose joint probability density function at x and y is $p(x, y)$; let $p(x)$ and $p(y)$ be their marginal probability densities at x and y, respectively. Then, by direct analogy with (5.51), the mutual information between X and Y is defined as:

$$I(X:Y) = E_{(X,Y)}\left[\log \frac{p(X, Y)}{p(X)p(Y)}\right],$$

where the expectation is with respect to the joint distribution of X and Y, and $p(X)$ denotes the density of X evaluated at X; similarly, $p(Y)$ and $p(X, Y)$. How can $I(X:Y)$ help us design an experiment?

An answer to the above question lies in the considerations which originally motivated an introduction of the notion of mutual information. In the coding theory framework, if X represents the message to be sent and Y the message received by passing a coded version of X through a channel corrupted by noise – such a channel being represented by $p(X|Y)$ – then one strives to find that channel for which $I(X:Y)$ is a maximum. Recall that $I(X:Y)$ is analogous to $g(n)$ of (5.47), and $g(n)$ is the expected gain in utility. The maximum of $I(X:Y)$ with respect to $p(X|Y)$ is the channel capacity. With the above in place, let us now turn our attention to the problem of experimental design. Here the experimenter controls the design variables in $\mathcal{A}$. Let $A \in \mathcal{A}$, and suppose that the information contained in Y about X varies with A. I denote this dependence of X on both Y and A by $p(X \mid Y, A)$. The quantity $p(X \mid Y, A)$ is like a channel in coding theory, and the experimenter can choose from a family of channels $\{p(X|Y, A); A \in \mathcal{A})$. Then, the criterion would be to choose that A, say A^*, for which the mutual information about X and Y is maximized. That is, we seek the A^* which yields

$$\max_{A\in\mathcal{A}} E_{(X,Y|A)}\left[\log\frac{p(X,Y|A)}{p(X)p(Y|A)}\right]. \tag{5.80}$$

An equivalent way of writing the above is

$$\max_{A\in\mathcal{A}} E_{Y|A}E_{X|Y,A}\left[\log\frac{p(X,Y|A)}{p(X)}\right]; \tag{5.81}$$

the details are left as an exercise for the reader.

To invoke the above principle in the context of an experimental design whose aim is predicting a future observation Y_{n+1} (which depends on the design variables in $\mathcal{A}$), given $\mathbf{Y}=(Y_1,\ldots,Y_n)$, we replace X by Y_{n+1} and Y by $\mathbf{Y}$ in (5.81) to write

$$\max_{A\in\mathcal{A}} E_{\mathbf{Y}|A}E_{Y_{n+1}|\mathbf{Y},A}\left[\log\frac{p(Y_{n+1}|\mathbf{Y},A)}{p(Y_{n+1})}\right]; \tag{5.82}$$

$p(Y_{n+1}\mid\mathbf{Y},A)$ and $p(Y_{n+1})$ are the conditional (on $\mathbf{Y}$ and A) and the unconditional predictive densities of Y_{n+1}, respectively.

This summarizes our discussion of the general principle for designing experiments when the aim of experimentation is predicting a future observable. If the object of experimentation is inference about some parameter θ which depends on the elements of $\mathcal{A}$ as well as $\mathbf{Y}$, then in (5.82) Y_{n+1} is replaced by θ. I now consider the special scenario of invoking this principle in the context of designing an accelerated test.

Maximizing Mutual Information in Accelerated Tests

The principle underlying (5.82) is easiest to implement when the underlying predictive distributions are Gaussian and when the dependence of the predictive random variable on A is linear. Thus it is best that we make some changes to the accelerated life-testing model of section 5.8.1. Specifically, let us suppose that the T_{jl}s, $j=1,\ldots,m$, $l=1,\ldots,n$, have a lognormal distribution with parameters μ_j and σ_j^2. This would imply that the $\ln T_{jl}$'s will have a Gaussian distribution with mean μ_j and variance σ_j^2. The lognormal as a chance distribution for lifetimes has sometimes been considered as an alternative to the Weibull. For the time transformation function, we consider as before, the Power Law model of the form

$$E(T_{jl}|\mu_j,\sigma_j^2)=\exp\left(\mu_j+\frac{\sigma_j^2}{2}\right)=\frac{c}{S_j^P},$$

where $c>0$ and $P\geq 0$ are unknown parameters.

If we let $\tilde{z}_{jl}=\ln T_{jl}$, $a=\ln c$, $b=-\ln P$ and $V_j=\ln S_j$, then the above Power Law relationship becomes $z_{jl}=a+bV_j-\sigma_j^2/2+\epsilon_{jl}$, where $\epsilon_{jl}\sim N(0,\sigma_j^2)$. Further simplification is achieved, but without a compromise in generality, if the σ_j^2's are assumed known, $j=1,\ldots,m$. Specifically, if $z_{jl}=\tilde{z}_{jl}+\sigma_j^2/2$, then the Power Law-based time transformation function becomes

$$z_{jl}=a+bV_j+\epsilon_{jl} \tag{5.83}$$

with $\epsilon_{jl}\sim N(0,\sigma_j^2)$, $j=1,\ldots,m$, $l=1,\ldots,n$.

If $\mathbf{z}$ denotes the vector of all the z_{jl}'s, the vector $\beta = (a, b)'$, and the design matrix $\mathbf{A}$

$$\mathbf{A} = \begin{bmatrix} 1, \ldots, 1 & 1, \ldots, 1 & \cdots & 1, \ldots, 1 \\ V_1, \ldots, V_1 & V_2, \ldots, V_2 & \cdots & V_m, \ldots, V_m \end{bmatrix}$$

then, in matrix form, (5.83) becomes $\mathbf{z} = \mathbf{A}\boldsymbol{\beta} + \boldsymbol{\epsilon}$, where the vector $\boldsymbol{\epsilon}$ is a multivariate Gaussian with mean vector $\mathbf{0}$ and covariance matrix $\boldsymbol{\Sigma}$; $\boldsymbol{\Sigma}$ is a diagonal matrix with diagonal terms $\{\sigma_1^2\mathbf{I}_n, \sigma_2^2\mathbf{I}_n, \ldots, \sigma_m^2\mathbf{I}_n\}$, and $\mathbf{I}_n$ is an $n \times n$ identity matrix.

For the prior on $\boldsymbol{\beta}$, we assume that $\boldsymbol{\beta} \sim N(\boldsymbol{\beta}_0, \boldsymbol{\Sigma}_0)$ where $\boldsymbol{\beta}_0 = (a_0, b_0)'$ is specified and so is

$$\boldsymbol{\Sigma}_0 = \begin{bmatrix} \sigma_a^2 & \sigma_{ab} \\ \sigma_{ab} & \sigma_b^2 \end{bmatrix},$$

the variance-covariance matrix of a and b.

Our aim is to predict T_u, the time of failure under S_u, and with σ_u^2 assumed known, interest will therefore be focused on $Z_u = \ln T_u + \sigma_u^2/2$. For doing so, we invoke (5.82) to find that $\mathbf{A}$, say $\mathbf{A}^*$ which maximizes

$$E_{\mathbf{Z}|\mathbf{A}} E_{Z_u|\mathbf{Z},\mathbf{A}} \left[\log \frac{p(Z_u|\mathbf{Z}, \mathbf{A})}{p(Z_u)} \right]. \tag{5.84}$$

In Verdinelli, Polson and Singpurwalla (1993), it has been shown that maximizing (5.84) reduces to minimizing the variance of the predictive distribution of Z_u given $\mathbf{Z}$ and $\mathbf{A}$, with respect to $\mathbf{A}$. That is, we need to minimize the quantity

$$s^2 = (1, V_u)(\mathbf{A}'\boldsymbol{\Sigma}^{-1}\mathbf{A} + \boldsymbol{\Sigma}_0^{-1})^{-1} \begin{pmatrix} 1 \\ V_u \end{pmatrix} + \sigma_u^2, \tag{5.85}$$

with respect to the design matrix $\mathbf{A}$.

A consideration of some special cases in which the result of (5.85) is invoked yields some interesting conclusions. For example, when $c = m = 1$, that is, it is possible to test at only one accelerated stress level S_1, then

$$s^2 = \frac{\sigma_1^2}{n} + \frac{V_u^2}{V_1^2 + \delta} + \sigma_u^2,$$

where $\delta = \sigma_1^2(n\sigma_b^2)^{-1}$. Thus s^2 is minimized when V_1 is as large as possible.

With C and P unknown, but with m still one,

$$s^2 = \frac{(V_1 - V_u)^2 + \delta_b + V_u^2\delta_a}{(1 + \delta_a)(V_1^2 + \delta_b) - V_1^2} + \sigma_u^2,$$

where $\delta_a = \sigma_1^2/n\sigma_a^2$ and $\delta_b = \sigma_1^2/n\sigma_b^2$. Thus, it is $V_1 = (1 + \delta_a)V_u$ that minimizes s^2.

The above special cases, though unrealistic, illustrate the kind of results that the maximization of mutual information principle yields. More research involving an application of these ideas to scenarios other than that considered here is needed. Polson (1992) provides a broad perspective and indicates several possibilities involving inference about parameters of time transformation models, censoring, step-stress testing and the treatment of binary data.

I conclude this chapter by acknowledging the fact that much of what I have presented has some way to go before it can be implemented for day-to-day use. This presents several opportunities for others to embark on more detailed investigations into the scenarios that interest them and under conditions that are more realistic than those considered by us. My purpose here has been to indicate how one may think about life-testing and reliability using some of the more modern ideas that probability and statistics have to offer.

Chapter 6

Composite Reliability: Signatures

Here lies Isaac Newton: A body at rest shall remain at rest.

6.1 INTRODUCTION: HIERARCHICAL MODELS

Hierarchical modeling was alluded to in section 2.6.1 as an expansion of the prior $P(\theta; \mathcal{H})$, using the law of total probability. Specifically, for some parameter, say $\boldsymbol{\alpha}$, we may invoke the law of total probability on $P(\theta; \mathcal{H})$ to write (2.4) in an expanded form as

$$P(X^{(n)}; \mathcal{H}) = \sum_{\theta} \prod_{i=1}^{n} f(X_i \mid \theta) \sum_{\alpha} P(\theta \mid \boldsymbol{\alpha}) P(\boldsymbol{\alpha}; \mathcal{H}), \tag{6.1}$$

where $P(\boldsymbol{\alpha}; \mathcal{H})$ is a prior on $\boldsymbol{\alpha}$. The scalar or vector α is known as a **hyperparameter**. The model of (6.1) is a two-stage hierarchical model because the probability law of $X^{(n)}$ entails the operation of extending the conversation two times, the first to θ and then to $\boldsymbol{\alpha}$. In principle, I may continue the above process by expanding $P(\boldsymbol{\alpha}; \mathcal{H})$ using another set of hyperparameters, so that the hierarchy of modeling gets enlarged from two stages to three, and so on. From a subjectivistic point of view, the motivation for constructing hierarchical models is ease of specifying $P(\theta; \mathcal{H})$ via a consideration of $\boldsymbol{\alpha}$. In other words, a hierarchical approach facilitates the specification of probability models by a step-by-step breakdown of the underlying uncertainties. In practice, one often does not go much beyond a two-stage model unless the physics of the situation is such that it becomes meaningful to do so. The Kalman Filter model of section 5.8.2 (more details about which can be found in Meinhold and Singpurwalla (1983)) is perhaps one of the best-known examples of a hierarchical model. The observation (5.70) encapsulates the first stage of the hierarchy, and the system (5.71) the second stage. From a practical point of view hierarchical models have, in many cases, provided a better predictive capability than single stage models, the success of the Kalman Filter being a prime example. A unifying perspective on statistical modeling via a step-by-step hierarchical expansion is described in Singpurwalla (1989).

Reliability and Risk: A Bayesian Perspective N.D. Singpurwalla
© 2006 John Wiley & Sons, Ltd

The purpose of this chapter is twofold. The first is to introduce the notion of 'composite reliability', and the second to introduce a technology called 'signature analysis', which reliability engineers and medical practioners find useful. Both entail hierarchical modeling.

6.2 'COMPOSITE RELIABILITY': PARTIAL EXCHANGEABILITY

By 'composite reliability', I do not mean the reliability of composite materials. Rather, what I have in mind here is an omnibus measure for characterizing the trustworthiness of a collection of unidentical, but similar, items. What has prompted me to introduce and to consider the notion of an overall reliability? It is that often decision makers in government, industry and public policy, and also the public at large, are more concerned about the collective trustworthiness of groups of similar items rather than the trustworthiness of a single or a particular group of items. Some examples to highlight this matter are: the reliability of automobiles made in the US, the safety of nuclear power plants, the quality of European medical equipment, etc. Indeed, it was the matter of nuclear power plant safety that motivated our interest in coming up with a way to describe overall reliability. Individual power plants may be safe and reliable, but what can we say about the collective behavior of the power plants given that they differ in design and operating procedures? Similarly, with automobiles. All compact cars could be judged exchangeable with each other, but not with mid-sized and full-sized cars. Yet regulators and consumer advocates may want to know if the automobile industry, taken as a whole, is providing reliable automobiles. Then of course there are claims such as cars made in Sweden are safer than their Korean counterparts, and that Japanese autos are more reliable than those made in the US.[1] How can we quantify and make precise statements such as these? Certainly, some brands of US-made cars are more reliable than certain brands of cars made in Japan. Similar is the case when it comes to quality and safety. Thus what I need here is a mechanism via which heterogenous but similar groups of items can be compared vis-à-vis their reliability. One such mechanism is via the notion of composite reliability described below. This notion has been introduced and discussed by Chen and Singpurwalla (1996), who were motivated by the problem of assessing the reliability of emergency diesel generators used in nuclear power plants. Some related work is in a paper by Arjas and Bhattacharjee (2004), who focus on inter-failure times encountered in a longitudinal analysis of value failures in nuclear power plants.

To appreciate the notion of 'composite reliability', we start by re-visiting the notion of reliability, or chance, discussed in section 4.4. However, we focus attention here on the case of Bernoulli random variables, X_{ij}, $i = 1, \ldots, k$; $j = 1, \ldots, n_i$. Here X_{ij} takes the value 1(0) if the j-th item in the i-th group survives (fails) its mission time. We suppose that for each i, the X_{ij}s, $j = 1, \ldots, n_i$ are exchangeable. That is, all the Xs within the i-th group are exchangeable. Then, given p_i, the reliability (or chance) of the items in the i-th group, $i = 1, \ldots, k$, the X_{ij}s, $j = 1, \ldots, n_i$ would be independent and identically distributed Bernoulli random variables with parameter p_i, so that

$$P(X_{i1} = x_{i1}, \ldots, X_{in_i} = x_{in_i}) = \int_0^1 \prod_{j=1}^{n_i} p_i^{x_{ij}} (1 - p_i)^{1 - x_{ij}} F(\mathrm{d}p_i); \tag{6.2}$$

see Theorem 3.2 of section 3.1 (de Finetti's Theorem). $F(\mathrm{d}p_i)$ describes our uncertainty about the p_i that generates $X_{i1}, X_{i2}, \ldots, X_{in_i}$; $0 < p_i < 1$.

[1] See, for example, an article in *The Washington Post* dated March 10, 2004, wherein it is stated that 'For the first time in 25 years, U.S. carmakers can say that they make more reliable cars than their competitors in Europe. Asian manufacturers still hold top bragging rights, however'.

To relate the $X_{i1}, \ldots, X_{in_i}$ to $X_{k1}, \ldots, X_{kn_k}$, for all $k \neq i$, we create a hierarchical architecture. This we do by supposing that the p_i's themselves are dependent random drawings from some joint distribution whose nature is described below, by (6.3). Specifically, we assume that given a (scalar or vector) parameter η, the p_i's are independent and identically distributed with a density $\pi_1(p \mid \eta)$ at p, $0 < p < 1$. Then unconditionally, the joint density of $p_1, \ldots, p_k$ at $p_1^*, \ldots, p_k^*$ will be of the form

$$\pi(p_1^*, \ldots, p_k^*) = \int_\eta \prod_{i=1}^{k} \pi_1(p_i^* \mid \eta) F(\mathrm{d}\eta), \tag{6.3}$$

where $F(\mathrm{d}\eta)$ describes our uncertainty about η.

The model of (6.3) makes the p_i's exchangeable, and the quantity $\pi_1(p \mid \eta)$ is the chance distribution that generates the p_i's. By analogy with the interpretation of (2.4) of Chapter 2, we call this chance distribution composite reliability. Thus:

Composite Reliability is a Chance Distribution

that generates the individual chances associated with an observable subsequence of exchangeable lifetimes in a partially exchangeable sequence.

The marginal distribution of p_i, obtained via (6.3), gives us the $F(\mathrm{d}p_i)$ of (6.2); thus the linkage between reliability and composite reliability. Under the models of (6.2) and (6.3), the X_{ij}s, $j = 1, \ldots, n_i$, provide information about the p_i that generated them, and collectively, the X_{ij}s, $i = 1, \ldots, k$, $j = 1, \ldots, n$ provide information about the η that generates the p_i's. Since the composite reliability is a (chance) distribution, we may compare the composite reliabilities of two groups of items by comparing their respective chance distributions. Specifically, suppose that $F_1(\mathrm{d}p)$ denotes the distribution function corresponding to a chance distribution $\pi_1(p \mid \eta)$ and $F_2(\mathrm{d}p)$ the distribution function corresponding to, say, $\pi_2(p \mid \eta)$. Then the composite reliability of the group of items whose lifetimes are a consequence of $\pi_1(p \mid \eta)$ is greater than the composite reliability of items whose lifetimes are a consequence of $\pi_2(p \mid \eta)$, if $F_1(\mathrm{d}p) \leq F_2(\mathrm{d}p)$ for all $p \in (0, 1)$.

6.2.1 Simulating Exchangeable and Partially Exchangeable Sequences

It was mentioned before that a linkage between reliability and composite reliability is achieved by putting together (6.2) and (6.3). The schemata of Figure 6.2 illustrates the mechanics of how this can be done. Indeed Figure 6.2 can be seen as a roadmap for simulating the partially exchangeable Bernoulli sequence X_{ij}, $i = 1, \ldots, k$, $j = 1, \ldots, n_i$. However, to better appreciate the anatomy of Figure 6.2, it is helpful if we first consider the simulation of an exchangeable Bernoulli sequence, say $X_1, \ldots, X_n$, constructed via the process of (6.2). That is, for any i, $i = 1, \ldots, n$, $x_i = 0$ or 1, and $p \in (0, 1)$,

$$P(X_i = x_i) = \int_0^1 p^{x_i}(1-p)^{1-x_i} F(\mathrm{d}p). \tag{6.4}$$

Figure 6.1 illustrates the mechanics of simulating the process corresponding to (6.4).

The $\{u_i\}$ and $\{v_i\}$ in Figure 6.1 denote sequences of mutually (and also contemporaneously over i) independent uniformly distributed random deviates on (0, 1). Each u_i in conjunction with the prior $F(\mathrm{d}p)$ generates a p_i, and in turn each p_i, in conjunction with the Bernoulli chance distribution $p_i^{x_i}(1-p_i)^{1-x_i}$ and a v_i, generates a Bernoulli x_i. This cycle is repeated n times.

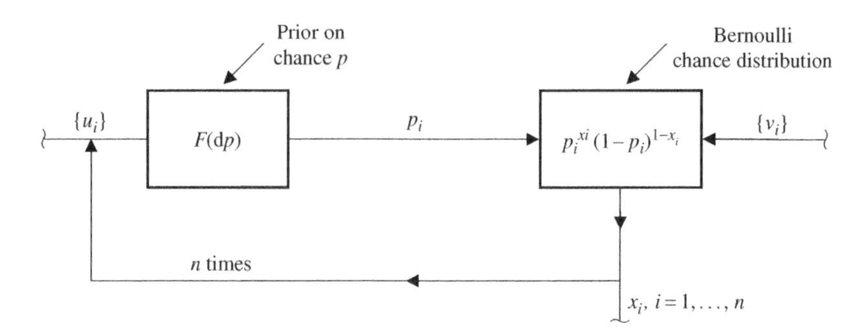

Figure 6.1 Simulating an exchangeable Bernoulli sequence.

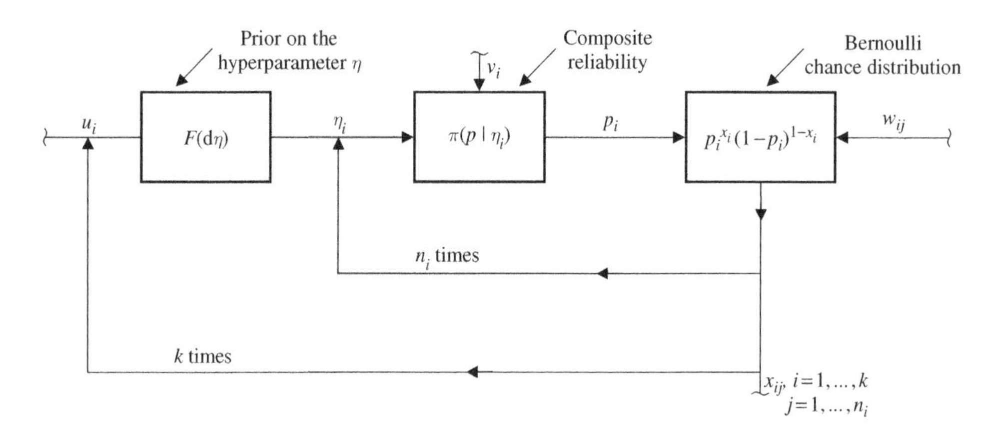

Figure 6.2 Simulating a partially exchangeable Bernoulli sequence.

The process of repeatedly generating a p_i becomes the equivalent of weighting $p^{x_i}(1-p)^{1-x_i}$ by $F(dp)$ of (6.4), for a range of values of p, $0 < p < 1$.

An expansion of Figure 6.1 to encompass the generation of a sequence $\{\eta_i\}$, $i = 1, \ldots, k$, results in Figure 6.2, which is a roadmap for generating the partially exchangeable sequence X_{ij}, $i = 1, \ldots, k$, and $j = 1, \ldots, n_i$. The sequence $\{w_i\}$ is like the sequences $\{u_i\}$ and $\{v_i\}$, a collection of independent uniformly distributed random variables on (0, 1). The role of composite reliability as being an intermediary between the prior $F(d\eta)$ and the Bernoulli chance distribution $p_i^{x_i}(1-p_i)^{1-x_i}$ is now transparent.

6.2.2 The Composite Reliability of Ultra-reliable Units

The purpose of this section is to illustrate the foregoing material by developing a specific model for the composite reliability, and its associated priors. The development here is based on Chen and Singpurwalla (1996), who were interested in a scenario involving highly reliable units, namely emergency diesel generators used in nuclear power plants. Let $\theta_i = 1 - p_i$ denote the unreliability of the units in the i-th group, $i = 1, \ldots, k$; i.e. $P(X_{ij} = x_{ij} \mid \theta_i) = \theta_i^{1-x_{ij}}(1-\theta_i)^{x_{ij}}$, for $x_{ij} = 0$ or 1. We suppose, as is reasonable to do so, that each θ_i is small. That is, the units in each group are highly reliable. Thus a model for the composite reliability should be such that the

θ_i's it generates should take small values. Our strategy for doing this comprises of two stages; this I shall describe in the sequel.

Our proposed model (for composite unreliability) has a density at θ which is β with parameters η and 1, i.e.

$$\pi_1(\theta \mid \eta) = \eta(\eta+1)\theta^{\eta-1}(1-\theta), \quad 0<\theta<1. \tag{6.5}$$

Figure 6.3 illustrates this density for different values of η.

The model $\pi_1(\theta \mid \eta)$ illustrated in Figure 6.3 is motivated in two stages. In stage one, we assume that the probability density of θ is L-shaped, with an upper bound γ, where $\gamma < 1$ (Figure 6.4). This is best expressed via

$$\pi_0(\theta \mid \eta, \gamma) = \eta\theta^{\eta-1}/\gamma^{\eta}, \quad \text{for } 0<\theta<\gamma,$$

where η and γ are hyperparameters. Verify that $\pi_0(\theta \mid \eta, \gamma)$ is a beta density on $(0, \gamma)$ with parameters η and 1. To reflect the belief that small values of θ are much more likely than large values, we require that $\eta \in (0, 1)$. Observe that $\pi_0(\theta \mid \eta, \gamma)$ becomes approximately uniform on $(0, \gamma)$ as $\eta \to 1$.

In stage two, I prescribe a model for γ, the upper bound on θ. Since it is unlikely that γ will take values that are very small, a meaningful model for γ is one wherein its probability density is uniform with a steep decrease in the vicinity of zero. A way to achieve this is via a beta density over (0, 1) with parameters β and 1, with β taking values greater than but close to one. That is, the probability density at γ is of the form

$$\pi(\gamma \mid \beta, 1) \propto \gamma^{\beta-1}, \quad 0<\gamma<1.$$

One way to ensure that β is greater than one, and yet close to it, is to let $\beta = 1+\eta$. An advantage of this choice is that all that remains to be assessed for the two-stage development described above is the single parameter η. With $\pi_0(\theta \mid \eta, \gamma)$ and $\pi(\gamma \mid \beta, 1)$ as specified, it

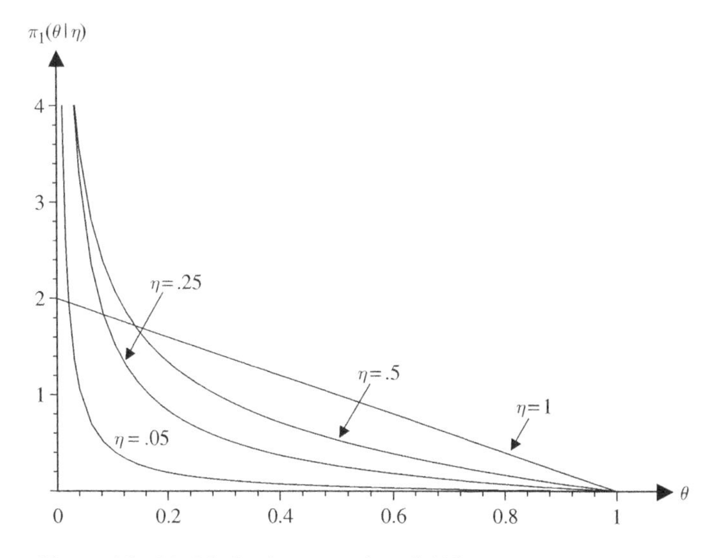

Figure 6.3 Models for the composite reliability of ultra-reliable items.

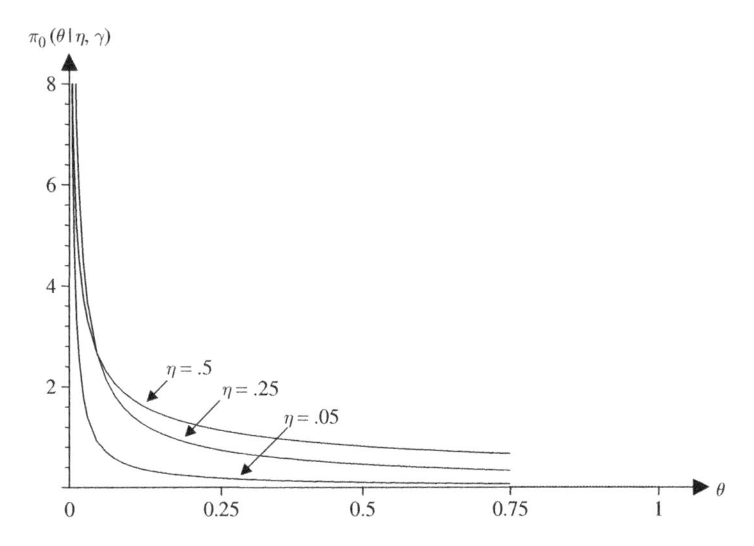

Figure 6.4 Probability density of θ for different values of η, with $\gamma = 0.75$.

is easy to verify that unconditionally (on γ), θ has the model of (6.5). More discussion on this choice as a model for the unreliability (or the composite unreliability) of ultra-reliable items is given in section 2.1 of Chen and Singpurwalla (1996).

With $\pi_1(\theta \mid \eta)$ specified, our next task is to obtain a joint distribution for $\theta_1, \ldots, \theta_k$. To do so, we follow the steps leading to (6.3). This entails specifying a prior distribution for η, $F(\mathrm{d}\eta)$. Since $\eta \in (0, 1)$, a suitable choice for $F(\mathrm{d}\eta)$ would be a beta density on $(0, 1)$ with hyperparameters β_1 and β_2, where β_1 and β_2 are so specified that values of η close to zero get emphasized (Figure 6.3). A natural choice would be to set $\beta_1 = \beta_2 = 1$, so that $F(\mathrm{d}\eta)$ is a uniform distribution on $(0, 1)$ (sections 5.2.3 and 5.3.2). With the above in place, the joint density of the θ_i's at $\theta_1, \ldots, \theta_k$ becomes the analogue of (6.3); it takes the form

$$\pi(\theta_1, \ldots, \theta_k \mid \beta_1, \beta_2) = \int_0^1 \left\{ \prod_{i=1}^{k} \eta(\eta + 1)\theta_i^{\eta-1}(1 - \theta_i) \right\} F(\mathrm{d}\eta | \beta_1, \beta_2). \tag{6.6}$$

6.2.3 Assessing Reliability and Composite Reliability

The aim of this section is to discuss an assessment of the p_i's, $i = 1, \ldots, k$, and the $\pi_1(p \mid \eta)$ of section 6.2, in the light of data. Since $p_i = 1 - \theta_i$, $i = 1, \ldots, k$, and since the θ_i's are viewed as independent drawings from $\pi_1(\theta \mid \eta)$ (section 6.2.2) assessing the θ_i's and $\pi_1(\theta \mid \eta)$ is equivalent to assessing the p_i's and $\pi_1(p \mid \eta)$. In what follows we shall discuss an assessment of the θ_i's and $\pi_1(\theta \mid \eta)$ because the models of section 6.2.2 have been motivated in terms of these quantities.

Suppose, then, that the observed data, say $\mathbf{d}$, consists of x_{ij}, $j = 1, \ldots, n_i$, $i = 1, \ldots, k$. Let $y_i = \sum_{j=1}^{n_i} x_{ij}$, $i = 1, \ldots, k$, and $\boldsymbol{\theta} = (\theta_1, \ldots, \theta_k)$. Since $P(X_{ij} = x_{ij} \mid \theta_i) = \theta_i^{1-x_{ij}}(1 - \theta_i)^{x_{ij}}$, the likelihood of $\boldsymbol{\theta}$ in the light of $\mathbf{d}$, assuming that the stopping rule is non-informative, is $\prod_{i=1}^{k} \theta_i^{n_i - y_i}(1-\theta_i)^{y_i}$. Consequently, the posterior distribution of $\boldsymbol{\theta}$ and η is given as

$$\pi(\boldsymbol{\theta}, \eta; \mathbf{d}, \beta_1, \beta_2) \propto \mathcal{L}(\boldsymbol{\theta}; \mathbf{d}) \prod_{i=1}^{k} \eta(\eta+1)\theta_i^{\eta-1}(1-\theta_i)F(\mathrm{d}\eta \mid \beta_1, \beta_2),$$

where $\mathcal{L}(\boldsymbol{\theta}; \mathbf{d})$ denotes the likelihood given above. An attractive feature of this posterior distribution is that it lends itself to an evaluation of the posterior distributions of the θ_i's, $i = 1, \ldots, k$, and of η, via the Markov Chain Monté Carlo method of Appendix A. The specifics of how this can be done is left as an exercise for the reader.

The posterior distribution of each θ_i, $i = 1, \ldots, k$, represents our assessment of the *unreliability* of the items in group i. The posterior distribution of η enables us to numerically obtain $E(\eta; \mathbf{d}, \beta_1, \beta_2)$, its mean. This mean, when plugged into (6.5) replacing η, provides an estimate of $\pi_1(\theta|\eta)$, which then becomes our assessment of the composite unreliability. See the end of section 5.4.4 for the rationale behind the substitution of $E(\eta; \mathbf{d}, \beta_1, \beta_2)$ for η. Comparing composite unreliabilities is isomorphic to the comparison of composite reliabilities.

6.3 SIGNATURE ANALYSIS AND SIGNATURES AS COVARIATES

Defects in certain physical and biological items sometimes manifest themselves as oscillary periodic motions, like vibrations, or as electrical signals, like leakage currents and voltages. For example, rotor imbalance in rotating machinery causes an excessive wear on the bearings and this in turn leads to vibrations that cause damage to electrical insulations and a general discomfort to humans. In the biomedical context, an electrocardiogram (ECG) – which measures changes in the electrical potential produced by contractions of the heart – is used by physicians to assess the condition of the heart muscle. Vibration signals are generally recorded in the time domain as a time series (section 5.8.2). Similarly, the ECG is a time domain plot of the voltage; the shape and the pattern of this plot reveal deficiencies. Such time domain plots are known as signature data, and signature analysis is an interpretation of the signature data. For reasons that will become clear in the sequel, it is common to interpret signature data in the frequency domain through its 'power spectrum' which is then referred to as the **signature**.[2] The power spectrum, which is introduced and discussed in Appendix B, is a graphical display of 'frequency' (on the horizontal axis) versus 'amplitude of the spectrum' (on the vertical axis). The goal of signature analysis is to identify particular types of defects, if any, and to assess their relative impact on an item's survival. In the context of an ECG, large amplitudes at high frequencies are indicators of potential medical problems. As an example, I illustrate via Figure 6.5 the estimated power spectrum of a patient's ECG-generated signature data, prior to and post heart surgery (Pierce *et al.*, 1989). The effect of surgery has been a dampening of the large amplitudes at all frequencies (expressed as Hz – see Appendix B), suggesting a positive impact of the surgical intervention.

Figure 6.5 is but one illustration of the usefulness of signatures, namely comparison and classification. More importantly, signatures provide a snapshot of several defects that otherwise might be difficult to observe directly. In the case of rotating machinery, the frequency at which the amplitude (of the spectrum) takes a large value indicates the nature of the defect, and the

[2] Not to be confused with the notion of the signature of a coherent system, discussed by Boland and Samaniego (2004).

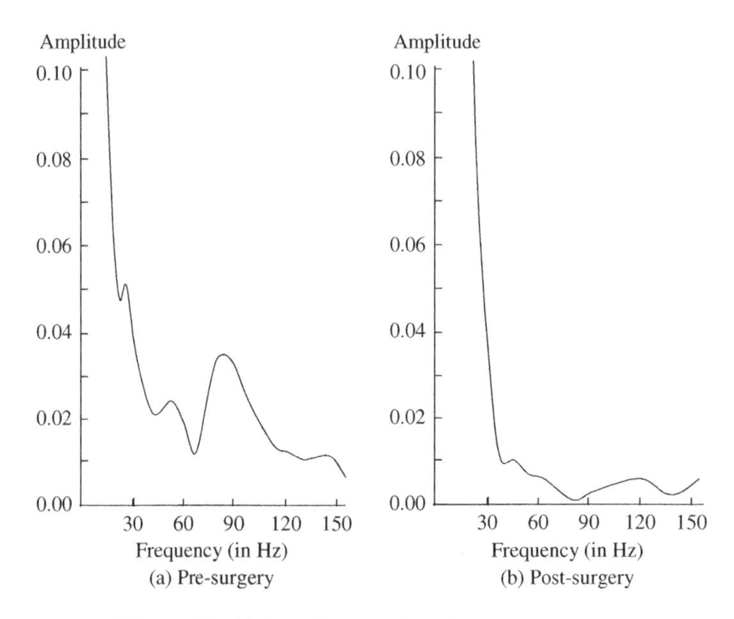

Figure 6.5 Estimated pre- and post-surgery signatures.

magnitude of the amplitude indicates the severity of the defect. Thus, for example, a large amplitude at a frequency corresponding to one times the motor running speed suggests an imbalance, whereas a large amplitude at two times the motor running speed suggests misalignment. Similarly, the signature of an ECG can reveal the nature of heart disease and the extent of the disease.

Whereas an examination of the signature can help one identify the nature and the magnitude of defects, it cannot by itself help assess the impact of the identified defects on an item's lifelength. To do so, we need to link signature analysis with the techniques of life-data analysis. This we are able to do by treating the signature as a covariate (section 4.9). The motivation here would be to explore the relative impact of the defects on an item's lifelength and to explore as to whether the defects tend to have an additive effect or if they tend to cancel out. For example, in the case of rotating machinery, would a small amount of imbalance be more deleterious to lifetime than a large amount of misalignment, or would imbalance and misalignment taken together cancel out the deleterious effect of each defect?

The aim of this section is to describe a hierarchical Bayesian approach for obtaining signatures and for harnessing the import of such signatures for lifelength prediction. It is our view that signatures provide a powerful diagnostic and assessment tool in survival analysis, a tool whose potential remains to be fully exploited in reliability theory and life-data analysis. What I attempt to describe here are a few preliminary efforts. But to do so, I need to highlight some standard material on 'Fourier series models', and the spectrum of a periodic function. This I do below in section 6.3.1 assuming that the reader is familiar with trigonometric functions, Fourier series models and the power spectrum, which for convenience are described in Appendix B. The material in Appendix B is abstracted from Anderson's (1971) classic treatise on time-series analysis.

6.3.1 Assessing the Power Spectrum via a Regression Model

Let $Y(t)$ denote the observed value of a variable of interest, such as the displacement due to vibration or the voltage of a heart muscle at time t, $t = 1, \ldots, T$. For purposes of discussion suppose that T is odd, and that $Y(t)$ is described via the regression model

$$Y(t) = f(t) + \epsilon(t), \tag{6.7}$$

where $f(t)$ is an unknown periodic function with a known period T. For example, T could be the time taken by a motor to complete a revolution or the time taken by the heart muscle to complete a pumping cycle. The error $\epsilon(t)$ is assumed to be Gaussian with mean 0 and variance σ^2; also, the sequence $\{\epsilon(t)\}$ is uncorrelated.

Our aim is to explore if there are hidden periodicities in $f(t)$, i.e. periodicities in addition to the one that is known to be T. As mentioned before, hidden periodicities could be deleterious to an item's survival. To identify the hidden periodicities we approximate $f(t)$ by a Fourier series model of the form:

$$f(t) = \alpha(0) + \sum_{j=1}^{q} \left(\alpha(m_j) \cos\left(\frac{2\pi}{T} m_j t\right) + \beta(m_j) \sin\left(\frac{2\pi}{T} m_j t\right) \right), \tag{6.8}$$

where $1 \leq q \leq (T-1)/2$, and $\{m_1, m_2, \ldots, m_q\}$ is a subset of $I = \{1, 2, \ldots, (T-1)/2\}$ – see equation (B.3) in Appendix B. The $\alpha(m_j)$ and $\beta(m_j)$ are unknown and have to be estimated from the $Y(t)$s; they are the amplitudes of the cosine and sine curves having periods T/m_j and frequencies m_j/T, $j = 1, \ldots, q$. The summand q and the subset $\{m_1, \ldots, m_q\}$ are chosen in accordance with our prior beliefs about the number and the order of the hidden periods. A plot of $\rho^2(m_j) \stackrel{\text{def}}{=} \alpha^2(m_j) + \beta^2(m_j)$ versus the frequency m_j/T, $j = 1, \ldots, q$, is the power spectrum of $f(t)$ (section B.4 of Appendix B). Our aim is to describe a Bayesian approach for assessing the power spectrum using the $Y(t)$s; this tantamounts to assessing $\alpha(0)$, the $\alpha(m_j)$'s, and the $\beta(m_j)$'s, $j = 1, \ldots, q$.

The problem described above has a rich legacy within the frequentist paradigm. For example, Anderson (1971, p. 105) describes a least-squares approach for estimating the unknowns $\alpha(0)$, $\alpha(m_j)$ and $\beta(m_j)$, and discusses the properties of the estimators. It makes for some fascinating reading and sets the tone for what can be done using a Bayesian approach. But before I prescribe our Bayesian approach, it may be helpful to remind the reader that what we are looking for are large values of $\rho^2(m_j)$ because these are a measure of how closely the trigonometric functions having period T/m_j describe the data. In essence, signature analysis pertains to identifying those periods T/m_j at which $\rho^2(m_j)$ is large, because it is such periods that give us clues about the nature and the magnitude of defects.

6.3.2 Bayesian Assessment of the Power Spectrum

For the model of section 6.3.1 as encapsulated via the regression model of (6.7) and (6.8), we start by defining the vectors

$$F_t' = \left(1, \cos\left(\frac{2\pi}{T} m_1 t\right), \sin\left(\frac{2\pi}{T} m_1 t\right), \ldots, \cos\left(\frac{2\pi}{T} m_q t\right), \sin\left(\frac{2\pi}{T} m_q t\right)\right),$$

$$\boldsymbol{\theta}' = \left(\alpha_0, \alpha(m_1), \beta(m_1), \ldots \alpha(m_q), \beta(m_q)\right),$$

and the scalar $\phi = 1/\sigma^2$.

Then (6.7) and (6.8) boil down to the statement that

$$(Y(t) \mid \boldsymbol{\theta}, \phi) \sim N(F_t'\boldsymbol{\theta}, 1/\phi), \tag{6.9}$$

where $N(\mu, \sigma^2)$ denotes a Gaussian distribution with mean μ and variance σ^2. The quantity ϕ is called the **precision** of the normal distribution; it is the reciprocal of the variance.

The vector $\boldsymbol{\theta}$ needs to be endowed with a prior, and so does the scalar ϕ. For this, we suppose that

$$(\boldsymbol{\theta} \mid \phi) \sim N(\boldsymbol{\Lambda}_0, \phi^{-1}\mathbf{R}_0), \quad \text{and that} \tag{6.10}$$

$$\phi \sim G\left(\frac{n_0}{2}, \frac{d_0}{2}\right),$$

where $G(\alpha, \beta)$ denotes a gamma distribution with shape α and scale β (section 5.4.6). The quantity $\boldsymbol{\Lambda}_0$ is the mean vector of the Gaussian distribution of $\boldsymbol{\theta}$, and $\phi^{-1}\mathbf{R}_0$ is its covariance matrix. The quantities $\boldsymbol{\Lambda}_0$, $\mathbf{R}_0$, n_0 and d_0 have to be specified by us and should reflect our prior beliefs about the order and the extent of the hidden periodicities. The elements of the covariance matrix $\phi^{-1}\mathbf{R}_0$ should reflect our strength of conviction about our specified $\boldsymbol{\Lambda}_0$. The motivation for multiplying $\mathbf{R}_0$ by ϕ^{-1} in the Gaussian distribution of $\boldsymbol{\theta}$ is to make the variances of the components of $\boldsymbol{\theta}$ free of the scale used for measuring the $Y(t)$. The motivation for using $n_0/2$ and $d_0/2$ as the shape and scale parameters of the gamma distribution is that this choice makes ϕd_0 have a chi-square distribution with n_0 degrees of freedom.

Specifying a Prior for the Power Spectrum

The following ground rules for specifying $\boldsymbol{\Lambda}_0$ and $\phi^{-1}\mathbf{R}_0$ may be helpful.

Should our prior belief be such that there are no hidden periodicities exhibited by the item in question, then all the elements of $\boldsymbol{\Lambda}_0$ should be zero. This could also be seen as a reasonable choice for a default prior. In the case of rotating machinery, if an imbalance is suspected then the second and third elements of $\boldsymbol{\Lambda}_0$, i.e. the elements corresponding to $\alpha(m_1)$ and $\beta(m_1)$, would be assigned the same non-zero value, the value being reflective of the extent of imbalance that is perceived, similar is the case of misalignment. In essence, specifying a meaningful prior for $\boldsymbol{\theta}$ through a judicious choice for $\boldsymbol{\Lambda}_0$ will entail a good knowledge of how defects manifest themselves in the power spectrum, be it in the context of an ECG or a physical item.

With $\boldsymbol{\Lambda}_0$ set at its default value $\mathbf{0}$, the elements of the covariance matrix $\phi^{-1}\mathbf{R}_0$, when non-zero, should be large and decreasing in value with the harmonics of the fundamental frequency H_1 – (section B of Appendix B). This is because it is often the case that the observed spectra of many time series tend to reveal no amplitudes at large frequencies. It may also be meaningful to suppose that covariances of the coefficients associated with any two harmonics is zero. For the paired coefficients within each harmonic, the corresponding variances are set equal, and the correlation set to be a neutral 0.5. That is, $\text{Var}[\alpha(m_j)] = \text{Var}[\beta(m_j)] = V(m_j)$, and $\text{Cov}[\alpha(m_j), \beta(m_j)] = 0.5V(m_j)$, $j = 1, \ldots, q$. Since $V(m_j)$ is to be a decreasing function of m_j, one plausible choice would be to let $V(m_j) = \phi^{-1}\exp(-km_j)$ for some positive $k > 0$. I denote the above as $V(m_j) = \phi^{-1}r_0(m_j)$. Thus, our proposed default values for $\boldsymbol{\Lambda}_0$ and $\phi^{-1}\mathbf{R}_0$ are: $\boldsymbol{\Lambda}_0 = (0, \ldots 0)'$ and

$$\phi^{-1}\mathbf{R}_0 = \begin{bmatrix} r_0(0) & 0 & 0 & & & \\ 0 & r_0(m_1) & 0.5r_0(m_1) & & \mathbf{0} & \\ 0 & 0.5r_0(m_1) & r_0(m_1) & & & \\ & & & \ddots & & \\ & \mathbf{0} & & & r_0(m_q) & 0.5r_0(m_q) \\ & & & & 0.5r_0(m_q) & r_0(m_q) \end{bmatrix}.$$

For parameters n_0 and d_0 of the gamma prior on ϕ, the choice $n_0 = 2$ and $d_0 = 3$ would ensure a strong belief about the rationality of (6.7); this would tend to make ϕ have a small variance. Another possibility would be to let $n_0 = d_0 = 0$, in which case the prior on ϕ is non-informative. However, this non-informative prior does not adversely affect the updating mechanism.

Posterior Power Spectrum Under the Regression Model

Let the observed data be denoted by the vector $\mathbf{Y}_t = (Y(1), \ldots, Y(t))'$ and let $\mathbf{F}_t = (F_1', \ldots, F_t')$ be a $(2q+1) \times t$ matrix. Then upon observing $\mathbf{Y}_t$, a standard application of Bayes' law to (6.9) and (6.10) leads to the claim (cf. DeGroot 1970, p. 249) that the posterior distribution of $\boldsymbol{\theta}$, given ϕ, is of the form

$$(\boldsymbol{\theta} \mid \phi; \mathbf{Y}_t) \sim N\left(\boldsymbol{\Lambda}_t, \phi^{-1}\mathbf{R}_t\right), \tag{6.11}$$

where $\boldsymbol{\Lambda}_t = \mathbf{R}_t(\mathbf{F}_t\mathbf{Y}_t + \mathbf{R}_0^{-1}\boldsymbol{\Lambda}_0)$, and $\mathbf{R}_t = (\mathbf{F}_t\mathbf{F}_t' + \mathbf{R}_0^{-1})^{-1}$.

Also, the posterior distribution of ϕ is

$$(\phi; \mathbf{Y}_t) \sim G\left(\frac{n_t}{2}, \frac{d_t}{2}\right),$$

where $n_t = n_0 + t$, and

$$d_t = d_0 + \mathbf{Y}_t'\mathbf{Y}_t + \boldsymbol{\Lambda}_0'\mathbf{R}_0^{-1}\boldsymbol{\Lambda}_0 - (\mathbf{F}_t\mathbf{Y}_t + \mathbf{R}_0^{-1}\boldsymbol{\Lambda}_0)'\mathbf{R}_t(\mathbf{F}_t\mathbf{Y}_t + \mathbf{R}_0^{-1}\boldsymbol{\Lambda}_0).$$

Averaging out ϕ in (6.11) with respect to its posterior distribution leads us to the result that

$$(\boldsymbol{\theta}; \mathbf{Y}_t) \sim T_{n_t}\left(\boldsymbol{\Lambda}_t, \mathbf{R}_t\frac{d_t}{n_t}\right); \tag{6.12}$$

i.e. the posterior distribution of $\boldsymbol{\theta}$ is a multivariate Student's t distribution in dimension $(2q+1)$ with $n_0 + t$ degrees of freedom with a mode $\boldsymbol{\Lambda}_t$ and a scale matrix $\mathbf{R}_t d_t / n_t$ (DeGroot, 1970, p. 59) for a description of the multivariate Student's-t distribution. A consequence of (6.12) is that the i-th element of $\boldsymbol{\theta}$ has a univariate Student's-t distribution whose mode is the i-th element of $\boldsymbol{\Lambda}_t$ and whose scale is the i-th diagonal entry of the scale matrix $\mathbf{R}_t d_t / n_t$.

For a posterior assessment of the power spectrum what we need is the posterior distribution (given $\mathbf{Y}_t$) of the vector

$$\left[\left(\alpha^2(m_1) + \beta^2(m_1)\right), \ldots, \left(\alpha^2(m_q) + \beta^2(m_q)\right)\right].$$

Obtaining this posterior distribution turns out to be a formidable task unless one resorts to a MCMC-based simulation (Appendix A); and even this would be a cumbersome endeavor. However, in Campodónico and Singpurwalla (1994b) an expression for the marginal posterior distribution of $\rho^2(m_j) \stackrel{\text{def}}{=} \alpha^2(m_j) + \beta^2(m_j)$ is obtained (Appendix B of Campodónico and Singpurwalla, 1994b) from which $E(\rho^2(m_j); \mathbf{Y}_t)$, the mean of this posterior, can be assessed. This turns out to be

$$E(\rho^2(m_j); \mathbf{Y}_t) = \boldsymbol{\Lambda}_t\left(\alpha(m_j)\right)^2 + \boldsymbol{\Lambda}_t\left(\beta(m_j)\right)^2 + \frac{d_t}{n_t - 2}\left[r_t(\alpha(m_j)) + r_t(\beta(m_j))\right], \tag{6.13}$$

$j = 1, \ldots, q$, where $\boldsymbol{\Lambda}_t\left(\alpha(m_j)\right)$ and $\boldsymbol{\Lambda}_t\left(\beta(m_j)\right)$ are the elements of the vector $\boldsymbol{\Lambda}_t$ which correspond to the posterior expected values of $\alpha(m_j)$ and $\beta(m_j)$, and $r_t(\alpha(m_j))$ and $r_t(\beta(m_j))$ are the diagonal elements of the matrix $\mathbf{R}_t$ associated with $\alpha(m_j)$ and $\beta(m_j)$. A plot of $E(\rho^2(m_j);\mathbf{Y}_t)$ versus $m_j/T,\ j=1,\ldots,q$, could be used as a posterior Bayes assessment of the power spectrum. In principle of course, a proper Bayesian assessment of the power spectrum would be a q-dimensional surface representing the posterior distribution of the vector $[\rho^2(m_1),\ldots,\rho^2(m_q)]$, but this would be of little practical value in terms of linkage with survival analysis.

6.3.3 A Hierarchical Bayes Assessment of the Power Spectrum

As an alternative to the model prescribed by (6.7) and (6.8), we may entertain a hierarchical Bayes–Kalman Filter-type model with an observation equation:

$$Y(t)=\alpha_t(0)+\alpha_t(m_1)+\alpha_t(m_2)+\cdots+\alpha_t(m_q)+\epsilon(t),$$

and a collection of system equations:

$$\begin{aligned}
\alpha_t(0)&=\alpha_{t-1}(0)+v_t(0)\\
\alpha_t(m_1)&=\alpha_{t-1}(m_1)\cos\left(\frac{2\pi}{T}m_1\right)+\beta_{t-1}(m_1)\sin\left(\frac{2\pi}{T}m_1\right)+v_t(m_1)\\
\beta_t(m_1)&=-\alpha_{t-1}(m_1)\sin\left(\frac{2\pi}{T}m_1\right)+\beta_{t-1}(m_1)\cos\left(\frac{2\pi}{T}m_1\right)+w_t(m_1)\\
&\ \vdots\\
\alpha_t(m_q)&=\alpha_{t-1}(m_q)\cos\left(\frac{2\pi}{T}m_q\right)+\beta_{t-1}(m_q)\sin\left(\frac{2\pi}{T}m_q\right)+v_t(m_q)\\
\beta_t(m_q)&=-\alpha_{t-1}(m_q)\sin\left(\frac{2\pi}{T}m_q\right)+\beta_{t-1}(m_q)\cos\left(\frac{2\pi}{T}m_q\right)+w_t(m_q).
\end{aligned}$$

In matrix notation, the observation and system equations are written as

$$Y(t)=\mathbf{F}'\boldsymbol{\theta}(t)+\epsilon(t),\quad \text{and} \tag{6.14}$$

$$\boldsymbol{\theta}(t)=\mathbf{G}\boldsymbol{\theta}(t-1)+\mathbf{W}(t), \tag{6.15}$$

where $\mathbf{F}$, $\boldsymbol{\theta}(t)$ and $\mathbf{W}(t)$ are vectors of dimension $(2q+1)$ taking the form:

$$\begin{aligned}
\mathbf{F}'&=(1,1,0,1,0,\ldots,1,0)\\
\boldsymbol{\theta}'(t)&=\left[\alpha_t(0),\alpha_t(m_1),\beta_t(m_1),\ldots,\alpha_t(m_q),\beta_t(m_q)\right],\ \text{and}\\
\mathbf{W}(t)&=\left[v_t(0),v_t(m_1),w_t(m_1),\ldots,v_t(m_q),w_t(m_q)\right];
\end{aligned}$$

$\mathbf{G}$ is a $(2q+1)\times(2q+1)$ matrix of the form

$$\mathbf{G}=\begin{bmatrix}
1 & 0 & 0 & & & \\
0 & \cos\left(\frac{2\pi}{T}m_1\right) & \sin\left(\frac{2\pi}{T}m_1\right) & & \mathbf{0} & \\
0 & -\sin\left(\frac{2\pi}{T}m_1\right) & \cos\left(\frac{2\pi}{T}m_1\right) & & & \\
 & & & \ddots & & \\
 & \mathbf{0} & & & \cos\left(\frac{2\pi}{T}m_q\right) & \sin\left(\frac{2\pi}{T}m_q\right)\\
 & & & & -\sin\left(\frac{2\pi}{T}m_q\right) & \cos\left(\frac{2\pi}{T}m_q\right)
\end{bmatrix}.$$

The formulation of (6.15) and (6.15) results in a Fourier series model of the form discussed in section 6.3.1. This can be verified by first invoking the relationship of (6.15) t times, to observe that

$$\boldsymbol{\theta}(t) = \mathbf{G}^{(t)}\boldsymbol{\theta}(0) + \Omega(t), \tag{6.16}$$

where $\mathbf{G}^{(t)}$ is the matrix $\mathbf{G}$ multiplied by itself t times (i.e. $\mathbf{G}^{(t)} = \mathbf{G} \times \mathbf{G} \times \cdots \times \mathbf{G}$), $\boldsymbol{\theta}(0)$ is a vector of initial values whose $(2q+1)$ elements are

$$\boldsymbol{\theta}'(0) = \left[\alpha_0(0), \alpha_0(m_1), \beta_0(m_1), \ldots, \alpha_0(m_q), \beta_0(m_q)\right],$$

and the error term $\Omega(t)$ is

$$\Omega(t) = \mathbf{G}^{(t-1)}\mathbf{W}(1) + \mathbf{G}^{(t-2)}\mathbf{W}(2) + \cdots + \mathbf{G}\mathbf{W}(t-1) + \mathbf{W}(t).$$

If in (6.14) $\boldsymbol{\theta}(t)$ is replaced by the version shown in (6.16) above, then $Y(t)$ can also be written as a Fourier series model of the form

$$Y(t) = \alpha_0(0) + \sum_{j=1}^{q}\left(\alpha_0(m_j)\cos\left(\frac{2\pi}{T}m_j t\right) + \beta_0(m_j)\sin\left(\frac{2\pi}{T}m_j t\right)\right) + \Omega(t) + \epsilon(t). \tag{6.17}$$

A comparison of (6.8) and (6.17) suggests that the $\alpha(0)$, $\alpha(m_j)$ and $\beta(m_j)$ of the former have been replaced by $\alpha_0(0)$, $\alpha_0(m_j)$ and $\beta_0(m_j)$, $j = 1, \ldots, q$, and that the $\epsilon(t)$ of (6.7) gets supplanted by $\epsilon(t) + \Omega(t)$. This latter feature suggests a larger variability in the model for $Y(t)$ in section 6.3.2 than the one in section 6.3.1.

To proceed further we need to make some error theory assumptions and specify a prior distribution for $\boldsymbol{\theta}(0)$. This we do via the following:

$$\begin{aligned}
(\epsilon(t)|\phi) &\sim N(0, 1/\phi);\\
(\mathbf{W}(t)|\phi) &\sim N(0, \Sigma(t)/\phi);\\
(\theta(0)|\phi) &\sim N(\boldsymbol{\Lambda}_0, \mathbf{R}_0/\phi), \quad \text{and as before}\\
\phi &\sim G\left(\frac{n_0}{2}, \frac{d_0}{2}\right).
\end{aligned}$$

The quantities $\Sigma(t)$, $\boldsymbol{\Lambda}_0$, $\mathbf{R}_0$, n_0 and d_0 are to be specified by us as prior parameters; i.e. the starting values of a Kalman Filter model. For $\boldsymbol{\Lambda}_0$, $\mathbf{R}_0$, n_0 and d_0 considerations analogous to those discussed in section 6.3.2 under the sub-heading 'Specifying a prior for the power spectrum' would apply. The role of these parameters parallels that of the parameters $\boldsymbol{\Lambda}_0$, $\mathbf{R}_0$, n_0 and d_0 of section 6.3.1. A general rule of thumb for specifying $\Sigma(t)$ would be to set $\Sigma(t) = k\mathbf{R}_0$, for all t, with $k \in (0, 1)$. When the observations $Y(t)$, $t = 1, 2, \ldots,$ are closely spaced, k would be small, in the vicinity of 0. The motivation for choosing a small value of k is that the goodness of a Fourier series approximation increases as the time between observations decreases. This completes our specification of a three-stage hierarchical model for assessing the power spectrum. The observation and systems equations constitute the first two stages of the hierarchy, and the gamma distribution of ϕ, the third stage.

Posterior Power Spectrum Under the Hierarchical Bayes Model
Upon observing $\mathbf{Y}_t = (Y(1), \ldots, Y(t))'$, we use standard Bayesian updating techniques (in this case Kalman Filtering) to obtain the posterior distribution of $\boldsymbol{\theta}(t)$ as:

$$(\boldsymbol{\theta}(t); \mathbf{Y}_t) \sim T_{n_t}(\boldsymbol{\Lambda}_t^*, \ \mathbf{R}_t^*(d_t^*/n_t^*)), \tag{6.18}$$

where the quantities $\boldsymbol{\Lambda}_t$, $\mathbf{R}_t$, n_t and d_t are prescribed below (West and Harrison, 1989, pp. 118–119). Observe that (6.18) parallels (6.12), save for the fact that the parameters take different values. These are of the form:

$$\boldsymbol{\Lambda}_t^* = \mathbf{G}\boldsymbol{\Lambda}_{t-1}^* + \mathbf{A}_t e_t,$$

where

$$\mathbf{A}_t = (\mathbf{G}\mathbf{R}_{t-1}^*\mathbf{G}' + \Sigma(t))\mathbf{F}(1 + \mathbf{F}'(\mathbf{G}\mathbf{R}_{t-1}^*\mathbf{G}' + \Sigma(t))\mathbf{F})^{-1}, \quad \text{and}$$
$$e_t = Y(t) - \mathbf{F}'\mathbf{G}\boldsymbol{\Lambda}_{t-1}^*.$$

Also

$$n_t^* = n_{t-1}^* + 1,$$
$$d_t^* = (e_t^*)^2(1 + \mathbf{F}'(\mathbf{G}\mathbf{R}_{t-1}^*\mathbf{G}' + \Sigma(t))\mathbf{F})^{-1} + d_{t-1}^*, \quad \text{and}$$
$$\mathbf{R}_t^* = (\mathbf{G}\mathbf{R}_{t-1}^*\mathbf{G}' + \Sigma(t)) - \mathbf{A}_t\mathbf{A}_t'(1 + \mathbf{F}'(\mathbf{G}\mathbf{R}_{t-1}^*\mathbf{G}' + \Sigma(t))\mathbf{F})^{-1}.$$

In the above recursions, $n_0^* = n_0$, $d_0^* = d_0$, $\boldsymbol{\Lambda}_0^* = \boldsymbol{\Lambda}_0$ and $\mathbf{R}_0^* = \mathbf{R}_0$; a strategy for specifying them was discussed before.

The isomorphism between (6.12) and (6.18) allows us to use the methodology following (6.12) to assess the power spectrum for this section's hierarchical Bayes set-up. The main difference is that $\boldsymbol{\Lambda}_t^*$, $\mathbf{R}_t^*$, n_t^* and d_t^* are used in place of $\boldsymbol{\Lambda}_t$, $\mathbf{R}_t$, n_t and d_t, respectively.

6.3.4 The Spectrum as a Covariate Using an Accelerated Life Model

Since the power spectrum can be seen as a snapshot of potential defects, it makes sense to use it as a covariate for assessing lifetimes. The purpose of this section is to propose an approach for doing the above. With this in mind, suppose that T is the lifelength of an item, and suppose that the vector

$$\rho^* \stackrel{\text{def}}{=} [\rho^*(m_1), \rho^*(m_2), \ldots, \rho^*(m_q)]',$$

where $\rho^*(m_j) = E(\rho^2(m_j); \mathbf{Y}_t)$ (equation (6.13) is our assessment of the $\rho^*(m_j)$ for this item. Note that $\boldsymbol{\rho}^*$ could be obtained via either the regression model of section 6.3.2, or via the Kalman Filter model of section 6.3.3. Recall that a plot of $\rho^*(m_j)$ versus m_j/T, for $j = 1, \ldots, q$ represents our assessment of the power spectrum.

The linkage between ρ^* and T can be made through what is known as an 'accelerated life model'. One such version of this model – the version used by Campodónico and Singpurwalla (1994b) – is to assume that

$$\ln T = -\sum_{j=1}^{q} C_j \rho^*(m_j) + \ln W, \tag{6.19}$$

where $C_1, \ldots, C_q$ are unknown constants, and $\ln W \sim N(0, \tau^2)$, with τ^2 unknown. Equation (6.19) describes what is known as a **log-linear model** for T.

Suppose that n units are observed to failure and their lifetimes $t_1, \ldots, t_n$ noted. Also recorded, just prior to their failure, are their respective signatures $\boldsymbol{\rho}_i^*$, $i = 1, \ldots, n$. Then, as per the dictates of the log-linear model,

$$\ln t_i = -\sum_{j=1}^{q} C_j \rho_i^*(m_j) + \epsilon_i, \tag{6.20}$$

where the ϵ_i's are independent and are such that $\epsilon_i \sim N(0, \tau^2)$. Given the t_i's and their corresponding $\rho_i^*(m_1), \rho_i^*(m_2), \ldots, \rho_i^*(m_q)$, $i = 1, \ldots, n$, standard Bayesian analysis of the log-linear model (cf. Box and Tiao, 1992, p. 113) enables us to assess the posterior distribution of $(\mathbf{C}, \tau^2)$, where $\mathbf{C} = (C_1, \ldots, C_q)$. The natural conjugate prior for $\mathbf{C}$ is a multivariate normal distribution whose parameters reflect our opinions about the relative impact of each defect on the item's lifetime. A neutral choice would be to assume that the prior means of the C_is are zero, that their variances are large and their covariances are small. A natural conjugate prior for $1/\tau^2$ would be a gamma, independent of the prior for $\mathbf{C}$. The mode of the posterior distribution of $\mathbf{C}$ provides a sense of the relative importance of each $\rho(m_j)$, $j = 1, \ldots, q$ vis-à-vis the item's lifetime.

A flavor of the above is best illustrated by an overview of the traction motor example considered by Campodónico and Singpurwalla (1994b). Here, the 'leakage current' is a proxy for motor's lifetime, and a signature of the motor's vibration during rotation is an indicator of its underlying defects.

Illustrative Example – Vibration of Traction Motors

Traction motors for locomotives are normally tested for defects using vibration signals taken at different running speeds. We shall restrict attention here to the case of the 900 revolutions per minute (rpm) speed. A large amplitude – i.e. $\rho^2(m_j)$ – at $m_j = 1$ suggests an imbalance or a bent shaft, whereas a large amplitude at $m_j = 2$ suggests a misalignment of the motor. Similarly, large amplitudes at $m_j = 3(4)$ suggest looseness of the bearing on the shaft (housing). If the gear in a motor has k teeth, then a large amplitude at $m_j = k$ suggests problems having to do with gearmesh. Finally, if the number of ball-bearings in the motor is n, then large amplitudes at $m_j = 0.4n$ $(0.6n)$ suggest defects on the outer (inner) races of the ball-bearings. In our particular case, $k = 18$ and $n = 13$.

Thus, for this scenario, it makes sense to choose the frequencies of 1 through 10 (times the motor running speed) and the frequency of 18 (times the motor running speed). That is, in the notation of section 6.3.1, $\mathcal{J} = \{1, 2, \ldots, 10, 18\}$ and $q = 11$. We then use the regression model of (6.7) and (6.8), with $Y(t)$ as the lateral displacement of the motor, to obtain a Bayesian assessment of the power spectrum. For the prior specifications, $\boldsymbol{\Lambda}_0$ was set to its default value 0 and the diagonal elements of the matrix $\mathbf{R}_0$ of Section 6.3.2 were taken to be (1, 0.95, 0.95, 0.9, 0.9, 0.85, 0.85, 0.8, 0.8, 0.75, 0.75, 0.7, 0.7, 0.65, 0.65, 0.6, 0.6, 0.55 and 0.55); the parameters n_0 and d_0 were set equal to 0. Equation (6.13) was used to assess the spectrum for $m_j \in \mathcal{J}$. Figure 6.6 shows a plot of the assessed spectrum (or a signature) of a motor labeled #4. The leakage current for this motor was 1.975.

Signatures similar to those of Figure 6.6 were also obtained for eight other traction motors, making the total number of motors tested as nine; also recorded were their associated leakage currents, t_i, $i = 1, \ldots, 9$. The log-linear model of (6.20) was invoked on these data and the C_j, $j = 1, \ldots, 11$ assessed. A plot of the assessed C_js, rescaled so that they are all between 0 and 1, is shown in Figure 6.7. We refer to such rescaled C_js as **spectral weights**. For details on the priors for $\mathbf{C} = (C_1, \ldots, C_{11})$ and τ^2, see (Campodónico and Singpurwalla 1994b).

An examination of Figure 6.7 suggests that defects associated with other frequencies (4, 6, 7, 9 and 10) are more deleterious to the motor's lifetime than those corresponding to the other frequencies. A comparison of Figures 6.6 and 6.7 suggests that even though the amplitudes associated with the frequencies at 6, 7, 8, 9 and 10 are small, their impact on lifetimes is large. Recall that large amplitudes at frequencies 5–10 correspond to defects in the inner and outer race of the bearings, whereas large amplitudes at frequencies 1–3 suggest an imbalance. Thus, it appears that good bearing races and tightness of the bearing on the housing is more important to a traction motor's lifetime than misalignment and imbalance.

6.3.5 Closing Remarks on Signatures and Covariates

Hopefully the above realistic but simplified example underscores the relevance of signature analysis in reliability and survival analysis. It enables us to assess the relative importance of the types of defects on an item's lifelength. It also provides us with a means for predicting

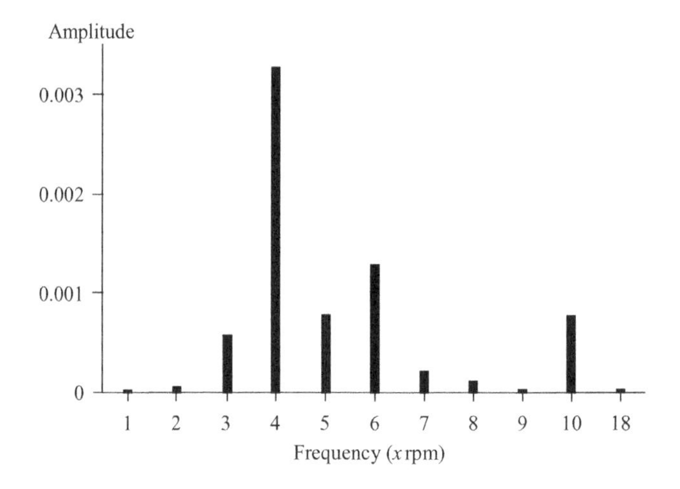

Figure 6.6 Assessed signature of lateral displacement data for motor #4.

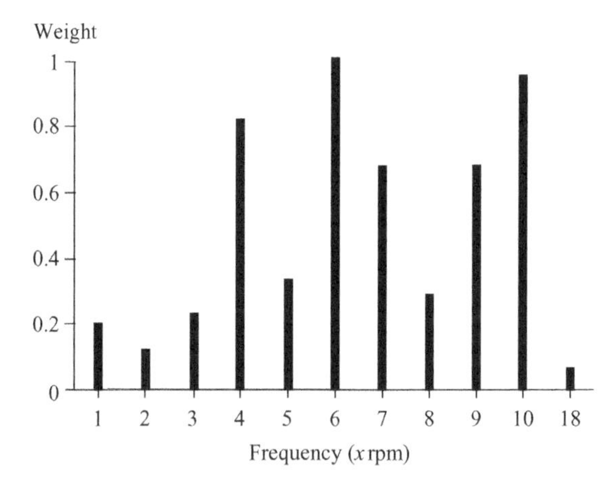

Figure 6.7 The spectral weights C_j for traction motor data.

lifetimes, given an item's signature. Recently, much has been written under the general rubric of degradation modeling (for example, Whitmore (1995); also section 7.5). In principle, using signatures as covariates can also be seen as a form of degradation modeling, provided that one can relate amplitudes of the spectra with defects. As mentioned before, the full potential of using signatures as covariates needs to be fully explored. For instance, is the accelerated life model of (6.19) the only meaningful way of relating signatures and lifetimes? Also needed is the development of efficient computational tools. The formulae in sections 6.3.2 and 6.3.3 look cumbersome and intimidating. In actuality, they are not; they are nothing more than the equations of Kalman Filtering, successfully used by engineers using control theory. Modern computing technologies such as the MCMC will enable us to deal with such expressions in a rather straightforward manner.

By a way of concluding to this chapter on hierarchical modeling, I remark that hierarchical models are not only useful with respect to their traditional role of better encapsulating observed data, as most Bayesians seem to view, but they can also play a role in terms of developing new ideas, or expanding upon the existing ones. A case in point is our notion of composite reliability. This can be seen as a way of embedding the current notion of reliability (articulated in the early 1950s) within a more general conceptual framework.

Chapter 7

Survival in Dynamic Environments

Kolmogorov: He almost surely lies here

7.1 INTRODUCTION: WHY STOCHASTIC HAZARD FUNCTIONS?

The actual environment in which most items generally operate is dynamic. That is, factors which go to constitute an environment, such as temperature, humidity, usage rate, etc., change with time. Moreover, the manner in which the changes occur is not deterministic, in the sense that the times of change and the amount of change cannot be pre-specified. Consequently, the behavior of such factors, actually covariates, is best described via stochastic process models, such as the Poisson process of section 4.7.4, or the gamma process of section 4.9.1. Static environments are rare. They are generally encountered in controlled life-tests wherein the conditions of the test environment are kept constant (section 5.4.1). In the biomedical scenario, the changing covariates are sometimes referred to as **markers**, and their random nature described via what are known as **marker processes** (cf. Whitmore, Crowder and Lawless, 1998).

One possible way of capturing the effect of a random environment on an item's lifelength is to make its failure rate function (be it predictive or model) a stochastic process. This is what we mean by a **stochastic hazard**. The standard models of reliability derived from failure rate functions that are deterministic (Chapter 4) do not account for the dynamics of the observed and/or the unobserved covariates. The idea of making the failure rate function a stochastic process can be traced back to Gaver (1963); this was followed up by the works of Antelman and Savage (1965) and of Harris and Singpurwalla (1967). However, it is only recently, subsequent to the papers of Arjas (1981) and Kebir (1991), that the merits of the idea are beginning to be better appreciated (cf. Singpurwalla, 1995b). Stochastic hazards have already been introduced in sections 4.7.6 and 4.9.1, but these were embedded within the context of a broader class of problems. The basic features of hazard rate processes remain to be articulated and this is what I endeavor to do below.

Reliability and Risk: A Bayesian Perspective N.D. Singpurwalla
© 2006 John Wiley & Sons, Ltd

7.2 HAZARD RATE PROCESSES

A stochastic process was introduced in section 4.7.6 as a collection of random variables whose marginal and joint distributions can be fully specified. Specifically, if for every $t \in [0, \infty)$, $r(t)$ is a random variable with a specified distribution, and if for every collection of $k \geq 1$ points $0 < t_1 < t_2 < \cdots < t_k$, the joint distribution of the k random variables $r(t_1), \ldots, r(t_k)$ is also specified, then $\{r(t); t \geq 0\}$ is a stochastic process. Furthermore, if for all $t \in [0, \infty)$, $r(t)$ is non-negative, real-valued and continuous in t from the right, then $\{r(t); t \geq 0\}$ is said to be a non-negative, real-valued and right-continuous stochastic process. Suppose that $\{r(t); t \geq 0\}$ is such a process, and suppose that T denotes the lifelength of an item. Then $\{r(t); t \geq 0\}$ is said to be the **hazard rate process** of T if

$$P(T \geq t | r(s), 0 \leq s < t) = \exp\left[-\int_0^t r(s)\,\mathrm{d}s\right]. \tag{7.1}$$

The quantity $R(t) \stackrel{\text{def}}{=} \int_0^t r(s)\,\mathrm{d}s$ also defines a stochastic process $\{R(t); t \geq 0\}$, which is known as the **cumulative hazard process**. Such a process has already been encountered by us in section 4.7.6, wherein it was taken to be a gamma process. Thus in terms of a cumulative hazard process, the right-hand side of (7.1) can be written as $\exp[-R(t)]$. A consequence of the above is that for any $t \geq 0$

$$P(T \geq t) = E\{\exp(-R(t))\}, \tag{7.2}$$

where E denotes the expectation. The arguments leading us from (7.1) to (7.2) are subtle (cf. Pitman and Speed, 1973) even though the result is intuitive. Before discussing meaningful choices for the hazard rate and the cumulative hazard process, the following relationship between the hazard rate process and the intensity function of a doubly stochastic Poisson process is useful to note.

Relationship with Doubly Stochastic Poisson Processes

Recall that homogenous and non-homogenous Poisson processes and their intensity functions were introduced in section 4.7.4, and that doubly stochastic Poisson processes and their stochastic intensity functions introduced in section 5.2.4. Now, for any lifetime T having a survival function whose failure rate is $r(t)$, $t \geq 0$, we have

$$P(T > t) = \exp\left[-\int_0^t r(s)\,\mathrm{d}s\right]. \tag{7.3}$$

But the right-hand side of (7.3) is also the distribution function of U, the time of the first jump in a non-homogenous Poisson process, say $\{N(t); t \geq 0\}$, with intensity function $r(s)$, $s \geq 0$. That is

$$P(U > t) = P(N(t) = 0) = \exp\left[-\int_0^t r(s)\,\mathrm{d}s\right]. \tag{7.4}$$

Thus the failure rate function can also be interpreted as the intensity function of a non-homogenous Poisson process in the sense of (7.3) and (7.4) above. In the same vein it can be argued (cf. Kebir, 1991) that a hazard rate process can be interpreted as the intensity process of a doubly stochastic Poisson process. This connection between Poisson processes and the hazard rate serves as a motivation for the point process approach to reliability and survival analysis discussed in Chapter 8. For now we shall concentrate on some specific choices for the hazard rate process $\{r(t); t \geq 0\}$ and cumulative hazard rate process $\{R(t); t \geq 0\}$.

7.2.1 Hazard Rates as Shot-noise Processes

Shot-noise processes were introduced in section 4.7.6 for describing the effect of a certain kind of environment on the model failure rates of multi-component systems. Here we suppose that the failure rate function itself can be described by a shot-noise process. For this we will adhere to the notation and terminology of section 4.7.6. Much of what is said below parallels the material of section 4.7.6, save for some minor variations. Specifically, for $t \geq 0$, we write

$$r(t) = \sum_{k=1}^{\infty} X_k h(t - T_{(k)}),$$

where X_k is the size of the jump in the failure rate at time $T_{(k)}$, and $h(\bullet)$ is the attenuation function (Figure 4.16); but bear in mind that here $X_0 \equiv 0$, so that $r(0) = 0$. The X_ks are assumed to be independent of each other and also of the $T_{(k)}$s. Also, the X_ks have a common distribution G.

A motivation for describing the failure rate function via a shot-noise process is based on the premise that the shocks (or events) of the process induce stresses of unknown magnitude on a component, that the component experiences a residual stress that may or may not decay over time and that the stresses are additive. We also suppose that stress is the cause of failure, so that the pattern of the failure rate function is determined by the (time indexed) pattern of the stress. If, as in section 4.7.6, we let $M(t) = \int_0^t m(u)\,du$, where $m(u)$ is the intensity function of the Poisson process that generates the jump times of the failure rate, $H(t) = \int_0^t h(u)\,du$, and G^* the Laplace transform of G, then

$$P(T \geq t) = \exp\left[-M(t) + \int_0^t G^*(H(u))m(t-u)\,du\right]; \tag{7.5}$$

T is the time to failure of an item experiencing the shot-noise hazard rate process described above (for details, see section 4.7.6).

Note, from the material discussed before, that T is also the time to occurrence of the first event in a doubly stochastic Poisson process whose intensity function is described by a shot-noise process.

Special cases of $h(\bullet)$, G and $m(\bullet)$ will yield specific forms of the survival function $P(T \geq t)$. For example, when G is exponential with a scale $b > 0$, $m(u)$ a constant $m > 0$, and $h(u) = 1$, so that $r(t)$ is non-decreasing,

$$P(T \geq t) = \exp(-mt)\left(\frac{b+t}{b}\right)^{mb}, \tag{7.6}$$

a Pareto distribution of the third kind. The above result parallels that of (4.49) save for the fact that here $\lambda_i = 1$ and that $\left(\frac{b+t}{b}\right)$ is raised to the power mb instead of $mb - 1$. The reason for this difference is that in the present case $r(t) = 0$ until the occurrence of the first jump.

As a second example, suppose that the jump sizes are a constant d, where d is set to one, and that $h(u) = (1+u)^{-1}$, signaling a slow decay of the failure rate until the next jump. Then, it can be verified that for $m(u) = m$,

$$P(T \geq t) = \exp(-mt)\,(1+t)^m, \tag{7.7}$$

again a Pareto distribution of the third kind.

As a third and final example, suppose that the attenuation function reflects an exponential decay of the form $h(u)=\exp(-au)$, for some $a>0$, that $M(t)=mt$, and that G is an exponential distribution with scale $b>0$. Then

$$P(T\geq t)=\exp\left[-\frac{mabt}{1+ab}\right]\left\{\frac{1+ab-e^{-at}}{ab}\right\}^{\frac{mb}{1+ab}}. \tag{7.8}$$

I leave it as an exercise for the reader to show that (7.6)–(7.8) are true.

7.2.2 Hazard Rates as Lévy Processes

A Lévy process, though mentioned in section 4.9.1, remains to be formally defined. Processes with independent increments are called **additive processes**, and the Lévy process is a special type of additive process. Specifically, a Lévy process is a continuous process with stationary independent increments; i.e. the increments of the process have the same distribution. Lévy processes encompass a large family of processes, the Poisson, the compound Poisson, the gamma process (of the type discussed in section 4.9.1) and the 'Brownian motion process' (to be discussed in section 7.2.3) are some noteworthy examples. **Increasing Lévy processes** (also known as 'non-negative Lévy processes' or as 'subordinators') are Lévy processes whose stationary independent increments are non-negative. Examples are the Poisson, the compound Poisson and the gamma process mentioned before.

Lévy processes have turned out to be a valuable modeling tool in reliability and survival analysis, not only because of their generality but also because they produce, in many cases, computationally tractable results. This computational tractability is the consequence of a celebrated decomposition of increasing processes with independent increments, namely, the Itô decomposition (cf. Itô, 1969). The decomposition takes a particularly simple form when the underlying process is an increasing Lévy process (equation (7.9). Increasing Lévy processes are viable candidates for describing failure rate functions that are increasing (and also cumulative hazard functions; section 7.3.2). Suppose then that the hazard rate process $\{r(t); t\geq 0\}$ is an increasing Lévy process. Then, for some $\alpha\geq 0$, where α is known as the drift rate

$$r(t)=\alpha t+\int_0^t\int_0^\infty zN(du,dz), \tag{7.9}$$

where N has a Poisson distribution in two dimensions on $[0,\infty)\times[0,\infty)$, with mean $n(du,dz)$. That is, for any subset, say B, of $[0,\infty)\times[0,\infty)$, $P(N(B)=j)=\exp[-n(B)]\frac{(n(B))^j}{j!}$. Furthermore $n(du,dz)=du\nu(dz)$, where $\nu(dz)$ does not depend on u; $\nu(dz)$ is known as the **Lévy measure**. Intuitively, the Lévy measure characterizes both the expected frequency and the size of jumps in a Lévy process. The essence of the above result (which is also known as a **Lévy representation**) is best appreciated via the specific examples that are discussed below. When $\alpha=0$, the underlying Lévy process does not have an upward drift. Consequently, it does not have any non-random parts and so $r(t)$ increases only by jumps. The αt term can be replaced by the more general $\alpha(t)$, where $\alpha(t)$ is a non-decreasing continuous function of time with $\alpha(0)=0$.

Based on the decomposition of (7.9), Kebir (1991), in his Corollary 3.2, is able to show that the survival function of the time to failure T is:

$$P(T\geq t)=\exp\left(-\frac{\alpha t^2}{2}\right)\exp\left[-\int_0^t ds\int_0^\infty[1-\exp(-(t-s)x)]\right]\nu(dx), \tag{7.10}$$

where $\nu(\mathrm{d}x)$ is the Lévy measure of the underlying Lévy process. Equation (7.10) is striking. Striking, because its right-hand side is also the survival function of an item which has a deterministic failure rate function of the form:

$$r(t) = \alpha t + \int_0^\infty [1 - \exp(tx)]\, \nu(\mathrm{d}x). \tag{7.11}$$

Thus an item experiencing an increasing Lévy process as a hazard rate process has a survival function that is identical to the survival function of an item experiencing a deterministic failure rate function of the form given by (7.11). A broader interpretation of this result has been that a continuous increasing stochastic process with stationary independent increments is essentially deterministic (cf. Çinlar, 1979; also see section 7.3.2).

Specific choices for the Lévy measure $\nu(\mathrm{d}x)$ yield specific expressions for the survival function of (7.10). For example, the Lévy measure of a gamma process with shape at, $a > 0$, and scale b is

$$\nu(\mathrm{d}x) = \frac{a}{x}\exp(-bx)\,\mathrm{d}x, \qquad x > 0. \tag{7.12}$$

Consequently, with the drift rate $\alpha = 0$, (7.10) gives the survival function as:

$$P(T \geq t) = \exp\left[-\int_0^t a\log\left(\frac{b+x}{b}\right)\mathrm{d}x\right]; \tag{7.13}$$

its deterministic failure rate function is $a\log(b^{-1}(b+t))$.

In section 9.4.2, the extended gamma process, which is a weighted gamma process, will be introduced. Equation (9.11) pertains to the survival function generated by such a process; it complements (7.13) above.

As a second example, suppose that the hazard rate process is assumed to be a homogenous compound Poisson process (section 4.9.1), with an intensity function $m(t) = m > 0$, and with G as the distribution of the increments. Then the Lévy measure of this process is $\nu(\mathrm{d}x) = mG(\mathrm{d}x)$. Consequently, its survival function is

$$P(T \geq t) = \exp\left[-\int_0^t m(1 - G^*(x))\,\mathrm{d}x\right], \tag{7.14}$$

where G^* is the Laplace transform of G. If G is assumed to be a gamma distribution with shape parameter $\alpha > 0$, and scale parameter b, then the Lévy measure of the process is

$$\nu(\mathrm{d}x) = m\frac{b^a}{\Gamma(a)}x^{a-1}\mathrm{e}^{-bx}\mathrm{d}x, \quad \text{and}$$

$$G^*(x) = (b/(b+s))^a.$$

Consequently, its survival function is

$$P(T \geq t) = \begin{cases} \exp(-mt)((b+t)/b)^{mb}, & \text{if } a = 1 \\ \exp(-m)\left(t + \frac{b}{a-1}\left(\frac{b}{b+t}\right)^{a-1} - \frac{b}{a-1}\right), & \text{if } a \neq 1 \end{cases}$$

Finally, if the hazard rate process $\{r(t); t \geq 0\}$ is a **stable process**, i.e. if for any $t > 0$, its Laplace transform $E\{\exp(-sr(t))\}$ is of the form $\exp(-tbs^a)$, for $b > 0$ and $0 < a < 1$, then the Lévy measure of the process is

$$\nu(\mathrm{d}x) = \frac{abx^{-(1+a)}}{\Gamma(1-a)}. \tag{7.15}$$

The above when plugged in (7.10) gives us the survival function of an item experiencing a stable process for its failure rate.

7.2.3 Hazard Rates as Functions of Diffusion Processes

The material of this section is largely based on the work of Myers (1981), who induces a hazard rate process by assuming a Brownian motion process for the covariates (or markers) that influence failure. This is in contrast to the material of sections 7.2.1 and 7.2.2 wherein stochastic process models were directly prescribed for the failure rate function. Myer's work appears to have been inspired by ideas on modeling human mortality and aging by Woodbury and Manton (1977). Yashin and Manton (1997) give an authoritative overview of this and related issues. A justification for following the modeling strategy proposed by the above is that prescribing stochastic process models on unobserved entities like the failure rate function is unnatural. Rather, one should describe the stochastic behavior of observable entities, like covariates (and markers), by a stochastic process model, and to then induce a stochastic process for the failure rate function by assuming a relationship between the covariate(s) and the failure rate. Indeed, this was the strategy used by us in section 4.7.6 wherein the intrinsic failure rate $\lambda(u) = \lambda$, for $u \geq 0$, was modulated by the covariate-related shot-noise process $\{\eta(u);\ u \geq 0\}$ to produce the hazard rate process $\{\lambda\eta(u);\ u \geq 0\}$. The material to follow is based on the above line of thinking, save for the fact that a Brownian motion process is used to describe the evolution of covariate(s), and the relationship between the covariate(s) and the failure rate is more elaborate than the multiplicative one mentioned above. We start with an overview of the Wiener process and Brownian motion, but to set the stage for this I need to introduce some notation. Let the vector $\mathbf{Z}(t)$ denote the values at t of a collection of covariates that are assumed to influence lifelength. Suppose that $\mathbf{Z}(t)$ is of dimension $m \geq 1$, and that it is possible to observe and to measure each element of $\mathbf{Z}(t)$, for any $t \geq 0$. Let $Z_i(t)$, $t \geq 0$, be the i-th element of $\mathbf{Z}(t)$, $i = 1, \ldots, m$.

Brownian Motion, Wiener Processes and Diffusion

A stochastic process $\{Z_i(t);\ t \geq 0\}$ is said to be a **Wiener process**, starting from $Z_i(0) = z$, say, where $z \geq 0$, if:

(i) $Z_i(t)$ is continuous in t, for $t \geq 0$;
(ii) $Z_i(t)$ has independent increments; and
(iii) $(Z_i(s+t) - Z_i(s))$ has a Gaussian distribution with mean 0 and variance $\sigma^2 t$, for all $s, t \geq 0$, with σ^2 a positive constant.

When $z = 0$ and $\sigma^2 = 1$, $\{Z_i(t);\ t \geq 0\}$ is called the **standard Wiener process**; with $\sigma^2 \neq 1$, it is the process $\{Z_i(t)/\sigma;\ t \geq 0\}$ that is a standard Wiener process. It can be verified that for $0 \leq s \leq t$, the autocovariance function of $\{Z_i(t);\ t \geq 0\}$, $E\{(Z_i(t) - Z_i(0))(Z_i(s) - Z_i(0))\} = \sigma^2 s$; thus for any $s, t \geq 0$, the autocovariance function is $\sigma^2 \min(s, t)$. If $E(Z_i(t)) = \eta t$, for some $\eta > 0$, then the Wiener process is said to have a drift parameter η. In general, if $\{Z(t);\ t \geq 0\}$ is a standard Wiener process, then the process $\{\tilde{Z}(t);\ t \geq 0\}$, where $\tilde{Z}(t) = \eta t + \sigma^2 W_t$ will be a Wiener process with drift η and diffusion parameter σ^2. Furthermore, $\tilde{Z}(t)$ will have a Gaussian distribution with mean μt and variance $\sigma^2 t$, and for $s \leq t$ its covariance will be $\sigma^2 s(1 + \mu^2 t)$.

The vector stochastic process $\{\mathbf{Z}(t);\ t \geq 0\}$ is said to be a **Brownian motion process** if:

(i) $\mathbf{Z}(0) = \mathbf{0}$, and
(ii) $\{Z_i(t);\ t \geq 0\}$, $i = 1, \ldots, m$ are independent and identically distributed Wiener processes.

Thus a collection of independent and identically distributed Wiener processes, all starting at 0, constitute a Brownian motion process. It is for this reason that the Wiener process is also referred to as Brownian motion. A **standard Brownian motion process** is a Brownian motion process in which all the constituent processes $\{Z_i(t); t \geq 0\}$ are standard Wiener processes. The Brownian motion and the Wiener process are archetypal examples of diffusion processes. That is, processes whose sample paths are continuous and which possess 'the strong Markov property' (see Karlin and Taylor (1981), p. 149 for a definition).

Hazard Rate as Squared Brownian Motion

For the covariate vector $\{\mathbf{Z}(t); t \geq 0\}$, Myers (1981) assumes a diffusion process via the Itô stochastic differential equation

$$\mathrm{d}\mathbf{Z}(t) = \mathbf{a}(t)\mathbf{Z}(t) + \mathbf{b}(t)\,\mathrm{d}\mathbf{W}(t),$$

where $\mathbf{a}(t)$ and $\mathbf{b}(t)$ are continuous, and $\{\mathbf{W}(t); t \geq 0\}$ is a standard Brownian process of dimension m. The failure rate function is assumed to be a quadratic function of $\mathbf{Z}(t)$. For the hazard rate process thus induced, Myers (1981) obtains an expression for the survival function (the theorem in section 2.2 of Myers' paper). The expression for this survival function is not in a closed form and involves the solution of a matrix differential equation (called the 'Riccati equation'). However, in the one-dimensional case, i.e. for $m = 1$, an interesting closed-form expression for the survival function can be obtained; this is discussed below.

Let $\{Z(t); t \geq 0\}$ be a one-dimensional covariate process whose dynamic is defined via the relationship

$$Z(t) = Z(0) + \sigma W(t),$$

where $W(t)$ is a one-dimensional standard Wiener process. Thus $Z(t)$ is a Wiener diffusion process starting at $Z(0) > 0$, and having scale $\sigma > 0$. For some constant $q > 0$, let the hazard rate process $\{r(t); t \geq 0\}$ be such that $r(t) = q(Z(t))^2$. Then the survival function of T, the time to failure, will be

$$P(T \geq t) = E\left[\exp\left(-q\int_0^t Z^2(u)\mathrm{d}u\right)\right];$$

see (7.2). The failure rate function of T, namely $-\frac{\mathrm{d}}{\mathrm{d}t}\log P(T \geq t)$ turns out to be

$$q(Z(0))^2 \mathrm{sech}^2(At) + A\tanh(At),$$

where $A = (2\sigma^2 q)^{1/2}$, $\mathrm{sech}(x) = \dfrac{2}{\mathrm{e}^x + \mathrm{e}^{-x}}$, and $\tanh(x) = \dfrac{\mathrm{e}^x - \mathrm{e}^{-x}}{\mathrm{e}^x + \mathrm{e}^{-x}}$.

When $Z(0) = 0$ and $\sigma^2 = 1$, the survival function of T is given as:

$$P(T \geq t) = (\cosh(\sqrt{2qt}))^{1/2},$$

where $\cosh(x) = (\mathrm{e}^x + \mathrm{e}^{-x})/2$.

The above expression for the survival function is the Laplace transform of a definite integral of squared Brownian motion (in dimension one), and is known as the **Cameron–Martin Formula**, after Cameron and Martin (1944).

7.3 CUMULATIVE HAZARD PROCESSES

The theme of section 7.2 was to obtain expressions for the survival function of items that operate in a dynamic environment and therefore experience stochastic hazard rate functions. The exponentiation formula of reliability and survival, averaged over the hazard rate process (equation (7.2)), was the key operational tool. The cumulative hazard process $\{R(t); t \geq 0\}$, where $R(t) = \int_0^t r(u)\,du$, was introduced in section 7.2. This process comes about via the argument that it is a consequence of assuming a failure rate function that is a stochastic process. Different kinds of stochastic processes were assumed for $\{r(t); t \geq 0\}$, such as the shot-noise, the Lévy and the diffusion (namely, a squared Brownian motion). The assumed stochastic process for $\{r(t); t \geq 0\}$ induces a stochastic process for $\{R(t); t \geq 0\}$. Alternatively, one may find it more convenient to directly prescribe a stochastic process for $\{R(t); t \geq 0\}$ itself, and in what follows this is the point of view that will be adopted here.

The aim of this section is to describe a different approach for obtaining the survival function of items operating in a dynamic environment. The approach starts by assuming a stochastic process model for $\{R(t); t \geq 0\}$, and instead of using (7.2) to obtain $P(T \geq t)$, it uses the hitting time of this process to a random threshold, say X, to obtain a survival function. The random threshold X is the hazard potential of the item, and X has an exponential distribution with scale parameter one (section 4.6). Durham and Padgett (1997) take a similar approach for obtaining survival functions. Their motivation, however, is different from ours; it is based on cumulative damage and initial strength, assumed known. Lee and Whitmore (2006) consider the scenario of hitting times with covariate information.

Hitting times of stochastic processes to a specified threshold (or barrier) is a well-studied topic in probability; for example, Barndorff-Nielsen, Blaesid and Halgreen (1978). Thus, as an example, if the process $\{R(t); t \geq 0\}$ is assumed to be a Wiener process, then the hitting time of this process to a specified barrier will have an inverse Gaussian distribution. But using the above argument as a justification for the inverse Gaussian as a failure model is flawed. This is because the Wiener process is not monotonically increasing in time, whereas $R(t)$ needs to be non-decreasing in t. However, the process $\sup_{0<u\leq t} \{W(t); t \geq 0\}$, where $\{W(t); t \geq 0\}$ is a standard Wiener process, is continuous, non-negative and non-decreasing in t. This process is called a **Brownian Maximum Process**. Thus it makes sense to describe $\{R(t); t \geq 0\}$ by a Brownian maximum process. The hitting time of a Brownian maximum process to a specified threshold will continue to have an inverse Gaussian distribution. That is, if T_x denotes the hitting time of a Brownian maximum process to a threshold x, then

$$P(T_x \leq t) = 2(1 - \Phi(x/\sqrt{t})),$$

where $\Phi(s) = \int_{-\infty}^{s} \frac{1}{\sqrt{2\pi}} e^{-\frac{u^2}{2}}\,du$.

In our particular scenario, the threshold x is not specified. Rather, it is an exponentially distributed random variable with scale parameter one. This means that T, the time to failure, will have a distribution that is a scale mixture of inverse Gaussian distributions, the mixing distribution being exponential (1). Specifically, it can be verified that the survival function is of the form

$$P(T > t) = 2e^{t/2}\Phi(-\sqrt{t}).$$

In what follows let us consider some other non-decreasing stochastic processes for $\{R(t); t \geq 0\}$, such as a compound Poisson, an increasing Lévy, an integrated geometric Brownian motion and a Markov additive process to obtain some families of survival functions using the above line of development. The important question of which process one must use and when to use it remains to be satisfactorily addressed. In some cases, the choice of a process would be natural, in others, the choice would be based on subjective (or personalistic) considerations. An empiricist answer would be to choose that process which best describes the failure data at hand, if any. My aim here is to offer the reader a menu of possibilities.

7.3.1 The Cumulative Hazard as a Compound Poisson Process

A compound Poisson process with an intensity function $m(u) = m > 0$, and having jumps of size $X_1, X_2, \ldots,$ with X_is independent and identically distributed with distribution function G, is perhaps one of the simplest models for describing the process $\{R(t); t \geq 0\}$. Observe that $R(t)$ is right continuous and non-decreasing in t. If the X_is are also independent of the hazard potential X, as is reasonable to assume, then if T denotes the hitting time of this process to the random threshold X, the survival function of T, conditional on m known, is

$$P(T > t \mid m) = \sum_{k=0}^{\infty} \frac{e^{-mt}(mt)^k}{k!} \int_0^{\infty} G^{(k)}(u) e^{-u} \, du, \tag{7.16}$$

where $G^{(k)}$ is the k-fold convolution of G with itself. The details are easy to verify. When G is such that $G(u) = 0$, $u < 0$ (i.e. the jump sizes are non-negative), Esary, Marshall and Proschan (1973) show that (7.16) simplifies as

$$P(T > t \mid m) = e^{-mt}, t \geq 0, m > 0.$$

Averaging out m with respect to any reasonable distribution for it would result in a survival function having a decreasing failure rate function. In particular, if m has a gamma distribution, then the distribution of T will be a Pareto.

7.3.2 The Cumulative Hazard as an Increasing Lévy Process

An omnibus way to describe $\{R(t); t \geq 0\}$ is via an increasing Lévy process (section 7.2.2). Such a process encompasses the compound Poisson, the gamma, and the stable process. There is another advantage to using this process in the context of its hitting time distribution to an exponential threshold. Specifically, the hitting time distribution of an increasing Lévy process with Lévy measure ν is the Laplace transform of the process, and the Laplace transform of the process is given by the well-known **Lévy–Khinchin formula** given below. Specifically, I start by noting that

$$P(T > t) = P(X > R(t)) = \int_0^{\infty} e^{-u} B_t(du),$$

where B_t is the distribution function of $R(t)$. Thus $P(T > t) = E[\exp(-R(t))]$, which is the Laplace transform of $R(t)$. When $\{R(t); t \geq 0\}$ is an increasing Lévy process with Lévy measure ν, the Laplace transform is of the form

$$E[\exp(-R(t))] = \exp\left[-\alpha(t) - \int_0^{\infty} (1 - e^{-u})\nu(du)\right], \tag{7.17}$$

which is the Lévy–Khinchin formula; see (for example Gjessing, Aalen and Hjort, 2003). As before, $\alpha(t)$, a continuous non-decreasing function of t with $\alpha(0) = 0$, is the non-random part of the process.

By choosing different forms for $\nu(du)$, different processes for $\{R(t); t \geq 0\}$ can be prescribed. For example, if $\nu(du)$ is of the form given by (7.12) or (7.15), the resulting process for $\{R(t); t \geq 0\}$ will be a gamma or a stable process, respectively, whereas if $\nu(du) = mG(du)$ then the resulting process will be a homogenous compound Poisson process. Once $\nu(du)$ is specified, (7.17) can be evaluated to obtain $P(T > t)$, the survival function.

The Case of a Continuous Increasing Strong Markov Process
A striking feature of such processes is that they are essentially deterministic (cf. Blumenthal, Getoor and McKean, 1962). For the special case of a continuous increasing Lévy process, $R(t)$ can be written as

$$R(t) = \alpha t + \int_0^\infty (1 - e^{-xt})\nu(dx)$$

where $\alpha \geq 0$ and $\nu(dx)$ is the Lévy measure of the process. When $\nu(dx)$ is of the form

$$\nu(dx) = \frac{a}{x}\exp(-bx)\,dx,$$

the continuous increasing Lévy process for $R(t)$ will be a gamma process with shape (scale) at (b), and

$$R(t) = \alpha t + a\log\left(\frac{b+t}{b}\right), \quad t \geq 0. \tag{7.18}$$

Note the parallel between (7.18) and (7.13), the latter being a consequence of prescribing a gamma process for the hazard rate process.

With $\alpha = 0$, the $R(T)$ given above hits a threshold X, at $T = b[\exp(X/a) - 1]$. Since X is the hazard potential, it has an exponential (1) distribution. Consequently

$$P(T \geq t) = \left(1 + \frac{t}{b}\right)^{-a},$$

which is a Pareto distribution.

7.3.3 Cumulative Hazard as Geometric Brownian Motion

When $\{R(t);\, t \geq 0\}$ is described by a Brownian maximum process, as was done in the introduction to section 7.3, it is implied that the process for $R(t)$ increases in t by jumps. As another functional of the Brownian motion process, and one which ensures that the process for $R(t)$ is strictly increasing in t, I consider the relationship

$$R(t) = \int_0^t \exp(2W(u))du, \tag{7.19}$$

where $\{W(u);\, u \geq 0\}$ is a standard Wiener process; the scalar 2 is chosen for technical convenience. The quantity $\exp(W(u))$ defines what is known as a **geometric Brownian motion**, so the process $\{R(t);\, t \geq 0\}$ defined by (7.19) is an integrated geometric Brownian motion process. Such a process has been investigated by Yor (1992), who has obtained results that are useful for obtaining its hitting time distribution. The details are cumbersome and therefore not spelled out here. It can be seen (cf. Singpurwalla, 2006a,b) that the survival function of T is of the form

$$P(T > t) = \int_0^\infty \int_0^\infty P(H(t) \in d\nu)e^{-x}dx,$$

where $P(H(t) \in d\nu)/d\nu$

$$= \sqrt{\frac{2\pi}{\upsilon}}\int_0^\infty \exp\left(-\frac{y^2}{2t} + \frac{\upsilon}{2}\cosh^2 y\right)\sinh y \sin\left(\frac{\pi y}{t}\right)(1 - \Phi(\sqrt{\upsilon}\cosh y))dy,$$

with $\Phi(s) = \int_{-\infty}^{s} \frac{1}{\sqrt{2\pi}}e^{-\frac{u^2}{2}}du$, $\cosh(x) = \frac{e^x + e^{-x}}{2}$, $\sinh(x) = \frac{e^x - e^{-x}}{2}$.

7.3.4 The Cumulative Hazard as a Markov Additive Process

In section 7.2.3 I described a strategy for incorporating the effect of a covariate on the survival function by making the failure rate a function of the covariate. The particular scenario that we discussed involved a Wiener process for the covariate $Z(t)$, and the failure rate was $q(Z(t))^2$, for some constant $q > 0$. With $Z(0) = 0$ and $\sigma^2 = 1$, the above set-up resulted in the survival function being given by the Cameron–Martin formula. There is another strategy for incorporating the effect of a stochastically varying covariate on the survival function, and this is through the cumulative hazard function. This strategy is implemented via the structure of a bivariate stochastic process, called the Markov additive process, so named because one member of the bivariate process is a Markov process (see below), and the other an increasing additive process. Such processes were introduced by Çinlar (1972); I start with an overview of their salient features.

Markov and Markov Additive Processes

For us to discuss Markov additive processes (MAP), it is necessary that I first introduce Markov processes and the Markov property. To relate these to the material at hand, we shall adhere to the notation we have been using. Specifically, the covariate process $\{Z(t); t \geq 0\}$ will be used to denote a Markov process, and the cumulative hazard process $\{R(t); t \geq 0\}$ to denote an increasing additive process. These two processes constitute a MAP.

A stochastic process $\{Z(t); t \geq 0\}$, with $Z(t)$ taking values in a set $\mathcal{E}$, where $\mathcal{E} \subset (-\infty, +\infty)$, is said to be a **Markov process** with **state space** $\mathcal{E}$ if it possesses the **Markov property**. That is, for any $\mathcal{A} \subseteq \mathcal{E}$, and any s, t

$$P(Z(t+s) \in \mathcal{A} | Z(u), 0 \leq u \leq s) = P(Z(t+s) \in \mathcal{A} \mid Z(s)).$$

This means that the future course of the process depends only on its present value, the entire history of the process not being relevant. If the above relationship holds for all values of t, then the Markov process is said to be time homogenous. When such is the case, we may suppress t, and write that for any $s > 0$, and $i, j \in \mathcal{E}$

$$P(Z(t+s) = j \mid Z(s) = i) = p_s(i, j). \tag{7.20}$$

The conditional probability $p_s(i, j)$ is known as the **transition function** of the Markov process. Note that $p_s(i, j) \geq 0$, $\sum_{j \in \mathcal{E}} p_s(i, j) = 1$, and for any $v, s \geq 0$, $p_{v+s}(i, j) = \sum_{k \in \mathcal{E}} p_v(i, k) p_s(k, j)$.

Suppose that $0 \equiv T_0 < T_1 < T_2 < \ldots$, are the times at which a Markov process makes a transition from one state, say i, to the next, say j. The spacings $T_{n+1} - T_n$, $n = 0, 1, 2, \ldots$, are known as the **sojourn times** of the process. An important feature of Markov processes is that their sojourn times have an exponential distribution with a scale parameter that depends only on the current state of the process, and not the state to which the process is transitioning to. Specifically,

$$P(T_{n+1} - T_n \geq u \mid Z(T_n) = i, Z(T_{n+1}) = j) = e^{-\lambda(i)u}, \tag{7.21}$$

for some $u \geq 0$ and $\lambda(i) > 0$; $\lambda(i)$ does not depend on j. We will be making use of this feature in an illustration that will be given later.

Turning attention to the MAP, we say that a bivariate stochastic process $\{Z(t), R(t); t \geq 0\}$ is a **Markov additive process** if:

(i) $Z(t)$ and $R(t)$ are right continuous with left-hand limits;
(ii) $R(t) \geq 0$ and increases with t;
(iii) $Z(t)$ takes values in a set, say $\mathcal{E}$, where $\mathcal{E}$ is either countable or is a subset of the real line;

(iv) For each $e \in \mathcal{E}$, there is a well-defined probability (measure), say P^e, with the feature that

$$P^e(Z(0) = e, R(0) = 0) = 1, \quad \text{and}$$

(v) For all $e \in \mathcal{E}$, s, t, $\mathcal{A} \subset \mathcal{E}$, and $B \subset (-\infty, +\infty)$,

$$P^e(Z(t+s) \in \mathcal{A}, (R(s+t) - R(s)) \in \mathcal{B} | \mathcal{H}(s)) = P^{Z(s)}(Z(t) \in \mathcal{A}, R(t) \in \mathcal{B}),$$

where $\mathcal{H}(s)$ is the entire history of the process until time s.

Note that (v) above prescribes the conditions under which the two probability measures P^e and $P^{Z(s)}$ give identical answers. The essential import of a MAP is encapsulated in Theorem 2.22 of Çinlar (1972), which says that if $\{Z(t), R(t); t \geq 0\}$ is a MAP, then:

(a) $\{Z(t); t \geq 0\}$ is a Markov process with a state space $\mathcal{E}$ and a transition function $P^e(Z(t) \in \mathcal{A}, R(t) \in [0, \infty))$, and
(b) Given $Z(t)$, the (conditional) law of $R(t)$ is that of an increasing additive process (i.e. a process that can be represented as the sum of independent non-negative random variables).

Using (a) and (b) above, Çinlar (1972) was able to argue that during a small time interval $(t, t + \mathrm{d}t)$, the probability law of $R(t)$ is that of an increasing Lévy process whose parameters depend on the value $Z(t)$. Thus when $\{Z(t); t \geq 0\}$ represents a covariate process and $\{R(t); t \geq 0\}$ the cumulative hazard process, a MAP for $\{Z(t), R(t); t \geq 0\}$ is able to describe the dependence of a covariate on the cumulative hazard. When $\mathcal{E}$ consists of a single state, then the MAP reduces to a univariate process, namely a Lévy process. This completes our overview of the Markov process and the MAP. We now consider some specific examples of the MAP to illustrate the kind of results that can be obtained for the problem of relating the cumulative hazard to a covariate.

Random Usage as a Covariate

The simplest scenario is one wherein a unit is intermittently used and which experiences an increase in its cumulative hazard whenever it is being used. Thus, usage is the covariate which influences the item's lifelength. Alternatively put, the item operates in an environment of usage that is dynamic. Such a covariate can be described by a two-state Markov process, with the states ω denoting working, and r denoting rest; i.e. $\mathcal{E} = \{\omega, r\}$. Thus we assume that the usage process $\{Z(t); t \geq 0\}$ is a Markov process with a state space $\mathcal{E} = \{\omega, r\}$. Recall (Equation (7.21)) that for such a process the (random) amount of time the unit is being used has an exponential distribution with scale parameter $\lambda(\omega) > 0$, and the amount of time for which it is out of use also has an exponential distribution with scale parameter $\lambda(r) > 0$. The assumption of a two-state Markov process for usage may or may not be meaningful. It is sometimes the case that rest times have distributions other than the exponential, such as the lognormal when non-usage (rest) is for the purpose of repair. Similarly, the usage time distributions could be other than an exponential, such as say a Weibull signaling the feature that long usage times accentuate the termination of use. When such is the case, the MAP model to be described below will not be appropriate.

To continue with the modeling process, suppose that when the unit is in state ω, its cumulative hazard $R(\bullet)$ increases by jumps of random size at random points in time (within any period of usage). Assume that the jump sizes are independent of each other and also of the jump times, and have a distribution function, say G. Suppose also that the jump times are determined by a homogenous Poisson process with rate m. In other words, within any interval of usage, say $\mathcal{J}$, the cumulative hazard process $\{R(t); t \in \mathcal{J}\}$ is described by a homogenous compound

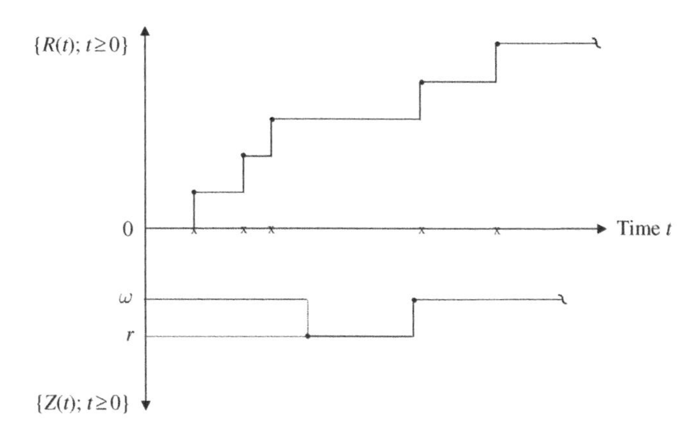

Figure 7.1 The sample path of a MAP for $\{R(t), Z(t); t \geq 0\}$.

Poisson process with rate m and jump distribution G. When the item is in state r, the cumulative hazard stays put at its current value; i.e. it does not increase. In other words, the cumulative hazard is in a state of suspended animation whenever the unit is in state r. Since a homogenous compound Poisson process is an increasing Lévy process with no drift and a finite Lévy measure $\nu(\mathrm{d}x) = mG(\mathrm{d}x)$ (section 7.2.2) $\{R(t); t \geq 0\}$ is an increasing additive process whose parameter m depends on the state of the usage process $\{Z(t); t \geq 0\}$, which is a Markov process. Therefore, it makes sense to describe the scenario of conceptualizing random usage as a covariate, via the structure of a MAP for the bivariate process $\{Z(t), R(t); t \geq 0\}$. The sample path of the process constructed above is shown in Figure 7.1. The crosses on the time axis t denote the jump times of the $\{R(t); t \geq 0\}$ process shown on the upper part of Figure 7.1. The lower part shows the sample path of the two-state usage process $\{Z(t); t \geq 0\}$. Note that $R(t)$ does not experience any jumps when $Z(t)$ is in state r.

A unit operating under certain circumstances wherein the bivariate process $\{Z(t), R(t); t \geq 0\}$ is a MAP of the kind described above will fail when $\{R(t); t \geq 0\}$ hits a random threshold X, where X has an exponential distribution with scale parameter one. I have seen before, in section 7.3.2, that $P(T > t)$, the survival function of this unit, is the Laplace transform of $R(t)$. Specifically, $P(T > t) = E[\exp(-R(t))]$. To obtain this quantity, I appeal to a result of Çinlar (1977), who shows that given $\{Z(u); 0 \leq u \leq t\}$ the history of the usage process up to and until time t, the quantity

$$E[\exp(-R(t))|Z(u); 0 \leq u \leq t] = \exp\left[-A(t)\int_0^\infty (1 - \mathrm{e}^{-u})mG(\mathrm{d}u)\right], \tag{7.22}$$

where $A(t)$ is the total amount of time that the covariate process $\{Z(u); 0 \leq u \leq t\}$ spends in state ω. That is, $A(t)$ is the total time spent by the unit in a working state during the time interval $[0, t]$.

A comparison of (7.22) with (7.17) with $\nu(\mathrm{d}u)$ in the latter equation replaced by $mG(\mathrm{d}u)$ is instructive. The main difference is that when the $A(t)$ of (7.22) gets replaced by t, (7.17) comes about. This makes sense because were the unit to operate in only one state, namely the state ω, $A(t) = t$, and I have said before that when $\mathcal{E}$ consists of a single state, the MAP reduces to a univariate increasing additive process. To obtain $P(T > t) = E[-\exp(R(t))]$, we need to average out the right-hand side of (7.22) with respect to the distribution of $A(t)$. It is unlikely that a closed form expression for $P(T > t)$ can be obtained for any specified G; we will need to resort to a use of numerical techniques. An exception could be the case wherein the jump sizes are a constant, say d, so that $G(\mathrm{d}u) = 1$ when $u = d$ and zero otherwise. When such is the case

$$P(T > t) = E\left[\mathrm{e}^{-m(1-\mathrm{e}^{-d})A(t)}\right],$$

where now the expectation is with respect to the distribution of $A(t)$. Note that with $d = 1$, the MAP discussed here is also the hidden Markov model mentioned in section 5.2.4.

As another illustration of the scenario of random usage, consider the situation wherein the nature of usage is germane. For example, usage under different climatic conditions such as average, cold, freezing and hot or such as aggressive, benign and normal. Suppose that the environment of usage is such that $\{Z(t);\, t \geq 0\}$ is a multi-state Markov process. The sojourn times of $\{Z(t);\, t \geq 0\}$ in each state will continue to be exponentially distributed with scale parameters that are state dependent. Instead of supposing that the cumulative hazard process $\{R(t);\, t \geq 0\}$ increases by jumps, we now assume that it increases continuously, so that locally (i.e. during the sojourn time in a particular state) $\{R(t);\, t \geq 0\}$ is a gamma process whose shape and scale parameters are state dependent; they are $\alpha(Z(t))$ and $\beta(Z(t))$, respectively. Then, as before (equation (7.22)) it can be shown (cf. Çinlar, 1977) that

$$E[\exp(-R(t))|Z(u);\, 0 \leq u \leq t] = \exp\left[-\int_0^t \alpha(Z(u)) \log\left(1 + \frac{u}{\beta(Z(u))}\right) \mathrm{d}u\right]. \tag{7.23}$$

Here again, it is instructive to compare (7.23) above with (7.13) and note the parallel between the two. When the usage process $\{Z(t);\, t \geq 0\}$ consists of a single state, $\alpha(Z(u)) = a$ and $\beta(Z(u)) = b$, and (7.23) will reduce to (7.13).

7.4 COMPETING RISKS AND COMPETING RISK PROCESSES

Loosely speaking, by the term 'competing risks', it is meant competing causes of failure. The notion of competing risks has its origin in the biomedical context wherein interest centers around the cause of death of a biological unit, given that there are several agents that compete for its lifetime. The issue is quite complex since the causes do not operate in isolation of each other, it often being the case that one cause acerbates the effect of the other. Some celebrated references on this topic are Makeham (1873), who introduces the problem, and Cornfield (1957), who articulates the issues that the problem spawns.

A simple way to conceptualize a model for competing risks is through the scenario of a two-component series system, the failure of the first of the two components being labeled as the cause of failure of the system. Thus, for example, in the case of an automobile the cause of failure could be a mechanical defect or an electrical defect. In the interest of simplicity, here we do not distinguish between a prima facie cause and a genuine cause (section 4.8.1). Accordingly, let T_1 and T_2 be the lifetimes of a two-component system, and T the lifetime of the system. Let $R_1(t)$ and $R_2(t)$ be the cumulative hazard functions of T_1 and T_2, respectively; $t \geq 0$. The component labeled 1 will fail when $R_1(t)$ crosses its hazard potential X_1 (section 4.6), similarly with component 2, whose hazard potential is X_2. The system fails when $R_1(t)$ crosses X_1, or $R_2(t)$ crosses X_2, whichever occurs first. The cause of failure of the system is $R_1(t)$ if $R_1(t)$ crosses X_1 prior to $R_2(t)$ crossing X_2, and vice versa. Thus we may interpret the cumulative hazards as the risks to the system that compete with each other for the system's lifelength. Recall (section 4.6) that the cumulative hazard encapsulates an assessor's judgment about the manner in which an item's quality and the environment in which the item operates interact, vis-à-vis the item's lifelength. From an actuarial and an operational point of view, it is the system's time to failure T that is of greater interest than the cause of failure. From the point of view of system design, system maintenance and a postmortem of system failure, it is the cause of failure that is germane. We start with a consideration of deterministic $R_1(t)$ and $R_2(t)$, and in section 7.4.2 discuss the scenario of stochastic cumulative hazard functions, independent and dependent.

7.4.1 Deterministic Competing Risks

With the two-component series system as a model for competing risks, the survival function of T, given by $P(T \geq t)$, is of the form

$$\begin{aligned} P(T \geq t) &= P(R_1(t) \leq X_1, R_2(t) \leq X_2) \\ &= P(X_1 \geq R_1(t), X_2 \geq R_2(t)) \\ &= \exp(-(R_1(t) + R_2(t))), \end{aligned} \tag{7.24}$$

if X_1 and X_2, the hazard potentials, are independent. This means that the cumulative hazard experienced by the system of the two competing cumulative hazards, $R_1(t)$ and $R_2(t)$. In other words, under the assumption of independent hazard potentials of each component, the risk to the system is an additive function of the component risks.

There is another way to look at (7.24). Specifically, this equation can also be seen as the survival function of a single item that experiences a cumulative hazard of $R(t) \stackrel{\text{def}}{=} R_1(t) + R_2(t)$. But when such is the case, how may we interpret $R_1(t)$ and $R_2(t)$? More generally, in the case of a single item experiencing a cumulative hazard of $R(t)$, can there be a decomposition of $R(t)$, and if so could it be additive? To address these questions, one possible strategy would be to associate each $R_i(t)$, $i = 1, 2, \ldots,$ with a covariate, such as a factor of the environment, say the temperature, and to suppose that were the item to experience the factor i alone, then its cumulative hazard would be $R_i(t)$. The item fails when $R_i(t)$ crosses the item's hazard potential X. With the item simultaneously experiencing m factors that constitute its environment, its survival function will be

$$\begin{aligned} P(T \geq t) &= P(R_1(t) \leq X, R_2(t) \leq X, \ldots, R_m(t) \leq X) \\ &= P(X \geq \max\{R_1(t), \ldots, R_m(t)\}) \\ &= \exp(-\max\{R_1(t), \ldots, R_m(t)\}). \end{aligned} \tag{7.25}$$

Thus far the scenario described above, the decomposition of $R(t)$, is not additive; rather, here $R(t) = \max\{R_1(t), \ldots, R_m(t)\}$.

The above two perspectives on competing risk modeling, one entailing the scenario of a series system and the other the case of a single unit experiencing several failure-causing agents, lead to different answers. Which of these two perspectives best encapsulates a competing risk mechanism? Do the two formulae of (7.24) and (7.25) bear some relationship to each other? In what follows, I attempt to address the above questions.

Relating the Two Formulae of Competing Risk Models

The existing literature on competing risk models is based on a multi-component series system with independent or dependent component lifelengths $T_1, \ldots, T_m$, say. Under this conceptualization, the m lifetimes are assumed to compete with each other for the system's lifetime. When the lifetimes are assumed independent, a generalization of (7.24) with additive cumulative hazards (risks) results. By contrast, we assume independent (and identically distributed) hazard potentials to obtain (7.24) or its generalization. Since independent hazard potentials imply independent lifetimes (Theorem 4.2) there is a parallel between the manner in which we have approached the problem and the manner in which it has been traditionally done. However, this parallel loses its transparency when we consider the scenario of a single unit experiencing k competing hazards, because now we obtain the formula of (7.25), which for the competing risk scenario appears to be new. Looking at the competing risk problem from the point of view leading us to (7.25) seems to be more natural than the one involving a series system. But not too surprisingly, the two scenarios are related, in the sense that the single unit case is diametrically opposite to the independent hazard potential case. This is because a single hazard potential X can be seen as a representation of several identical hazard potentials (namely, the case of total dependence). Recall that any random variable is totally dependent on itself. As a consequence, for the series system model of competing risks, when the hazard potentials are positively dependent, the survival function $P(T \geq t)$ is bounded as:

$$\exp\left(-\sum_{i=1}^{m} R_i(t)\right) \leq P(T \geq t) \leq \exp(-\max\{R_1(t), \ldots, R_m(t)\}). \tag{7.26}$$

Thus, the two perspectives on competing risk modeling can be reconciled via the notion of independent and dependent hazard potentials, the left-hand side of (7.26) reflecting the former and the right-hand side the latter. It is interesting to note that (7.26) reaffirms a well-known result in reliability, namely that the reliability of a series system whose lifetimes are 'associated' (loosely speaking, positively dependent) is bounded below by the reliability of a series system, and above by the reliability of a parallel system (cf. Barlow and Proschan, 1975, p. 34).

7.4.2 Stochastic Competing Risks and Competing Risk Processes

The material of section 7.4.1 assumed that the cumulative hazard functions $R_i(t), t \geq 0, i = 1, 2, \ldots,$ were known and specified. Thus, the question of independent or dependent competing risks was not germane to the material of that section. The issue of independence and dependence was embodied in the context of hazard potentials. All the same, and for the record, it is useful to mention that in the existing literature on competing risks, the issue of dependent competing risks tantamounts to a consideration of dependent lifetimes in the series system model of section 7.4.1. This I think is a misnomer! A proper framework for describing dependent competing risks would be to assume that the $R_i(t)$s are random functions, and for this it is best to describe the cumulative hazard functions via stochastic process models of the form $\{R_i(t); t \geq 0\}, i = 1, 2, \ldots$ The purpose of this section is to develop a foundation for discussing dependent competing risks by prescribing a joint stochastic process model for $\{R_1(t), R_2(t), \ldots; t \geq 0\}$. The issue of independent competing risk processes will fall out as a special case. To keep the modeling simple, I restrict attention to the case of a bivariate stochastic process $\{R_1(t), R_2(t); t \geq 0\}$ and label it a **dependent competing risk process**. Dependence between the processes $\{R_1(t); t \geq 0\}$ and $\{R_2(t); t \geq 0\}$ will induce dependence between the corresponding lifetimes T_1 and T_2, and this will ensure that the current view of what constitutes dependent competing risks is encompassed within the broader framework of dependent competing risk processes. Thus to summarize, in what follows we consider the scenario of a single item experiencing two dependent competing risk processes encapsulated via the bivariate stochastic process $\{R_1(t), R_2(t); t \geq 0\}$, where $R_1(t)$ and $R_2(t)$, the cumulative hazards, are non-decreasing in t. What kind of models shall we prescribe for $R_1(t)$ and $R_2(t)$, and how shall we create interdependence between the two processes $\{R_1(t); t \geq 0\}$ and $\{R_2(t); t \geq 0\}$? A simple strategy is the one described below. It is based on the ideas proposed by Lemoine and Wenocur (1985), and by Wenocur (1989), though in a context different from ours. Note that because of the non-negative and non-decreasing nature of the $R_i(t)$s, a bivariate Wiener process for $\{R_1(t), R_2(t); t \geq 0\}$ is unsuitable, and neither is the two-state MAP of Figure 7.1.

A Bivariate Stochastic Process for the Cumulative Hazards

For purposes of motivation I start with a discrete time version of one of the two processes mentioned above, say $\{R_1(t); t \geq 0\}$. Suppose that this process can be monitered only at every h units of time, so that what is at stake here is the sequence $\{R_1(j)\}, j = 0, h, 2h, \ldots, nh, \ldots$ We assume that if at time nh, the state of the process is $R_1(n)$, then at time $(n+1)h$ its state is determined by a first-order stochastic difference equation:

$$R_1(n+1) - R_1(n) = \alpha(R_1(n))\epsilon(n) + \beta(R_1(n))h, \tag{7.27}$$

where $\{\epsilon(n)\}$ is a sequence of independent and identically distributed random variables. Since $R_1(t)$ is the cumulative hazard at time t, we need to assume that $R_1(0) = 0$, and to ensure that $R_1(t) > 0$ for $t > 0$, we require that $\alpha(R_1(0))$ or $\beta(R_1(0))$, or both $\alpha(R_1(0))$ and $\beta(R_1(0))$ are non-zero. Equation (7.27) states that over h units of time, and starting from some state say ω, $R_1(\bullet)$ will increase on average by an amount $\alpha(\omega)E(\omega) + \beta(\omega)h$, and that the variance of this average increase is $(\alpha(\omega))^2 V(\omega)$, where $E(\omega)$ and $V(\omega)$ are the mean and variance of $\epsilon(n)$. To guarantee that $R_1(\bullet)$ is strictly increasing with time, we require that $\alpha(\bullet), \beta(\bullet)$ and $\epsilon(n)$ are non-negative. In particular, we assume that the $\epsilon(n)$'s have a gamma distribution with scale one and shape h.

As an aside, we note that a diffusion process is a continuous approximation of (7.27), which in turn can also be seen as a discrete approximation to a diffusion process. When the $\epsilon(n)$'s are assumed to be normally distributed we are able to utilize the enormous literature of diffusion processes. But this we are unable to do here because when the $\epsilon(n)$'s are normal, $R_1(t)$ cannot be non-decreasing in t.

A unit experiencing the process $\{R_1(t); t \geq 0\}$ alone, will fail when this process reaches the unit's hazard potential X. If in addition to $\{R_1(t); t \geq 0\}$ the unit also experiences the process $\{R_2(t); t \geq 0\}$, then the unit will fail when either process first hits X. The two processes in question compete with each other for the item's lifelength. With the above in place we need to make the process $\{R_2(t); t \geq 0\}$ dependent on the process $\{R_1(t); t \geq 0\}$. This after all is what would genuinely constitute a dependent competing risk process. To do so, a simple strategy would be to assume that the sample path of $\{R_2(t); t \geq 0\}$ is an impulse function of the form $R_2(t) = 0$, for all $t \neq t^*$, and $R_2(t^*) = \infty$, for some $t = t^* > 0$, where the rate of occurrence of the impulse depends on the state of the process $\{R_1(t); t \geq 0\}$. The idea here is that it is a traumatic event that creates the process $\{R_2(t); t \geq 0\}$, and that the rate of occurrence of trauma depends on the state of the process $\{R_1(t); t \geq 0\}$. This assumption makes sense if $\{R_1(t); t \geq 0\}$ is interpreted as degradation or wear [see, for example Yang and Klutke (2000)]. Specifically, we assume that

$$P(\text{a traumatic event occurs in } [t, t+h) | R_1(t) = z) = 1 - \exp(-k(z)h),$$

where $k(\bullet)$ is some function of z.

If the hazard potential of the item takes the value x, that is if $X = x$, then given $R_1(0), R_1(h), \ldots, R_1(nh)$, the state of the $\{R_1(t); t \geq 0\}$ process at times $0, h, \ldots, nh$, the probability that the unit is surviving at nh is of the form

$$\exp\left(-h \sum_{j=0}^{n-1} k(R_1(jh))\right) I_{[0,x)}(R_1(nh)), \tag{7.28}$$

where $I_A(\bullet)$ is the indicator function of a set A.

Conditioning on $R_1(0)=0$ alone, and taking the limits in (7.28) so that $nh \to t$, the conditional probability that an item experiencing the dependent competing risk process $\{R_1(t), R_2(t); t \geq 0\}$, and having its hazard potential set at x, will survive to t is of the form

$$p(0, x, t) = E^0 \left[\exp\left(\int_0^t -k(R_1(s))\, ds \right) I_{[0,x)}(R_1(t)) \right]. \tag{7.29}$$

The right-hand side of (7.29) above can be more generally written as

$$E^\omega \left[\exp\left(\int_0^t -k(R_1(s))\, ds \right) f(R_1(t)) \right] \stackrel{\text{def}}{=} p(\omega, t),$$

for some $f(\bullet)$; it is known as the Kac Functional of the state process. This functional satisfies the Feynman–Kac Equation, namely,

$$\frac{\partial}{\partial t} p(\omega, t) = -k(\omega) p(\omega, t) + Up(\omega, t)$$

for some U. Under certain conditions on f (namely, that f be bounded and satisfy the Lip_1 condition) the Feynman–Kac Equation has a unique solution. Thus, for example, with $\alpha(u) = 1, \beta(u) = 0$, making $\{R(t); t \geq 0\}$ a gamma process, Wenocur (1989) shows that when $f(u) = 1$ and $k(u) = u$,

$$p(\omega, t) = \exp(-t(\omega - 1) - (1 + t)\log(1 + t)),$$

which for $\omega = 0$, takes the form

$$p(0, t) = \exp(-(1 + t)\log(1 + t) + t) = P(T \geq t). \tag{7.30}$$

Similarly, when $\alpha(u) = 1, \beta(u) = 1, f(u) = 1$, and $k(u) = u$,

$$P(T \geq t) = \exp(-(1 + t)\log(1 + t) + t - t^2/2). \tag{7.31}$$

When $f(\bullet)$ is the indicator function of (7.29) it is bounded but not continuous and therefore does not satisfy the Lip_1 condition. Thus closed-form results analogous to those of (7.30) and (7.31) cannot be had. All the same if the threshold x is set to a very large value, conceptually infinite, we may replace the indicator function by the function $f(u) = 1$, for all $u > 0$, and then use (7.30) and (7.31). In doing so, the effect of x is of course lost.

7.5 DEGRADATION AND AGING PROCESSES

Recently, much is being written on what is known as ‘degradation modeling’, and reliability assessment using ‘degradation data’; see, (for example, Nair, 1998). According to Lu, Meeker and Escobar (1996), reliability assessments based on degradation data tend to be superior to those that are solely based on lifelength data that is either censored and/or truncated. The thinking here is that the cause of all failures is some degradation mechanism at work, such as the progression of a chemical reaction, and that failure occurs when the level of degradation hits some threshold. This is the kind of argument that has been used to motivate the inverse Gaussian as a failure model (cf. Doksum, 1991). When degradation data is used to assess reliability, the understanding is that degradation is a phenomenon that can be observed and measured. But this line of thinking may need a reconsideration because the meaning of degradation seems to suggest that degradation may not be something that can be directly observed and hence measured. Specifically, our review of the engineering literature (cf. Bogdanoff and Kozin, 1985) suggests that degradation is regarded as the irreversible accumulation of damage throughout life, that ultimately leads to failure. Whereas the term 'damage' itself is not defined, it is claimed that damage manifests as corrosion, cracks, physical wear (i.e. the depletion of material), etc. With regards to aging, our review of the literature on longevity and mortality indicates that aging pertains to a unit's position in a state space wherein the probabilities of failure are greater than in a former position.

Thus it appears that both aging and degradation are abstract constructs that cannot be directly observed and thus cannot be measured. However, these constructs serve to describe a process which leads to failure. By contrast, what can be observed and measured are manifestations of damage, such as crack growth, corrosion and wear. The question therefore arises as to how one can mathematically model the degradation phenomenon and how can one relate it to the observables mentioned above. We need to model degradation because it could help us predict, prolong and manage lifetimes. In what follows, I propose a conceptual framework which links the unobservable degradation with observable agents such as crack growth and corrosion. This framework capitalizes on the material of this chapter, in particular, the structure discussed in section 7.3.4.

7.5.1 A Probabilistic Framework for Degradation Modeling

With degradation seen as an unobservable abstract construct that triggers failure upon hitting a threshold, and with the observables such as crack growth seen as manifestations of damage, our proposed framework treats the former as the cumulative hazard and the latter as a covariate or a marker that influences the former. Recall that covariates and markers are influential variables that are precursors to failure. In some scenarios, a failure occurs when an influential variable crosses a threshold. For example, the blood pressure or the pulse rate of a human. When such is the case, the relationship between a marker and the failure time is deterministic. In some other situations, failure is not deterministically related to an observable marker. For example, stock price is a precursor to business failure but there is not a one-to-one relationship between the two. Some other examples are the tracking of AIDS death by the CD4 cell count and failure due to fatigue by crack growth. Here, the stock price, CD4 dell counts and crack growth are markers that provide a diagnostic to impending failure. Markers are often functions of time t, and as such are best described by stochastic process models, namely, by marker processes. Thus the phenomenon of degradation (and aging), which we in our framework have identified with the cumulative hazard, will also be described by a stochastic process $\{R(t); t \geq 0\}$, where $R(t)$ is non-decreasing in t. The item fails when $R(t)$ hits its hazard potential X, where X has an exponential distribution with scale one. This fits in well with the view that the failure of an item occurs when the degradation level hits a threshold.

Thus to summarize, our framework for modeling degradation and an observable that is a manifestation of damage, is to regard the two as the elements of a bivariate stochastic process $\{Z(t), R(t); t \geq 0\}$ where $Z(t)$ represents an observable marker, so that $\{Z(t); t \geq 0\}$ is a marker process, and $R(t)$ represents the unobserved degradation. The behavior of the $\{R(t); t \geq 0\}$ process parallels that of the cumulative hazard process discussed in section 7.3. With $Z(t)$ and $R(t)$ interpreted as above, I shall call the bivariate process $\{Z(t), R(t); t \geq 0\}$ a **degradation process**.

7.5.2 Specifying Degradation Processes

What are some possible choices for the probabilistic structure of the degradation process $\{Z(t), R(t); t \geq 0\}$? A natural one, given the parallel between the basic structure of the degradation process and the material of section 7.3.4, is the Markov additive process. However, this would require that the process $\{Z(t); t \geq 0\}$ which represents the observeables such as crack growth, corrosion and wear be a Markov process. This seems to be a reasonable assumption for which there is some precedence (cf. Sobczyk, 1987). Should $\{Z(t), R(t); t \geq 0\}$ be taken to be a Markov additive process, and should one wish to use observations on $\{Z(t); t \geq 0\}$ for inferences about the process $\{R(t); t \geq 0\}$, an exercise that parallels the use of 'degradation data' for reliability assessment, then a statistical theory for inference on degradation processes needs to be developed. This important issue remains to be addressed and is an open problem.

Another possibility, and one that has been proposed by Whitmore, Crowder and Lawless (1998) is to use a bivariate Wiener process, actually a Brownian motion process, for $\{Z(t), R(t); t \geq 0\}$. The correlation between the two processes in question determines the usefulness of the marker for tracking progress of the failure causing process $\{R(t); t \geq 0\}$. When a Wiener process is assumed for $\{R(t); t \geq 0\}$, $R(t)$ will not be non-decreasing in t. Consequently, looking at degradation as a cumulative hazard will not make sense and the advantage of interpreting the threshold as an exponentially distributed (with scale one) hazard potential will not be there. The ability to do statistical inference using data on the marker process as well as data on failure times is a key advantage of using a bivariate Wiener process.

The third possibility for linking the degradation and marker processes is through Cox's (1972) proportional hazards model. But here we do not specify a direct linkage between $R(t)$ and $Z(t)$, rather, a linkage made between the processes $\{r(t); t \geq 0\}$ and $\{Z(t); t \geq 0\}$, where $R(t) = \int_0^t r(s)\,ds$. With the degradation $R(t)$ viewed as the cumulative hazard, $\{r(t); t \geq 0\}$ is the hazard rate process of section 7.2. Two strategies for linking $r(t)$ and $Z(t)$ have been proposed in the literature: the proportional hazards model and the additive hazards model (section 4.9.1).

Under the former

$$r(t) = r_0(t) \exp(\beta Z(t)), \tag{7.32}$$

where β is an unknown constant and $r_0(t)$ is a component of $r(t)$ that is not related to the marker $Z(t)$. Note that in (7.32), $r(t)$ depends on the value of the marker at time t only, and not the previous values of the marker. With $r(t)$ interpreted as the hazard rate, $r_0(t)$ is known as the baseline hazard (section 4.9.1). Since $r(t)$ is (informally) the risk that a failure will occur at t, it is reasonable to attribute the level or risk in terms of the level of a marker, and this is the motivation behind (7.32).

Under the additive hazards model, proposed by Aalen (1989),

$$r(t) = r_0(t) + \beta Z(t), \tag{7.33}$$

where $r_0(t)$ and β have the interpretation given above. The additive hazards model has been considered by us before (equation (4.59)) albeit in a different context.

Conceptually, the models of (7.32) and (7.33) are a break from the tradition of regression models in the sense that it is the hazard function, rather than the conditional expectation, that is the basis of regression. The models are quite flexible and can be generalized in several ways. For example, $Z(t)$ can be replaced by its lagged value, say $Z(t-a)$ for some $a \geq 0$, or by its integrated value $Z(t) = \int_0^t Z(u)\,du$. Another possibility would be to extend our consideration to the case of several markers, say k, so that

$$\begin{aligned} r(t) &= r_0(t) \exp(\beta \mathbf{Z}(t)), \quad \text{or} \\ &= r_0(t) + \beta \mathbf{Z}(t), \end{aligned} \tag{7.34}$$

where $\beta = (\beta_1, \ldots, \beta_k)$ and $\mathbf{Z}(t) = (Z_1(t), \ldots, Z_k(t))'$.

Processes for $\{Z(t); t \geq 0\}$ that have been considered before have been a jump process (cf. Jewell and Kalbfleisch, 1996) and a diffusion process (cf. Woodbury and Manton, 1977) (section 7.2.3). Recall that in the latter case, the link function used is a quadratic function of the marker.

A more concrete and manageable possibility is to describe the observable marker $\{Z(t); t \geq 0\}$ by a Wiener process with drift and diffusion (section 7.2.3) and the degradation process $\{R(t); t \geq 0\}$ by a Brownian Maximum Process (section 7.3), derived from the Wiener process for $\{Z(t); t \geq 0\}$. Since the two processes are related to each other, they provide the link between cumulative hazard and degradation. The details of this kind of modeling and the issues of inference that it spawns are in Singpurwalla (2006b). The approach leads to a mixture of inverse Gaussian distributions as the lifetime of a unit experiencing failure due to degradation.

There could be other approaches for modeling the degradation phenomenon discussed here. One possibility is to consider a Kalman filter type model of section 5.8.2 with $R(t)$ as the unknown state of nature that is non-decreasing in t and an observation equation governed by the observable $Z(t)$. This line of thinking remains to be explored.

Chapter 8

Point Processes for Event Histories

Here lies David Hilbert: He no longer has problems.

8.1 INTRODUCTION: WHAT IS EVENT HISTORY?

The material of Chapter 7 was based on the premise that to capture the effects of dynamic environments, one may model the hazard function and the cumulative hazard function by a suitable stochastic process. However, stochastic process models can be made to play a larger role in reliability theory when one looks at event history data, particularly the life history of maintained systems, and systems that experience multiple modes of failure, and also when tracking the condition of degrading units, known to engineers as **condition monitoring**. Point process models, which are indeed stochastic processes (section 4.7.6), offer a convenient platform for doing so. Furthermore, point process models can bring into play the modern technology of martingales and their associated limit theorems, via the framework of dynamic stochastic process models (section 8.4). Such theorems facilitate an analysis of point process data, albeit under asymptotic conditions. Point process data are generated when one continuously tracks the life history of units and systems.

The aim of this chapter is to give the reader a flavor of the above. But before doing so, it is incumbent on me to say a few words about what we mean by the terms 'life-history data' and 'event histories'. This I do next.

By life history, we mean here the sample path of a stochastic process as it moves from one state to another (section 7.3.4). The transition from one state to another creates an event, and by **event history**, we mean the collection of points over time, each point representing a time of transition. Thus, point process models, like the Poisson counting process of section 4.7.4, come naturally into play. The simplest case is when there are only two states to consider, say 'functioning' or 'failed', with the latter being an absorbing state (i.e. once the process enters this state, it is unable to exit it). Figure 8.1 illustrates a two-state system with Figure 8.1(a) depicting the two states, shown as boxes labeled State 1 and State 0, and Figure 8.1(b) showing the sample path (or life history) and the event history of the system. The model of Figure 8.1 encapsulates

Reliability and Risk: A Bayesian Perspective N.D. Singpurwalla
© 2006 John Wiley & Sons, Ltd

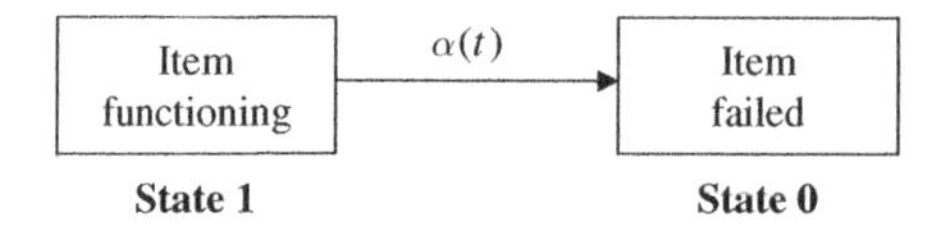

(a) State-space of the two-state process

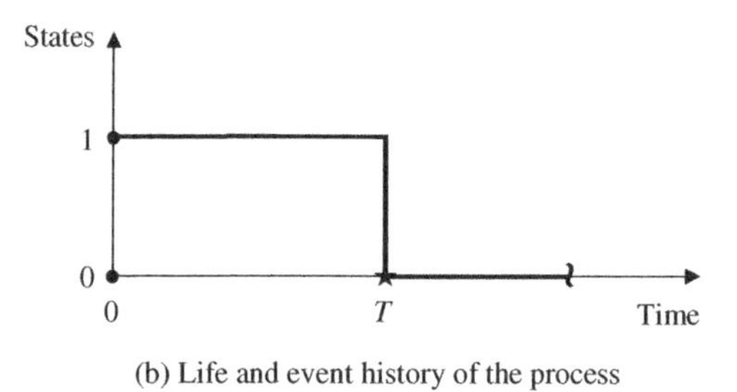

(b) Life and event history of the process

Figure 8.1 State space and sample path of a two-state stochastic process.

much of what we have been discussing thus far under the labels of 'survival analysis', 'life-table analysis' and 'reliability analysis'.

In Figure 8.1 the stochastic process starts in state 1 at time 0, and makes a permanent transition to state 0 at time T. Since T is unknown at time 0, T is a random variable. The arrow connecting the boxes in Figure 8.1(a) shows the direction of the transition, and the label $\alpha(t)$ on the arrow is known as the intensity of the process. It is the rate at which a transition from state 1 to state 0 occurs, and it is akin to the intensity function of a Poisson process (section 4.7.4). In this two-state scenario, the transition creates a single event at time T, this event being denoted by a star on the time axis in Figure 8.1(b). With point process models, much interest centers around the quantity $\alpha(t)$. I therefore need to first articulate the meaning of $\alpha(t)$.

Our interpretation of $\alpha(t)$, the intensity function of a point process, can be motivated by the material of section 4.7.4 on the intensity function $\lambda(t)$ of the non-homogenous Poisson process. We recall from there that $\lambda(t) = \mathrm{d}[\Lambda(t)]/\mathrm{d}t$, where $\Lambda(t) = E[N(t)]$, and $N(t)$ is the number of events up until time $t \geq 0$. Thus, $\lambda(t)$ can also be written as

$$\lambda(t) = \lim_{\mathrm{d}t \downarrow 0} \frac{E\left[N(t+\mathrm{d}t) - N(t)\right]}{\mathrm{d}t}. \tag{8.1}$$

The independent increments property of the Poisson process mandates that $\lambda(t)$ does not depend on the history of the process up until t. Suppose now that I generalize the above by conditioning on a quantity $\mathcal{F}_t$, where $\mathcal{F}_t$ denotes the collection of all that can be observed in $[0, t]$, and then define

$$\alpha(t) = \lim_{\mathrm{d}t \downarrow 0} \frac{1}{\mathrm{d}t} E\left[N(t+\mathrm{d}t) - N(t) | \mathcal{F}_t\right], \tag{8.2}$$

as the intensity function of a point process. In doing so we must assume that the limit exists. It is this generalization that gives birth to the notion of a dynamic model, that is, a model in which the future development is explained in terms of the past.

In the case of the two-state process of Figure 8.1, $\alpha(t)$ takes a familiar form, namely,

$$\alpha(t) = \lim_{\mathrm{d}t \downarrow 0} \frac{P(t \leq T \leq t + \mathrm{d}t | T \geq t)}{\mathrm{d}t};$$

thus $\alpha(t)\mathrm{d}t$ is (approximately) the probability that the process makes a transition from state 1 to state 0 in the interval $(t, t+\mathrm{d}t)$, given that it was in state 1 at time t. Alternatively seen, $\alpha(t)$ is also the failure rate at t of the survival function of T. This type of a connection between the failure rate function and the intensity function (of a non-homogenous Poisson process) has been seen by us before vis-à-vis (7.3) and (7.4) of section 7.2. It is important to bear in mind that the $\alpha(t)$ of (8.2) has been shown to be identical to the failure rate function of $P(T \geq t)$ when the stochastic process is a two-state process of the kind described in Figure 8.1. This in general need not always be the case.

8.1.1 Parameterizing the Intensity Function

In certain scenarios, especially the medical ones, there may be explanatory variables (or covariates) upon which T depends. Such variables could be qualitative, such as sex, geographical region, name of manufacturer, or quantitative, such as blood pressure, time in storage, etc. When such is the case, the explanatory variables become a part of $\mathcal{F}_t$, the observable history, and so their effect needs to be incorporated in $\alpha(t)$. One strategy for doing so is to parameterize $\alpha(t)$ via the multiplicative proportional hazards model of Cox (1972) – equation (7.32). When such is the case, one writes

$$\alpha(t) = \alpha_0(t)\exp(\beta'\mathbf{Z}(t)), \tag{8.3}$$

where $\beta = (\beta_1, \ldots, \beta_p)'$ is a vector of parameters and $\mathbf{Z}(t) = (Z_1(t), \ldots, Z_p(t))'$ is a vector of time-dependent covariates; $\alpha_0(t)$, known as the **baseline intensity**, is the intensity of transition when $\mathbf{Z}(t) = \mathbf{0}$, for $t \geq 0$.

There is another, more encompassing, approach for parameterizing $\alpha(t)$. This approach, due to Aalen (1987), is known as the **multiplicative intensity model**. The framework of this model includes the parameterization specified in (8.3), as well as the intensity functions appropriate to the several scenarios that will be described in section 8.3. Besides its generality, a key advantage of the multiplicative intensity model is that it enables us to invoke the martingale theory of point processes. Because of the importance of this model, its role will be articulated in separate sections – sections 8.3 and 8.4 – once the practical scenarios of section 8.2 are put forth. Peña and Hollander (2004) give an exhaustive and up to date overview of the various parameterizations for $\alpha(t)$ with a particular focus towards applications in biomedicine.

8.2 OTHER POINT PROCESSES IN RELIABILITY AND LIFE-TESTING

Whereas the two-state point process discussed before serves as a convenient vehicle for illustrating the basic ideas and notions about point processes, a better appreciation of the usefulness of such processes can be had by looking at some other, more general problems in reliability and survival analysis. I have in mind four archetypical scenarios: the first pertains to items experiencing multiple modes of failure; second, to items experiencing degradation prior to failure; and third involving items experiencing maintenance and repair. My last scenario pertains to life-testing wherein censoring and withdrawals come into play. Each of these is described below.

8.2.1 Multiple Failure Modes and Competing Risks

In general, it is difficult to pinpoint a single incident, or a collection of incidents, as the true (i.e. a genuine) cause of an item's failure. However, it may be possible to describe the manner in which an item can fail, assuming that it can fail in several ways. For example, a series system of k units can fail in at least k different ways, each corresponding to the failure of a particular unit or the simultaneous failure of more than one unit. Engineers refer to the different ways of failure as **failure modes**, and often these are taken as the (prima facie) causes of failure. When an investigation of failure (such as a postmortem) entails specifying the mode of failure, the exercise is known as **failure mode and effects analysis**, or simply FMEA. The scenario of multi-mode failure is not unlike that of competing risks discussed in section 7.4.1. There, the matter was conceptualized in terms of the hitting times of several cumulative hazard functions (deterministic or stochastic) to a threshold. The strategy described here is different; it is based on intensities of transition from working state to the failed state, the failed state being indexed by a mode of failure (Figure 8.2).

In Figure 8.2(a), I show the possible transitions of a unit that can experience k modes of failure, from the functioning state labeled 1, to one of the k failed states labeled $0^{(1)}, \ldots, 0^{(k)}$. The $\alpha_i(t)$'s denote the intensities of transition from state 1 to state $0^{(i)}$, $i = 1, \ldots, k$. Figure 8.2(b) shows the life and event history of the unit were a transition to state $0^{(i)}$ at time T_i be the only mode of failure.

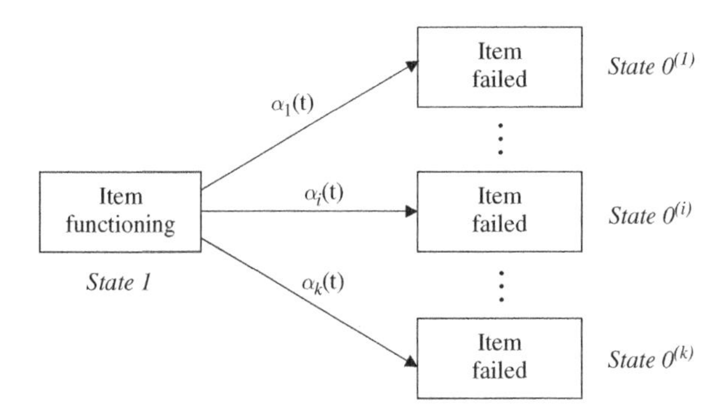

(a) State space of a unit with k failure modes

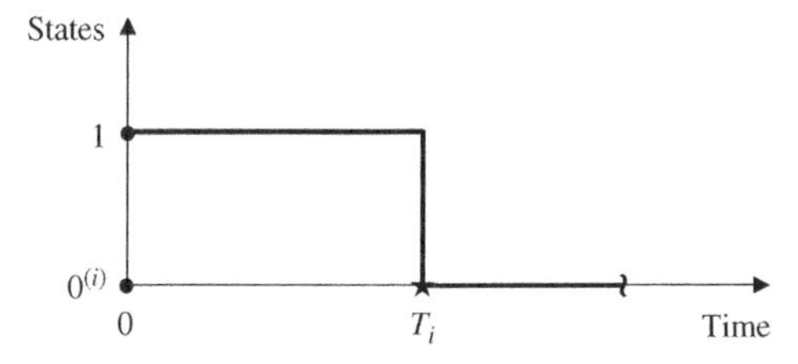

(b) Life and event history of unit transitioning to state 0

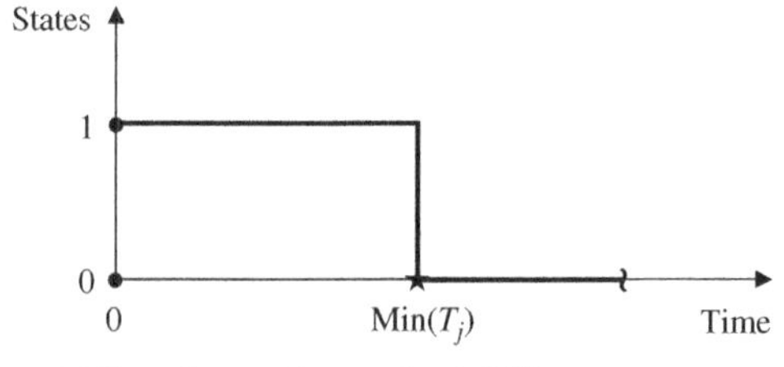

(c) Life and event history of unit failing

Figure 8.2 State space and sample paths of a unit with k failure modes.

Note that for the scenario of Figure 8.2, the underlying stochastic process for the life history of the unit has a state space of size $(k+1)$, and even though k different events are possible – with each illustrated via a sample path of the kind shown in Figure 8.2(b), – only one will be observed, namely, the one corresponding to the time of the first transition. This is shown in Figure 8.2(c). The challenge here is to prescribe a meaningful model for each of the $\alpha_i(t)$'s. One strategy would be to assume that an $\alpha_i(t)$ depends on the history of T_i alone (i.e. supposing that $0^{(i)}$ is the only mode of failure), so that $\alpha_i(t)$ is the failure rate function of $P(T_i \geq t)$, for $i = 1, \ldots, k$. Were we to conglomerate the k-failed states into one (failed) state, say state 0, and denote by $\alpha(t)$ the intensity of transition from state 1 to state 0, then it is only logical to suppose that $\alpha(t)$ would depend on the history of all the k T_i's, so that $\alpha(t)$ must be a function of $\alpha_1(t), \ldots, \alpha_k(t)$. With that in mind, one possibility is to set $\alpha(t) = \sum_{i=1}^{k} \alpha_i(t)$, making $\alpha(t)$ the sum of the failure rates of the survival functions of the k T_is. But $\alpha(t)$ is itself the failure rate of the survival function of $\min_j T_j$ (Figure 8.2(c)). Thus, the assumption $\alpha(t) = \sum_{i=1}^{k} \alpha_i(t)$ yields a result parallel to that of (7.24) on competing risks with independent hazard potentials. Alternatively put, the assumption of additive intensities can be justified from the point of view of independent hazard potentials. Similarly, the choice $\alpha(t) = \max_i \alpha_i(t)$ can be justified via (7.25) of section 7.4.1 wherein we considered a common hazard potential.

8.2.2 Items Experiencing Degradation and Deterioration

We next turn attention to the scenario of items experiencing degradation prior to failure. Suppose that a unit can exist in $(k+1)$ states, states 1 to k being functioning states, with state 1 being a perfect (non-degraded) state, and state 0 being a failed state. In Figure 8.3(a) I show the state space and one possible transition protocol for this unit. Here, the unit can transition from its current state to either its neighboring degraded state or the failed state. Other protocols are possible. In Figures 8.3(b) and (c) I show two possible sample paths for the said process. In the former the unit traverses all the k degraded states prior to failure; in the latter the unit fails after existing in only three of the k possible degraded states. The sample path of Figure 8.3(b) shows that the process can generate k events, each denoted by a star on the time axis. Figure 8.3(c) illustrates the scenario wherein less than k events are generated by the process.

The arrows between the states have labels involving α_{ij}, $i = 1, \ldots, k$; $j = 0, 2, \ldots, k$. These encapsulate the intensity of transition from state i to state j. When the probability of transition from state i to state j depends only on the time, then α_{ij} will be indexed by t and written $\alpha_{ij}(t)$, where

$$\alpha_{ij}(t) = \lim_{dt \downarrow} \frac{1}{dt} P_{ij}(t, t + dt). \tag{8.4}$$

Here $P_{ij}(s, t)$ is the probability that a unit in state i at time s will transition to state j at time $t \geq s$. With $\alpha_{ij}(t)$ as written above, the underlying process is Markovian. When $\alpha_{ij}(t) = \alpha_{i,j}$ for all $i, j = 1, \ldots, k$, and $t \geq 0$, the underlying process is time homogenous and Markovian.

More generally we may want to assume a semi-Markovian structure wherein the transition intensities depend not only on the states in question but also on the time t and the time spent in the current state. Thus, a transition form state i to the state j at time t will depend on i, j, t and T_i, the time spent in state i. This makes sense because a long time spent in the working state could increase the probability of transitioning to the failed state. When such is the case the α_{ij}'s will be indexed by both t and T_i, and written $\alpha_{ij}(t, T_i)$.

8.2.3 Units Experiencing Maintenance and Repair

I now consider the scenario of units that are subjected to scheduled maintenance and unscheduled repairs due to random failures. Often scheduled maintenance occurs at preset times; however,

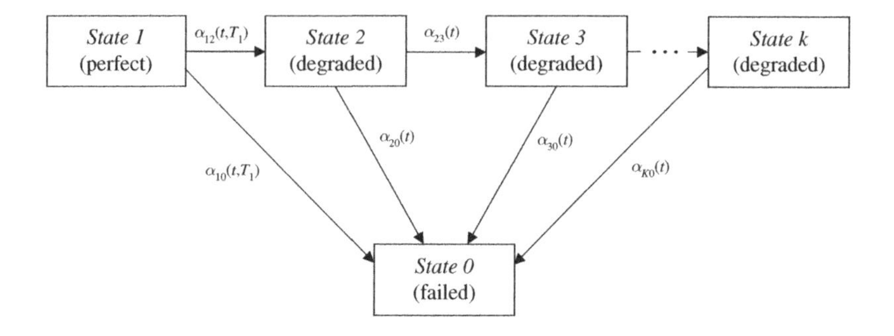

(a) State-space of a unit experiencing degradation

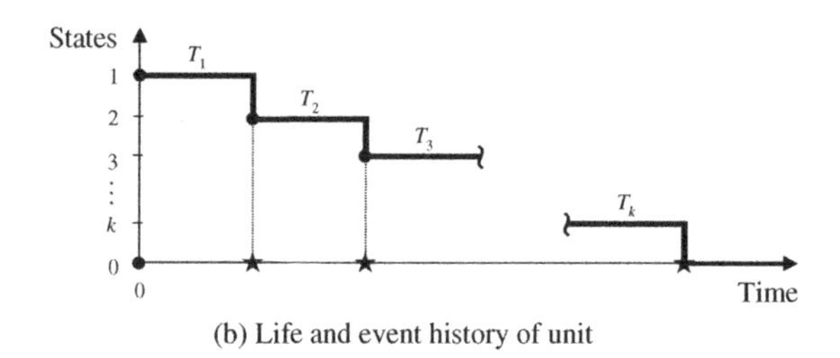

(b) Life and event history of unit

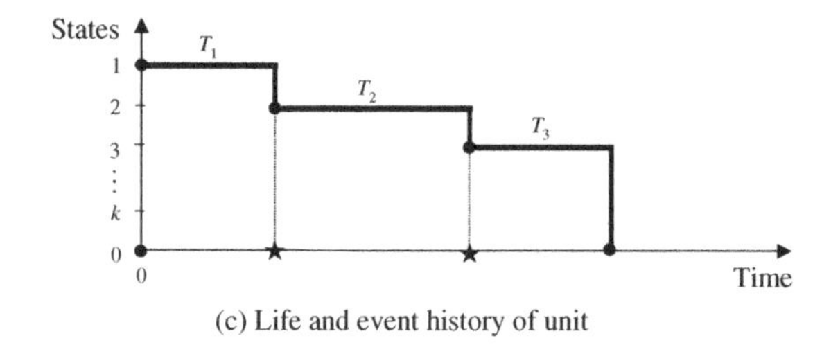

(c) Life and event history of unit

Figure 8.3 State space and sample paths of a unit in degraded states.

in actuality it happens at times in the vicinity of the schedule so that planned maintenance could also be random. Subsequent to any maintenance/repair action, a unit may be restored to a working state, or it may be permanently discarded. Units that are restored to a working state may experience a minimal (or imperfect) repair (Brown and Proschan, 1982). The state space of the above-described maintenance/repair process is shown in Figure 8.4(a). Note that there could be a transition from maintenance to repair; this is because routine maintenance could uncover the need for repairs to avoid impending failures. A possible point process generated by this maintenance/repair phenomenon is shown in Figure 8.4(b). It is for illustrative purposes only; other life histories are also possible.

The T_i's of Figure 8.4(b) indicate the up-times (i.e. the functioning times) of the unit; the D_i's are the down-times, be they due to maintenance or due to repair. Here again the arrows between the states have labels involving α_{ij}, $i, j = 0, 1, 2, 3$ with each α_{ij} indexed by a t alone or by a t and a T_i. When indexed by t alone the life- (or event-) history process is Markovian; when indexed by a t and a T_i, the process is semi-Markovian. The simplest life-history process is a time homogenous Markov process wherein the α_{ij}'s are not indexed by time at all.

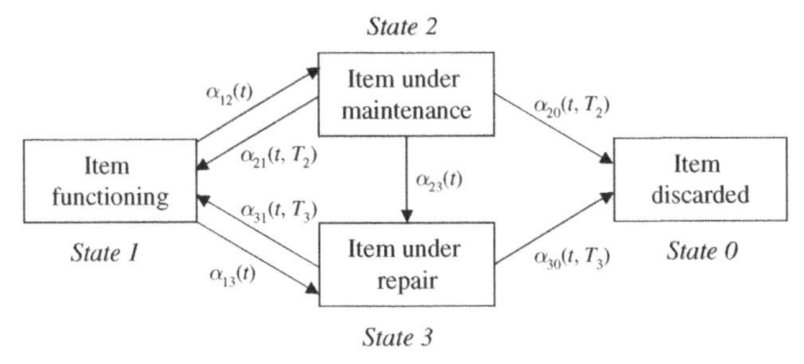

(a) State space of an item experiencing maintenance/repair

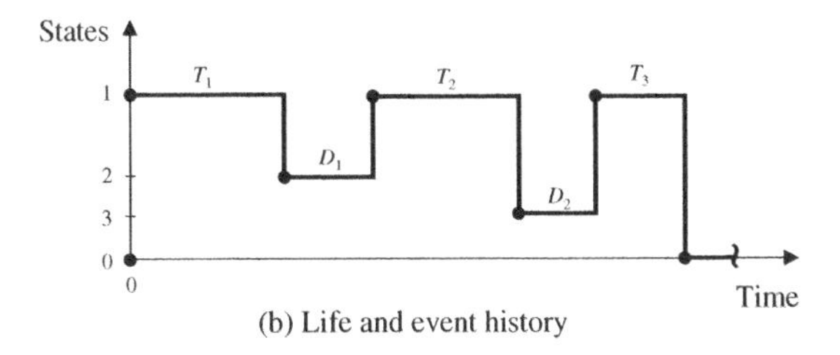

(b) Life and event history

Figure 8.4 State space and sample path under maintenance and repair.

In all the scenarios discussed hereto, including the one of section 8.1, interest generally centres around availability; i.e. the proportion of time (over some time horizon) the unit is in a functioning state.

8.2.4 Life-testing Under Censorship and Withdrawals

The notion of censoring was introduced by me in section 5.4.1 wherein strategies for designing industrial life-tests were discussed. Whereas censorship can be a matter of design, often it is not, and particularly so when we are confronted with retrospective failure data; i.e. data that is generated in the field. The matter of censorship is quite important in the context of medical survival studies or studies pertaining to drug approval. Here, a group of (presumably identical) individuals are observed over some period of time with the aim of assessing the distribution of the time to a certain event, say the remission of a disease, or relating this distribution to individual characteristics. Since individuals may opt to withdraw from a study, or the study itself may be terminated at some arbitrary time, censorship and withdrawals are a facet of medical survival studies that is outside an experimenter's control.

Thus, a typical feature of medical survival data with interest centered around the lifetime T_i of the i-th individual, $i = 1, \ldots, n$, is that one is not able to observe a T_i for every one of the n individuals in the study. For some individuals I may know that T_i exceeds some quantity c_i; i.e. T_i is right-censored (or time-truncated) at c_i. Any particular c_i could be a consequence of an individual's withdrawal from the study, or an individual's record being lost to a follow-up or the termination of the study. I may therefore describe the event history of i-th individual by a process which starting from state 1 (surviving) can take at most one downward jump to state 0 (failed), the occurrence or not of the jump depending on whether $T_i \leq c_i$, or not (Figure 8.5). The state space of this process is identical to that of Figure 8.1(a).

There is another aspect of medical survival data that I will not discuss here in the context of point process models. This pertains to postmortem information wherein the actual value

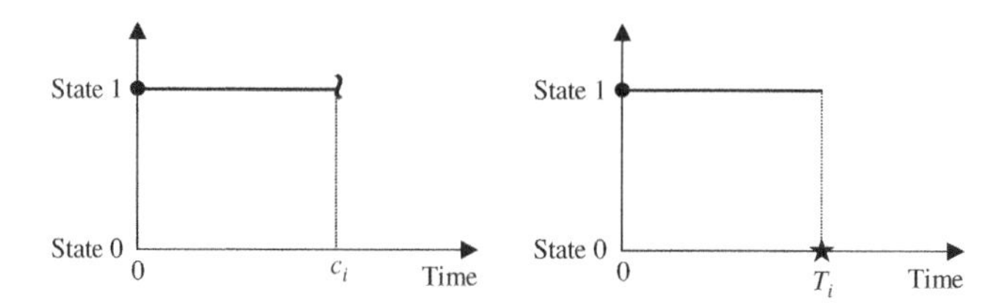

Figure 8.5 Event history with censoring (failure) at $c_i(T_i)$.

of T_i is not observed, but it is known that $T_i < \tau_i$, where τ_i is the time of the postmortem (section 4.4.3).

8.3 MULTIPLICATIVE INTENSITY AND MULTIVARIATE POINT PROCESSES

In section 8.1.1 it was mentioned that there is another, more encompassing, approach for parameterizing $\alpha(t)$, and that was via Aalen's (1987) multiplicative intensity model. Here one assumes that the intensity function $\alpha(t)$ is of the general form

$$\alpha(t) = \alpha_0(t) Z(t), \qquad t \geq 0, \tag{8.5}$$

where $\alpha_0(t)$ is some unknown deterministic function of time, and $Z(t)$ is an observable stochastic process over $[0, t)$. Aalen and Huseby (1991) use this kind of an approach to model several scenarios in biomedicine. Since $Z(t)$ is a stochastic process, so will be $\alpha(t)$; consequently, the process $\{\alpha(t); t \geq 0\}$ is known as the **intensity process**. The notion of an intensity process has already been introduced by me, in section 5.2.4, via the context of a doubly Stochastic Poisson process (or the Cox Process). But the idea can be made more general; see Yashin and Arjas (1998) and section 8.3.1.

To see how the multiple failure mode scenario of section 8.2.1 can be made to fit into the framework of equation (8.5), consider the relationship $\alpha(t) = \sum_{i=1}^{k} \alpha_i(t)$. Suppose, for now, that $\alpha_1(t) = \alpha_2(t) = \ldots = \alpha_k(t) = \alpha_0(t)$, where $\alpha_0(t)$ is unknown. Then I may write

$$\alpha(t) = \alpha_0(t) Z(t), \tag{8.6}$$

where the stochastic process $\{Z(t); t \geq 0\}$ is determined by the relationship $Z(t) = k$, for $t < \min_i T_i$, and $Z(t) = 0$, otherwise. Here $Z(t)$ can interpreted as the number of failure modes that are at the risk of transition to state 0 at the time t.

When the $\alpha_i(t)$'s cannot be assumed equal, and also when a unit experiences degradation, or maintenance and repair, the univariate point process model discussed thus far is inadequate. It needs to be generalized to the multivariate case; this is described next.

8.3.1 Multivariate Counting and Intensity Processes

A univariate counting (or point) process was informally introduced in section 4.7.4 under the venue of a Poisson process, and also in the materials of sections 8.1 and 8.2. More formally, a counting process $\{N(t); t \geq 0\}$ has a sample path that is an integer-valued step function with jumps of size $+1$, and is assumed here to be right continuous, with $N(0) = 0$. Thus $N(t)$ denotes the random number of points (i.e. the events of interest) that have occurred in the time interval $[0, t]$ (Figure 8.6).

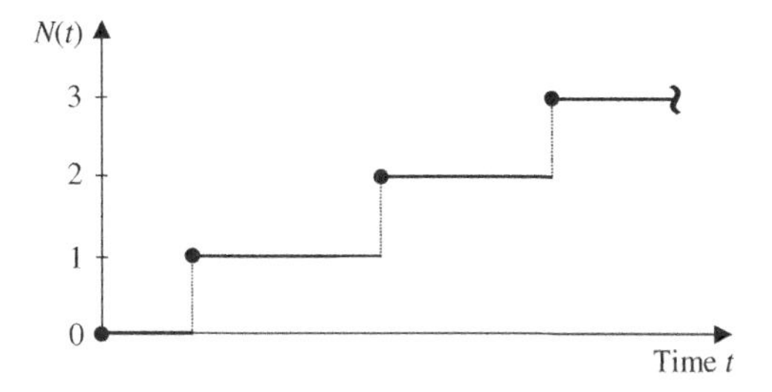

Figure 8.6 Right-continuous sample path of a counting process.

Associated with any counting process $\{N(t); t \geq 0\}$ is its intensity process $\{\alpha(t); t \geq 0\}$, where $\alpha(t)$ is defined as:

$$\alpha(t)\,\mathrm{d}t = P\{N(t) \text{ jumps in an interval of length } \mathrm{d}t \text{ around time } t | \mathcal{F}_{t^-}\}, \tag{8.7}$$

and, as before, $\mathcal{F}_{t^-}$ denotes all that can possibly happen until just before time t (equation (8.2)).

A collection of n univariate counting processes $\{N_i(t); t \geq 0,\ i = 1, \ldots, n\}$ constitutes a multivariate counting process, denoted $\{\mathbf{N}(t); t \geq 0\}$. Associated with a multivariate counting process $\{\mathbf{N}(t); t \geq 0\}$ is an intensity process $\{\alpha(t); t \geq 0\}$ where $\boldsymbol{\alpha}(t) = (\alpha_1(t), \ldots, \alpha_n(t))$, is a collection of n intensity processes. There is an additional caveat about what constitutes a multivariate counting process, and that is the requirement that no two components of the process jump at the same time. The components of a multivariate counting process may depend on each other.

With the notion of a multivariate counting process in place, it is easy to see how the scenario of the multiple mode failure with the $\alpha_i(t)$'s distinct can be made to fit into its framework. To see this, let the counting process $\{N_i(t); t \geq 0\}$ be associated with the failure mode i, $i = 1, \ldots, k$. This process can take at most one jump, and its intensity process is $\{\alpha_i(t)Z(t); t \geq 0\}$, where the observable process $\{Z(t); t \geq 0\}$ is such that $Z(t) = 1$, if $\min_i T_i \geq t$; $Z(t) = 0$ otherwise. Thus, the multiple mode failure phenomenon can be modeled as a multivariate counting process $\{\mathbf{N}(t); t \geq 0\}$, where $\mathbf{N}(t) = (N_1(t), \ldots, N_k(t))$, with intensity process $\{\boldsymbol{\alpha}(t); t \geq 0\}$ where $\boldsymbol{\alpha}(t) = (\alpha_1(t)Z(t), \ldots, \alpha_k(t)Z(t))$.

The scenarios of Figures 8.3 and 8.4 can also be modeled via a multivariate counting process $\{N_{ij}(t);\ t \geq 0,\ i, j \in \mathcal{J}\}$, where $N_{ij}(t)$ denotes the number of transitions from state i to state j in time $[0, t]$, and $\mathcal{J}$ is the state space of the process. For the degrading unit scenario, $\mathcal{J} = \{0, 1, \ldots, k\}$, whereas for the maintenance/repair scenario, $\mathcal{J} = \{0, 1, 2, 3\}$. For defining a multiplicative intensity function associated with each of the above component processes, we let the indicator function $Z_i(t) = 1$, if the device is in state i just prior to time t (i.e. the device is at the risk of transition out of state i), and $Z_i(t) = 0$, otherwise. Then, the intensity process associated with the counting process $\{N_{ij}(t); t \geq 0\}$ assuming that the process is Markovian is of the form $\{\alpha_{ij}(t)Z_i(t); t \geq 0\}$. When the process is semi-Markov, the $\alpha_{ij}(t)$ gets replaced by $\alpha_{ij}(t, T_i)$.

Perhaps the best known, and possibly the most discussed, application of multivariate point processes and their intensity processes pertains to the treatment of censored survival data (section 8.2.4). Here one defines the i-th component of a multivariate counting process via $N_i(t) = I(T_i \leq t, D_i = 1)$, where $I(\bullet)$ is the indicator function, and $D_i = 1$, if T_i is an observed survival time; $D_i = 0$ otherwise. With an exchangeable population of n individuals under observation, $i = 1, \ldots, n$, and in what follows we assume that the n pairs (T_i, D_i) are independent. The process $\{N_i(t); t \geq 0\}$ jumps to 1 at T_i, if T_i is a time to failure; otherwise the process does not jump at all.

Since the n individuals in the study are judged to be exchangeable, the survival times T_i, $i = 1, \ldots, n$, have a common distribution. Let $\alpha(t)$ denote the failure rate of this common distribution. Then, $\{\alpha_i(t); t \geq 0\}$, the intensity process of $\{N_i(t); t \geq 0\}$, is given as $\alpha_i(t) = \alpha(t) Z_i(t)$, where $Z_i(t) = 1$ if $T_i \geq t$, and $c_i \geq t$; $Z_i(t) = 0$, otherwise. To see why, note that at time t, we know that either the i-th individual is surviving and still under observation or the individual is no more under observation because of censorship or failure. In the latter two cases, the probability of $N_i(t)$ taking an upward jump (in the vicinity of t) is zero. In the former case it is the failure rate at t.

Thus to summarize, the life-testing with the censorship scenario of section 8.2.4 can be described via a multivariate counting process $\mathbf{N}(t) = (N_1(t), \ldots, N_k(t))$ with an intensity process $\{\boldsymbol{\alpha}(t); t \geq 0\}$, where $\boldsymbol{\alpha}(t) = (\alpha(t)Z_1(t), \ldots, \alpha(t)Z_k(t))$, with $Z_i(t) = 1$, if the i-th individual is surviving at t; $Z_i(t) = 0$, otherwise. Alternatively, we may also describe the said scenario via a univariate counting process $\{N(t); t \geq 0\}$ with an intensity process $\{\alpha(t) Z(t); t \geq 0\}$ where $N(t) = \sum_{i=1}^{t} N_i(t)$, and $Z(t) = \sum_{i=1}^{t} Z_i(t)$. Here $Z(t)$ would connote the number of individuals at risk, just prior to t.

When the population under observation cannot be judged homogenous, the intensity process of the multivariate counting process takes the form $\boldsymbol{\alpha}(t) = (\alpha_1(t)Z_1(t), \ldots, \alpha_k(t)Z_k(t))$, where $\alpha_i(t)$ is the failure rate of the distribution of T_i.

I close this section by affirming the generality of the multiplicative intensity model vis-à-vis its relevance to a wide range of scenarios in reliability modeling, and for life-testing under censorship. The next task is to show how the structure of this model facilitates an analyses of random failures via martingales and their associated limit theorems. This is outlined in section 8.5, but to better appreciate the material therein it is necessary to give an informal overview of a dynamic stochastic process model, its martingale properties and the Doob–Meyer decomposition of stochastic processes. This is done next in section 8.4.

8.4 DYNAMIC PROCESSES AND STATISTICAL MODELS: MARTINGALES

The purpose of this section is to provide an overview of dynamic stochastic process models and their martingale properties. This I do in an informal setting by considering a discrete time stochastic process $\{X_t; t = 1, 2, \ldots\}$. The material here is largely based on Aalen (1987). The essence of a dynamic process is that the present course of the process depends on its past history. This section also enables us to introduce terminology that is specific to the theory of martingales, terminology that will facilitate the reading of the material of section 8.5.

Assuming X_0 known, I start by focusing attention on $(X_t - X_{t-1})$, the change experienced by the process between time $(t-1)$ and t. By assuming that the time interval $[t-1, t]$ is small, $(X_t - X_{t-1})$ could be seen as the instantaneous change experienced at time t. We then ask what would be our 'best' prediction of $(X_t - X_{t-1})$, were we to know $X_1, \ldots, X_{t-1}$? A meaningful answer to this question would be the conditional expectation

$$E(X_t - X_{t-1} \mid X_1, \ldots, X_{t-1}) \overset{\text{def}}{=} V_t.$$

With V_t as a predictor of $(X_t - X_{t-1})$, the error of prediction would be $X_t - E(X_t \mid X_1, \ldots, X_{t-1})$.

Motivated by the fact that $\sum_j^t (X_j - X_{j-1}) = X_t$, I may consider the sums $\sum_j^t V_j \overset{\text{def}}{=} U_t$, and $\sum_j^t X_j - E(X_j \mid X_1, \ldots, X_{j-1}) \overset{\text{def}}{=} M_t$, and declare that

$$X_t = U_t + M_t, \quad t = 1, 2, \ldots \tag{8.7}$$

This relationship implies that the sum of changes in X_t equals the sum of all the predicted changes plus the sum of prediction errors. More generally, we may say that the process $\{X_t; t = 1, 2, \ldots\}$ has been split into two processes $\{U_t; t = 1, 2, \ldots\}$ representing the sum of our best predictions of the instantaneous changes, and the process $\{M_t; t = 1, 2, \ldots\}$ representing the sum of our errors of prediction. Equation (8.7) is known as the **Doob Decomposition** of the process $\{X_t; t = 1, 2, \ldots\}$.

Innocuous as it may appear, the Doob decomposition is endowed with some striking properties. These properties enable such decompositions to play a fundamental role in probability theory. To appreciate why, I start by noting that

$$E(M_t \mid X_1, \ldots, X_{t-1}) = M_{t-1}, \tag{8.8}$$

a feature that can be verified via (8.7) and the fact that $E(X_t \mid X_1, \ldots, X_{t-1}) - X_{t-1} = V_t$. The relationship (8.8) implies that the expectation of M_t given the past equals its value at time $(t-1)$. A process $\{M_t; t = 1, 2, \ldots\}$ satisfying the properties of (8.8) is called a **martingale with respect to the process** $\{X_t; t = 1, 2, \ldots\}$. The term 'martingale' is an acronym that derives from a French term that describes a gambling strategy of doubling one's bet until a win is secured. This gambling strategy captures the notion of a fair game.

In probability and statistics, martingales arise quiet often, in a large variety of contexts. Some examples are sums of random variables, branching processes, urn schemes, Markov chains, and likelihood ratios. In other words, martingales are all around us. One of their biggest virtues is that subject to a condition on their moments, they always converge – this is Doob's convergence theorem. Roughly speaking, this theorem says that if $\{M_t; t = 1, 2, \ldots\}$ is a martingale with $E\left(M_t^2\right) < K < \infty$, for some K and for all t, then there exists a random variable M such that M_t converges to M (almost surely and in mean square).[1] Theorems like this and other martingale-related theorems such as the Doob–Kolmogorov inequality[2] enable one to study estimators, test statistics, and other likelihood based quantities when they are given a martingale representation. Thus the attractiveness of martingales. But how do martingales differ from the traditional scenario of independent, identically distributed random variables, and processes with independent increments? To address these questions let us focus attention on the martingale differences $\{M_t - M_{t-1}\}$ and explore their properties.

Using the relationship of (8.8) and the fact that for any two random variables X and Y, $E(X) = E_Y(E(X|Y))$, we can show that:

$$E(M_t - M_{t-1}) = 0, \tag{8.9}$$

and that

$$E[(M_t - M_{t-1})(M_{t-1} - M_{t-2})] = 0. \tag{8.10}$$

Thus, the sequence of martingale differences $\{M_t - M_{t-1}\}$ can be seen as a generalization of the usual error theory assumptions made in statistical modeling. The errors are usually described via zero mean, independent and identically distributed random quantities. The generalization

[1] Let $M_1, M_2, \ldots, M_n, \ldots$ be a sequence of random variables, and M another random variable, with the property that $E|M_n^r| < \infty$ for all n and for some $r \geq 1$, and $E(|M_n - M|^r) \downarrow 0$ as $n \uparrow \infty$. Then the sequence is said to **converge in mean (mean square)** if $r = 1(2)$.

[2] If $\{M_t; t = 1, 2, \ldots\}$ is a martingale with respect to $\{X_t; t = 1, 2, \ldots\}$ then $P(\max_{1 \leq i \leq n} |M_i| \geq \epsilon) \leq \frac{1}{\epsilon^2} E(M_n^2)$ whenever $\epsilon > 0$.

here is suggested by the property of (8.9), which implies uncorrelatedness (but not necessarily independence). The relationship of (8.9) is known as the property of **orthogonal increments** and this is what distinguishes the martingale property from the independent increments property.

The martingale differences serve the same function as the errors (in predicting the changes $X_t - X_{t-1}$ by V_t), and are also known as **innovations** since they represent the new development in the process. Because the U_t's are a function of the X_t's up until X_{t-1}, the process $\{U_t; t = 1, 2, \ldots\}$ is called a **predictable process**. Some other properties of martingales and martingale differences will be described later, once we discuss the decomposition of continuous time processes, especially the counting process, which are the focus of our discussion in Chapter 8.

8.4.1 Decomposition of Continuous Time Processes

There is a generalization of the Doob decomposition to continuous time stochastic processes; it is called a **Doob–Meyer decomposition**. Here, as a continuos time analogue to (8.7), we write

$$X_t = U_t + M_t, \qquad \text{for } t \geq 0, \tag{8.11}$$

where the process $\{M_t; t \geq 0\}$ is a martingale with respect to the process $\{X_t; t \geq 0\}$, and the process $\{U_t; t \geq 0\}$ is determined by an integral of the 'best' predictions; i.e.

$$U_t = \int_0^t V_s \, \mathrm{d}s. \tag{8.12}$$

The process $\{V_t; t \geq 0\}$ is such that each V_t is known just before t; it is called a **local characteristic**. The process $\{U_t; t \geq 0\}$ is known as a **compensator** of the process $\{X_t; t \geq 0\}$.

By a **dynamic statistical model** we mean any parametrization of the predictable process $\{V_t; t = 1, 2, \ldots \text{ or } t \geq 0\}$. An archetypal example is the linear model $V_t = \sum_t^k \beta_t Z_t$, where the β_i's are unknown coefficients and each of the kZ_t's are observable and predictable stochastic processes. Another example is an autoregressive process with random coefficients

$$X_t - X_{t-1} = \alpha_t X_{t-1} + (M_t - M_{t-1}),$$

obtained by setting $V_t = \alpha_t X_{t-1}$ and combining it with (8.7).

In the continuous setting, expressions that are analogous to (8.8) and (8.9) of the discrete case take the forms

$$E(M_t \mid \mathcal{F}_{t^-}) = M_{t^-},$$

and

$$E(\mathrm{d}M_t \mid \mathcal{F}_{t^-}) = 0,$$

respectively; these expressions have a role to play in describing a central limit for martingales. I next introduce the notion of a predictable variation process of a martingale, namely, the process $\{\langle M_t \rangle; t \geq 0\}$, where

$$\mathrm{d}\langle M_t \rangle \stackrel{\text{def}}{=} E\left[\mathrm{d}M_t^2 \mid \mathcal{F}_{t^-}\right] = \mathrm{Var}[\mathrm{d}M_t \mid \mathcal{F}_{t^-}].$$

The above process is predictable and non-decreasing. It can be seen as the sum of the (conditional) variances of the increments $\{M_t; t \geq 0\}$ over small intervals that partition $[0, t]$ (Gill, 1984), from which the material below has been abstracted.

For two martingales $\{M_t; t \geq 0\}$ and $\{M'_t; t \geq 0\}$ the predictable covariance process, $\{\langle M_t, M'_t \rangle ; t \geq 0\}$ can be analogously defined.

The ideas and notions described above can be made concrete via the scenario of a univariate counting process $\{N_t; t \geq 0\}$ of section 8.3.1. I start with the Doob–Meyer decomposition of this process via the template of (8.11) and (8.12) so that the martingale $\{M_t; t \geq 0\}$ is asserted as

$$N_t = \int_0^t V_s \, \mathrm{d}s + M_t, \tag{8.13}$$

where the local characteristic V_s is the intensity process $\{\alpha(s); s \geq 0\}$ of the point process. This intensity process can also be written as

$$\alpha(t) \, \mathrm{d}t = E(\mathrm{d}N_t | \mathcal{F}_{t^-}); \tag{8.14}$$

this is because in a small interval of time $\mathrm{d}t$, N_t either takes a jump or does not.

As an aside, a dynamic statistical model underlying this counting process can be had by parameterizing $\alpha(t)$ via one of the several multiplicative intensity processes described in section 8.3.1, or via the Cox model of (8.3), depending on the scenario at hand.

Since $(\mathrm{d}N_t | \mathcal{F}_{t^-})$ is a zero or one random variable with expectation $\alpha(t) \, \mathrm{d}t$, its conditional variance is $\alpha(t) \, \mathrm{d}t(1 - \alpha(t) \, \mathrm{d}t) \approx \alpha(t) \, \mathrm{d}t$. Consequently, the quantity $\langle M_t \rangle$ is of the form

$$\langle M_t \rangle = \int_0^t \alpha(s) \, \mathrm{d}s,$$

so that the predictable variance process of the counting process $\{M_t; t \geq 0\}$ is $\left\{\int_0^t \alpha(s) \, \mathrm{d}s; t \geq 0\right\}$. To illustrate the nature of the predictable covariance process spawned by two counting processes, suppose that there is another counting process $\{N'_t; t \geq 0\}$ whose intensity process is $\{\alpha'(t); t \geq 0\}$. Suppose that the two counting process under consideration constitute a bivariate counting process, so that N_t and N'_t do not simultaneously experience a jump. That is, $(\mathrm{d}N_t)(\mathrm{d}N'_t)$ is always zero, and hence the conditional covariance between $\mathrm{d}N_t$ and $\mathrm{d}N'_t$ is $(\alpha(t) \, \mathrm{d}t)(\alpha'(t) \, \mathrm{d}t) \approx 0$. Consequently the predictable covariance process $\{\langle M_t, M'_t \rangle ; t \geq 0\} = 0$. When such is the case the two martingale processes $\{M_t; t \geq 0\}$ and $\{M'_t; t \geq 0\}$ are said to be **orthogonal**.

8.4.2 Stochastic Integrals and a Martingale Central Limit Theorem

Suppose that $\{M_t; t \geq 0\}$ is a continuous time martingale and $\{H_t; t \geq 0\}$ some predictable process. Define a new process, say $\{M^*_t; t \geq 0\}$, which is a transformation of $\{M_t; t \geq 0\}$ by the stochastic integral

$$M^*_t = \int_0^t H(s) \, \mathrm{d}M_s,$$

so that $\mathrm{d}M^*_t = H(s) \, \mathrm{d}M_s$. Then it is easy to verify that the process $\{M^*_t; t \geq 0\}$ is also a martingale and that its predictable variance process is

$$\langle M^*_t \rangle = \int_0^t (H(s))^2 \, \mathrm{d} \langle M_t \rangle .$$

In a similar manner, we may obtain the predictable covariance process of the integrals of two predictable processes with respect to two martingales.

The martingale central limit theorem prescribes conditions under which a martingale converges to a Brownian motion process $\{Z_t; t \geq 0\}$ (section 7.2.3). Loosely speaking, consider a sequence of martingale processes $\{M_t^{(n)}; t \geq 0 \text{ and } n = 1, 2, \ldots\}$ such that the jumps of $M_t^{(n)}$ becomes smaller as n increases; that is, $M_t^{(n)}$ gets small as n increases; that is, $M_t^{(n)}$ becomes more and more continuous. Furthermore, suppose that as n increases, the predictable variance process of $\left\{M_t^{(n)}; t \geq 0\right\}$ becomes deterministic; i.e. $\left\{\left\langle M_t^{(n)}\right\rangle; t \geq 0\right\} \to A(t)$, where $A(\bullet)$ is a deterministic function. Then $M_t^{(n)}$ converges in distribution to $Z(t)$ as $n \to \infty$. Specifically, for each t, $M_t^{(n)}$ has a Gaussian distribution with mean 0 and variance $A(t)$, and the increments of $M_t^{(n)}$ are independent (and no more merely orthogonal!).

8.5 POINT PROCESSES WITH MULTIPLICATIVE INTENSITIES

To set things in perspective, I start this section with an overview of what has been said so far in sections 8.1–8.3. Section 8.1 pertained to articulating what one means by event- and life-history data, and the role of a point process for describing these data. Also introduced therein was the notion of the intensity function and the parameterization of this function. Section 8.2 pertained to looking at several scenarios in reliability modeling, such as competing risks, degradation and deterioration, multi-mode failures and maintenance (replacement/repair) problems, as point processes. Also considered therein was the scenario of life-testing with censoring and withdrawals. In section 8.3, I introduced multivariate counting processes and their associated intensity processes. A key feature of the material here was a parameterization of the intensity function in a multiplicative form involving an unknown function and an observable stochastic process; (equation (8.5)). The observable stochastic process is a consequence of the nuances associated with the scenarios considered in sections 8.1 and 8.2. In some sense, parameterizing the intensity function in a multiplicative form is the crux of the material in section 8.3.

Overall, the main message underlying all that has been said thus far is the important role played by point processes and their multiplicative intensity functions for describing a wide range of scenarios that arise in reliability modeling and survival analysis. Having done so, we now divert our attention to the material of section 8.4. The essential import of this material is that a point process spawns, via the Doob–Meyer decomposition, a martingale process which enjoys, among other things, convergence to a Brownian motion process, via a central limit theorem. Section 8.4 also describes the transformation of a martingale to another martingale via a stochastic integral. Such transformations come into play when one wishes to do inference involving point process data.

The aim of this section is to show how the material of sections 8.1 through 8.3, which is specific to reliability modeling and survival analysis, can be connected with the material of section 8.4 on martingales to obtain results that provide meaningful descriptions of the several stochastic failure processes. The key to doing so is to link counting processes and martingales. Accordingly, suppose that $\{N(t); t \geq 0\}$ is a univariate counting process with an intensity process $\{\alpha(s); s \geq 0\}$. Suppose further that $\alpha(t) = \alpha_0(t)Z(t)$, where $\alpha_0(t)$ is an unknown function of time and $Z(t)$ is an observable stochastic process over $[0, t)$ (equation (8.5)). Then, by the Doob–Meyer decomposition of the counting process, we may write, for $t \geq 0$

$$M(t) = N(t) - \int_0^t \alpha(s)\,\mathrm{d}s, \tag{8.15}$$

where $\{M(t); t \geq 0\}$ is a martingale and the process $\left\{\int_0^t \alpha(s)\,\mathrm{d}s; s \geq 0\right\}$ is a compensator of the counting process. Because $\{M(t); t \geq 0\}$ is a martingale $E[M(t) \mid \mathcal{F}(u)] = M(u)$, $u < t$.

Equation (8.15) is the basis for a unified approach to studying the behavior of items experiencing the types of issues described in sections 8.1 and 8.2. Since the compensator is a predictable process, $N(t)$ inherits the martingale properties of $\{M(t); t \geq 0\}$, in particular the martingale central limit theorem and the Doob convergence theorems. Thus, we are able to discuss the asymptotic behavior of the univariate and multivariate counting process introduced in sections 8.1–8.3. These counting processes provide insight about the stochastic behavior of items and systems under study.

Chapter 9

Non-parametric Bayes Methods in Reliability

Here lies Stephen Banach: with plenty of space.

9.1 THE WHAT AND WHY OF NON-PARAMETRIC BAYES

By the term 'non-parametric Bayes methods in reliability', I mean an analysis of life history data without the use of any of the chance distributions mentioned in Chapter 5, or for that matter any other parametric family of chance distributions. Thus the methods of this chapter serve the same purpose as those of section 5.4. Furthermore, as it will become clear in the sequel, there are elements of commonality between the strategies to be proposed here and those discussed in Chapter 7 on stochastic hazards – thus the placement of the material of this chapter subsequent to that of Chapter 7 (and also that of Chapter 8).

A motivation for considering non-parametric Bayes methods in reliability and survival analysis is prompted by the often expressed view that, in any Bayesian analysis, the specification of a realistic likelihood is more important than the kind of priors endowed on its parameters. Chance distributions almost always form the basis of constructing likelihoods and thus their judicious choice is a matter of importance. Non-parametric Bayes methods come into play when one wishes to perform a Bayesian analysis of failure time data using a likelihood function that is not suggested by any standard distribution used in reliability, such as the exponential and the Weibull. However, there is a price to be paid for doing so, a price similar to that paid when one uses binary success/failure data in lieu of the actual time to failure data.

In non-parametric Bayes, the general strategy is to place a prior distribution on the class of all probability distribution functions, say F, and using life-history data to obtain a posterior distribution on the class of all probability distributions. The unknown distribution function F is the 'parameter' in non-parametric problems. A variant of the above strategy is to place a prior distribution on the class of all cumulative hazard functions, or on the hazard rates themselves. Endowing prior distributions on the hazard rate function is an attractive option because hazard rates are easy to interpret and can be generalized to more complicated models like Markov chains.

Reliability and Risk: A Bayesian Perspective N.D. Singpurwalla
© 2006 John Wiley & Sons, Ltd

What constitutes a meaningful and workable family of prior distributions is a challenging issue that has generated an array of imaginative and impressive ideas, the papers by Ferguson (1973) and by Doksum (1974) being landmark events. A striking feature of these approaches is the realization that the prior distributions for the underlying distribution functions constitute a stochastic process, and so the issue boils down to identifying those processes that lend themselves to a convenient prior to posterior transformation. Connection with the material of Chapter 7 is now clear, even though the motivations are different; modeling in the case of Chapter 7, and inference in the case of Chapter 9. The novelty here is that we are considering a prior to posterior conversion of a stochastic process using life-history data.

In what follows, I give an account of the basic set-up and a broad-brushed perspective of the key developments. The details have been deliberately kept to a minimum, the hope being that interested readers will go to the original references cited to fill in the gaps. The **Dirichlet distribution** and its variants – the **ordered** and the **generalized Dirichlet** – play a fundamental role, and so does the Lévy process of section 7.2.2. The Dirichlet distribution spawns the **Dirichlet process**, which is a candidate prior process on the class of all distribution functions. Thus, section 9.2 is devoted to a review of the Dirichlet distribution and its variants. In the context of the theme of this chapter, the Dirichlet distribution has entered into the arena via the papers of Kraft and Van Eeden (1964) and of Ramsey (1972). These pertain to the development of Bayesian methods for bioassay (or dose–response experiments). Note that the potency curve mentioned in section 5.8.1 being non-increasing and bounded between zero and one behaves like a survival function. Thus, from a chronological point of view, one may see the works of Kraft and Van Eeden, and Ramsey as a starting point for the highly sophisticated and more recent work in non-parametric Bayes methods. Because of its foundational position, and also its intuitive appeal, Ramsey's (1972) work on a Bayesian estimation of the potency curve is overviewed in section 9.3. Ramsey's material can be viewed as the non-parametric Bayes analogue of the parametric Bayes approach to dose–response and damage assessment discussed in section 5.8.

The remainder of this chapter is organized as follows. Section 9.4 focuses attention on prior distributions for the hazard rate functions, whereas Section 9.5 considers priors for the cumulative hazard functions. Section 9.6 is devoted to a consideration of priors for the cumulative distribution functions, or equivalently, the survival function.

9.2 THE DIRICHLET DISTRIBUTION AND ITS VARIANTS

The Dirichlet distribution is a generalization of the beta distribution to the multivariate case. In Bayesian inference, the attractiveness of the beta and the Dirichlet distributions stems from their conjugacy property (section 5.4.7) with respect to sampling from the Bernoulli and the multinomial distributions, respectively. The multinomial distribution, though mentioned in section 3.3, has not been properly introduced. I, therefore, start this section by describing the multinomial.

The multinomial is a generalization of the Bernoulli, in the sense that a random variable X can take k possible mutually exclusive and exhaustive values, say $a_1, \ldots, a_k$, with respective probabilities $p_1, \ldots, p_k$, where $\sum_{i=1}^{k} p_i = 1$. Interest centers around the probability that in n replications of X, the event $X = a_i$ occurs N_i times, $i = 1, \ldots, k$, with $\sum_{i=1}^{k} N_i = n$. Then, given $p_1, \ldots, p_k$, it is easy to see that the joint distribution of $(N_1, \ldots, N_k)$ is the **multinomial distribution**:

$$P(N_1 = n_1, \ldots, N_k = n_k \mid p_1, \ldots, p_k; n) = \frac{n!}{n_1! \cdots n_k!} p_1^{n_1} \cdots p_k^{n_k}. \tag{9.1}$$

Verify that $E(N_i) = np_i$, and that $\text{Var}(N_i) = np_i(1 - p_i)$ and that $\text{Cov}(N_i, N_j) = -np_ip_j$, for $i \neq j$. Also, for any i, N_i has a binomial distribution with parameters n and p_i, $i = 1, \ldots, k$.

The connection between the multinomial and the Dirichlet is described later, once the latter is introduced. To do so, we start by considering a random variable X having a beta distribution with parameters ν_1 and ν_2, whose density at x, $0 < x < 1$, denoted $\text{Beta}(\nu_1, \nu_2)$, is given as:

$$f(x; \nu_1, \nu_2) = \frac{\Gamma(\nu_1 + \nu_2)}{\Gamma(\nu_1)\Gamma(\nu_2)} x^{\nu_1 - 1}(1 - x)^{\nu_2 - 1};$$

see (5.11). The expectation and the variance of X are: $E(X) = \nu_1/(\nu_1 + \nu_2)$ and $\text{Var}(X) = \nu_1\nu_2/((\nu_1 + \nu_2)^2(\nu_1 + \nu_2 + 1))$. Furthermore, if Z_1 and Z_2 are independent with Z_i having a gamma distribution with scale (shape) parameter $1(\nu_i)$, denoted $\mathcal{G}(1, \nu_i), i = 1, 2$, then $X \stackrel{\text{def}}{=} Z_1/(Z_1 + Z_2)$ will have the above beta distribution. Details about these and related matters can be found in Wilks (1962, p. 173).

Since a (univariate) beta distribution has two parameters ν_1 and ν_2, its k-variate analogue for $X_1, \ldots, X_k$, with $k \geq 2$, will have $(k + 1)$ parameters $\nu_1, \ldots, \nu_k, \nu_{k+1}$. This k-variate analogue is called a **Dirichlet distribution** with parameters $(\nu_1, \ldots, \nu_k, \nu_{k+1})$, denoted $\mathcal{D}(\nu_1, \ldots, \nu_k, \nu_{k+1})$. For any point $\mathbf{x} = (x_1, \ldots, x_k)$ in the set S_k, where $S_k = \{(x_1, \ldots, x_k) : x_i \geq 0, i = 1, \ldots, k, \sum_{i=1}^{k} x_i \leq 1\}$, the distribution $\mathcal{D}(\nu_1, \ldots, \nu_k, \nu_{k+1})$ has at $\mathbf{x}$, density

$$f(\mathbf{x}; \nu_1, \ldots, \nu_{k+1}) = \frac{\Gamma(\nu_1 + \cdots + \nu_{k+1})}{\Gamma(\nu_1) \cdots \Gamma(\nu_{k+1})} x_1^{\nu_1 - 1} \cdots x_k^{\nu_k - 1}(1 - x_1 - x_2 - \cdots - x_k)^{\nu_{k+1} - 1}, \quad (9.2)$$

for $\nu_i \geq 0$. Observe that when $k = 1$, $\mathcal{D}(\nu_1, \ldots, \nu_{k+1}) = \text{Beta}(\nu_1, \nu_2)$; thus the Dirichlet is seen as a generalization of the beta.

Furthermore, if $Z_1, \ldots, Z_{k+1}$ are independent with Z_i having the distribution $\mathcal{G}(1, \nu_i), i = 1, \ldots, k + 1$, then the k ratios $Z_i/(Z_1 + \cdots + Z_{k+1}), i = 1, \ldots, k$, will have $\mathcal{D}(\nu_1, \ldots, \nu_{k+1})$ as their joint distribution.

The mean, variance and the covariance of the X_is are given below:

$$E(X_i) = \frac{\nu_i}{\nu_1 + \cdots + \nu_{k+1}},$$

$$V(X_i) = \frac{\nu_i(\nu_1 + \cdots + \nu_{k+1} - \nu_i)}{(\nu_1 + \cdots + \nu_{k+1})^2(\nu_1 + \cdots + \nu_{k+1} + 1)},$$

$$E(X_i, X_j) = -\frac{\nu_i\nu_j}{(\nu_1 + \cdots + \nu_{k+1})^2(\nu_1 + \cdots + \nu_{k+1} + 1)}, \quad i \neq j.$$

Observe that $E(X_i, X_j)$ does not account for the proximity between i and j, and this in some scenarios can be a disadvantage, also note that X_i and X_j are negatively correlated.

Some attractive properties of the Dirichlet are the behavior of its marginal and conditional distributions, and its property of closure under additivity.

To appreciate these suppose that $(X_1, \ldots, X_k)$ has the distribution $\mathcal{D}(\nu_1, \ldots, \nu_{k+1})$. Then

(*a*) For any $k_1 < k$, the random variables $X_1, X_2, \ldots, X_{k_1}$ have the distribution

$$\mathcal{D}(\nu_1, \ldots, \nu_{k_1}, (\nu_{k_1+1} + \cdots + \nu_{k+1}));$$

(*b*) Given $(X_1, \ldots, X_{k-1})$, the conditional distribution of the random variable $X_k/(1 - X_1 - X_2 - \cdots - X_{k-1})$ is $\text{Beta}(\nu_k, \nu_{k+1})$;

(*c*) The sum $(X_1 + \cdots + X_k)$ has the distribution $\text{Beta}((\nu_1 + \cdots + \nu_k), \nu_{k+1})$.

Finally, the Dirichlet distribution's natural conjugacy property vis-à-vis sampling from the multinomial distribution can be stated by saying that if the $(p_1, \ldots, p_{k-1})$ of (9.1) have the distribution $\mathcal{D}(\nu_1, \ldots, \nu_k)$ as a prior, then upon observing $n_1, \ldots, n_k$, the posterior distribution of $(p_1, \ldots, p_{k-1})$ will also be a Dirichlet, and it will be of the form $\mathcal{D}(\nu_1 + n_1, \ldots, \nu_k + n_k)$. This result is easy to verify (cf. De Groot, 1970, p. 174).

9.2.1 The Ordered Dirichlet Distribution

The ordered Dirichlet distribution has proved to be useful for addressing problems in bioassay, and in reliability assessment, when the failure rate function is assumed to be increasing. This distribution arises when we consider successive sums of Dirichlet variables. Specifically, suppose that the vector $(X_1, \ldots, X_k)$ has the distribution $\mathcal{D}(\nu_1, \ldots, \nu_{k+1})$. Consider the random variables $Y_1 = X_1, Y_2 = X_1 + X_2, \ldots, Y_k = X_1 + \cdots + X_k$. Observe that $0 \equiv Y_0 < Y_1 < \cdots < Y_k < Y_{k+1} \equiv 1$; i.e. the successive Y_is are ordered. It can be seen (cf. Wilks, 1962, p. 182) that the vector $(Y_1, \ldots, Y_k)$ has a joint distribution, with density at $(y_1, \ldots, y_k)$, where $y_1 < y_2 < \cdots < y_k$ is of the form

$$\frac{\Gamma(\nu_1 + \cdots + \nu_{k+1})}{\Gamma(\nu_1) \cdots \Gamma(\nu_{k+1})} y_1^{\nu_1 - 1} (y_2 - y_1)^{\nu_2 - 1} \cdots (y_k - y_{k-1})^{\nu_k - 1} (1 - y_k)^{\nu_{k+1} - 1}. \tag{9.3}$$

This distribution of $(Y_1, \ldots, Y_k)$ is known as an **ordered Dirichlet distribution**; it will be denoted $\mathcal{D}^*(\nu_1, \ldots, \nu_{k+1})$. For future reference, when in section 9.6 we consider the Dirichlet process, it is useful to note that because the X_is are correlated, the increments $Y_1, (Y_2 - Y_1), (Y_3 - Y_2), \ldots, (Y_i - Y_{i-1})$, are not independent. That is, the Dirichlet process does not have independent increments.

If we set $y_0 \equiv 0$ and $y_{k+1} \equiv 1$, then (9.3) can be compactly written as being proportional to

$$\prod_{i=1}^{k+1} (y_i - y_{i-1})^{\nu_i - 1}. \tag{9.4}$$

The above form of the ordered Dirichlet distribution has a constructive appeal whose nature will be explained in section 9.3. For now I will introduce a generalization of the Dirichlet distribution, a generalization that is based on the notion of 'neutrality'.

9.2.2 The Generalized Dirichlet – Concept of Neutrality

Connor and Mossiman (1969) introduce a generalization of the Dirichlet distribution that is motivated by the problem of how to randomly divide an interval. The matter of randomly dividing an interval occurs in several contexts. An archetypal scenario is the proportions (say by weight) of the chemical compounds that constitute a substance. As an example, let $P_1, \ldots, P_k$ denote the (random) proportions of the k chemical compounds of a rock, with P_1 pertaining to silica. Then a question of interest to geologists is the effect of eliminating silica on the remaining proportions. In other words, the behavior of the random vector $\boldsymbol{P}_{-1} = \left\{\frac{P_2}{1-P_1}, \ldots, \frac{P_k}{1-P_1}\right\}$, where the P_is ≥ 0, $i = 1, \ldots, k$, and $\sum_{i=1}^{k} P_i = 1$.

One possibility could be the assumption that the vector $\boldsymbol{P}_{-1}$ is independent of P_1. This means that P_1 does not influence (or is *neutral* to) the manner in which the remaining proportions $P_2, \ldots, P_k$ divide the remainder of the interval $(P_1, 1)$. This also means that P_1 is independent of the vectors $\left\{\frac{P_3}{1-P_1-P_2}, \ldots, \frac{P_k}{1-P_1-P_2}\right\}, \left\{\frac{P_4}{1-P_1-P_2-P_3}, \ldots, \frac{P_k}{1-P_1-P_2-P_3}\right\}$, and so on. When such is the case P_1 is said to be **neutral**.

The notion of a neutral P_1 when extended to all the k proportions yields a **completely neutral** vector $(P_1, \ldots, P_k)$. The notion of a completely neutral vector, also introduced by Connor and

Mossiman (1969), provides a mechanism for randomly dividing an interval. To see how, we start by first generating a random variable P_1 from some distribution with support (0,1). We then generate P_2, independently of P_1, but with some distribution having support $(1 - P_1, 1)$, and then a third random variable P_3, independently of the vector (P_1, P_2), but over the interval $(1 - P_1 - P_2, 1)$, and so on. The vector $(P_1, \ldots, P_k)$ so generated will be completely neutral.

Completely neutral vectors exhibit some interesting properties, one of which leads us to a generalization of the Dirichlet distribution. In particular, $(P_1, \ldots, P_k)$ is a completely neutral random vector if and only if $(Z_1, \ldots, Z_k)$ are mutually independent, where $Z_1 = P_1, Z_i = P_i/(1 - P_1 - P_2 - \cdots - P_{i-1}), i = 1, \ldots, k$, and $Z_k = 1$ is degenerate (cf. Theorem 9.2 of Connor and Mossiman, 1969). Note that $0 \le Z_i \le 1$. To obtain a generalization of the Dirichlet, we assume that the Z_is, $i = 1, \ldots, k - 1$, which are independent, have a univariate beta distribution on (0,1) with parameters ν_{1i} and ν_{2i}; that is, each Z_i is (ν_{1i}, ν_{2i}), $i = 1, \ldots, k - 1$. Then, it can be shown (Connor and Mossiman, 1969, or Lochner, 1975) that the joint distribution of $(P_1, \ldots, P_{k-1})$ has density at $(p_1, \ldots, p_{k-1})$ of the form

$$\prod_{i=1}^{k-1} \frac{\Gamma(\nu_{1i} + \nu_{2i})}{\Gamma(\nu_{1i})\Gamma(\nu_{2i})} p_i^{\nu_{1i}-1}(1 - p_1 - p_2 - \cdots - p_{k-1})^{\gamma_i}, \tag{9.5}$$

where $\gamma_i = \nu_{2i} - \nu_{1(i+1)} - \nu_{2(i+1)}$, for $i = 1, 2, \ldots, k - 2$ and $\gamma_{k-1} = \nu_{2(k-1)} - 1$.

Equation (9.5) is a generalization of the Dirichlet distribution – a generalization because setting $\gamma_1 = \gamma_2 = \cdots = \gamma_{k-2} = 0$ gives us a Dirichlet distribution. A consequence is that a vector of proportions following the standard Dirichlet distribution of (9.2) is completely neutral. Our motivation for discussing this generalization of the Dirichlet distribution is prompted by two considerations: one is to introduce the notion of neutrality, and the other is that Lochner (1975) has proposed a use of this distribution for a Bayesian analysis of life-test data (section 9.4). The notion of neutrality, which was extended by Doksum (1974) for stochastic processes, plays a role in the context of priors for the cumulative hazard rate functions and also for the construction of priors on the cumulative distribution functions (sections 9.5 and 9.6).

9.3 A NON-PARAMETRIC BAYES APPROACH TO BIOASSAY

A parametric Bayes approach to the bioassay problem based on the Weibull survival function as a model for the response curve was presented in section 5.8, under the label of dose–response testing or damage assessment studies. The focus there was on a use of filtering techniques with the double lognormal as the underlying distribution. In this section, I describe a non-parametric Bayes approach pioneered by Kraft and Van Eeden (1964), and developed by Ramsey (1972).

Adhering to the notation of section 5.8, we let X denote the level of a dose, and suppose that X takes values x, $0 \le x \le \infty$. Let $Y(x)$ denote the response of a subject to a dose x, with $Y(x) = 1$, if the subject survives a dose x, and $Y(x) = 0$, otherwise (i.e. the subject responds to dose x). It makes sense to suppose that $Y(x)$ is non-increasing in x, with $Y(0) = 1$ and $Y(\infty) = 0$. It is common to administer a dose x_i to n_i subjects, $i = 1, \ldots, m$, where n_i, m and x_i are pre-chosen. The $n_i \ge 1$ subjects experiencing dose x_i are assumed to be exchangeable. Let $P(x) = P(Y(x) = 0)$; i.e. $P(x)$ is the probability that a subject responds to the dose x. We suppose that $P(x)$, the response probability function, or the potency curve, increases in x so that $P(x)$ behaves like a distribution function with $P(0) = 0$ and $P(\infty) = 1$. Recall that in section 5.8.1 a prior probabilistic structure was imposed on $Y(x)$ via the model of (5.69). By contrast, in this section we endow a probabilistic structure on $P(x)$ via an ordered Dirichlet distribution. But before doing so we first write out a likelihood for the $P(x_i)$s given the pair (n_i, s_i), $i = 1, \ldots, m$, where s_i denotes the total number of subjects that do produce a response out of the n_i subjects that were administered the dose x_i. That is, s_i is the total number of $Y(x_i)$s that take the value 0. Clearly $\mathcal{L}$, the likelihood of $P(x_1), \ldots, P(x_m)$ with (n_i, s_i), $i = 1, \ldots, m$ fixed and known, is

$$\mathcal{L} \propto \prod_{i=1}^{m} (P(x_i))^{s_i} (1 - P(x_i))^{n_i - s_i}; \tag{9.6}$$

we assume that given the $P(x_i)$s the responses $Y(x_i)$ are independent.

Our aim is to assess the function $P(x)$, $x \geq 0$, based on prior information about $P(x)$ and the data (n_i, s_i), $i = 1, \ldots, m$. The reasons for doing so could be many, one of which is to know $P(x^*)$ – the probability of a response at some $x^* \neq x_i$, $i = 1, \ldots, m$.

9.3.1 A Prior for Potency

Ramsey (1972) has proposed a version of the ordered Dirichlet as a prior for $P(x_1), \ldots, P(x_m)$, and has given a constructive mechanism that lead to this choice. His prior boils down to the supposition that the differences in potency at the observational doses have a Dirichlet distribution whose density, say π, is of the form

$$\pi \propto \left[\prod_{i=1}^{m+1} \{P(x_i) - P(x_{i-1})\}^{\alpha_i} \right]^{\beta}, \tag{9.7}$$

with $P(x_0) = 0$ and $P(x_{m+1}) = 1$. The constants β and α_i, $i = 1, \ldots, m + 1$ are non-negative and $\sum_{i=1}^{m+1} \alpha_i = 1$.

The density π given above can be seen to be identical to that described by (9.4), provided that when averaging with π, we integrate with respect to

$$\frac{\prod_{i=1}^{m} \mathrm{d}P(x_i)}{\prod_{i=1}^{m+1} \{P(x_i) - P(x_{i-1})\}},$$

instead of $\prod_{i=1}^{m} \mathrm{d}P(x_i)$, which is what we would do were we to integrate with respect to (9.4). Thus Ramsey's (1972) choice for a prior on $P(x_i)$s, $i = 1, \ldots, m$, is said to be a version of the ordered Dirichlet. To motivate this choice Ramsey (1972) gives the construction described below; but first I need the following.

Some Preliminaries

For $i = 1, \ldots, m$, let $A_i = \sum_{j=1}^{i} \alpha_j$, and $a_i = 1 - A_i = \sum_{j=i+1}^{m+1} \alpha_j$; recall that α_j's are non-negative constants that sum to one, for $j = 1, \ldots, m + 1$. Consider a random variable Y whose density at y is the following version of the beta density on (0,1), with parameters $\beta\nu_1$ and $\beta\nu_2$:

$$\frac{\Gamma(\beta(\nu_1 + \nu_2))}{\Gamma(\beta\nu_1)\Gamma(\beta\nu_2)} y^{\beta\nu_1} (1 - y)^{\beta\nu_2}, \quad 0 < y < 1, \tag{9.8}$$

with $\nu_1, \nu_2, \beta \geq 0$. When averaging with respect to (9.8) we integrate with respect to $\mathrm{d}y/[y(1 - y)]$. I denote this density by Beta $(\nu_1, \nu_2, \beta; 0, 1)$.

If $Z = a + (b - a)Y$, and Y is as defined by (9.8), then Z will have a translated beta distribution over (a, b) denoted Beta $(\nu_1, \nu_2, \beta; a, b)$.

With the above preliminaries in place, we start by supposing that the distribution of $P(x_1)$ is Beta$(\alpha_1, a_1, \beta; 0, 1)$, and continue by assuming that for $i > 1$, the distribution of $P(x_i)$ given $P(x_1), \ldots, P(x_{i-1})$ depends only on $P(x_{i-1})$, and is Beta$(\alpha_i, a_i, \beta; P(x_{i-1}), 1)$. That is, the $P(x_i)$s, $i > 1$ have a Markov character, with the distribution of $P(x_i)$ being a translated beta over $(P(x_{i-1}), 1)$. A consequence of the above construction is that the potencies $P(x_1), \ldots, P(x_m)$ have a joint distribution, this joint distribution being identical to that which would result from π, the assumed Dirichlet of the differences. We denote this joint distribution by $\mathcal{D}_m(\alpha_1, \ldots, \alpha_{m+1}; \beta)$. This joint distribution has the following three attractive properties:

(*a*) If $(P(x_1), \ldots, P(x_m))$ has the distribution $\mathcal{D}_m(\alpha_1, \ldots, \alpha_{m+1}; \beta)$ then $(P(x_1), \ldots, P(x_{m-1}))$ will have the distribution $\mathcal{D}_{m-1}(\alpha'_1, \ldots, \alpha'_m; \beta)$ where $\alpha'_i = \alpha_i$ for $i = 1, \ldots, m-1$ and $\alpha'_m = \alpha_m + \alpha_{m+1}$.

(*b*) Under (*a*) above, the distribution of $P(x_i)$ for any $i, i = 1, \ldots m$, is $\beta(A_i, 1 - A_i, \beta; 0, 1)$; the parameter A_i is both the mean and the mode. Also, $(A_1, \ldots, A_m)$ is the mode of the joint density π.

Thus if $P^*(x)$ is our best guess of $P(x)$, then our choices for the α_i's would be: $\alpha_1 = P^*(x_1), \alpha_i = P^*(x_i) - P^*(x_{i-1}), i = 2, \ldots, m$, and $\alpha_{m+1} = 1 - P^*(x_m)$.

(*c*) If $P(x_{i-1}) = p_1$ and $P(x_{i+1}) = p_2$, then given p_1 and p_2, the distribution of $P(x_i)$ is Beta$(\alpha_i, \alpha_{i+1}, \beta; p_1, p_2)$.

A consequence of the above is that as $\beta \downarrow 0$, the beta conditional density of $P(x_i)$ becomes uniform over (p_1, p_2), and it becomes concentrated within this interval around a point determined by α_i and α_{i+1}, as β increases. In other words, β controls the probability that $P(x_i)$ is close to $P(x_{i-1})$ or $P(x_{i+1})$, and in so doing controls the smoothness in the prior/posterior process. Alternatively viewed, β controls an assessor's strength of belief about the prior guess $P^*(x)$.

9.3.2 The Posterior Potency

The joint density of the posterior is given by the product of the likelihood (equation (9.6)) and the prior (equation (9.7)). A Bayes estimate of the potency curve is the mode of the joint posterior. That is, the k-dimensional point $(\widehat{P}(x_1), \ldots, \widehat{P}(x_m))$ which maximizes the posterior density of $(P(x_1), \ldots, P(x_m))$. For values of $\beta \in (0, \infty)$, Ramsey (1972) has argued that the posterior density is convex and unimodal, so that a unique modal function exists. Also outlined by Ramsey (1972) are numerical and optimization-based approaches for obtaining the modal function. The details of these are not pursued here, because the main point of this section is to show how the Dirichlet and the ordered Dirichlet enter the arena of non-parametric Bayes in the context of reliability and survival analysis via the scenarios of bioassay and damage assessment. The use of the Dirichlet as a prior distribution was further developed and extended by Ferguson (1973) who formally introduced the Dirichlet process as a prior for the response function $P(x)$. This and related developments will be discussed in section 9.6.

Before closing this section it may be useful to add that the Dirichlet specification mentioned here does not capture features of the dose–response curve that may be known a priori. For example, it may be felt that $P(x)$ is either concave, convex or ogive (i.e. changes from convex to concave). Priors that capture such features were developed by Singpurwalla and Shaked (1990), and further extended by Ramgopal, Laud and Smith (1993). Smythe (2004) gives a recent overview.

9.4 PRIOR DISTRIBUTIONS ON THE HAZARD FUNCTION

To set the stage for the material of this section, as well as the following two sections, we start with some preliminaries.

Suppose that $P(T \geq t)$ the predictive survival function of T, the lifelength of an item, is $(1 - F(t))$; thus $F(t) = P(T < t)$ is the distribution function of T. Our aim is to assess $F(t)$ using prior information and life-history data of n items judged exchangeable with the item of interest. We start by partioning the time axis into $(k + 1)$ intervals $0 < t_1 < t_2 < \cdots < t_k < \infty$, and let $p_1 = F(t_1), p_i = F(t_i) - F(t_{i-1}), i = 2, \ldots, k$. Since T is a lifelength, $F(0) = 0$ and $F(\infty) = 1$. Let the vector $\mathbf{p} = (p_1, \ldots, p_k)$, and let n_i denote the number of observed lifetimes in the interval $[t_{i-1}, t_i)$, with n_1 denoting the number of observed lifetimes less than t_1 and n_{k+1}, the number $\geq t_k$; thus $n = \sum_{i=1}^{k+1} n_i$. Given $\mathbf{p}$, the vector $\mathbf{d} = (n_1, \ldots, n_{k+1})$ has a multinomial distribution

$$\frac{n!}{n_1! \cdots n_{k+1}!} p_1^{n_1} \cdots p_k^{n_k} (1 - p_1 - p_2 - \cdots - p_k)^{n_{k+1}}. \tag{9.9}$$

If we were to assign a prior distribution to the vector $\mathbf{p}$, then this prior and the above likelihood would yield the posterior of $\mathbf{p}$ from which I could assess $F(t)$.

The scenario here is not unlike that of section 9.3 on bioassay, since the behavior of $F(x)$ is identical to that of the response curve $P(x)$. The key difference was that in section 9.3 our data was quantal (response or no response), whereas here we look at the number of failures in different intervals. In section 9.3, a Dirichlet prior distribution was assigned to the differences in potency (equation (9.7)) and in principle, one could proceed here along similar lines using the multinomial of (9.9) as a likelihood instead of the Bernoulli. One could also justify a use of the Dirichlet prior following the constructive scheme of Ramsey (1972). However, the priors that we discuss in this section will be different. Different, because a disadvantage of the Dirichlet distribution is that its covariance dose not account for proximity, so that the relationship between p_i and p_j is not a function of how close i and j are to each other. Thus the priors that we consider here are the generalized Dirichlet of section 9.2.2, and a generalization of the gamma process (introduced in section 4.9.1). The generalized Dirichlet, motivated by Connor and Mossiman's (1969) construction of completely neutral vectors, is obtained by assuming that the piecewise constant segments of the hazard rate function have independent beta distributions. Thus it is the piecewise constant hazard rate function that is endowed with a generalized Dirichlet distribution; this is described in section 9.4.1. By contrast, the material of section 9.4.2 pertains to endowing a continuous non-decreasing hazard rate function with the generalized gamma as a prior.

9.4.1 Independent Beta Priors on Piecewise Constant Hazards

Let $Z_1 = p_1$ and $Z_i = p_i/(1 - p_1 - p_2 - \cdots - p_{i-1})$, for $i = 2, \ldots, k$. Then Z_i is the failure rate of $F(t)$ over the interval $[i - 1, i)$, and the concatenated collection of Z_is, $i = 1, \ldots, k$ constitute the piecewise constant hazard rate function of $F(t)$ over the interval $[0, t_k)$. The Z_is would be between 0 and 1. Since the hazard rate function cannot be observed, we treat it as a random function, or a stochastic process, and assign probabilities to its sample paths. One such strategy is to assume that the Z_is are independent, with Z_i having a beta distribution $\text{Beta}(\nu_{1i}, \nu_{2i}), i = 1, \ldots, k$. Then, the vector $\mathbf{p}$ will have as its prior distribution a generalized Dirichlet, with density at $(p_1, \ldots, p_k)$ of the form

$$\prod_{i=1}^{k} \frac{\Gamma(\nu_{1i} + \nu_{2i})}{\Gamma(\nu_{1i})\Gamma(\nu_{2i})} p_i^{\nu_{1i}-1} (1 - p_1 - p_2 - \cdots - p_k)^{\gamma_i}, \tag{9.10}$$

where $\gamma_i = \nu_{2i} - \nu_{1(i+1)} - \nu_{2(i+1)}$, for $i = 1, 2, \ldots, k - 1$ and $\gamma_k = \nu_{2k} - 1$; see equation (9.5).

Given the counts data **d**, the prior of (9.10) when combined with the likelihood of (9.9) yields the posterior distribution of **p**, with density (cf. Lochner, 1975)

$$\left[\prod_{i=1}^{k} \frac{\Gamma(\nu_{1i}+n_i+\nu_{2i}+n_{i+1}+\cdots+n_{k+1})}{\Gamma(\nu_{1i}+n_i)\Gamma(\nu_{2i}+n_{i+1}+\cdots+n_{k+1})} p_i^{\nu_{1i}+n_i-1}(1-p_1-\cdots-p_i)^{\gamma_i}\right] \cdot (1-p_1-p_2-\cdots-p_k)^{n_k+1}. \tag{9.11}$$

Properties of this posterior distribution and approaches for using it to obtain an assessment of $F(x)$ are given by Lochner (1975). A disadvantage in using the generalized Dirichlet as a prior for **p** is the excessive number of parameters that one needs to specify. Some strategies for simplifying this task are also given by Lochner (1975). However, a key drawback is that the generalized Dirichlet lacks the additive property that a Dirichlet has. That is if $\mathbf{p}=(p_1,\ldots,p_k)$ has the generalized Dirichlet distribution, then $(p_1, p_2+p_3, p_4,\ldots,p_k)$ will not be a generalized Dirichlet. This leads to an inconsistency with respect to inference. Specifically, our inferential conclusions will be dependent on whether we combine cells at the prior stage or at the posterior stage. Another disadvantage is that in using count data we sacrifice information that the actual failure times provide.

Because of the above obstacles, one is motivated to consider other families of stochastic processes as prior distributions for the hazard rate function. The need for seeking alternatives becomes particularly germane if one has additional information about the behavior of the hazard rate function, such as its monotonicity, that needs to be incorporated into the analyses. The assumption of complete neutrality upon which the generalized Dirichlet is based upon is unable to account for any structural patterns in the hazard rate function. Of the alternatives proposed, the ones by Padgett and Wei (1981) and by Arjas and Gasbarra (1994) take the prior on the hazard rate function to be a (non-decreasing) jump process, a Poisson process with a constant jump size, and the one by Mazzuchi and Singpurwalla (1985) takes the prior on a function of the hazard rate function to be an ordered Dirichlet. In all the above cases the hazard rate function is assumed to be non-decreasing (actually monotone in the case considered by Mazzuchi and Singpurwalla (1985)). Two other alternatives are the use of an extended gamma process, introduced by Dykstra and Laud (1981), and the Markov Beta and Markov Gamma processes considered by Nieto-Barajas and Walker (2002). In what follows, I overview the first of the above two developments; I focus on the extended gamma case because the extended gamma process possesses features that could be of a broader interest.

9.4.2 The Extended Gamma Process as a Prior

Suppose that $r(t)$, the hazard rate function of $F(t)$ is non-decreasing in $t \geq 0$. My aim here is to specify a prior probability distribution over the collection of all possible non-decreasing hazard rate functions. This is done by choosing an appropriate stochastic process whose sample paths are non-decreasing functions over $(0,\infty)$. Dykstra and Laud (1981) have introduced and proposed the extended gamma process for doing so. To define this process, we start by recalling (section 4.9.1) that a stochastic process $\{M(t); t \geq 0\}$ is a gamma process with a shape function $a(t)$ and a scale parameter 1, if for any $t \geq s \geq 0$ and $M(0)=0$, $M(t)$ has independent increments and $(M(t)-M(s))$ has a gamma distribution with a scale parameter 1 and a shape parameter $(a(t)-a(s))$. The function $a(t)$ is non-decreasing, left-continuous and real-valued, with $a(0)=0$. Ferguson (1973) shows that such a process exists, and that the process has non-decreasing, left-continuous sample paths.

Suppose now that $\beta(t), t \geq 0$ is a positive, right-continuous and real-valued function that is bounded away from zero, and which has left-hand limits. Then, the stochastic process $\{M^*(t); t \geq 0\}$ is an **extended gamma process** where $M^*(t)$ is defined as the integral

$$M^*(t) = \int_0^t \beta(s)\mathrm{d}M(s);$$

we denote such a process as $\Gamma(a(t), \beta(t))$.

Dykstra and Laud (1981) show that the extended gamma process $\Gamma(a(t), \beta(t))$ has independent increments with

$$E[M^*(t)] = \int_0^t \beta(s)\mathrm{d}a(s), \quad \text{and} \quad \mathrm{Var}[M^*(t)] = \int_0^t \beta^2(s)\mathrm{d}a(s).$$

Both the gamma process and the extended gamma process are **jump processes**; i.e. they increase only by taking jumps. Finally, in the spirit of the material of section 7.2, if the hazard rate process is taken to be an extended gamma process $\Gamma(a(t), \beta(t))$, then the survival function is (cf. Theorem 3.1 of Dykstra and Laud (1981)) of the form

$$P(T \geq t) = \exp\left[-\int_0^t \log(1 + \beta(s)(t - s))\mathrm{d}a(s)\right]; \tag{9.12}$$

this equation complements the likes of (7.13).

Equation (9.12), besides being useful as a modeling tool, also facilitates one to obtain the joint survival probability of n observations $T_1, \ldots, T_n$ from $F(t)$. Specifically,

$$P(T_1 \geq t_1, \ldots, T_n \geq t_n) = \exp\left[-\int_0^\infty \log\left(1 + \beta(s)\sum_{i=1}^n (s - t_i)^*\right)\mathrm{d}a(s)\right], \tag{9.13}$$

where $a^* = \max(a, 0)$.

The above enables us to write out the likelihood of n observations truncated to the right at $t_1, \ldots, t_n$. This likelihood, when used in conjunction with $\Gamma(a(s), \beta(s))$ as a prior on the hazard rate process yields $\Gamma(a(s), \widehat{\beta}(s))$ as an extended gamma process for the posterior of the hazard rate process. Here

$$\widehat{\beta}(s) = \frac{\beta(s)}{1 + \beta(s)\sum_{i=1}^n (t_i - s)^*}.$$

The posterior of the hazard rate process when $t_1, \ldots, t_n$ are the actual times to failure is a mixture of extended gamma processes whose form is cumbersome to write out (Theorem 3.3 of Dykstra and Laud (1981)). However, Laud, Smith and Damien (1996) approximate the posterior hazard rate process by approximating its random independent increments via a Monté Carlo method involving the Gibbs sampler. Once the posterior hazard rate process is obtained by approximation, one can obtain an approximation for the survival function. A use of the Gibbs sampler for estimating the posterior hazard rate process has also been considered by Arjas and Gasbarra (1994), who, we recall, assume a jump process structure as a prior for the hazard rate process.

9.5 PRIOR DISTRIBUTIONS FOR THE CUMULATIVE HAZARD FUNCTION

Section 9.4 was devoted to a consideration of prior distributions for the hazard rate function, assumed to be a piecewise constant function in section 9.4.1, and continuous and non-decreasing in section 9.4.2. In the first case, the prior was assumed to be a collection of independent beta distributions, and in the second case an extended gamma process. The generalized Dirichlet on the increments of $F(t)$ was a consequence of the independent beta assumption. In this section, we consider prior distributions on the cumulative hazard rate function $R(t) = \int_0^t r(u)\mathrm{d}u$, for $t \geq 0$. A consideration of priors for the cumulative hazard is attractive because the cumulative hazard can be defined even when F has no density, as $R(t) = \int_0^t \mathrm{d}F(u)/(1 - F(u))$. Here again, since $R(t), t \geq 0$, cannot be observed, we treat it as a random function (or a stochastic process) and assign probabilities to its sample paths. I describe here two constructions, one due to Kalbfleisch (1978) involving the gamma process, and the other by Hjort (1990), who introduces a new process, which he calls the beta process. Kalbfleisch's (1978) construction entails the assumption that the piecewise constant segments of the hazard rate function have independent gamma distributions (section 9.5.1). In this sense it parallels the construction of Lochner (1975) discussed in section 9.4.1. By contrast, Hjort's (1990) beta process construction, which will be discussed in section 9.5.2, proceeds by discretizing the time axis and assuming that the resulting point mass hazard rates have independent beta distributions.

9.5.1 Neutral to the Right Probabilities and Gamma Process Priors

As a preamble to our discussion of a gamma process prior for the cumulative hazard functions, I need to introduce the notion of a neutral to the right-random probability. To do so, I follow the notation and set-up laid out at the start of section 9.4, by partitioning the time axis into a finite number k of disjoint intervals $[0 \equiv t_0, t_1), [t_1, t_2), \ldots, [t_{i-1}, t_i), \ldots, [t_{k-1}, t_k = \infty)$, and letting $Z_i = P(T \in [t_{i-1}, t_i) | T \geq t_{i-1}), i = 1, \ldots, k$, be the hazard rate of $F(t)$ over the i-th interval. Thus the concatenated collection of Z_is constitutes the piecewise constant hazard rate function of $F(t)$ over $[0, t_k)$.

Since $1 - F(t) = \exp(-R(t))$, it can be seen that

$$R(t_i) = \sum_{j=1}^{i} -\log(1 - Z_j) \stackrel{\text{def}}{=} \sum_{j=1}^{i} r_j, \tag{9.14}$$

where $r_i = -\log(1 - Z_i), i = 1, \ldots, k$ and $R(t_0) = 0$.

If for every partition of the type $[t_{i-1}, t_i), i = 1, \ldots, k$, a joint distribution for the collective $(Z_1, \ldots, Z_k)$, or equivalently the collective $(r_1, \ldots, r_k)$, can be specified, subject to consistency conditions, then a probability distribution is also specified for the set of all cumulative hazard functions $\{R(t)\}$, where $R(t)$ is non-decreasing in $t \geq 0$. When the Z_is (or equivalently the r_i's) are judged independent, then $\{R(t); t \geq 0\}$ is a non-decreasing stochastic process with independent increments. Under these circumstances, the random distribution function F and its probability law will have the property of being neutral to the right. The notion of neutral to the right random probabilities was introduced by Doksum (1974); it can be seen as an extension of Connor and Mossiman's (1969) notion of complete neutrality of random probability vectors, to stochastic processes. When F is neutral to the right and the process $\{R(t); t \geq 0\}$ such that $R(t)$ has no non-random part (i.e. the α of (7.9) and the $\alpha(t)$ of (7.17) are zero), then F is necessarily a discrete distribution function (Doksum, 1974, Corollary 3.2). Futhermore, if the independent increments process $\{R(t); t \geq 0\}$ has stationary increments (i.e. when the Z_is or the r_i's are independent and identically distributed), then the process $\{R(t); t \geq 0\}$ is referred to by Kingman (1975) as a **subordinator**.

On Neutral to the Right (Left) Probabilities

Neutral to the right-random probabilities are of interest in Bayesian inference because of their closure properties. Specifically, the posterior distribution of a random probability neutral to the right, is also neutral to the right (Doksum, 1974, Theorem 4.2). A consequence is that is if the cumulative hazard rate function is, a priori, a non-decreasing function with independent increments, then it will also be, a posteriori, a non-decreasing function with independent increments. This feature is made more explicit via the example of a gamma process that will be discussed in the sub-section that follows; but first, a definition of neutral to the right distribution.

We say that a random distribution function F is **neutral to the right** if its normalized increments are independent. That is, for $t_1 < t_2 < \cdots < t_k$,

$$F(t_1), \frac{F(t_2) - F(t_1)}{1 - F(t_1)}, \ldots, \frac{F(t_k) - F(t_{k-1})}{1 - F(t_{k-1})},$$

are independent. Observe that the above definition parallels that of a completely neutral vector $(P_1, \ldots, P_k)$ (section 9.2.2) If I was to set $P_1 = F(t_1)$, $P_i = F(t_i) - F(t_{i-1})$, and $Z_i = P_i/(1 - P_1 - P_2 - \ldots - P_{i-1})$, for $i = 2, \ldots, k$. Thus the claim that the notion of neutral to the right-random probabilities is an extension of the notion of completely neutral vectors to the case of processes. Doksum (1974) has also introduced the notion of **neutral to the left**. Here, for any $t_1 < t_2 < \cdots < t_k$, the proportional increments

$$F(t_k), \frac{F(t_k) - F(t_{k-1})}{F(t_k)}, \ldots, \frac{F(t_2) - F(t_1)}{F(t_2)},$$

are independent.

With the above as a background on neutral to the right (and left) probabilities, we are now ready to introduce the gamma process prior for the cumulative hazard function as an example of a neutral to the right probability. Additional material on neutral to the right processes is given later in section 9.6.2.

Gamma Process Priors for the Cumulative Hazard

Reverting to the scenario leading to (9.14), Kalbfleisch (1978) has proposed a gamma process as a prior for $\{R(t); t \geq 0\}$ by assuming that the r_i's are independent with r_i having a gamma distribution with shape $(\alpha_i - \alpha_{i-1})$ and scale c, where $\alpha_i = cR^*(t_i)$, $i = 1, \ldots, k$. The function $R^*(t), t \geq 0$, is interpreted as one's best guess of $R(t), t \geq 0$; equivalently, one may interpret $\exp(-R^*(t))$ as one's best guess about $1 - F(t)$. The constant c is a measure of one's strength of conviction (or credibility) about the guess $R^*(t)$. Large values of c indicate a strong conviction. The assignment of independent gammas to the r_i's is akin to Lochner's (1975) choice of independent beta distributions for the Z_is.

The structure of the gamma distribution given above satisfies the requirement of consistency, in the sense that the distribution assigned to $r_i + r_{i+1}$ for the concatenated interval $[t_{i-1}, t_{i+1})$ will be the same as that deduced from the gamma distributions assumed for r_i and r_{i+1}. A consideration of the partition $[0, t)$, $[t, \infty)$ shows that the aforementioned construction results in a gamma process as a prior for $R(t)$; its shape function is $cR^*(t)$ and scale c. Consequently, for any $t \geq 0$, $E(R(t)) = R^*(t)$ and $\text{Var}(R(t)) = R^*(t)/c$.

Since the gamma process for $\{R(t); t \geq 0\}$ has independent increments, the random probability $F(t)$ that it induces is neutral to the right. Consequently, given n failure times, say $\tau_1, \ldots, \tau_n$, the posterior cumulative hazard function will also be a process with independent increments. Kalbfleisch (1978) has shown that the increment at τ_i of the posterior cumulative hazard rate function has a distribution with density $A(c+A_i, c+A_{i+1})$ at u, where $A_i = n-i, i=1, \ldots, n$, and $A(a, b)$ is of the form

$$\frac{u^{-1}\{\exp(-bu) - \exp(au)\}}{\log(a/b)}.$$

Between τ_{i-1} and τ_i, the increments are prescribed by a gamma process with shape function $cR^*(\bullet)$ and scale $c+A_i$. The survival function $1 - F(t)$ can be recovered from the cumulative hazard rate process either via simulation or by approximation using the expected value of the process. However, one can foresee the following concerns about using the gamma process prior on the cumulative hazard function.

The first concern pertains to motivation. In particular, assuming a gamma distribution on $r_i = -\log(1-Z_i)$, where Z_i is the failure rate of $F(t)$ over the interval $[t_{i-1}, t_i)$ lacks intuition. The second concern could stem from the fact that the independent increments property of $\{R(t); t \geq 0\}$ may not be meaningful. Under aging and wear, the successive Z_is would be judged to be increasing and thus interdependent. By contrast, the extended gamma process of section 9.4.2 on the hazard rate function would lead to a process for the cumulative hazard function that will not have independent increments and would thus be more desirable. Finally, since $F(t)$ being neutral to the right would result in the feature that any realization of the process $\{R(t); t \geq 0\}$ will be discrete, difficulties will be encountered when ties are present in the failure time data, and such ties will occur with a non-zero probability. The nature of such difficulties and how they can be overcome is described by Kalbfleisch (1978) in section 5.5 of his paper.

9.5.2 Beta Process Priors for the Cumulative Hazard

Hjort (1990) has introduced a new class of stochastic processes for the analysis of life-history data; he calls such processes 'beta processes'. These processes, which constitute a large and flexible family, are also of value for inference connected with time inhomogeneous Markov chains. When used as priors on the cumulative hazard function they exhibit the property of conjugacy and enable one to obtain easy-to-interpret and easy-to-compute Bayes estimators of the cumulative hazard. Hjort's (1990) development proceeds by starting with the discrete time scenario and then taking appropriate limits. In doing so, Hjort (1990) also develops a discrete time framework for the analysis of failure data.

By way of preliminaries and notation, suppose that a lifetime T is discrete, taking values $0, b, 2b, \ldots,$ for some $b > 0$. Later on, we let b go to zero so that the discrete time results can be extended to the continuous case. For $j = 0, 1, \ldots,$ let $p(jb) = P(T = jb)$, $F(jb) = P(T \leq jb) = \sum_{i=0}^{j} p(ib)$, $r(jb) = P(T = jb | T \geq jb) = p(jb)/F[jb, \infty)$, where $F[jb, \infty) = 1 - F((j-1)b)$, and $R(jb) = \sum_{i=0}^{j} r(ib)$. Note that $r(jb)$ and $R(jb)$ are the point mass hazard rate and the point mass cumulative hazard rate of F at jb, respectively. Also, (cf. (4.3) and (4.1)), F and p can be recovered from a knowledge of r via the relationships:

$$F(jb) = 1 - \prod_{i=0}^{j}(1 - r(ib)) \quad \text{and} \quad p(jb) = r(jb)\left[\prod_{i=0}^{j-1}(1 - r(ib))\right]. \tag{9.15}$$

The aim here is to construct a class of priors for the cumulative hazard function R such that given failure time data $T_1, \ldots, T_n$ (with or without censoring) one can conveniently obtain the posterior distribution of R, and thence a non-parametric Bayes estimate of F. Hjort (1990) motivates his class of priors by looking at the likelihood of the data in terms of the point mass hazard function. To do so, he considers the observables $(X_i, \delta_i), i = 1, \ldots, n$, where $X_i = \min(T_i, c_i)$, c_i being the right censoring time of T_i, and $\delta_i = 1$ if $T_i \leq c_i$; $\delta_i = 0$ if $T_i > c_i$. Thus

$\delta_i = 1$ corresponds to a failure at or before c_i, the convention here being that at c_i failure takes precedence over censorship. The set-up described above gives rise to two counting processes, $\{N(jb); j = 0, 1, 2, \ldots\}$ and $\{Y(jb); j = 0, 1, 2, \ldots\}$, where

$$N(jb) = \sum_{i=1}^{n} \mathcal{I}(X_i \leq jb \text{ and } \delta_i = 1),$$

$$Y(jb) = \sum_{i=1}^{n} \mathcal{I}(X_i \geq jb);$$

$\mathcal{I}(a)$, the **identity function**, is such that $\mathcal{I}(a) = 1(0)$ if the event a occurs (does not occur). The process $\{N(jb); j = 0, 1, \ldots\}$ counts the number of failures at or before time jb, and the process $\{Y(jb); j = 0, 1, \ldots\}$ counts the number of items at risk of failure at time jb. The increment in the $\{N(jb); j = 0, 1, \ldots\}$ process at time jb, namely $N(jb) - N((j-1)b)$, will be denoted as $\mathrm{d}N(jb)$.

Then it is easy to verify that given the observables x_i and δ_i, $i = 1, \ldots, n$, the likelihood of p (or F) is

$$\left[\prod_{i:\delta_i=1} p(x_i)\right]\left[\prod_{i:\delta_i=0} F(x_i, \infty)\right], \tag{9.16}$$

where $F(x_i, \infty) = P(T > x_i)$.

Invoking the relationship of (9.15), and then simplifying the resulting algebra, the likelihood becomes:

$$\prod_{i=1}^{n}\prod_{j=0}^{\infty}\{1 - r(jb)\}^{\mathcal{I}(j<x_i \text{ or } (j=x_i \text{ and } \delta_i=0))}\{r(jb)\}^{(j=x_i \text{ and } \delta_i=1)}$$

$$= \prod_{j=0}^{\infty}\left[\{1 - r(jb)\}^{Y(jb)-\mathrm{d}N(jb)}\{r(jb)\}^{\mathrm{d}N(jb)}\right]. \tag{9.17}$$

Note that we now have the likelihood of r with the data (x_i, δ_i), $i = 1, \ldots, n$, as being fixed.

Suppose now that the point mass hazard rates $r(jb)$, $j = 0, 1, \ldots,$ are assumed to be independent with $r(jb)$ having a prior density $h_{jb}(s)$, for $0 \leq s \leq 1$. Then given the data (x_i, δ_i), $i = 1, \ldots, n$, the $r(jb)$'s will continue to be independent with $r(jb)$ having a posterior density $h^*_{jb}(s)$ where

$$h^*_{jb}(s) \propto s^{\mathrm{d}N(jb)}(1 - s)^{Y(jb)-\mathrm{d}N(jb)} h_{jb}(s). \tag{9.18}$$

This result is a consequence of (9.17).

With the above in place, let

$$R \sim \mathrm{Beta}(c, R_o) \tag{9.19}$$

denote the feature that the cumulative hazard function when viewed as a stochastic process $\{R(jb); j = 0, 1, \ldots\}$ has independent increments $r(jb)$ where, following the notation of section 9.2,

$$r(jb) \sim \mathrm{Beta}((c(jb)r_0(jb), c(jb))(1 - r_0(jb))); \tag{9.20}$$

i.e. $r(jb)$ has a beta density with the parameters $c(jb)$ and $r_0(jb)$.

Clearly, since $E[r(jb)] = r_0(jb) = \mathrm{d}R_0(jb)$, I have $E[R(jb)] = R_0(jb)$; similarly, $\mathrm{Var}[r(jb)] = r_0(jb)(1 - r_0(jb))/(c(jb) + 1)$. Consequently, given data, the posterior cumulative hazard process has the feature [cf. Equation (9.18)] that

$$(R|\text{data}) \sim \text{Beta}\left\{c + Y, \sum \frac{c\mathrm{d}R_0 + \mathrm{d}N}{c + Y}\right\}, \tag{9.21}$$

in notation that should be obvious. Thus a conjugacy of the process of (9.19). The mean of the posterior cumulative hazard process, say $\widehat{R}(jb) = E[R(jb)|\text{data}]$, is of the form

$$\widehat{R}(jb) = \sum_{i=0}^{j} \frac{c(ib)r_0(ib) + \mathrm{d}N(ib)}{c(ib) + Y(ib)}; \tag{9.22}$$

which in compact notation can be written as

$$\widehat{R}(jb) = \sum_{0}^{jb} \frac{c\mathrm{d}R_0 + \mathrm{d}N}{c + Y};$$

its variance is $\sum_0^{jb} \mathrm{d}\widehat{R}(1 - \mathrm{d}\widehat{R})/(c + Y + 1)$.

The quantity $\widehat{R}(jb)$ can be seen as a non-parametric Bayes estimate of the cumulative hazard function.

Relationship to the Kaplan–Meier and Nelson–Aalen Estimators

The parameters $c(jb), j = 0, 1, \ldots,$ can be interpreted as those reflecting a *strength of belief* in the prior guess R_0. This is because, in (9.22), the weight given to $r_0(ib)$ decreases in $c(ib)$, to the extent that as $c(ib)$ tends to zero, $\widehat{R}(jb) = \sum_{i=0}^{j} \mathrm{d}N(ib)/Y(ib)$ – the 'Nelson–Aalen' (non-Bayesian) estimator of $R(jb)$. As $c(ib)$ grows large, $\widehat{R}(jb)$ becomes $R_0(ib)$, the prior guess.

An easy spin-off of the above is the development of a non-parametric Bayes estimator of the distribution function of F. A consequence of (9.15) and (9.22) is that

$$E[F(jb)|\text{ data}] \stackrel{\text{def}}{=} \widehat{F}(jb) = 1 - \prod_{0}^{j}(1 - E[r(ib)|\text{ data}]) = 1 - \prod_{0}^{jb}\left[1 - \frac{c\mathrm{d}R_0 + \mathrm{d}N}{c + Y}\right], \tag{9.23}$$

which when $c(\bullet)$ tends to zero becomes the 'Kaplan–Meier' non-parametric Bayes estimator of F, namely

$$F^*(jb) = 1 - \prod_{0}^{jb}\left[1 - \frac{\mathrm{d}N}{Y}\right].$$

When $c(\bullet)$ increases to infinity, the non-parametric Bayes estimator of $F(jb)$ is simply $1 - \prod_{i=0}^{j}(1 - r_0(ib))$, the prior guess.

Thus to conclude, large values of $c(\bullet)$ imply a strong belief about the prior guess $r_0(\bullet)$, and vice-versa for small values of $c(\bullet)$.

Extension to the Continuous Case: Beta Processes

Suppose now that T can take values in $[0, \infty)$, with $F(t) = P(T \leq t)$, and $F(0) = 0$. We will allow $F(\bullet)$ to have jumps, the effect of which is that the well-known exponentiation formula of reliability and survival analysis (equation (4.9)) will not hold. The cumulative hazard function of F is a non-negative, non-decreasing and right-continuous function F on $[0, \infty)$, which by analogy with the case of T discrete should satisfy the requirement that $\mathrm{d}R(s) = \mathrm{d}F(s)/F[s, \infty)$. Consequently, we define

$$R[a, b) = \int_{[a,b)} \mathrm{d}F(s)/F[s, \infty),$$

as the cumulative hazard rate of F. With $R[a, b)$ as defined above, it is immediate that

$$F[a, b) = \int_{[a,b)} F[s, \infty)\, \mathrm{d}R(s), \quad 0 \leq a \leq b < \infty.$$

A solution of the above equation is the *product integral formula* (section 4.3.1)

$$F(t) = 1 - \prod_{[0,t]} \{1 - \mathrm{d}R(s)\}, \quad t \geq 0; \tag{9.24}$$

thus F is uniquely determined by R.

The quantity $\prod_{[a,b]}\{1 - \mathrm{d}R(s)\} = \exp(-R[a, b])$, if and only if R is continuous, and thus $-\log(1 - F) = R$, only under the assumption of the said continuity. In general, since $1 - x \leq \mathrm{e}^{-x}$, we may claim that $-\log(1 - F(t)) \geq R(t)$.

With the above as background we turn to the construction of prior distributions for R, but require that the class of all cumulative hazard rates be such that for any R in this class, (9.24) would lead to an F on $[0, \infty)$, with $F(0) = 0$. I denote this class by $\mathcal{A}$. Since the class of priors for $R(jb)$ in the discrete case has paths that lie in $\mathcal{A}$, and whose independent increments are governed by (9.20), we are motivated to consider a process on $[0, \infty)$ with paths in $\mathcal{A}$ whose infinitesimal independent increments are governed by

$$\mathrm{d}R(s) \sim \mathrm{Beta}((c(s)\mathrm{d}R_0(s), c(s))(1 - \mathrm{d}R_0(s))). \tag{9.25}$$

The existence of such a process has been established by Hjort (1990) in his Theorem 3.1 which goes as follows:

Theorem 9.1 (Hjort (1990)). *Let R_0 in $\mathcal{A}$ be continuous and let c_0 be a piecewise continuous non-negative function. Then there exists a Lévy process $R(\bullet)$ whose paths fall in $\mathcal{A}$ and whose Lévy representation*

$$E[\exp -\theta R(t)] = \exp\left[-\int_0^1 (1 - \mathrm{e}^{-\theta s})\mathrm{d}L_t(s)\right],$$

has Lévy measure $\mathrm{d}L_t(s)$ of the form

$$\mathrm{d}L_t(s) = \left[\int_0^t c(z)s^{-1}(1 - s)^{c(z)-1}\mathrm{d}R_0(z)\right]\mathrm{d}s,\ t \geq 0, 0 < s < 1.$$

The process R given above is called a **beta process** with parameters $c(\bullet)$ and $R_0(\bullet)$, and like (9.19) for the discrete case denoted $R \sim \mathrm{Beta}(c, R_0)$.

The beta process defined above conforms to the requirement of (9.25) in a certain sense and preserves some of the beta characteristics. Specifically, $E\{R(a, b]\} = R_0(a, b]$, making R_0 the prior guess, and (cf. Hjort, 1990, p. 1272)

$$\begin{aligned}\mathrm{Var}\{R(a, b]\} &= \int_{(a,b]} \mathrm{d}R_0(s)(1 - \mathrm{d}R_0(s))/(c(s) + 1) \\ &= \int_{(a,b]} \mathrm{d}R_0(s)/(c(s) + 1).\end{aligned}$$

An extension of the beta process defined above consists in allowing R_0 to have a finite number of jumps. This is necessitated by the fact that when considering life-history data, the posterior distribution of R will have jumps at times of observation. The specifics of this extension are at the end of section 3.3 of Hjort (1990); they will not be stated here because the results to be given below are not affected by the extension.

With the beta process as a prior for R, the posterior of R given the life-history data $(X_1, \delta_1), \ldots, (X_n, \delta_n)$ is also a beta process of the form

$$(R|\text{data}) \sim \text{Beta}\left[c(\bullet) + Y(\bullet), \int_0^{(\bullet)} \frac{c\,\mathrm{d}R_0 + \mathrm{d}N}{c + Y}\right];$$

see Corollary 4.1 of Hjort (1990).

The Bayes estimator of $R(t)$ is the mean of the above posterior, and is given as:

$$\widehat{R}(t) = \int_0^t \frac{c\,\mathrm{d}R_0 + \mathrm{d}N}{c + Y}, \tag{9.26}$$

and its variance is $\int_0^t \frac{\mathrm{d}\widehat{R}(1-\mathrm{d}\widehat{R})}{c+Y+1}$; (9.26) is the continuous time analogue of the expression following (9.22). Finally, as a parallel to (9.23), a non-parametric Bayes estimator of F is:

$$\widehat{F}(t) = 1 - \prod_{[0,t]}\left[1 - \frac{c\,\mathrm{d}R_0 + \mathrm{d}N}{c + Y}\right];$$

see Theorem 4.3 of Hjort (1990).

As before, as $c(\bullet)$ decreases to zero, $\widehat{R}$ and $\widehat{F}$ tend to the non-parametric Nelson–Aalen and Kaplan–Meier estimators of R and F, respectively.

9.6 PRIORS FOR THE CUMULATIVE DISTRIBUTION FUNCTION

Chronologically, the material here should precede that of the previous two sections. This is because much of what is said in sections 9.4 and 9.5 has its roots in the material of this section. In what follows, I overview some key aspects of Ferguson's (1973) Dirichlet processes, and Doksum's (1974) processes neutral to the right (NTR). The NTR processes were introduced in section 9.5.1, but within the context of prior distributions for the cumulative hazard function. The intent now is to cast these processes in their original setting. However, to do so, it is desirable to use the framework of measure theoretic probability. This I do, but with some hesitation. Readers not at ease with the abstractions of measure theoretic probability may choose to concentrate only on the essence of the theorems given below. But before doing so, the following preamble may be useful vis-à-vis an overall perspective.

There appear to be two general approaches through which the problem of making inferences about an unknown survival function $\overline{F} = 1 - F$ have been explored. The first one, best summarized in Ferguson and Phadia (1979), starts by noting that any absolutely continuous survival function $\overline{F}(t)$ can always be written as $\overline{F} = \exp(-B(t))$, where $B(t)$ is some non-negative, non-decreasing and continuous function on $[0, \infty)$ with $B(0) = 0$. One then supposes that $\{B(t); t \geq 0\}$ is a Lévy process with non-negative increments and the property that $B(0) = 0$. With the above in place, it can be shown that given the data, $\{B(t); t \geq 0\}$ continues to be a Lévy process and that the expectation of $\exp(-B(t))$ provides inference about $\overline{F}(t)$. The Dirichlet and the NTR processes discussed in sections 9.6.1 and 9.6.2 below are in conformance with this approach.

The second approach centers around viewing the cumulative hazard rate function as a stochastic process $\{R(t); t \geq 0\}$. But when $\overline{F}$ is not absolutely continuous, one is not free to choose any stochastic process for $\{R(t); t \geq 0\}$. For example, we cannot choose a gamma process (which is a special Lévy process), when $\overline{F}$ is not absolutely continuous, because doing so will not produce a proper cumulative distribution function F (cf. Hjort, 1990, p. 1269). Thus every Lévy process cannot be used as a prior process for $\{R(t); t \geq 0\}$ unless of course $\overline{F}$ is absolutely continuous in which case $B(t) = R(t)$. It is because of this line of reasoning that Hjort (1990) has introduced the 'beta process' of section 9.5.2. For such processes given the failure data the posterior process continues to be a beta process.

9.6.1 The Dirichlet Process Prior

To discuss the Dirichlet process, we find it helpful to adopt a framework prescribed by Sethuraman (1994) in his famous paper on a constructive definition of Dirichlet processes. Accordingly, suppose that T is a random variable that generates 'data' t. Assume that T takes values in a measurable space $(\mathcal{X}, \mathcal{B})$, and let P be an unknown probability measure on $(\mathcal{X}, \mathcal{B})$. By 'measure' we mean non-negative, 'σ-additive' function. In the kind of applications that are of interest to us here, $\mathcal{X}$ will be the real line $\mathbb{R}$, and $\mathcal{B}$ the σ-algebra of Borel subsets of $\mathbb{R}$. Suppose that the unknown P takes values in $\mathcal{F}$, where $\mathcal{F}$ is the collection (or space) of all probability measures on $(\mathcal{X}, \mathcal{B})$. Let $\mathcal{C}$ be the smallest σ-algebra generated by sets of the form $\{P: P(A) < \omega\}$, where $A \in \mathcal{B}$ and $\omega \in [0, 1]$. Let ν be a probability measure on $(\mathcal{F}, \mathcal{C})$. Such a probability measure can be seen as a prior distribution of P. The Bayesian solution is to obtain a posterior distribution of P given the data t; we denote this posterior distribution by ν^t.

When the sample space $\mathcal{X}$ is a finite set, say $\mathcal{X} = \{1, 2, \ldots, k\}$, then every probability measure P on $\mathcal{X}$ will be given by the vector $(p_1 = P(1), p_2 = P(2), \ldots, p_k = P(k))$. This vector takes values in the simplex $\Delta_k = \left\{(p_1, \ldots, p_k): 0 \leq p_1 \leq 1, \ldots, 0 \leq p_k \leq 1, \sum_{j=1}^{k} p_j = 1\right\}$, and Δ_k is a subset of $\mathbb{R}_k$. In such cases a natural choice for P would be the $(k-1)$ variate Dirichlet distribution $\mathcal{D}(\gamma_1, \ldots, \gamma_k)$, where the parameters of this distribution are such that $\gamma_i \geq 0, i = 1, \ldots, k$ and $\sum_{i=1}^{k} \gamma_i > 0$ (equation (9.2)). Thus defining probability measures on Δ_k is relatively straightforward. When $\mathcal{X}$ is countably infinite the situation becomes much trickier because now we need to construct probability measures on $\Delta_\infty \subseteq \mathbb{R}_\infty$, with the requirement that $p_i \geq 0$ and $\sum_{i=1}^{\infty} p_i = 1$. This scenario and the attending difficulties that it creates have been discussed by Kingman (1975).

When $\mathcal{X}$ is some arbitrary measurable space, Bayesian non-parametrics becomes feasible only if one can define a large class of prior distributions on $(\mathcal{F}, \mathcal{C})$ for which the posterior distributions can be easily obtained. One such family of prior distributions is the Dirichlet process prior. There are a large number of Dirichlet processes on $\mathcal{X}$, one for each $\alpha \in \mathcal{M}$, where $\mathcal{M}$ is the class of all non-zero finite measures on $(\mathcal{X}, \mathcal{B})$. Specifically,

Definition 9.1. *A probability measure* ν *on* $(\mathcal{F}, \mathcal{C})$ *is said to be a* **Dirichlet process** *with parameter* α, *if for every measurable partition* $\{B_1, \ldots, B_k\}$ *of* $\mathcal{X}$ *(i.e. the* B_i*s are disjoint and* $\bigcup_{i=1}^{k} B_i = \mathcal{X}$*), the distribution of* $(P(B_1), \ldots, P(B_k))$ *under* ν *is the finite dimensional Dirichlet distribution* $\mathcal{D}(\alpha(B_1), \ldots, \alpha(B_k))$.

In particular, when $\mathcal{X}$ is $\mathbb{R}$, every $B \in \mathcal{B}$ will be such that $P(B)$ is Beta$(\alpha(b), \alpha(\mathbb{R}) - \alpha(B))$, and so $E[P(B)] = \alpha(B)/\alpha(\mathbb{R})$. The distribution $\mathcal{D}(\alpha(B_1), \ldots, \alpha(B_k))$, when it can be demonstrated to exist, will be denoted $\mathcal{D}_\alpha$. Since the measure ν is indexed by the elements of $\mathcal{B}$, it is a stochastic process; and $\mathcal{X}$ is the parameter space of the process.

Apart from the fact that the marginals of the Dirichlet measures have finite dimensional Dirichlet distributions, the Dirichlet measures have other properties that make them attractive in Bayesian non-parametrics. The most important of these properties is that of conjugacy. This property, plus much of what has been said above can be best articulated by

Theorem 9.2 (Ferguson (1973)). *Let $X_1, \ldots, X_n$ be a sample from P, where $P \in \mathcal{F}$, and the space $(\mathcal{F}, \mathcal{C})$ is (a priori) endowed with a probability measure ν that is a Dirichlet process $\mathcal{D}_\alpha$. Then given $X_1, \ldots, X_n$, the posterior probability measure on $(\mathcal{F}, \mathcal{C})$ will also be a Dirichlet process with parameter $\alpha + \sum_{i=1}^{n} \delta(X_i)$, where $\delta(X)$ is a probability measure that is degenerate at X.*

Ferguson (1973) gives two proofs for the existence of the Dirichlet process $\mathcal{D}_\alpha$ on $(\mathcal{F}, \mathcal{C})$, one using Kolmogorov consistency conditions and the other a constructive one showing that ν is the sum of countable number of point masses, whatever be the value of α. Sethuraman and Tiwari (1982) give an alternative construction of the Dirichlet process which is simpler than that of Ferguson (1973) and which has been advantageously used by several authors to obtain new results involving the Dirichlet measure. Because of its simplicity and intuitive import, this constructive definition of the Dirichlet distribution is given below:

Theorem 9.3 (Sethuraman and Tiwari (1982)). *Let $Y_1, Y_2, \ldots,$ be independent and identically distributed with Y_i having a* Beta$(M, 1)$ *distribution, where $M > 0$. Let $Z_1, Z_2, \ldots,$ be independent and identically distributed with Z_i having a distribution F_0. Suppose that the sequences $\{Y_i\}$ and $\{Z_i\}$ are independent. Define $P_1 = (1 - Y_1)$ and $P_n = Y_1 Y_2 \cdots Y_{n-1}(1 - Y_n)$ for $n \geq 2$. Then $P = \sum P_j \delta(Z_j)$ is a Dirichlet process with parameter $\alpha = MF_0$.*

To gain a better feel of this constructive definition, we write the $\mathcal{D}_\alpha$ given above as $\mathcal{D}(M, F_0)$, and call M the 'coarseness' parameter. Then the construction of P can be conceptualized via the following steps:

(*a*) Choose F_0 in a manner that reflects our best guess about the distribution P that generates the data we wish to analyze.
(*b*) Draw an observation Z_1 from F_0.
(*c*) Draw an observation Y_1 from a Beta$(M, 1)$ distribution on (0,1).
(*d*) Assign the probability mass Y_1 to the observation Z_1.
(*e*) Now draw another observation Z_2 from F_0, and an observation Y_2 from a Beta$(M, 1)$ distribution on $(0, (1 - Y_1))$. Assign the probability mass Y_2 to Z_2.
(*f*) Repeat step (e) m times, with Y_i drawn from a Beta$(M, 1)$ distribution on $(0, (1 - Y_1 - Y_2 - \ldots - Y_{i-1}))$, and Z_i assigned mass Y_i. Observe that the Y_is, $i = 1, \ldots, m$, sum to one and that their joint distribution is a Dirichlet. Hence
(*g*) We may construct an empirical distribution (Z_i, Y_i), $i = 1, \ldots, m$, with jumps of size Y_i at Z_i. The empirical distribution so constructed will be the Dirichlet process $\mathcal{D}(M, F_0)$ (Figure 9.1).

Note that a small M would result in small values of Y_i, and small Y_is would yield a finer approximation of F_0 by the empirical distribution. Thus, the parameter M controls the coarseness of the approximation of F_0 by the Dirichlet process, where F_0 encapsulates one's best prior guess of P. Note that there is some parallel between the constructive steps described above and Ramsey's (1972) construction of a prior distribution for the potency curve mentioned in section 9.3.1.

There is one other noteworthy point about the Dirichlet measures discussed here. This pertains to the fact that $\mathcal{D}_\alpha$ gives probability one to the subset of all discrete probability measures on the sample space $(\mathcal{X}, \mathcal{B})$. This point seems to have been first noticed by Blackwell (1973). The implication of this property is that we would expect to see some observations repeated exactly, and this in some scenarios may not be true. To avoid such limitations one tries to find workable priors that choose continuous distributions with probability one. One such choice is the class of processes that are neutral to the right. These will be discussed later in the section that follows,

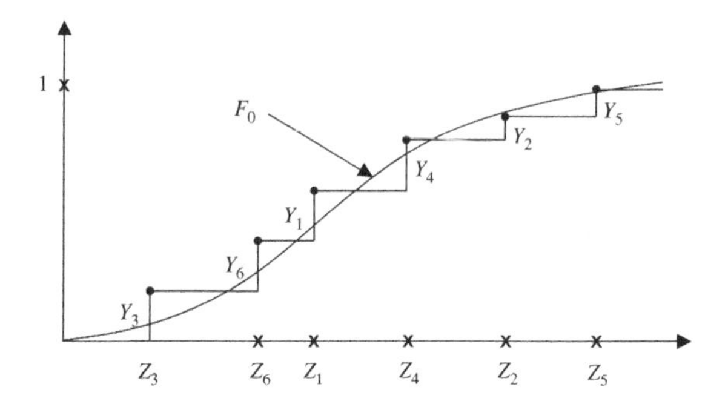

Figure 9.1 Constructing a Dirichlet process.

but first we shall illustrate how Theorem 9.2 given above can be put to work for addressing some standard statistical problems like estimating means, variances and distribution functions.

Example 9.1. Suppose that F is an unknown life distribution function whose estimate is desired. Let $\widehat{F}$ be a (Bayes) estimate of F based on the Bayes' risk which is taken to be a squared error loss function $\int_0^\infty E[F(t) - \widehat{F}(t)]^2 \mathrm{d}W(t)$, where $W(t)$ is some specified non-random weight function. If we assume that $F \in \mathcal{F}$, where $(\mathcal{F}, \mathcal{C})$ is endowed with a probability measure ν, and that ν is $\mathcal{D}_\alpha$, then $F(t), t \geq 0$, has distribution Beta$[\alpha(t), \alpha(\mathbb{R}^+) - \alpha(t)]$; here $\alpha(t)$ is a shorthand for $\alpha([0, t])$. Consequently, our Bayes estimate of $F(t)$ based on no data will be

$$\widehat{F}(t) = E[F(t)] = \frac{\alpha(t)}{\alpha(\mathbb{R}^+)} \stackrel{\text{def}}{=} F_0(t);$$

see Definition 9.1. We may interpret $F_0(t)$ as out prior guess about $F(t)$.

Suppose now that $T_1, \ldots, T_n$ are observations generated by F, and our aim is to update $F_0(t)$ using $T_1, \ldots, T_n$. Then, our Bayes estimate would be $E[F(t)|T_1, \ldots, T_n] \stackrel{\text{def}}{=} \widehat{F}_n(t)$, which by virtue of Theorem 9.2 will be

$$\begin{aligned}\widehat{F}_n(t) &= \frac{\alpha(t) + \sum_{i=1}^n \mathcal{I}_{[T_i, \infty)}(t)}{\alpha(\mathbb{R}^+) + n} \\ &= p_n F_0(t) + (1 - p_n) F_n(t), \end{aligned} \tag{9.27}$$

where $p_n = \alpha(\mathbb{R}^+)/(\alpha(\mathbb{R}^+) + n)$, and $F_n(t) = \sum_{i=1}^n \mathcal{I}_{[T_i, \infty)}(t)/n$ is the empirical distribution function of the sample.

The Bayes estimate (equation (9.27)) is therefore a mixture of the prior guess $F_0(t)$ and the sample distribution function $F_n(t)$. The mixing coefficient $p_n(t)$ gives a large weight to the prior guess $F_0(t)$ if $\alpha(\mathbb{R}^+)$ is large compared to n. When $\alpha(\mathbb{R}^+)$ is small compared to n, it is the empirical distribution function that receives a large weight. When $\alpha(\mathbb{R}^+)$ tends to zero, the case of the 'non-informative' Dirichlet process prior, $\widehat{F}$, is simply the empirical distribution function. Thus $\alpha(\mathbb{R}^+)$ encapsulates one's strength of belief in the prior guess $F_0(t)$ measured in units of the sample size n. Note that the parameter M of Theorem 9.3 corresponds to the $\alpha(\mathbb{R}^+)$ of this example. More generally, we may let M correspond to $\alpha(x)$. Hence, to put Theorem 9.2 to work a user need specify only two quantities, F_0 and $\alpha(\mathbb{R}^+)$, one's best guess about F, and a measure of the strength of belief in the guess F_0.

Filtering Under the Dirichlet Process

Whereas estimating the unknown distribution function F would encapsulate as assessment of all the uncertainty that is inherent in the scenario that generates the data $T_1, \ldots, T_n$, sometimes one may be satisfied with an assessment of only the mean and the variance of the unknown F. This is particularly so in the context of filtering wherein one endeavors to update the mean and the variance of F in the light of evolving information that comes in the form of data generated over time. In such scenarios, F is generally assumed to be a Gaussian distribution when the context of application is signal processing for target tracking, or F is the double lognormal distribution that was discussed in section 5.8.3. With F so specified, filtering will be parametric. But what if one chooses not to specify F and prefers to do filtering in a non-parametric Bayes framework? The methods of this section offer the necessary mechanism for doing so.

With the above in mind, consider the scenario of Example 1 and let

$$\mu_0 = \int_0^\infty t \mathrm{d}F_0(t) = \int_0^\infty \frac{t \mathrm{d}\alpha(t)}{\alpha(\mathbb{R}^+)}, \tag{9.28}$$

be the mean of our prior guess F_0 of F. The prior guess F_0 could be one of the standard distributions used in filtering. Ferguson (1973), in his Theorem 9.3, claims that under the Dirichlet process framework for F described before, the mean of F exists. Then, given the sample mean $\overline{T}_n = \sum_{i=1}^n T_i/n$, the Bayes estimate of the mean of F, again assuming a squared error loss, is of the form

$$p_n \mu_0 + (1 - p_n)\overline{T}_n, \tag{9.29}$$

which is a weighted combination of the prior guess μ_0 of the mean F (indeed in the absence of data, the mean of F_0 is also the mean of F) and the sample mean of F, $\overline{T}_n$.

A Bayes estimate of the variance of the unknown F is slightly more complicated to obtain. To see why, we first note that the variance of F,

$$\mathrm{Var}(F) = \int_0^\infty t^2 \,\mathrm{d}F(t) - \left[\int_0^\infty t \,\mathrm{d}F(t)\right]^2,$$

is a random variable whose expectation is the Bayes estimate of $\mathrm{Var}(F)$, when there is no data. This is given as

$$\begin{aligned} E[\mathrm{Var}(F)] &= E\left[\int_0^\infty t^2 \,\mathrm{d}F(t)\right] - E\left[\int_0^\infty t \,\mathrm{d}F(t)\right]^2 \\ &= \frac{\alpha(\mathbb{R}^+)}{\alpha(\mathbb{R}^+) + 1}\sigma_0^2, \end{aligned} \tag{9.30}$$

where σ_0^2 is the variance of our prior guess F_0. It is given as (cf. Ferguson (1973))

$$\sigma_0^2 = \frac{1}{\alpha(\mathbb{R}^+)} \int_0^\infty t^2 \,\mathrm{d}\alpha(t) - \mu_0^2.$$

Given a sample of size n with mean $\overline{T}_n$ and variance s_n^2, the Bayes estimate of $\mathrm{Var}(F)$ is

$$\frac{\alpha(\mathbb{R}^+) + n}{\alpha(\mathbb{R}^+) + n + 1} \left(p_n \sigma_0^2 + (1 - p_n) s_n^2\right) + p_n((1 - p_n)(\mu_0 - \overline{T}_n))^2.$$

The above can also be written as a mixture of three different estimates of the variance (cf. Ferguson, 1973) as

$$\frac{\alpha(\mathbb{R}^+)+n}{\alpha(\mathbb{R}^+)+n+1}\left[p_n\sigma_0^2+(1-p_n)\left\{\frac{p_n}{n}\sum_{i=1}^{n}(T_i-\mu_0)^2+(1-p_n)s_n^2\right\}\right]; \tag{9.31}$$

recall that, as before, $p_n=\alpha(\mathbb{R}^+)/[\alpha(\mathbb{R}^+)+1]$.

Thus filtering the mean and variance of the unknown F, under the Dirichlet process structure, starts with (9.28) and (9.30), and would get updated sequentially (upon receipt of new data) via (9.29) and (9.30).

The Case of Censored Data

The problem of using a Dirichlet process prior to obtain a non-parametric Bayes estimator of the F of Example 1, when the observed data consists of a right-censored sample, was considered by Susarla and Van Ryzin (1976). Under censoring the Dirichlet process prior loses its property of conjugacy and the results become cumbersome. These are given below.

Following the notation of section 9.5.2, let $(X_i, \delta_i), i=1,\ldots,n$, be the observables with $X_i=\min(T_i, c_i)$, c_i being the right-censoring time of T_i; $\delta_i=1$ if $T_i\le c_i$ and $\delta_i=0$, if $T_i>c_i$. Censoring is said to be **exclusive (inclusive)** censoring when the information given is of the form $T_i>(\ge)c_i$.

Let $u_1<u_2<\cdots<u_k$ be the distinct values among $X_1,\ldots,X_n$, and let λ_j be the number of censored observations at u_j. Let $k(t)$ denote the number of u_j's that are less than or equal to t and h_k, the number of X_js greater than u_k. Then given the exclusive right-censored data, and supposing that $F\in\mathcal{F}$, where $(\mathcal{F},\mathcal{C})$ is endowed with a probability measure ν that is $\mathcal{D}_\alpha$, the Bayes estimator of the survival function $(1-F(t))$ is, for $M=\alpha(\mathbb{R}^+)$ and $\alpha(t)=\alpha((t,\infty))$, of the form (Susarla and Van Ryzin, 1976):

$$\frac{\alpha(t)+h_k(t)}{M+n}\prod_{j=1}^{k(t)}\frac{\alpha(u_j)+h_j+\lambda_j}{\alpha(u_j)+h_j}; \tag{9.32}$$

this Bayes estimator is the expectation of the posterior survival function. This estimator reduces to the estimator of (9.27) for F when there is no censoring; the product term vanishes. As M goes to zero, the estimator reduces to the Kaplan–Meier estimator.

Ferguson, Phadia and Tiwari (1992) describe several other scenarios wherein the Dirichlet process priors have played a useful role for addressing several practical problems in statistics. Also noteworthy is a paper by Kumar and Tiwari (1989) wherein reliability estimation for an item operating in a random environment that is governed by a Dirichlet process has been discussed. This work can be seen as a way to generalize the model of section 4.7.5 on the bivariate Pareto.

9.6.2 Neutral to the Right-prior Processes

There are two limitations of the Dirichlet process priors. One is that such priors always choose discrete distributions with probability one. The other is that Dirichlet process priors are not well suited for the treatment of right-censored data as the material leading up to (9.32) indicates. To avoid these limitations one endeavors to find workable prior processes that choose continuous distributions with probability one. One such choice is the class of prior measures that are neutral to the right (NTR). These measures generalize the Dirichlet process priors and fit well with right-censored data, both exclusive and inclusive. Specifically (cf. Doksum, 1974) if the prior is NTR, then so is the posterior, irrespective of whether the data is a complete sample or a censored sample (exclusive or inclusive). By contrast, under censoring, a Dirichlet process prior will not lead to a Dirichlet process posterior; rather the posterior will be NTR.

Constructive Definition of NTR Distributions

We have stated before (in section 9.5.1) that a random distribution function F is NTR if for $t_1 < t_2 < \cdots < t_k$, the following are independent:

$$F(t_1), \frac{F(t_2) - F(t_1)}{1 - F(t_1)}, \ldots, \frac{F(t_k) - F(t_{k-1})}{1 - F(t_{k-1})}.$$

Because of the possibility that the denominator terms could be zero, an alternative definition of NTR is preferred. Accordingly,

Definition 9.2. *A random distribution function on the real line is said to be NTR, if for every m and $t_1 < t_2 < \cdots < t_k$, there exist independent random variables $V_1, \ldots, V_m$, such that the vector $(1 - F(t_1), 1 - F(t_2), \ldots, 1 - F(t_m))$ has the same distribution as the vector $(V_1, V_1 V_2, \ldots, \prod_{i=1}^{m} V_i)$.*

Different choices for the distributions of the V_is lead to different NTR distributions F. For example, if V_i has a $\beta(\alpha_i, \beta_i)$ distribution for $i = 1, \ldots, m$, with $\beta_i = \sum_{j=i+1}^{m-1} \alpha_j$, then F will be a Dirichlet process (cf. Doksum, 1974). The NTR family of distributions is therefore more general than the Dirichlet process family. Indeed, a characterization of the Dirichlet process is that it is both neutral to the right and also neutral to the left (cf. James and Mosimann, 1980). Besides avoiding the possibility of division by zero, Definition 9.2 has the virtue of being a constructive device. It gives us the ability to generate several families of NTR processes, depending on our choice for the distributions of the V_is.

Characterization of NTR and Relationship with Lévy Processes

Some properties of the NTR family of distributions have been given before, in section 9.5.1. In what follows, I elaborate on these properties, and give a few additional ones as well.

A key property of the NTR family of distributions is that if F is NTR, then the cumulative hazard function $R(t) = -\log(1 - F(t))$, when viewed as a stochastic process, will have independent increments. This feature is true even when $F(t)$ has discrete components, in which case the cumulative hazard function will be defined as $R^*(t) = \int_0^t dF(s)/(1 - F(s))$; (Dey, Erickson and Ramamoorthi, 2003). Note that $R(t)$ and $R^*(t)$ will be identical when F has no discrete components. The converse of the above property forms the basis for a simple characterization of NTR processes. Specifically, if a stochastic process, say Y_t, has independent increments, is non-decreasing and is right-continuous, then $F(t) = 1 - \exp(-Y_t)$ is a random distribution function that is NTR (Doksum, 1974).

The connection between the NTR property of F and the non-decreasing, right-continuous and independent increments property of Y_t is fortuitous. This is because we can import results from the theory of increasing Lévy processes (sections 7.2.2 and 7.3.2) to obtain features about the NTR class of distributions. One such feature would be the Lévy–Khinchin formula of (7.17), which would give us $P(T > t)$ when F, the distribution of T, is NTR. Another consequence of the aforementioned connection stems from the fact that when the $\alpha(t)$ of (7.17) is zero, the independent increments process Y_t will have non-random parts and thus Y_t will increase only by jumps. This means that $F(t)$ will also increase by jumps alone, and so with $\alpha(t)$ equal to zero, a neutral to the right $F(t)$ will be discrete with probability one.

Finally, the Lévy representation of Y_t also enables us to classify the NTR family of distributions into two classes; the **homogenous** and the **non-homogenous NTR processes**. If the Lévy measure of a right-continuous, non-decreasing, independent increments process Y_t is independent of t, then F is a homogenous NTR process; if the Lévy measure depends on t, then the corresponding NTR process is non-homogenous. Accordingly, a gamma process is a homogenous NTR process, whereas the Dirichlet process is a non-homogenous NTR (as well as neutral to the left) process.

Homogenous NTR processes have an advantage in that expressions for the posterior expectation of F given the data simplify; this is exemplified via (9.33) below.

The Posterior Distribution and Conjugacy of NTR Processes

We have stated before (in section 9.5.1) that NTR random probabilities are attractive from a Bayesian point of view because of Doksum's (1974) main result that if $X_1, \ldots, X_n$ is a sample from F, and if F is neutral to the right, then the posterior distribution of F given the sample is also neutral to the right. Thus the NTR family of random distributions is closed under a prior to posterior transformation; this is their property of conjugacy. In what follows, we describe the nature of the posterior distribution, first considering an uncensored sample of size one, and then the cases of inclusive and exclusive right censoring.

Theorem 9.4 (Doksum (1974)) . *Let F be a random distribution function that is NTR, and let X be a sample of size one from F. Then the posterior distribution of F given $X = x$ is NTR. The posterior distribution of an increment in $Y_t \stackrel{\text{def}}{=} -\log(1 - F(t))$ to the right of x is the same as its prior distribution. The posterior distribution of an increment in Y_t to the left of x is obtained by multiplying the prior density of the increment by e^{-y} and then renormalizing.*

Thus, for example, if the increment $Y_t - Y_s$, for $s < t < x$, has prior density $\mathrm{d}G(y)$, then the posterior density of the increment given $X = x$ is $\mathrm{e}^{-y}\,\mathrm{d}G(y) / \int_0^\infty \mathrm{e}^{-y}\,\mathrm{d}G(y)$. To complete a description of the posterior distribution, let $S = Y_x - Y_{\overline{x}}$ denote the jump in Y_t at x, and let $H_x(s)$ denote the distribution function of S at s. Then, if Y_t is a homogenous NTR process with Lévy measure $\nu(\mathrm{d}y)$,

$$H_x(s) = \frac{\int_0^s (1 - \mathrm{e}^{-y})\nu(\mathrm{d}y)}{\int_0^\infty (1 - \mathrm{e}^{-y})\nu(\mathrm{d}y)}; \tag{9.33}$$

observe that $H_x(s)$ is independent of x (cf. Ferguson and Phadia (1979), Theorem 5). This result suggest that even if the prior process is free from fixed points of discontinuity, the posterior will contain jumps at those points on the time axis where data are observed.

When X is censored at x, the following theorem describes the behavior of the posterior.

Theorem 9.5 (Ferguson and Phadia (1979), Theorem 3). *Let F be an NTR distribution function and let X be a sample of size one from F; let x be a real number.*

(a) Given $X > x$, the posterior distribution of X is also NTR. The posterior distribution of an increment to the right of x is the same as the prior; the posterior distribution of an increment to the left of x, or including x, is found by multiplying the prior density by e^{-y} and renormalizing, as in Theorem 9.4.

(b) Given that $X \geq x$, the posterior distribution of F is also NTR. The posterior distribution of an increment to the right of x or including x is the same as the prior distribution; the posterior distribution of an increment to the left of x is the prior density multiplied by e^{-y} and then normalized as in Theorem 9.4.

In both the Theorems 9.4 and 9.5 it is assumed that at x there is no a-priori assumption of discontinuity of F. The censored case is simpler to treat than the uncensored case because the jump at x does not have to be treated separately. In the case of exclusive (inclusive) censoring, the increment at x is treated as if it were to the left (right) of x. The general case of arbitrary sample sizes can, in principle, be treated by a repeated application of the above two theorems. The mechanics of implementing the above developments for practical applications gets cumbersome, and the best that one is able to achieve is the expected value of the posterior distribution expressed in terms of the moment-generating function (Theorem 4 of Ferguson and Phadia (1979)).

Simulation based approaches for a full Bayesian analysis involving the NTR processes have been proposed by Damien, Laud and Smith (1995) and by Walker and Damien (1998). The former exploit the structure of Lévy processes to generate (approximate) samples from 'infinitely divisible distributions'. The latter approach entails a hybrid of sampling methods and draws upon the Beta-Stacy process introduced by Walker and Muliere (1997). The Beta-Stacy process is an NTR process; it is a generalization of the Dirichlet process. Indeed, under censoring, the posterior distribution spawned by a Dirichlet process prior is a Beta-Stacy process.

Chapter 10

Survivability of Co-operative, Competing and Vague Systems

Euclid: His spirit is gone but here lie his elements.

10.1 INTRODUCTION: NOTION OF SYSTEMS AND THEIR COMPONENTS

A point of view that dates back to Greek philosophers is that, at a microscopic level, every item is an ensemble of smaller items that are linked together in some ordained manner. This hierarchical view of matter is carried further to smaller and smaller items, so that on an atomic scale every item is a collection of other items. However, such a regression to the infinitesimal must stop at some point, and the level at which this happens leads us to the notion of an element (a component or unit) of a system. The collection of elements constitutes a **system**, and the manner in which the elements get linked goes to define the **structure** (or the architecture) of the system. The simplest architectures are those of series and parallel systems, also known as **competing** and **co-operative** systems, respectively. Such architectures have already been touched upon in sections 4.7.3, 4.7.4 and 7.4. The architecture of networks, which can be seen as a combination of co-operative and competing systems, tends to be more intricate and therefore poses issues that could be challenging to analyze. The aim of this chapter is to describe how the survivability of each item and the manner in which they are connected translate to the survivability of the structure. System survivability analysis is therefore a study of the behavior of its parts, the metric for behavior here being survival. But there is more to this topic than merely the passive act of studying behavior. One is also interested in matters pertaining to making decisions that optimize behavior. Thus what is also germane here is the efficient design of system's architecture and the apportionment of reliability to each of its elements in order to achieve this optimum.

10.1.1 Overview of the Chapter

An overview of this chapter is given below; it provides a roadmap to a wide spectrum of issues that I endeavor to address. In section 10.2, I introduce our notation, state some preliminaries, describe the structure of some commonly encountered systems and provide some results pertaining to the

Reliability and Risk: A Bayesian Perspective N.D. Singpurwalla
© 2006 John Wiley & Sons, Ltd

reliability of systems in relationship to the reliability of their components. Much of the material of section 10.2 is old and honorable; it has been admirably covered in books such as those by Barlow and Proschan (1975), Aven and Jensen (1999), and Hoyland and Rausand (2004). The purpose of introducing section 10.2 is mainly to set the stage for describing the more recent material of subsequent sections, though in section 10.2.1, I give some new material on the nature of challenges posed by interdependencies, and families of multivariate Bernoulli distributions for modeling them. Section 10.3 constitutes the bulk of this chapter. It articulates on the nature of system survivability and introduces a framework for its formal treatment. Sections 10.3.2 and 10.3.3 describe the hierarchical nature of interdependencies and dependencies that our proposed framework for system survivability spawns. Section 10.3.3 concludes with a brief discussion of Monté Carlo integration that is useful for addressing the computational issues that the material of the above sections generates. Sections 10.3.4 and 10.3.5 can be seen as a coupling of two parts. In section 10.3.4, the first part, I introduce several continuous distributions on the unit hypercube. These are used in section 10.3.5 to address the inferential issues that arise when failure data on components and systems becomes available. Such issues are treated from a Bayesian point of view and involve an extensive use of the Markov Chain Monté Carlo (MCMC) method. In essence, the material of sections 10.3.4 and 10.3.5 constitutes a package for the commonly encountered practical problem of assessing a system's performance using data on the performance of its individual units and on the system itself; the material of these sections is based on Lynn, Singpurwalla and Smith (1998). Associated with section 10.3.4 is Appendix C, which points out circumstances in system survivability assessment wherein Borel's paradox can arise.

The focus of section 10.4 is on using techniques that are purely classified as 'machine learning' – like neural nets. These can be seen as a way to bypass the several challenges and difficulties of sections 10.2 and 10.3 that one encounters with modeling and computation in the context of large networks with layers of interdependencies. Sections 10.5 and 10.6 have a decision theoretic flavor. In section 10.5, I set up a framework for optimizing system reliability subject to cost constraints via a judicious allocation of reliability to each of its components. Motivated by the fact that when it comes to the matter of reliability, some individuals tend to be risk averse so that cost alone is not the driving criterion for making decisions, we discuss in section 10.6 the matter of the utility of reliability. Section 10.6 can be seen as a companion to section 10.5, since the material here is germane to making decisions about choosing between alternate system designs. Finally, in section 10.7 we discuss some recent work on characterizing the behavior of components and systems that can simultaneously exist in multiple states. The material here, motivated by concerns of realism, signals a change in the paradigm of how one may look at the relationship between a system and its components. In effect, this material takes an outward excursion from the foundational material of section 10.2, upon which the current theory of system reliability is based.

10.2 COHERENT SYSTEMS AND THEIR QUALITATIVE PROPERTIES

By way of preliminaries, let us freeze (i.e. fix) time at some value $\tau > 0$, and consider the state of each component of an n-component system. Define an indicator variable $X_i = 1(0)$, if the i-th component is functioning (failed) at τ. I suppose here that at any time τ, every component can be declared to exist in only one of two states, functioning or failed. In actuality components can be declared to exist in one of several states, such as perfect, degraded or failed. Furthermore, it is sometimes difficult to classify an item as being in any one of the two states so that the item is declared to simultaneously exist in multiple states. Issues of this kind are taken up in section 10.7.

Corresponding to the indicator variables $X_i, i = 1, \ldots, n$, for each component, we also define an indicator variable X for the entire system, where $X = 1(0)$ if the system is surviving (failed) at τ. Knowing the disposition of each $X_i, i = 1, \ldots, n$, we can determine the disposition of X if we know the architecture of the system. In other words, X is a function, say Φ, of the X_is. The function Φ is called the **structure function** of the system. In what follows, I let $\mathbf{X} \equiv (X_1, \ldots, X_n)$, so that $X = \Phi(\mathbf{X}) = 1$ or 0. When $\Phi(\mathbf{X}) = \prod_{i=1}^{n} X_i = \min_i X_i$, the architecture of the system is said to be a **series system**. In a series system, the system survives only when all its n components survive. The system fails as soon as one of its n components fails; thus one may say that the n components compete with each other to bring about a failure of the system (section 7.4.1). Series systems are therefore also known as 'competing systems'. When $\Phi(\mathbf{X}) = 1 - \prod_{i=1}^{n} X_i \stackrel{\text{def}}{=} \coprod_{i=1}^{n} X_i = \max_i X_i$, the system's architecture is said to be **parallel redundant**. The system fails only when all of its n components fail. Parallel systems are also known as 'co-operative systems'. This is because the system has, upon inception, $(n-1)$ components in redundancy, and these redundant components can be seen as co-operating with each other to keep the system afloat.

There is another aspect of systems with redundant components that needs mention. This has to do with what is known as **standby redundancy**. Under parallel redundancy, all the surviving components of the system are made to function (i.e. put to use) even though all that may be needed is a single functioning component. An example of parallel redundancy is an airplane that requires only two engines to fly, but which for added security has four engines, all of which are turned on during flight. Under standby redundancy, the redundant components are waiting to be turned on, until the failure of the active component at which time one of the surviving redundant component is commissioned for use. An electric lamp with $(n-1)$ spare light bulbs is an example of standby redundancy. Upon the failure of the first bulb, one of the $(n-1)$ spare surviving bulbs is put into service, and this process is repeated until all the n bulbs fail and darkness transcends. The mathematics governing the behavior of systems in standby and parallel redundancy is different. In the former case 'convolutions' of distribution functions come into play; in the latter it is the distribution of the 'n-th order statistic' that matters. Therefore a distinction between the two types of redundancies is germane.

When the structure function Φ is such that $\Phi(\mathbf{X}) = 1(0)$ if $\sum_{i=1}^{n} X_i \geq (<)k$, for $1 \leq k \leq n$, the system is said to be a ***k*-out-of-*n* system**. Its special case is the series (parallel redundant) system with $k = n(1)$. If an airplane with four engines needs at least any two of the four engines to function, then its underlying architecture is a two-out-of-four system; i.e. $k = 2$ and $n = 4$. In actuality many systems are a combination of series and parallel redundant systems. In Figure 10.1, we illustrate such a mixed system which is a series connection of two systems in parallel and two solo components. One of the parallel systems (also known as **modules**) has three components, and the other has two components in series that are connected in parallel with a single component.

It is evident that the structure function of this mixed system is of the form

$$\Phi(\mathbf{X}) = (X_1 \amalg X_2 \amalg X_3) \cdot X_4 \cdot X_5 \cdot [(X_6 \cdot X_7) \amalg X_8],$$

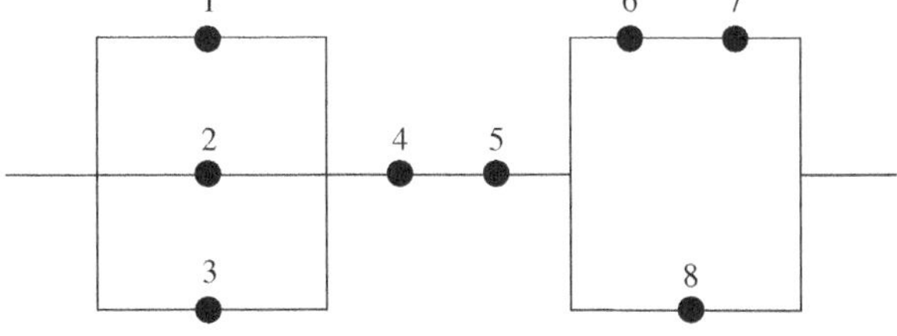

Figure 10.1 A mixed (series-parallel) system.

where $\mathbf{X} \equiv (X_1, \ldots, X_8)$. Upon expansion, the above expression becomes

$$\Phi(\mathbf{X}) = [1 - (1 - X_1)(1 - X_2)(1 - X_3)]X_4 \cdot X_5[1 - (1 - X_6 \cdot X_7)(1 - X_8)].$$

The above example underscores the fact that the writing out the structure of mixed systems is relatively straightforward using the Π and the $\amalg$ operators, so long as the components and the system are assumed to exist in binary [0,1] states. As a further support for this claim of simplicity, consider the bridge structure network of Figure 10.2.

The network of Figure 10.2 has a source and a sink; it is designed to transport material (or communicate information) from the source to the sink. The arrows in the figure show the direction of flow; it is unidirectional save for the link containing component (or node) 3, which is bi-directional. The bridge structure network can also be represented as a parallel connection of series systems (Figure 10.3) or as a series connection of parallel systems. Verify that its structure function can be written as:

$$\Phi(\mathbf{X}) = \begin{cases} (X_1 \cdot X_2) \amalg (X_4 \cdot X_5) \amalg (X_1 \cdot X_3 \cdot X_4) \amalg (X_4 \cdot X_3 \cdot X_2), \text{ or} \\ (X_1 \amalg X4) \cdot (X_2 \amalg X_5) \cdot (X_1 \amalg X_3 \amalg X_5) \cdot (X_4 \amalg X_3 \amalg X_2) \end{cases} \tag{10.1}$$

The components labelled (1 and 2), (4 and 5), (1 and 3 and 5), and (4 and 3 and 2) constitute what are known as the **min. path sets** of the network. These are the smallest set of components that need to function for the network to function (assuming that the remaining components have failed). If any component in a min. path set fails, the system fails.

What distinguishes the structure function of the system of Figure 10.1 from that of Figure 10.2 is that in the latter there is a replication of indicator functions in each of the four terms. Thus, with respect to the first equation above, the indicator variable X_1 appears in the term $(X_1 \cdot X_2)$ as well as the term $(X_1 \cdot X_3 \cdot X_4)$; similarly, X_4 appears in $(X_4 \cdot X_5)$ as well as $(X_4 \cdot X_3 \cdot X_2)$. This replication, though it may seem innocuous, will create an obstacle when we wish to assess the reliability of the network. But before getting into the nature of such obstacles, we need to articulate the notion of a **coherent system** and point out some inequalities connected with the behavior of such systems.

Loosely speaking, a coherent system is one wherein the structure function Φ is non-decreasing in each of its arguments and in which every component exists because it has a role to play. That is, the system does not contain components that are not essential (i.e. irrelevant). For example, with respect to mobility, the radio in an automobile is an irrelevant component. Similarly, seat belts are irrelevant vis-à-vis the automobile's ability to provide transportation, but are relevant if transportation with safety is a matter of interest.

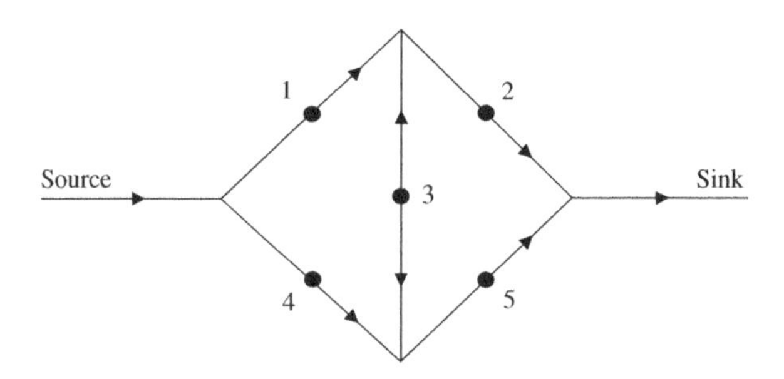

Figure 10.2 Bridge structure network.

With the X_is taking binary values 1 or 0, it can be seen that for any n-component coherent system

$$\prod X_i \leq \Phi(\mathbf{X}) \leq \coprod X_i. \tag{10.2}$$

This result suggests that any coherent system cannot be weaker (stronger) than a series (parallel) connection of its components.

There are two other aspects of coherent systems that warrant mention: duality and pivoting. The first defines $\Phi^D(\mathbf{X})$, the dual of a coherent system $\Phi(\mathbf{X})$ as

$$\Phi^D(\mathbf{X}) = 1 - \Phi(\mathbf{1} - \mathbf{X}),$$

where $(\mathbf{1} - \mathbf{X}) = (1 - X_1, \ldots, 1 - X_n)$. Verify that the dual of a series (parallel) system is a parallel (series) system and the dual of a k-out-of-n system is a $(n - k + 1)$-out-of-n system.

Dimension Reduction by Pivoting

Pivoting pertains to decomposing a coherent system of n components in terms of a coherent system of $(n - 1)$ components by pivoting on component i. Specifically, we can always write, that for any $i = 1, \ldots, n$,

$$\Phi(\mathbf{X}) = X_i \Phi(1_i, \mathbf{X}) + (1 - X_i)\Phi(0_i, \mathbf{X}), \tag{10.3}$$

where

$$\Phi(1_i, \mathbf{X}) = \Phi(X_1, \ldots, X_{i-1}, 1_i, X_{i+1}, \ldots, X_n)$$

and

$$\Phi(0_i, \mathbf{X}) = \Phi(X_1, \ldots, X_{i-1}, 0_i, X_{i+1}, \ldots, X_n).$$

Here $\Phi(1_i, \mathbf{X})$ represents the structure of a system for which X_i is fixed at 1, and the remaining $(n - 1)X_i$s are free to take the values 0 or 1; similarly with $\Phi(0_i, \mathbf{X})$.

Decomposition by pivoting is a useful tool for assessing the structure function of large and complex networks with many nodes. The main caveat here is picking the right node to pivot upon. One starts by pivoting on node i to create a structure function involving $(n - 1)$ nodes, and then one pivots on node j of the $(n - 1)$ component system to create a structure function involving $(n - 2)$ nodes, and so on until one arrives upon $\Phi(1, 1, \ldots, 1)$ and $\Phi(0, 0, \ldots, 0)$ whose values are 1 and 0, respectively. In actuality, it may not be necessary to keep pivoting until one arrives at $\Phi(\mathbf{1})$ and $\Phi(\mathbf{0})$; one may be able to specify the indicator function of a structure function at hand early on. I illustrate these issues by considering the bridge structure network of Figure 10.2 with pivoting on component 3. We have

$$\Phi(\mathbf{X}) = X_3 \Phi(1_3, \mathbf{X}) + (1 - X_3)\Phi(0_3, \mathbf{X}).$$

Clearly, $\Phi(0_3, \mathbf{X}) = (X_1 \cdot X_2) \amalg (X_4 \cdot X_5)$ and $\Phi(1_3, \mathbf{X}) = (X_1 \amalg X_4) \cdot (X_2 \amalg X_5)$. Hence

$$\Phi(\mathbf{X}) = X_3[(X_1 \amalg X_4) \cdot (X_2 \amalg X_5)] + (1 - X_3)[(X_1 \cdot X_2) \amalg (X_4 \cdot X_5)], \tag{10.4}$$

and there is no need to further decompose $\Phi(1_3, \mathbf{X})$ and $\Phi(0_3, \mathbf{X})$.

It now follows from (10.1) and (10.4) that the structure function of the bridge structure network of Figure 10.2 has three different versions, the two of (10.1) and the one of (10.4). The last equation has the virtue that there is no replication of any indicator variable in the ensuing terms. By contrast, there is a replication of indicator variables in several terms that go to constitute the two versions of $\Phi(\mathbf{X})$ in (10.1). The absence of replication is a consequence of pivoting on component 3. Another advantage of pivoting on component 3 is the ease with which we arrive upon (10.4). Had we started by pivoting on component 1, we would have had to pivot several times before which an expression for the structure function could be written out. Thus choosing the right component to pivot upon is an important task.

Time and 'Importance' Considerations

In our development thus far, the effect of time has been frozen, and I have focussed attention only on the X_is. The element of time may be introduced by noting that if $t_1, \ldots, t_n$ denote the actual times to failure of the n components, and if t denotes the time to failure of the system, then for a series system $t = \min_i t_i$, and for a parallel redundant system $t = \max_i t_i$. Consequently, by virtue of (10.2), we have the result that for any coherent system

$$\min_i t_i \leq t \leq \max_i t_i.$$

The case of random times to failure will be treated in sections 10.2.1 and 10.3.

A final point that I wish to discuss here pertains to the 'relative importance' of each component in a system. In any given system some components are more important (crucial) for the system's functioning than others. For example, with reference to the system of Figure 10.1, components 4 and 5 are more important than the remaining components because the failure of either would cause the system to immediately fail, even if all the other remaining components are functioning. With the above in mind, we would consider component i to be more important when $\Phi(1_i, \mathbf{X}) - \Phi(0_i, \mathbf{X}) = 1$ than when $\Phi(1_i, \mathbf{X}) = \Phi(0_i, \mathbf{X}) = 1$ or when $\Phi(1_i, \mathbf{X}) = \Phi(0_i, \mathbf{X}) = 0$; i.e. when component i becomes critical. Motivated by this observation, let

$$n(i) = \sum_{\{\mathbf{X}|X_i=1\}} [\Phi(1_i, \mathbf{X}) - \Phi(0_i, \mathbf{X})],$$

and define the **importance** of component i as $I(i) = n(i)/2^{n-1}$. The quantity $n(i)$ represents the number of times that component i can become critical to the system, and $I(i)$ is the proportion of times the component becomes critical. The notion of importance enables one to rank the components of a system. The notion of **system signature**, introduced by Samaniego (1985), enables one to rank entire systems from a structural point of view. Essentially, the signature of a system is a probability vector $\mathbf{S} = (S_1, \ldots, S_n)$, where S_i is the probability of system failure upon the occurence of the i-th component failure, $i = 1, \ldots, n$.

10.2.1 The Reliability of Coherent Systems

Reliability considerations come into play when T_i, the time to failure of the i-th component, is random (i.e. an unknown quantity). In what follows, we assume that all the T_is, $i = 1, \ldots, n$, are uncertain and that given 'chance' p_i, $P(T_i \geq \tau | p_i) = 1 - F(\tau | p_i) = p_i$, where p_i is assumed known. The case of p_i uncertain is taken up in section 10.3. Note that if p_i is the reliability of the i-th component for a mission of duration τ; it is also the expected value of X_i, since $P(T_i \geq \tau | p_i) = P(X_i = 1 | p_i)$. Similarly, let T denote the (uncertain) time to failure of the system, with $P(T \geq \tau | p) = p$, where p is the reliability of the system. Note that $p = E[\Phi(\mathbf{X}) | p]$, since $P(T \geq \tau | p) = P(\Phi(\mathbf{X}) = 1 | p)$. Knowing $\mathbf{p} = (p_1, \ldots, p_n)$, and the structure function of the system Φ, we can, in many cases, easily find p – provided the X_is are conditionally (given $\mathbf{p}$) independent. Specifically, for a series system, using multiplication rule of probability, we see that

$$P(\Phi(\mathbf{X})=1|\mathbf{p})=P(X_1=1,\ldots,X_n=1|\mathbf{p})=\prod_{i=1}^{n}P(X_i=1|p_i)=\prod_{i=1}^{n}p_i,$$

a function of the 'chances' p_i's alone. Similarly for a parallel system, we can verify that $P(\Phi(\mathbf{X})=1|\mathbf{p})=\coprod_{i=1}^{n}P(X_i=1|p_i)=\coprod_{i=1}^{n}p_i$, which again is a function of the p_i's alone.

As a consequence of the above relationships, plus (10.2), we have

$$\prod_{i=1}^{n}p_i\leq P(\Phi(\mathbf{X})=1|\mathbf{p})\leq\coprod_{i=1}^{n}p_i,$$

the implication of which is that the reliability of any coherent system whose lifetimes are independent is bounded from below (above) by reliability of a series (parallel) system.

When the system is a k-out-of-n parallel redundant systems, and when the known $\mathbf{p}$ is such that $p_1=p_2=\cdots=p_n=p^0$, then

$$P(\Phi(\mathbf{X})=1|\mathbf{p})=\sum_{j=k}^{n}\binom{n}{j}(p^0)^j(1-p^0)^{n-j}.$$

Thus for a series [parallel] system of three components each having component reliability p^0, $p=(p^0)^3$ $[1-(1-p^0)^3]$, whereas $p=3(p^0)^2(1-p^0)+(p^0)^3$, if the system is a two-out-of-three system. In the remainder of this section we shall drop the conditioning arguments p_i, $i=1,\ldots,n$ and p, since they are assumed known.

The pivotal decomposition of (10.3) leads us to an analogous formula for reliability, provided that the X_is are conditionally independent (given the p_i's). To see how, note that for any coherent system with structure function $\Phi(\mathbf{X})$

$$\begin{aligned}p&=E[\Phi(\mathbf{X})]=E[X_i]E[\Phi(1_i,\mathbf{X})]+(1-E[X_i])E[\Phi(0_i,\mathbf{X})],\\&=p_iE[\Phi(1_i,\mathbf{X})]+(1-p_i)E[\Phi(0_i,\mathbf{X})].\end{aligned}\tag{10.5}$$

It follows from (10.5) that the reliability of the system is linear in each p_i – the reliability of component i – for $i=1,\ldots,n$. Furthermore, when $p_1=p_2=\cdots=p_n=p^0$, p will be a polynomial in p^0. This polynomial is referred to as the **reliability polynomial**; its behavior has attracted the attention of computer scientists and network theorists; for example, Chari and Colbourn (1998).

In (10.5), if we consider the derivative of p with respect to p_i, we have

$$\frac{\mathrm{d}}{\mathrm{d}p_i}p=E[\Phi(1_i,\mathbf{X})-\Phi(0_i,\mathbf{X})];$$

since $[\Phi(1_i,\mathbf{X})-\Phi(0_i,\mathbf{X})]\geq 0$ with strict inequality holding for some vector, say $\mathbf{X}^0$, we see that the reliability of any coherent system is a strictly increasing function of each p_i.

To explore the shape of the system reliability function we start by considering a series [parallel] system of n components with $p_i=p^0$, for all i. Here $p=(p^0)^n$ $[1-(1-p^0)^n]$ lies entirely below [above] the diagonal connecting the points $(0,0)$ and $(1,1)$. However, for coherent systems that are mixtures of series and parallel systems, p as a function of p^0 starts off at zero crossing the diagonal mentioned above from below, and then culminates at one. In effect, p is like an S-shaped function of p^0.

We have seen that when the X_is are (conditionally) independent, therefore it is a straightforward issue to calculate the reliability of series, parallel, or any combination of series and parallel systems. The matter could get complicated in the case of networks, and this is even true when the X_is are assumed independent. To appreciate this point consider the bridge structure network of Figure 10.2, and focus attention on its structure function as given by the first equality of (10.1). Pictorially, this structure function can be depicted via Figure 10.3 below.

The bridge structure functions as long as any one of the four series-system branches of Figure 10.3 function. The nodes in each branch are the min. path sets. To facilitate calculation of the reliability of the bridge system, I define the indicator variables $Y_1 = X_1 \cdot X_2$, $Y_2 = X_4 \cdot X_5$, $Y_3 = X_1 \cdot X_3 \cdot X_5$, and $Y_4 = X_2 \cdot X_3 \cdot X_4$; then $\Phi(\mathbf{X}) = \coprod_{j=1}^{4} Y_j$.

Suppose that the X_is are conditionally (given the p_i's) independent, where $p_i = P(X_i = 1)$, $i = 1, \ldots, 5$. Then $P(Y_1 = 1) = p_1 p_2$, $P(Y_2 = 1) = p_4 p_5$, $P(Y_3 = 1) = p_1 p_3 p_5$, and $P(Y_4 = 1) = p_2 p_3 p_4$. However, to obtain $p = E[\Phi(\mathbf{X})]$, we need to know $P(\coprod_{j=1}^{4} Y_j = 1)$, and this will not be given by $\coprod_{j=1}^{4} \omega_j$, where $\omega_j = P(Y_j = 1)$, $j = 1, \ldots, 4$. The reason is that the Y_js are not independent even though the X_is are. For instance, since X_1 is common to both Y_1 and Y_3, Y_1 and Y_3 cannot be judged independent; similarly with the other Y_js. Thus in this formulation of $\Phi(\mathbf{X})$, it is the dependence between the Y_js that poses an obstacle for assessing the network's reliability.

There is a way to overcome the above obstacle, and this is via a judicious choice of pivoting. To get a sense for this consider (10.4), which is the structure function of the network obtained by pivoting on X_3; it was given as

$$\Phi(\mathbf{X}) = X_3[(X_1 \amalg X_4) \cdot (X_2 \amalg X_5)] + (1 - X_3)[(X_1 \cdot X_2) \amalg (X_4 \cdot X_5)].$$

If we let $Z_1 = (X_1 \amalg X_4) \cdot (X_2 \amalg X_5)$, then $P(Z_1 = 1) = (p_1 \amalg p_4) \cdot (p_2 \amalg p_5)$; similarly if $Z_2 = (X_1 \cdot X_2) \amalg (X_4 \cdot X_5)$, then $P(Z_2 = 1) = (p_1 \amalg p_2) \cdot (p_4 \amalg p_5)$. I may now invoke the relationship of 10.5 to obtain the network's reliability as

$$p = P(\Phi(\mathbf{X}) = 1) = p_3[(p_1 \amalg p_4) \cdot (p_2 \amalg p_5)] + (1 - p_3)[(p_1 \cdot p_2) \amalg (p_4 \cdot p_5)] \qquad (10.6)$$

Thus we see that pivoting is a way to circumvent the effect of dependencies caused by appearance of common nodes in the branches, the branches being determined by the min. path sets. But pivoting can also be seen as a way of computing network reliability by taking conditional expectations. This

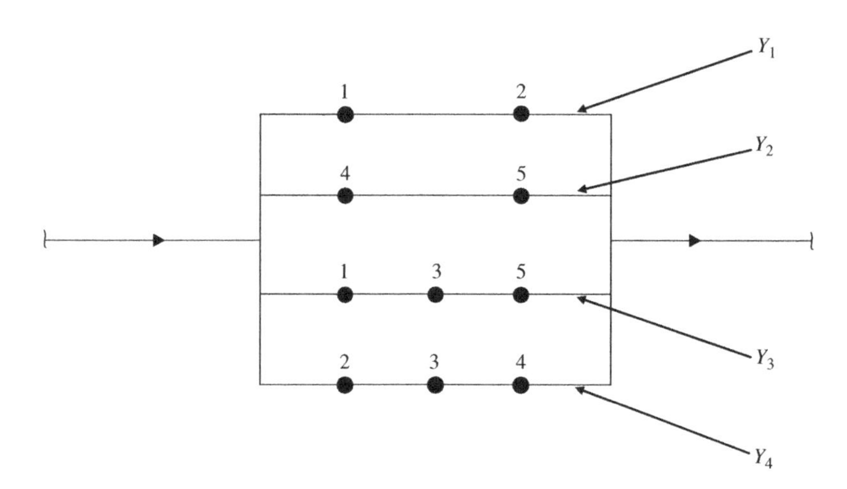

Figure 10.3 Series-parallel representation of the bridge structure of Figure 10.2.

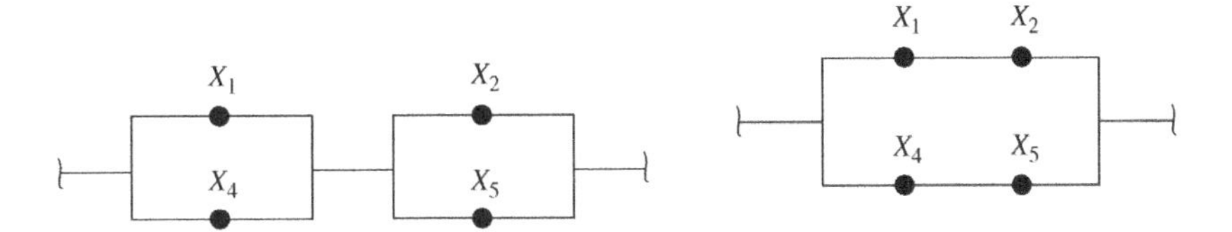

Figure 10.4 Decomposition of the bridge structure network of Figure 10.2 by pivoting on X_3.

is because (10.5) is, de facto, a consequence of conditioning on X_i and then taking expectations. Another way to look at pivoting is to examine (10.6) and to note that p is a p_3 weighted mixture of the reliability of two systems, one whose architecture is $(X_1 \amalg X_4) \cdot (X_2 \amalg X_5)$, and the other whose architecture is $(X_1 \cdot X_2) \amalg (X_4 \cdot X_5)$ (Figure 10.4).

Pivoting therefore breaks down a network's architecture into two sub-architectures and obtains the network's reliability as a mixture of the sub-architecture reliabilities. Thus, from a probabilistic point of view, a network can be seen as a mixture of co-operative and competing systems. In actual practice networks have interconnections that are more intricate than the one of Figure 10.2, and generally the size of a network tends to be large. This makes the task of identifying the right node to pivot upon challenging.

But the key matter that we need to come to terms with pertains to the issue of the assumed independence of the X_is (given the p_i's). This assumption is idealistic. Commonalities in manufacture, environment and operational policies would suggest that the lifetimes T_i, $i = 1, \ldots, n$, be judged dependent (section 4.7). This in turn would suggest that the X_is are also dependent, at least for certain sub-sets of $\{1, 2, \ldots, n\}$. In what follows I outline some strategies for incorporating dependence between the X_is and approaches for describing the nature of such dependencies.

The Challenge of Dependent Indicator Variables

We start with a reference to the system of Figure 10.1, and suppose that its eight indicator variables $(X_1, \ldots, X_8)$ are interdependent. In order to assess the reliability of this system incorporating the above interdependencies, the most direct and natural thing to do would be to introduce an eight-variate Bernoulli model. However, such high dimensional multivariate Bernoulli models are difficult to meaningfully specify, and when specified, can become cumbersome to work with (cf. Teugels (1990), whose use of the matrix calculus for representing multivariate Bernoullis is noteworthy). The main obstacle here is that the models are not parsimonious in the number of parameters, three in the bivariate case and seven in trivariate case.

In view of the above, it seems necessary to consider approximations to system reliability by reducing the dimensionality of dependence to say, two or at best three. One possibility would be to assume that dependence exists only among those components in each module of a system. Thus, for example, the system of Figure 10.3, which can be seen to have three modules comprising of components (1, 2, 3), components (4, 5), and components (6, 7, 8), would have $X_1 \triangledown X_2 \triangledown X_3$, $X_4 \triangledown X_5$, and $X_6 \triangledown X_7 \triangledown X_8$; here $\triangledown$ denotes 'dependence'. Once this is done, I may use a bivariate Bernoulli model to describe the joint behavior of (X_4, X_5), and trivariate Bernoulli models for (X_1, X_2, X_3) and (X_6, X_7, X_8). The particular form that these two- and three-dimensional Bernoulli models could take will be given in the sub-section that follows.

Whereas the strategy of restricting dependence to only the components in a module may get us going for modules with three or fewer components, the problem of high-dimensional dependency will continue to persist for modules having several components. Thus we seek the next level of approximation and one possibility would be to consider Bernoulli variables with Markov dependence. But to do this in a meaningful way, we need to label the components according to some logical order, and proximity seems like a meaningful way to proceed. In the case of a transportation or communication network, the direction of flow seems like a sensible thing to do. The labeling of components in Figure 10.1 for each of the three modules is already in order of proximity and thus no re-labelling is warranted. However, for the bridge structure network, the components in the branch labelled Y_4 of Figure 10.3 should be re-labelled as $(4, 3, 2)$ to correspond to the direction of the flow.

Under Markov dependence, given a collection of indicator variables $(X_1, \ldots, X_k)$, labeled according to some meaningful order, we assume that any X_i depends only on its predecessor X_{i-1}, for $i = 2, \ldots, k$. That is, dependence exists only among the adjoining indicator variables. A consequence is that

$$P(X_1, \ldots, X_k) = P(X_k|X_{k-1})P(X_{k-1}|X_{k-2}) \cdots P(X_2|X_1)P(X_1).$$

Now all that one needs to proceed further are a collection of bivariate Bernoulli distributions, each corresponding to the pair (X_i, X_{i-1}), $i = 2, 3, \ldots, k$.

When it comes to the bridge structure of Figure 10.2, with all the five nodes interdependent, issues do not become easy. We see here two possibilities, each carrying its own baggage. The first would be to start with a suitable five-variate Bernoulli distribution for $(X_1, \ldots, X_5)$ and use it to obtain the network's reliability via the inclusion-exclusion formula of probability (cf. Feller, 1968, p. 89), on the event $(\coprod_{j=1}^{4} Y_j = 1)$ of Figure 10.3. Accordingly,

$$P\left(\coprod_{j=1}^{4} Y_j = 1\right) = \sum_{j=1}^{4} P(Y_j = 1) - \sum_{i \neq j} P(Y_i \cdot Y_j = 1) + \sum_{i \neq j \neq k} P(Y_i \cdot Y_j \cdot Y_k = 1) - P\left(\prod_{1}^{4} Y_j = 1\right). \tag{10.7}$$

The five-variate Bernoulli distribution will enable us to obtain each of the above four probabilities. The cumbersome nature of such an assessment is obvious. A full five-variant Bernoulli distribution will entail 26 parameters, and assessing terms such as $P(\prod_{j=1}^{4} Y_j = 1) = P(X_1 = X_2 = X_3 = X_4 = X_5 = 1)$ will require of us a consideration of all the five variates. Simplification using Bonferroni's inequalities (Feller, 1957, p. 100), wherein $P(\coprod_{j=1}^{4} Y_j = 1) \geq \sum_{j=1}^{4} P(Yj = 1) - \sum_{i \neq j} P(Y_i \cdot Y_j = 1)$ will not help, because to assess $P(Y_1 \cdot Y_3 = 1)$, $P(Y_2 \cdot Y_4 = 1)$ and $P(Y_3 \cdot Y_4 = 1)$, I still need to specify $P(\prod_{j=1}^{5} X_i = 1)$.

Dimension reduction by assuming interdependence only among the components in each of the four min path sets $(1, 2)$, $(4, 5)$, $(1, 3, 5)$ and $(2, 3, 4)$ is again not possible because component replication relates the min. path sets to each other, and this in turn results in the need to consider interdependency between all the five X_is. Similarly, with an assumption of Markov dependence within each min. path set. Thus, the dimension reduction strategies that enabled us to approximate system reliability in the non-network scenario of Figure 10.1 do not work when it comes to the likes of networked systems.

Our proposed second strategy for assessing the reliability of the bridge structure network is to approximate the reliability by exploiting the pivotal decomposition shown in Figure 10.4. But to invoke this decomposition we need to assume that X_3 is independent of the remaining four X_is, and this may not be meaningful. Furthermore, doing so will reduce the dimensionality of dependence only by one; I will still need a four-variate Bernoulli model for (X_1, X_2, X_4, X_5) to obtain the network's reliability.

To conclude, the scenario of network reliability underscores the need to entertain multivariate Bernoulli distributions that are parsimonious and meaningful. In what follows I give examples of some possible choices.

Modelling Interdependence Using Multivariate Bernoullis

A bivariate Bernoulli for (X_1, X_2) is the simplest case to start with. The model has three natural parameters that characterize it. These are: $E(X_i) = p_i$, $i = 1, 2$ and $\sigma_{12} = \text{Cov}(X_1, X_2) = E((X_1 - p_1)(X_2 - p_2))$. The joint probabilities $p_{00} = P(X_1 = 0, X_2 = 0)$, $p_{10} = P(X_1 = 1, X_2 = 0)$, $p_{01} = P(X_1 = 0, X_2 = 1)$ and $p_{11} = P(X_1 = 1, X_2 = 1)$, with $p_{00} + p_{10} + p_{01} + p_{11} = 1$, can all be expressed in terms of the three natural parameters. To see how, let $\mu_{12} = E(X_1 X_2) = \sigma_{12} + p_1 p_2$; then $\mu_{12} = p_{11}$. Writing $q_i = 1 - p_i$, $i = 1, 2$, and solving for p_{00}, p_{10} and p_{01}, we obtain: $p_{00} = q_1 q_2 + \sigma_{12}$, $p_{10} = p_1 q_2 - \sigma_{12}$, and $p_{01} = q_1 p_2 - \sigma_{12}$. With $p_{11} = \sigma_{12} + p_1 p_2$, the above representations separate the independence components from the dependence ones; this is because $\sigma_{12} = 0$, when X_1 and X_2 are independent.

To fully specify the bivariate Bernoulli model we need to pin down on the natural parameters p_1, p_2 and σ_{12}. One way of streamlining the process for doing so would be to consider any suitable bivariate distribution of the lifetimes T_1 and T_2 discussed in section 4.1, and to associate T_1 and T_2 with X_1 and X_2, respectively. We can then deduce the required parameters from the parameters of the bivariate distribution of T_1 and T_2 via the relationship $P(X_1 = 1, X_2 = 1) = P(T_1 \geq \tau, T_2 \geq \tau)$ for any $\tau > 0$. Thus, for example, if the joint distribution of T_1 and T_2 is the BVE of section 4.7.4 with $P(T_1 \geq \tau, T_2 \geq \tau | \lambda_1, \lambda_2, \lambda_{12}) = \exp(-\lambda_1 \tau - \lambda_2 \tau - \lambda_{12} \tau)$ (equation (4.37)) then $p_i = \exp(-(\lambda_i + \lambda_{12})\tau)$, $i = 1, 2$, and $\sigma_{12} = \lambda_{12}/[\lambda(\lambda_1 + \lambda_{12})(\lambda_2 + \lambda_{12})]$, where $\lambda = \lambda_1 + \lambda_2 + \lambda_{12}$. Similarly with the other choices of T_1 and T_2.

The *trivariate Bernoulli* for (X_1, X_2, X_3) will have seven natural parameters $E(X_i) = p_i$, $i = 1, 2, 3$, $\sigma_{ij} = E((X_i - p_i)(X_j - p_j))$, $(i, j) = \{(1, 2), (1, 3), (2, 3)\}$, and $\sigma_{123} = E((X_1 - p_1)(X_2 - p_2)(X_3 - p_3))$. Whereas the structural import of a quantity such as σ_{123} is not clear, other than the sense that it encapsulates some form of dependence between X_1, X_2 and X_3, it can be seen (cf. Teugels, 1990) that quantities such as p_{000}, p_{100}, p_{010}, p_{001}, p_{110}, p_{101}, p_{011} and p_{111} can be expressed in terms of the seven natural parameters; here $p_{000} = P(X_1 = 0, X_2 = 0, X_3 = 0)$ and so on. Thus, for example, with $q_i = 1 - p_i$, $i = 1, 2, 3$, I can see that $p_{000} = q_1 q_2 q_3 + q_3 \sigma_{12} + q_2 \sigma_{13} + q_1 \sigma_{23} - \sigma_{123}$, and this representation separates the independence part, namely $q_1 q_2 q_3$, from the two- and three-way dependence as reflected by σ_{12}, σ_{13}, σ_{23} and σ_{123}, respectively. Dropping the three-way dependence by setting $\sigma_{123} = 0$ would make the model more parsimonious but still would require of us a consideration of six natural parameters. The representation of p_{000} given above not only separates the dependence and the independence parts, but it also shows how each dependent part contributes to the probabilities at hand.

It is clear from our discussion of the trivariate case described above that a consideration of higher order Bernoulli cases would become quiet cumbersome. Consequently, this line of multivariate Bernoulli model development will not be attractive from an applications point of view. Thus, we now turn attention to the case of Bernoulli models with Markov dependence.

Consider a sequence of n Bernoulli random variables $X_1, \ldots, X_n$, and assume that the successive quantities experience a Markovian dependence of the form:

$$P(X_i = 1) = 1 - P(X_i = 0) = p > 0, \quad i = 1, 2, \ldots, n,$$

and

$$P(X_i = 1 | X_{i-1} = 1) = \lambda, \quad i = 2, 3, \ldots, n.$$

It is evident that $P(X_i = 0 | X_{i-1} = 1) = 1 - \lambda$, $P(X_i = 1 | X_{i-1} = 0) = (1 - \lambda)p/q$, and that $P(X_i = 0 | X_{i-1} = 0) = (1 - 2p + \lambda p)/q$, where $q = 1 - p$. When $\lambda = p$, the model reduces to a collection of independent Bernoulli variables, and when $\lambda > p$ there will be a clustering among the ones and among the zeroes because of the Markov assumption for $x_i = 1$ or 0,

$$P(X_1 = x_1, \ldots, X_n = x_n) = P(X_1 = x_1) P(X_2 = x_2 | X_1 = x_1) \cdots P(X_n = x_n | X_{n-1} = x_{n-1}),$$

from which it follows that (cf. Klotz, 1973) that

$$P(X_1 = x_1, \ldots, X_n = x_n) = c_{1n}\eta_1^r \eta_2^s \eta_3^t, \tag{10.8}$$

where for $r = \sum_{i=2}^{n} x_{i-1}x_i$, $s = \sum_{i=1}^{n} x_i$ and $t = x_1 + x_m$,

$$\begin{aligned}
c_{1n} &= (1 - 2p + \lambda p)^{n-1}/q^{n-2},\\
\eta_1 &= \lambda(1 - 2p + \lambda p)/(p(1-\lambda)^2),\\
\eta_2 &= (1-\lambda^2)pq/(1 - 2p + \lambda p)^2, \text{ and}\\
\eta_3 &= (1 - 2p + \lambda p)/(q(1-\lambda)).
\end{aligned}$$

This multivariate Bernoulli model is parsimonious in the sense that it entails only two parameters p and λ. The simple functional form of (10.8) makes this model attractive from the point of view of applications. Its main negatives are the fact the variables $(X_1, \ldots, X_n)$ need to be such that their indices reflect some meaningful order, and that the relationship between the successive variables turns out to be the same; specifically, $\text{Cov}(X_i, X_j) = p(\lambda - p)$, for all $i \neq j$. Whereas the multivariate Bernoulli distributions introduced thus far are based on the modeling of pairwise and three-way dependencies, there remains the possibility of generating multivariate Bernoulli distributions via mixtures of independent Bernoullis. This approach to generating dependent lifelengths was considered by us before (section 4.7.5) in the context of items operating in a common environment. The bivariate Pareto distribution was motivated by arguments of this type. Thus in order to encapsulate interdependence among the X_is that are the consequence of systems and networks operating in a common environment, it seems natural to consider p-mixtures of independent (given p) identically distributed Bernoullis with $P(X_i = 1|p) = p$, for $i = 1, \ldots, n$. Consequently,

$$P(X_1 = x_1, X_2 = x_2, \ldots, X_n = x_n) = \int_0^1 p^s(1-p)^{n-s}\pi(p)\,\mathrm{d}p, \tag{10.9}$$

where $s = \sum_{i=1}^{n} x_i$, and $\pi(p)$ is the probability density function of p.

Note that (10.9) is in essence the infinite form of de Finetti's Theorem on exchangeable sequences (Theorem 3.2 of section 3.1.3). Different choices for $\pi(p)$ would result in different versions of the bivariate Bernoulli distribution for the sequence $(X_1, \ldots, X_n)$. Parsimony is assured because $\pi(p)$ would not contain an excessive number of parameters – two in the case of the beta distribution, the natural choice.

Suppose then that $\pi(p)$ is taken to be the beta distribution of (5.11) with parameters α and β. Then, in the bivariate case, (10.9) reduces to the form

$$P(X_1 = x_1, X_2 = x_2) = \frac{\Gamma(\alpha+\beta)\Gamma(\alpha + x_1 + x_2)\Gamma(\beta - x_1 - x_2 + 2)}{\Gamma(\alpha+\beta+2)\Gamma(\alpha)\Gamma(\beta)},$$

with marginals for X_i, $i = 1, 2$ given as

$$P(X_i = x_i) = \frac{\Gamma(\alpha+\beta)\Gamma(\alpha + x_i)\Gamma(\beta - x_i + 2)}{\Gamma(\alpha+\beta+2)\Gamma(\alpha)\Gamma(\beta)}.$$

Since $E(X_i) = E_p[E_{X_i}(X_i|p)] = E_p(p)$, $i = 1, 2$ and

$$\begin{aligned}\text{Cov}(X_1, X_2) &= E(X_1X_2) - (E(X_i))^2 \\ &= E_p[E_{X_1,X_2}(X_1X_2|p)] - [E_p(p)]^2 \\ &= E_p(p^2) - (E_p(p))^2 \\ &= \text{Var}_p(p),\end{aligned}$$

it follows that $E(X_i) = \alpha/(\alpha+\beta)$ and that the dependence between X_1 and X_2 is encapsulated by $\text{Cov}(X_1, X_2) = \alpha\beta/[(\alpha+\beta)^2(\alpha+\beta+1)]$. The above arguments generalize to the n-variate case, giving us a mechanism to generate higher order expectations such as $E[\prod_{i=1}^{n} X_i] = E_p(p^n)$ that are needed for system reliability assessment. The issue therefore boils down to obtaining $E_p(p^n)$ which, when $\pi(p)$ is a beta distribution, turns out to be

$$E_p(p^n) = \frac{\Gamma(\alpha+\beta)\Gamma(\alpha+n)}{\Gamma(\alpha+\beta+n)\Gamma(\alpha)}.$$

Since the nature of dependence between any two X_is, as encapsulated by $\text{Cov}(X_i, X_j)$, $i \neq j$, is the same over all i and j, this could be seen as a limitation of the bivariate Bernoulli model of (10.9). This limitation is not unlike that of the Markov dependent Bernoulli model of (10.8) wherein $\text{Cov}(X_i, X_j) = p(\lambda - p)$ for all $i \neq j$. In either case, both the Markov dependent and the mixture-based Bernoulli models offer a convenient way to compute the reliability of networks and systems whose indicator variables $(X_1, \ldots, X_n)$ can be assumed interdependent in the manner prescribed by (10.8) or (10.9).

10.3 THE SURVIVABILITY OF COHERENT SYSTEMS

Preamble and Overview: Reliability and Survivability

What do I mean by the term 'system survivability' and how does it differ from 'system reliability'? After all, the current literature in reliability theory does not make any such distinctions. To answer the above question I start by noting that the material of section 10.2.1 was centered around an assessment of p, a system's reliability, given its structure function Φ and assuming that the component reliabilities p_i, $i = 1, \ldots, n$ are known. We emphasize that p and the p_i's represent chance – not probabilities. As mentioned before (in section 4.4), by probabilities we would mean our personal uncertainties about the p_i's, the p, and also T, a system's lifetime. By the term 'System Survivability' we mean an assessment of T incorporating our uncertainties about the p_i's (and when appropriate p) as well. In actuality, the p_i's and the p are not known and are unlikely to be ever known, and thus the distinction between 'system reliability' and 'system survivability' is important and needs to be emphasized. Essentially, the former pertains to assessing uncertainty about T assuming the p_i's fixed and known; the latter assesses T incorporating uncertainties about the p_i's. By and large computer scientists and network analysts have focused attention on system reliability alone, leaving aside the matter of system survivability. In this section, I draw upon the former to address issues spawned by the latter. In so doing I provide a more complete picture about the assessment of a system's lifetime. I start by proposing (in section 10.3.1) a general architecture for looking into the issue of system survivability and follow this up by exploring survivability under the two scenarios of independence and interdependence – sections 10.3.2 and 10.3.3, respectively. In section 10.3.4, I describe some models on the unit hypercube that help implement the material of section 10.3.3. In section 10.3.5, I address, from a Bayesian point of view, the inferential issues that arise when failure data are available.

10.3.1 Performance Processes and their Driving Processes

We will continue to work with the framework and notation of section 10.2 but will slightly expand the latter to incorporate the additional generalities that we need to consider here. We start by replacing the p_i's and the p, by $p_i(\tau)$ and $p(\tau)$, respectively, for $i = 1, \ldots, n$. Since $\tau \geq 0$, $p_i(\tau)$ will be a decreasing (i.e. non-increasing) function of τ, with $p_i(0) = 1$ and $p_i(\tau) \downarrow 0$ as $\tau \uparrow \infty$, for $i = 1, \ldots, n$; similarly with $p(\tau)$. We assume that for any $\tau \geq 0$, $p_i(\tau)$ is unknown, and that our uncertainty about is described via our (personal) probability $\pi(p_i(\tau))$. Similarly, $\pi(p_1(\tau), \ldots, p_n(\tau))$ would encapsulate our joint uncertainty about the collective $(p_1(\tau), \ldots, p_n(\tau))$. But regarding $p_i(\tau)$ unknown for all $\tau \geq 0$, tantamounts to making $\{p_i(\tau); \tau \geq 0\}$ a decreasing stochastic process with state space $(0, 1]$ and parameter space $[0, \infty)$, for $i = 1, \ldots, n$. This in turn would make $\{p_1(\tau), \ldots, p_n(\tau); \tau \geq 0\}$ a joint stochastic process with state space $(0, 1]^n$ and parameter space $[0, \infty)$. The components of this joint stochastic process could be assumed independent or not depending on one's subjective assessment of the $p_i(\tau)$s. Similarly, $\{p(\tau); \tau \geq 0\}$ is also a stochastic process with state space $(0, 1]$ and parameter space $[0, \infty)$.

Analogous to what we did with the p_i's and p, we expand our notation to write $X_i(\tau) = 1(0)$, if the i-th component is functioning (failed) at τ, $\tau \geq 0$. We assume that $X_i(0) = 1$; thus, like $p_i(\tau)$, $X_i(\tau)$ is a decreasing (non-increasing) function of τ. Similarly $X(\tau) = 1(0)$, if the system is functioning (failed) at τ. We now assume that $P(X_i(\tau) = 1 | p_i(s); s \leq \tau) = p_i(\tau)$; i.e. the state of $X_i(\tau)$ depends only on the state of $p_i(\bullet)$ at time τ, and not on the past history of the process $\{p_i(s); s \leq \tau\}$. We continue to interpret $p_i(\tau)$ as the reliability of the i-th component at time τ. Furthermore, with the disposition of $X_i(\tau)$ uncertain over all τ, we look at $\{X_i(\tau); \tau \geq 0\}$ as a stochastic process that is driven by the process $\{p_i(\tau); \tau \geq 0\}$, and whose sample path is a decreasing step-function. We call the process $\{X_i(\tau); \tau \geq 0\}$ the **performance process** of the i-th component, $i = 1, \ldots, n$. Similarly, $\{X(\tau); \tau \geq 0\}$ is the performance process of the system, and $\{X_1(\tau), \ldots, X_n(\tau); \tau \geq 0\}$ the **joint performance process** of the n-components. With $P(X(\tau) = 1 | p(s); s \leq \tau) = p(\tau)$, the reliability of the system can be interpreted as $p(\tau)$.

Let $P_i(\tau) = P(X_i(\tau) = 1)$ denote the survivability of the i-th component to time τ, $i = 1, \ldots, n$, and $P(\tau) = P(X(\tau) = 1)$ the survivability of the system to τ. Then, by the law of total probability

$$P_i(\tau) = \int_0^1 P(X_i(\tau) = 1 | p_i(\tau)) \pi(p_i(\tau)) \, \mathrm{d}p_i(\tau)$$

$$= \int_0^1 p_i(\tau) \pi(p_i(\tau)) \, \mathrm{d}p_i(\tau) = E_\pi(p_i(\tau)).$$

Similarly, if $\pi(p(\tau))$ encapsulates the uncertainty about $p(\tau)$ for any $\tau \geq 0$, then

$$P(\tau) = E_\pi(p(\tau)).$$

Under certain circumstances (namely, the assumption of 'hierarchical independence', described in section 10.3.2), $P(\tau)$ will be a function, say h, of $P_1(\tau), \ldots, P_n(\tau)$ alone. The form of h will depend on the design of the system: series, parallel, or a mixture of series and parallel systems. When such is the case system survivability assessment is quite straightforward. Otherwise the issue becomes complicated and we then need to resort to approximations and simulations.

To summarize, our architecture for assessing system survivability analysis entails a consideration of two stochastic processes, a joint performance process $\{X_1(\tau), \ldots, X_n(\tau); \tau \geq 0\}$, and its **driving process** $\{p_1(\tau), \ldots, p_n(\tau); \tau \geq 0\}$. Note that the initial specification of the driving process is a subjective exercise, since each component of the process represents a chance, which is

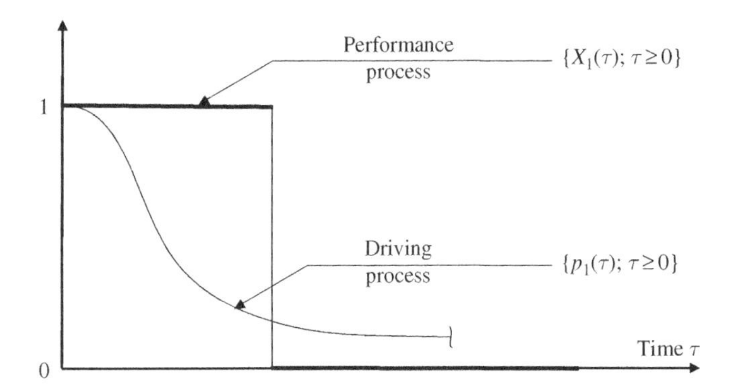

Figure 10.5 Sample paths of a performance process and its driving process.

unknown; the driving process can get updated when data is available. My process-based architecture is motivated by the fact that we have made a distinction between reliability and survivability and that we have not frozen time at some fixed τ, as is typically being done. In Figure 10.5 I illustrate, for the case $n=1$, the sample paths of the performance process $\{X_1(\tau);\ \tau \geq 0\}$ and its driving process $\{p_1(\tau);\ \tau \geq 0\}$.

Given the driving process, the performance process can be fully specified. Thus an issue that needs to be addressed pertains to some meaningful choices for the driving processes. But before getting into that matter, I first illustrate the workings of our framework by looking at the simple cases of n component series, parallel and other coherent systems; this is done below.

10.3.2 System Survivability Under Hierarchical Independence

Let $P^s(\tau)$ denote the survivability of a series system of n components. We start by considering the case $n=2$. Then

$$P^s(\tau) = P\,(X_1\,(\tau) = 1 \text{ and } X_2\,(\tau) = 1)\,,$$

which by the law of total probability can also be written as

$$P^s(\tau) = \int\limits_{P(\tau)} P(X_1\,(\tau) = 1,\, X_2\,(\tau) = 1 | p_1(\tau),\, p_2(\tau))\,\pi\,(p_1(\tau),\, p_2(\tau))\;\mathrm{d}p_1(\tau)\,\mathrm{d}p_2(\tau),$$

where $\pi\,(p_1(\tau),\, p_2(\tau))$ encapsulates our joint uncertainty about the component reliabilities $p_1(\tau)$ and $p_2(\tau)$, and $P(\tau)$ is the set of values that $p_1(\tau)$ and $p_2(\tau)$ can take; it is a subset of the unit square.

What should $\pi\,(p_1(\tau),\, p_2(\tau))$ be? This is an important question that needs to be meaningfully addressed. A question like this has not been raised before because a consideration of uncertainty about chances has not been a part of the toolkit of system reliability theory. A convenient starting point would be to assume that $p_1(\tau)$ and $p_2(\tau)$ are independent so that $\pi\,(p_1(\tau),\, p_2(\tau)) = \pi\,(p_1(\tau))\,\pi\,(p_2(\tau))$. The case of dependence between $p_1(\tau)$ and $p_2(\tau)$ will be considered next, in section 10.3.3. But assuming $p_1(\tau)$ and $p_2(\tau)$ independent tantamounts to the claim that a knowledge of the chance of failure of one component does not change our disposition about the chance of failure of the second; this may not be realistic because the components may have commonalities in their hazard potentials. Putting aside for now this matter of realism, the aforementioned independence assumptions lead us to the result that

$$P^s(\tau) = \int_{P(\tau)} P(X_1(\tau) = 1, X_2(\tau) = 1 | p_1(\tau), p_2(\tau)) \pi(p_1(\tau)) \pi(p_2(\tau)) \, \mathrm{d}p_1(\tau) \, \mathrm{d}p_2(\tau)$$

$$= \int_{P(\tau)} p_1(\tau) p_2(\tau) \pi(p_1(\tau)) \pi(p_2(\tau)) \, \mathrm{d}p_1(\tau) \, \mathrm{d}p_2(\tau),$$

provided that we also make the following additional independence assumptions:

(i) Given $p_1(\tau)$ and $p_2(\tau)$, $X_1(\tau)\, X_2(\tau)$ are independent;
(ii) Given $p_1(\tau)$, $X_1(\tau)$ is independent of $p_2(\tau)$; and
(iii) Given $p_2(\tau)$, $X_2(\tau)$ is independent of $p_1(\tau)$.

We now have in place a hierarchy of independence assumptions, (i), (ii) and (iii) above, constituting the first stage of hierarchy, and the independence of $p_1(\tau)$ and $p_2(\tau)$, being the second stage of the hierarchy. This is what we mean by the term **hierarchal independence**; first, the independence of $X_i(\tau)$s given the $p_i(\tau)$'s, and then the independence of $p_i(\tau)$'s themselves. Independence of the $X_i(\tau)$s is assured only when the $p_i(\tau)$'s are also independent. A consequence of the hierarchical independence is that now $P^s(\tau)$ can be written as

$$P^s(\tau) = P_1(\tau) P_2(\tau),$$

where $P_i(\tau)$ is the expected reliability of component i, $i = 1, 2$. Generalizing to the case of n components in series, we would have $P^s(\tau) = \prod_{i=1}^{n} P_i(\tau)$. Thus, under the assumption of hierarchical independence, the survivability of a series system is the product of the expected reliabilities of its components. Alternatively, the survivability of the system is also the expected value of its reliability.

The case of a parallel redundant system proceeds along similar lines. Here, if $P^p(\tau)$ denotes the survivability of n components in parallel redundancy then

$$P^p(\tau) = 1 - \prod_{i=1}^{n} (1 - P_i(\tau)) = \coprod_{i=1}^{n} P_i(\tau).$$

The expressions for $P^s(\tau)$ and $P^p(\tau)$ parallel those for the reliability of series and parallel systems, respectively, save for the fact that the component reliabilities get replaced by their expected values.

In general, for any coherent system of order n, for which the kind of assumptions of hierarchical independence described above can be made, it can be seen that the system's survivability is of the form

$$P(\tau) = h(P_1(\tau), \ldots, P_n(\tau)),$$

where h is some function of the component expected reliabilities.

10.3.3 System Survivability Under Interdependence

Note that by independence or interdependence, we are alluding to the behavior of the $X_i(\tau)$s. Unlike independence which had to be induced hierarchically, namely, independence of the $X_i(\tau)$s given the $p_i(\tau)$s and then an independence of the $p_i(\tau)$s themselves, interdependence can be brought into play at either or both stages of the hierarchy. This requires that we consider two scenarios: (a) independence of the $X_i(\tau)$s given the $p_i(\tau)$s and then dependence of the $p_i(\tau)$s; (b) dependence of the $X_i(\tau)$s irrespective of the disposition of the $p_i(\tau)$s. This is because under (a) dependence of the $p_i(\tau)$s induces a dependence of the $X_i(\tau)$s, even when the latter are conditionally (given the $p_i(\tau)$s) independent. We start with scenario (a).

Conditionally Independent Lifetimes with Dependent Reliabilities

Consider a coherent system of n components whose lifetimes $X_1(\tau), \ldots, X_n(\tau)$ are conditionally independent given their respective reliabilities $p_1(\tau), \ldots, p_n(\tau)$. We assume that $p_1(\tau), \ldots, p_n(\tau)$ are dependent with $\pi(p_1(\tau), \ldots, p_n(\tau))$ encapsulating our joint uncertainty about them; since $0 \le p_i(\tau) \le 1, i=1, \ldots, n$, $\pi(p_1(\tau), \ldots, p_n(\tau))$ resides on the unit hypercube. Section 10.3.4 is devoted to a discussion of some specific versions of $\pi(\bullet)$.

With the above set-up in place, it is possible to verify – via a repeated use of the pivotal decomposition formula – that the survivability of the system $P(\tau)$ will be of the form

$$P(\tau) = \int h(p_1(\tau), \ldots, p_n(\tau))\pi(p_1(\tau), \ldots, p_n(\tau))\,\mathrm{d}p_1(\tau)\cdots\mathrm{d}p_n(\tau), \tag{10.10}$$

where h is some function of the $p_i(\tau)$s. For example, in the case of series [parallel] systems

$$h(p_1(\tau), \ldots, p_n(\tau)) = \prod_{i=1}^{n} p_i(\tau) \left[\coprod_{i=1}^{n} p_i(\tau)\right].$$

The Case of Conditionally Dependent Lifetimes

We now turn attention to scenario (b) wherein the $X_i(\tau)$s are conditionally (given their respective $p_i(\tau)$s) dependent, and $\pi(p_1(\tau), \ldots, p_n(\tau))$ is arbitrary. When such is the case a probabilistic model for the $X_i(\tau)$s given the $p_i(\tau)$s needs to be specified. Possible choices are the multivariate Bernoullis given before or the multivariate failure models given in section 4.7, and depending on the scenario at hand any one of these can be selected. As an illustration, consider the case of a series system with two components discussed in section 10.3.2. Recall that the survivability of such a system is

$$P^s(\tau) = \int_{P(\tau)} P(X_1(\tau)=1, X_2(\tau)=1 | p_1(\tau), p_2(\tau))\pi(p_1(\tau), p_2(\tau))\,\mathrm{d}p_1(\tau)\,\mathrm{d}p_2(\tau).$$

With the $X_i(\tau)$s assumed (conditionally) dependent, we need to specify a probability model for the first term on the right-hand side of the above, and a possible choice is the BVE of (4.39). Under this choice $p_i(\tau) = \exp(-(\lambda_i + \lambda_{12})\tau)$, $i=1,2$, and $P(X_1(\tau)=1, X_2(\tau)=1 | p_1(\tau), p_2(\tau))$ becomes $P(X_1(\tau)=1, X_2(\tau)=1 | \lambda_1, \lambda_2, \lambda_{12}) = \exp(-\lambda\tau)$, where $\lambda = \lambda_1 + \lambda_2 + \lambda_{12}$. Consequently $\pi(p_1(\tau), p_2(\tau))$ would get replace by $\pi(\lambda_1, \lambda_2, \lambda_{12})$. Thus $P^s(\tau)$ would become

$$P^s(\tau) = \int_{(\lambda_1,\lambda_2,\lambda_{12})} \exp(-\lambda\tau)\pi(\lambda_1, \lambda_2, \lambda_{12})\,\mathrm{d}\lambda_1\,\mathrm{d}\lambda_2\,\mathrm{d}\lambda_{12}. \tag{10.11}$$

Similarly for parallel redundant systems and with other choices for $P(X_1(\tau) = 1, X_2(\tau) = 1 | p_1(\tau), p_2(\tau))$.

Computing Survivability by Monté Carlo Integrals

The two scenarios of interdependence considered here result in the need to evaluate integrals of the type given by (10.10) and (10.11). Often these cannot be analytically evaluated because of the complex and high-dimensional nature of the underlying $\pi(\bullet)$. One approach to bypassing such difficulties would be via a (crude) Monté Carlo integral wherein, in the case of (10.10), we approximate $P(\tau)$ by

$$\widehat{P}(\tau)=\frac{1}{M}\sum_{j=1}^{M}h(\mathbf{p}^{j}(\tau)), \tag{10.12}$$

where $\mathbf{p}^{j}(\tau)=(p_1^j(\tau),\ldots,p_n^j(\tau))$, $j=1,\ldots,M$, are M independent vector-valued samples generated from $\pi(\bullet)$. The ease with which such samples can be generated depends on the functional form of $\pi(\bullet)$. With a judicious choice of $\pi(\bullet)$, the samples can be straightforwardly generated via an MCMC-based simulation. Thus a task that remains to be addressed is the specification of $\pi(\bullet)$, and this is the topic of the next section. The matter of generating samples from $\pi(\bullet)$ will be taken up in section 10.3.5. The treatment of the relationship of (10.11) will proceed along similar lines.

Since $P(\tau)$ represents our personal probability that the system survives to τ, it connotes the amount that we are willing to bet on the event $(X(\tau)=1)$. Thus, should we be able to calculate $P(\tau)$ exactly, then the matter of assessing survivability will have been fully addressed once $P(\tau)$ is specified. However, when $P(\tau)$ is approximated by $\widehat{P}(\tau)$, it is $\widehat{P}(\tau)$ that represents the amount we are prepared to stake as our bet. But since $\widehat{P}(\tau)$ is obtained via a simulated set of data, it is subject to sampling error, and so it would make sense to specify upper and lower limits on it. More generally, the upper and lower probability paradigm for quantifying uncertainty, advocated by some such as Walley (1991), would now become germane.

A way to obtain the aforementioned probability limits would be to construct a histogram of the $h(\mathbf{p}^{j}(\tau))$s, $j=1,\ldots,M$, where each $h(\mathbf{p}^{j}(\tau))\in(0,1)$, and to pick two points on the histogram, say $\widehat{P}_L(\tau)$ and $\widehat{P}_U(\tau)$, such that the area of the histogram between these points is $1-\gamma$, for some $\gamma\in(0,1)$. The $\widehat{P}_U(\tau)$ and the $\widehat{P}_L(\tau)$ would represent the upper and lower betting probabilities. In the frequentist paradigm the upper and lower $100(1-\gamma)\%$ confidence limits on system reliability would be the analogues of $\widehat{P}_U(\tau)$ and $\widehat{P}_L(\tau)$, respectively.

10.3.4 Prior Distributions on the Unit Hypercube

The need for having at hand multivariate distributions on the unit hypercube has been pointed out in section 10.3.3, wherein it was understood that $\pi(p_1(\tau),\ldots,p_n(\tau))$ encapsulates our joint uncertainty about the component reliabilities $p_i(\tau)$, $i=1,\ldots,n$. Furthermore, since $\pi(\bullet)$ is a personal probability, it needs to be specified by taking into account subjectivistic considerations. The purpose of this section is to propose several possible candidates for $\pi(\bullet)$. The proposed candidates should not only be intuitively meaningful, but should also be such that their choice facilitates a Bayesian updating (when failure data are available) via Monté Carlo methods such as MCMC. The material of section 10.3.5 will be devoted to a discussion of the updating technology using MCMC, and survivability assessment via the Monté Carlo integral.

The construction of joint distributions on the unit hypercube has been of interest to statisticians concerned with a Bayesian analysis of Bernoulli data from similar but non-identical sources; see, for example, Danaher and Hardie (2005). Indeed, the method of copulas (discussed in section 4.7.1) is popular because it is able to fulfill such needs. Recall that a copula is a multivariate distribution on a unit hypercube whose marginal distributions are uniform. Thus one way to construct joint distributions on the unit hypercube with marginals having any general distribution could be via the method of copulas. But such methods tend to be mechanistic and lack the intuitive import that a subjectivistic specification of $\pi(\bullet)$ would mandate. We therefore need to consider approaches that are more fundamental and possess features that appear to be 'constructive'. One such approach has already been proposed in section 6.2.2 wherein we were interested in assessing the composite reliability of ultra reliable items. This resulted in the model of (6.6). In the subsections that follow, I revisit this construction and then propose a few additional ones that satisfy our requirements of meaningfulness and computability. Associated with the material of this section is the material in Appendix C on Borel's paradox in reliability assessment which interested readers may wish to visit.

Joint Priors for Highly Reliable and Related Units

There are many scenarios wherein it is reasonable to hold the belief that the $p_i(\tau)$s are related to each other but that the holder of such beliefs is unable to be specific about the nature of the relationship. One way to capture such scenarios is via the notion of exchangeability of the $p_i(\tau)$s, which to some such as Howson and Urbach (1989) is philosophically objectionable. Putting aside such objections we shall, in what follows, assume that the $p_i(\tau)$s are exchangeable. Furthermore, we shall also assume that the units are highly reliable, so that each $p_i(\tau)$ is in the vicinity of 1. The following two-stage construction, which parallels that given in section 6.2.2, yields a joint prior for $p_i(\tau)$s that encapsulates the features of similarity and high reliability.

To start off, suppose that given the parameters α and γ, the $p_i(\tau)$s, henceforth p_i if we suppress τ, are independent and have a beta distribution on $(\gamma, 1)$ with parameters $\alpha \in (0, 1)$ and 1, so that

$$\pi_1(p_i|\alpha, \gamma) = \frac{\alpha(1-p_i)^{\alpha-1}}{(1-\gamma)^{\alpha}}, \qquad \gamma < p_i < 1;$$

γ is a threshold on p_i.

For the second stage of our prior construction, we suppose that γ itself has a beta distribution on $(0, 1)$ with parameters $(\alpha + 1)$ and 1; i.e.

$$\pi_2(\gamma|\alpha) = (\alpha+1)(1-\gamma)^{\alpha}, \quad 0 < \gamma < 1.$$

It can now be seen that p_i has a beta distribution on (0,1) with parameters α and 2; i.e. for $i = 1, \ldots, n$,

$$\pi_1(p_i|\alpha) = \alpha(\alpha+1)p_i(1-p_i)^{\alpha-1}, \quad 0 < p_i < 1.$$

To obtain the joint distribution of the p_i's, namely $\pi(p_1, \ldots, p_n)$, I need to bring into the picture the smallest order statistic $p_{(1)} = \min(p_1, \ldots, p_n)$. Then it can be shown (cf. Chen and Singpurwalla, 1996) that

$$\pi(p_1, \ldots, p_n|\alpha) = \begin{cases} -\alpha^n \prod_{i=1}^{n}(1-p_i)^{\alpha-1}\ln(1-p_{(1)}), & \text{if } \alpha = (n-1)^{-1}, \\ \alpha^n \prod_{i=1}^{n}(1-p_i)^{\alpha-1}\frac{1-(1-p_{(1)})^{1-(n-1)\alpha}}{1-(n-1)\alpha}, & \text{otherwise.} \end{cases} \tag{10.13}$$

The attractiveness of this construction is that generating random samples from $\pi(\bullet)$ above is relatively straightforward. This matter will be taken up in section 10.3.5.

Joint Priors for Almost Identical Components

Many networks, especially those embedded in microcircuits, are made up of identical or almost identical units. This means that the p_i's should be nearly the same. Thus any joint distribution on the unit hypercube with its mass concentrated on the diagonal will capture the above feature. One such distribution, proposed by de Finetti (1972), takes, for $n=2$, the form

$$\pi(p_1, p_2|\theta_{12}) \propto \exp\left(-\frac{1}{2}(\theta_{12}(p_1 - p_2)^2)\right), \tag{10.14}$$

where θ_{12} dictates the nature of the marginal distributions of p_i, $i=1,2$, and the correlation between p_1 and p_2. Specifically, when $\theta_{12} > 0$, (10.14) is a truncated normal distribution on $(p_1 - p_2)$, and its marginal densities have an inverted U-shape. For $\theta_{12} < 0$, the bivariate density is cup-shaped and the marginals are U-shaped. Figures 4 and 5 of Lynn, Singpurwalla and Smith (1998) illustrate these shapes. A large positive correlation between p_1 and p_2 can be accounted for by letting $\theta_{12} \uparrow +\infty$. With $\theta_{12} < 0$, the correlation between p_1 and p_2 is negative.

In the trivariate case, (10.14) enlarges as

$$\pi(p_1, p_2, p_3|\bullet) \propto \exp\left(-\frac{1}{2}\{\theta_{12}(p_1 - p_2)^2 + \theta_{13}(p_1 - p_3)^2 + \theta_{23}(p_2 - p_3)^2\}\right),$$

with θ_{12}, θ_{13} and θ_{23} positive and large. Similarly, in the n-variate case we have

$$\pi(p_1, \ldots, p_n|\bullet) \propto \exp\left(-\frac{1}{2}Q\right), \tag{10.15}$$

where Q is a polynomial of degree two in $p_1, \ldots, p_n$. The approximate multivariate normality of the above form makes the generation of random samples from it relatively straightforward and so is its updating in the presence of failure data (section 10.3.5).

Joint Priors when Components are Ranked by Reliability

There could be circumstances wherein the components of a system can be ranked by their reliabilities. That is, it is possible to suppose that the components are labeled in such a way that

$$0 \equiv p_0 \le p_1 \le p_2 \le \cdots \le p_n \le p_{n+1} \equiv 1.$$

When such is the case we may resort to the strategy discussed in section 9.3 on bioassay, and arrive upon the ordered Dirichlet as a joint distribution for $(p_1, \ldots, p_n)$. Specifically, given the parameters $\alpha_i, \beta > 0$, with $\sum_{i=1}^{n+1} \alpha_i = 1$, we would have

$$\pi(p_1, \ldots, p_n|\boldsymbol{\alpha}, \beta) = \frac{\Gamma(\beta)}{\prod_{i=1}^{n+1} \Gamma(\beta\alpha_i)} \prod_{i=1}^{n+1} (p_i - p_{i-1})^{\beta\alpha_i - 1}, \tag{10.16}$$

where $\boldsymbol{\alpha} = (\alpha_1, \ldots, \alpha_{n+1})$. The essence of (10.16) is the assumption that $(p_i - p_{i-1}), i = 1, \ldots, n+1$, has a Dirichlet distribution.

It can be seen (cf. Mazzuchi and Soyer, 1993) that with $\alpha_i^* = \sum_{j=1}^{i} \alpha_j, i = 1, \ldots, n$, the marginal distribution of p_i is a beta distribution on (0, 1) of the form

$$\pi(p_i|\alpha, \beta) \propto p_i^{\beta\alpha_i^* - 1}(1 - p_i)^{\beta(1-\alpha_i^*)-1}. \tag{10.17}$$

Since $E(p_i|\bullet) = \alpha_i^*$ and $V(p_i|\bullet) = \alpha_i^*(1 - \alpha_i^*)/(\beta + 1)$, $\boldsymbol{\alpha}$ and β can be elicited via an elicitation of the mean and the variance of each p_i.

Generating samples from the $\pi(p_1, \ldots, p_n|\boldsymbol{\alpha}, \beta)$ of (10.16) is straightforward because of the constructive manner in which $\pi(\bullet|\bullet)$ is developed (section 9.3.1). More about this is said later in the section 10.3.5. Some other possible choices for $\pi(p_1, \ldots, p_n)$ are given in Lynn, Singpurwalla and Smith (1998). We now have at hand the necessary ingredients to assess system survivability with or without the incorporation of failure data on the system and its components. In the absence of data the arguments leading up to (10.12) provide a strategy. The details of how survivability can be assessed when data are available is given below in section 10.3.5. However, before doing so, it may be helpful to point out that there are circumstances wherein the scenario of assessing system reliability via multivariate distributions on the unit hypercube could lead to results that are paradoxical. This is because Borel's paradox could come into play here, with the consequence that we will end up with more than one answer for system survivability, all of them correct. A detailed discussion of this matter is in Singpurwalla and Swift (2001); for the sake of completeness, an overview is given in Appendix C.

Survivability Assessment Using Component and System Data

Perhaps one of the most commonly occurring practical problems in reliability and risk analysis is the assessment of system survivability given life test data on the components of the system, and sometimes the system itself. This problem has a long history, and it arises frequently in contexts such as the safety of nuclear power plants, weapons certification under test ban treaties, assessing technological risks of complex engineering systems, etc. Frequentist approaches for addressing this problem have often resulted in technical difficulties whose nature has been alluded to by Crowder *et al.* (1991). Bayesian approaches, proposed as early as the late 1970s (Mastran and Singpurwalla, 1978), have been hampered by the computational difficulties that high-dimensional posterior distributions pose. More important, the Bayesian methodology has not been based on the holistic architecture of section 10.3.1 described here. A consequence of this limited perspective is that the need to use one of the several multivariate distributions on the unit hypercube given in section 10.3.4 has not been necessitated. In what follows, we build on the material of the previous sections to provide a more complete answer to the problem of system survivability assessment. We start with some preliminaries and introduce an overall plan and conclude with a consideration of some special cases.

Preliminaries and the Overall Plan

Suppose that failure data on the system and its components is of the form (n, x) and (n_i, x_i), i=1,...,n, respectively. Here n_i denotes the number of copies of the i-th component that are tested for time τ, and x_i the number of copies that survive the test; similarly with n and x for the system as a whole. It is important to note that the (n_i, x_i) are not the result of a postmortem of (n, x). Let $\mathbf{d} = ((n_1, x_1), \ldots, (n_n, x_n))$. If for any particular component there are no test data available, then its corresponding n_i and x_i will be zero; similarly with n and x. With $\mathbf{d} = ((0, 0), \ldots, (0, 0))$ and $(n = 0, x = 0)$, system survivability is given by the $\widehat{P}(\tau)$ of (10.12), where $\widehat{P}(\tau)$ is the crude Monté Carlo integral of $P(\tau)$.

When the elements of $\mathbf{d}$ are not all $(0, 0)$, we need to update the $\pi(p_1, \ldots, p_n|\bullet)$s of section 10.3.4 via Bayes' law to obtain $\pi(p_1, \ldots, p_n|\bullet; \mathbf{d})$; this can be done analytically or by MCMC. The specifics would depend on the form chosen for $\pi(p_1, \ldots, p_n|\bullet)$. Once $\pi(p_1, \ldots, p_n|\bullet; \mathbf{d})$ is obtained we would generate M random samples from it to obtain, via the likes of (10.10), an updated version of $\widehat{P}(\tau)$, say $\widehat{P}(\tau; \mathbf{d})$.

If in addition to the $\mathbf{d}$ discussed above, failure data in the form of (n, x) is available on the system itself, then $\widehat{P}(\tau; \mathbf{d})$ needs to be updated to $\widehat{P}(\tau; \mathbf{d}, (n, x))$. This can be done numerically, using Bayes' law, as follows:

(a) Let $\mathbf{p}^j(\tau; \mathbf{d})$, $j = 1, \ldots, M$, denote M vector valued samples generated from $\pi(p_1, \ldots, p_n|\bullet; \mathbf{d})$.
(b) Let $h(\mathbf{p}^j(\tau; \mathbf{d}))$ denote the quantity obtained by plugging $\mathbf{p}^j(\tau; \mathbf{d})$ in $h(p_1(\tau), \ldots, p_n(\tau))$ of (10.10); note that $0 \leq h(\mathbf{p}^j(\tau; \mathbf{d})) \leq 1$.

(c) Let $\widehat{\pi}(\bullet)$ denote a histogram of the $h(\mathbf{p}^j(\tau; \mathbf{d}))$s, and let $\widehat{\pi}(p^{(k)}(\tau))$ be the value of the histogram at $p^{(k)}(\tau)$, where $0 \leq p^{(k)}(\tau) \leq 1$, and $k = 1, \ldots, H$.

(d) Update $\widehat{\pi}(p^{(k)}(\tau))$ to $\pi^*(p^{(k)}(\tau))$ in the light of (n, k) via Bayes' law as

$$\pi^*(p^{(k)}(\tau)) \propto (p^{(k)}(\tau))^x (1 - p^{(k)}(\tau))^{n-x} \widehat{\pi}(p^{(k)}(\tau)), k = 1, \ldots, H.$$

(e) Obtain $\widehat{P}(\tau; \mathbf{d}, (n, x))$ as

$$\widehat{P}(\tau; \mathbf{d}, (n, x)) = \sum_{k=1}^{H} p^{(k)}(\tau) \pi^*(p^{(k)}(\tau)).$$

Upper and lower betting probabilities can be constructed following the strategy outlined in the last part of section 10.3.3, save for the fact that the histogram frequencies will be prescribed by $\pi^*(p^{(k)}(\tau))$ as obtained in (d) above.

This completes our discussion on how to assess system survivability using failure data on the system and its components via a Bayesian approach. The key to implementing the plan proposed here is an ability to generate random vector valued samples from $\pi(\bullet)$, the joint distributions of section 10.3.4 on unit hypercubes, or its updated version. Strategies for doing this are discussed next.

Generating Random Samples from Distributions on Unit Hypercubes

We start with a consideration of the case of highly reliable and related units discussed in section 10.3.4. It resulted in the joint distribution $\pi(p_1, \ldots, p_n|\alpha)$ of (10.13) with α known. To generate samples from $\pi(p_1, \ldots, p_n|\alpha)$ we generate samples from $\pi(p_1, \ldots, p_n, \gamma|\alpha)$ and ignore the values of γ so generated. For doing so, we use the Gibbs sampling algorithm discussed in Appendix A. This requires that we have a knowledge of the full conditionals $\pi(p_1, \ldots, p_n|\alpha, \gamma)$ and $\pi(\gamma|p_1, \ldots, p_n, \alpha)$. The former is given as

$$\begin{aligned}\pi(p_1, \ldots, p_n|\alpha, \gamma) &= \prod_{i=1}^{n} \pi_1(p_i|\alpha, \gamma) \\ &= \prod_{i=1}^{n} \frac{\alpha(1-p_i)^{\alpha-1}}{(1-\gamma)^\alpha}, \quad \gamma < p_i < 1,\end{aligned}$$

and since γ is a lower threshold, the latter is simply

$$\pi(\gamma|p_1, \ldots, p_n, \alpha) = \pi(\gamma|p_{(1)}, \alpha),$$

which by Bayes' law (details omitted) is given as

$$\pi(\gamma|p_{(1)}, \alpha) = \begin{cases} -\dfrac{1}{(1-\gamma)\ln(1-p_{(1)})}, & \text{if } \alpha = (n-1)^{-1}, \\ \dfrac{1-(n-1)\alpha}{\left(1-(1-p_{(1)})^{1-(n-1)\alpha}\right)(1-\gamma)^{(n-1)\alpha}}, & \text{otherwise.} \end{cases}$$

When failure data $\mathbf{d}$ are available, we generate samples from $\pi(p_1, \ldots, p_n|\alpha; \mathbf{d})$, and this exercise follows the lines given above save for the fact that $\pi(p_1, \ldots, p_n|\alpha, \gamma)$ gets replaced by $\pi(p_1, \ldots, p_n|\alpha, \gamma; \mathbf{d})$ where

$$\pi(p_1, \ldots, p_n|\alpha, \gamma; \mathbf{d}) \propto \prod_{i=1}^{n} p_i^{n_i - x_i}(1-p_i)^{\alpha + x_i - 1},$$

for $\gamma < p_i < 1$. Also, $\pi(\gamma|p_1, \ldots, p_n, \alpha; \mathbf{d})$ is simply $\pi(\gamma|p_{(1)}, \alpha)$ since $\mathbf{d}$ has no effect on γ once $p_{(1)}$ is assumed known.

Moving on to the case of de Finetti's multivariate distribution that was deemed suitable for components that are almost identical, we first note that this distribution possesses nice closure properties. Specifically, under $\mathbf{d}$ with $N = (n_1 + \cdots + n_n)$ large, (10.15) becomes

$$\pi(p_1, \ldots, p_n|\bullet; \mathbf{d}) \propto \exp\left(-\frac{1}{2}(Q + NR)\right), \tag{10.18}$$

where $R = \sum_{i=1}^{n}[(p_i - \widehat{p}_i)/\alpha_i]^2$, with $\widehat{p}_i = x_i/n_i$ and $\alpha_i = (p_i(1 - p_i)/n_i N)^{1/2}, i = 1, \ldots, n$. The approximate multivariate normal form of (10.18) makes the task of generating samples from $\pi(p_1, \ldots, p_n|\bullet; \mathbf{d})$ straightforward.

In the case of an ordered Dirichlet suitable for components ranked by reliability, the constructive nature of (10.16) makes the generation of random samples from it relatively straightforward. Specifically, we start by generating an observation, say p_1^1, from $\pi(p_1|\alpha, \beta)$ given by (10.17) using the techniques prescribed in Atkinson (1979) for random number generation from a beta distribution. Then, given p_1^1, we generate an observation, say p_2^1, from a shifted beta distribution on $(p_1^1, 1)$, and so on, until we generate n observations $p_1^1, p_2^1, \ldots, p_n^1$. We repeat this cycle M times.

When failure data in the form of $\mathbf{d}$ are available, we need to generate samples from $\pi(p_1, \ldots, p_n|\alpha, \beta; \mathbf{d})$ where

$$\pi(p_1, \ldots, p_n|\boldsymbol{\alpha}, \beta; \mathbf{d}) \prod_{i=1}^{n} p_i^{x_i}(1 - p_i)^{n_i - x_i} \pi(p_1, \ldots, p_n|\boldsymbol{\alpha}, \beta).$$

Generating samples from $\pi(p_1, \ldots, p_n|\boldsymbol{\alpha}, \beta; \mathbf{d})$ turns out to be an onerous task, the details of which are given in Appendix B of Lynn, Singpurwalla and Smith (1998). The approach of these authors is based on the adaptation of an ingenious strategy proposed by Gelfand and Kuo (1991) in the context of Bayesian biassay.

10.4 MACHINE LEARNING METHODS IN SURVIVABILITY ASSESSMENT

By 'machine learning methods' we have in mind computationally intensive methods that appear to bypass probabilistic modeling and simulation requirements of sections 10.2 and 10.3, respectively. Recall that to analytically assess system survivability, we are required to prescribe three essentials: the structure function of the system; the uncertainties associated with the components' reliabilities; and probability models that encapsulate interdependencies between the component lifetimes. With large and complex systems, each of these becomes a onerous task. Furthermore, systems in actual use almost always experience maintenance (i.e. repairs and replacements), the effect of which is an enhancement of system survivability. Analytical approaches for the treatment of replacement and repair have not been discussed by me. Results from such approaches generally tend to be qualitative or at best asymptotic (i.e. long-run averages). Machine learning methods, like 'neural nets', circumvent the caveats and requirements mentioned above, and offer a user a viable alternative for assessing system survivability. As an aside, the useful role that machine learning methods can also play in statistical practice has recently been articulated by Brieman (2001) in a lively discussion article.

In what follows, our focus will be on neural net technology. In using this technology, one substitutes modeling and analysis with observed data. Thus one's ability to use neural nets depends on the availability of data. To gather such data, one needs to monitor a system for a modest period of time and this could be a drawback. In principle, survivability assessment using machine learning is only retrospectively feasible; i.e. for systems that are on and running. Thus, neural net-based methods cannot be used for systems that are under design and development, or systems that have yet to experience use. With machine learning one adopts an attitude that is akin to that taken by statisticians doing 'data analysis'. Here one tries to extract information out of data using specialized graphical and computational techniques, but without the benefit of an underlying stochastic model.

In what follows, I shall give an overview of the general architecture of a single-phased neural net with one or more hidden layers. For system survivability assessment we need a two-phased neural net. Once 'trained' (section 10.4.1) the first phase is designed to generate the binary states that a system's components will take in the future. The second phase uses as an input the output of the first phase, and generates the state of the system. In essence, the second phase of the neural net is necessitated by the feature that specifying the structure function of large and complex systems is too onerous a task for one to embark upon.

10.4.1 An Overview of the Neural Net Methodology

An artificial neural net (ANN) is a construct that is modeled after the human brain. Its building block is an idealized representation of a single nerve cell called the 'neuron', or in neural net terminology, a 'computational unit'. McCulloch and Pitts (1943) conceptualized the functioning of a neuron via the following construct.

A neuron receives as inputs $x_1, \ldots, x_p$ from p other neurons; each x_i is a binary, taking the values $+1$ or -1. The p inputs get transmitted to the neuron by links, called 'synapses'. The neuron transforms its p inputs to a binary output y, with $y = +1$ or -1 via the relationship

$$y = \begin{cases} +1, & \text{if } \sum_{i=1}^{p} w_i x_i + w_0 \geq 0, \\ -1, & \text{otherwise;} \end{cases}$$

the constants $w_1, \ldots, w_p$ are called the 'synaptic weights', and the constant w_0 is called the 'bias'. The output y now becomes an input to another neuron. The synaptic weights are the unknowns of the set-up, and the neural net methodology is centered around determining the w_i's via an iterative process that matches a neuron's output with an actual (observed) output. The synaptic weights encapsulate the importance of the input that gets passed from one neuron to the next. The iterative process, also known as the 'training process', is an empirical exercise about which much has been written and for which there is available an abundance of software; see for example, Lefevbre and Principe (1999). My aim here is not to dwell into the details of the training process; for this, see for example, the book by Haykin (1999). Rather, our purpose is to highlight the essence of an approach that is able to address an important practical problem in survival analysis. But before doing so, some additional details about the architecture of a neural net may be useful to know; these are given below.

In a multilayer net, the nodes are organized into layers; Figure 10.6 shows a neural net with a single layer – called the 'hidden layer'. In a feedforward net all the nodes send their output only to those nodes in the next layer, and not vice versa. In a fully connected net, every node is connected to each node in its adjacent layer. Figure 10.6 portrays a fully connected, feedforward net with one intermediate layer. In designing a neural net for any particular application, the key issues that need to be addressed are: the nature of interconnectivity (fully connected or otherwise), the direction of connectivity (feedforward or feedback) and the number of hidden layers. Computer scientists grapple with these issues, their aim being the approximation of a well-behaved function via a neural net methodology; see, for example, Funahashi (1989). Like

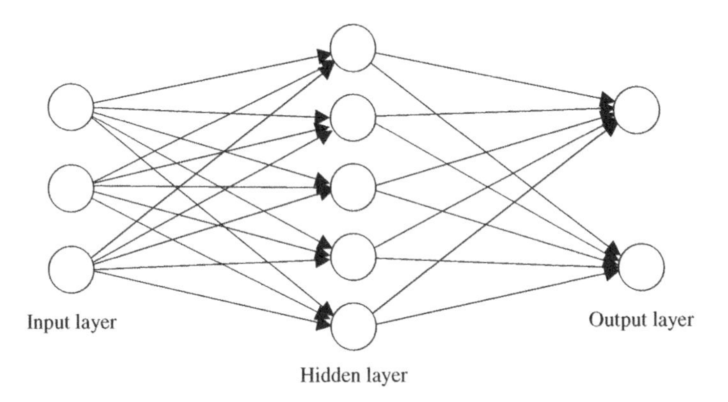

Figure 10.6 A fully connected feedforward net with one hidden layer.

the matter of training a neural net, a discussion on designing a neural net is beyond the scope of what we have in mind here, namely, how to use the neural net technology for assessing system survivability. Designing an efficient neural net, and training it for the purpose of system reliability assessment, is an open problem that warrants careful attention.

10.4.2 A Two-phased Neural Net for System Survivability

A two-phased neural net architecture is one wherein two separate neural nets are linked together in tandem so that the output of one net becomes an input to the second. Each net is designed to perform a different function. In this context, the first-phase net (henceforth ANN1) is designed to generate the binary states of the components of a multi-component system. The second-phase net (henceforth ANN2) takes as its input the output of ANN1 and generates the state of the system. To generate training data for ANN1 and ANN2, one observes a system of interest for a moderate amount of time, recording at some equally spaced time intervals the states of the components of the system and also the state of the system. For purposes of discussion, consider an n component system that is monitored at k equispaced time points $0 < \tau_1 < \tau_2 < \cdots < \tau_i < \cdots < \tau_k$. Let $\mathbf{X}(\tau_i) = (X_1(\tau_i), \ldots, X_n(\tau_i))$ and $X(\tau_i)$ denote the states of the components and the system, respectively at time τ_i, for $i = 1, \ldots, k$, with $\mathbf{X}(0) = (1, \ldots, 1)$ and $X(0) = 1$. That is, at time 0 all the n components of the system and the system itself are are supposed to be in the functioning state. The effects of component unreliabilities and their uncertainties, interdependence, and maintenance will be encapsulated in the manner in which $\mathbf{X}(\tau_i)$s change over i. The purpose ANN1 is to capture this change by learning about it; i.e. ANN1 trains itself by successively using $\mathbf{X}(0), \mathbf{X}(\tau_1), \ldots, \mathbf{X}(\tau_k)$ as data. Once trained, ANN1 is able to predict $\mathbf{X}(\tau_{k+j})$, $j = 1, 2, \ldots,$ the states of the components at the equispaced future times $\tau_{k+1}, \tau_{k+2}, \ldots,$ and so on. By the same token, ANN2 trains itself by learning about the relationship between the $\mathbf{X}(\tau_i)$ and $X(\tau_i)$, $i = 1, \ldots, k$, and is then able to predict $X(\tau_{k+j})$, using $\mathbf{X}(\tau_{k+j})$, $j = 1, 2, \ldots,$. Knowing the $X(\tau_{k+j})$, $j = 1, 2, \ldots J$ (say), we can assess system survivability via its 'availability'; i.e. the proportion of times out of J that $X(\tau_{k+j})$ takes the value one.

Whereas the design of ANN1 and its training are two major issues that need to be carefully addressed, one way to proceed would be to assume that ANN1 is a multi-layered, fully connected, feedforward net which generates its first set of synaptic weights $\mathbf{w}^{(1)}$ using $\mathbf{X}(0) = (1, \ldots, 1)$ as input data and $\mathbf{X}(\tau_1)$ as observed data. The synaptic weights $\mathbf{w}^{(1)}$ get updated to $\mathbf{w}^{(2)}$ with

$\mathbf{X}(\tau_1)$ now serving as input data and $\mathbf{X}(\tau_2)$ as observed data, and so on until we obtain $\mathbf{w}^{(k)}$. The $\mathbf{w}^{(k)}$ are then used to obtain $\widehat{\mathbf{X}}(\tau_{k+1})$ as a prediction of $\mathbf{X}(\tau_{k+1})$ via $\mathbf{X}(\tau_k)$ as input to ANN1. Similarly, $\widehat{\mathbf{X}}(\tau_{k+2})$ is obtained as prediction of $\mathbf{X}(\tau_{k+2})$ via ANN1 with $\widehat{\mathbf{X}}(\tau_k)$ as input, and so on.

Analogously, we may suppose that ANN2 is a multi-layered, fully connected, feedforward neural net which starts off with $X(0) = 1$ as input and $X(\tau_1)$ as data and successively uses $X(\tau_i)$ and $X(\tau_{i+1})$, $i = 1, \ldots, k-1$, as input and data, respectively, to obtain $\widehat{X}(\tau_{k+j})$ as a predictor of $X(\tau_{k+j})$, $j = 1, 2, \ldots$ Here again, to obtain $\widehat{X}(\tau_{k+2})$ I use $\widehat{X}(\tau_{k+1})$ as an input. Once the $\widehat{X}(\tau_{k+j})$, $j = 1, 2, \ldots, J$ (say) are obtained the survivability of the system over the time period $[\tau_{k+1}, \tau_{k+J}]$ is the proportion of times out of J that $\widehat{X}(\tau_{k+j})$, $j = 1, 2, \ldots, J$ takes the value 1. This completes our discussion on using machine learning for survivability assessment.

10.5 RELIABILITY ALLOCATION: OPTIMAL SYSTEM DESIGN

A key aspect of system design is the attainment of high survivability. For networks and systems with several inter-connected components, high survivability is achieved in one of two ways: by introducing redundant components, or by increasing the survivability of each component (i.e. strengthening the components). Either strategy increases costs and thus one feature of engineering design is to increase survivability subject to cost constraints. This tantamounts to making decisions in the face of uncertainty (or risk). When survivability is enhanced by redundancy, the decision problem boils down to determining the location of redundancies and the number of redundant components to install at each location. This class of problems has been studied under the label 'redundancy allocation', about which much has been written; see, for example, Barlow and Proschan (1975, p. 209) and the references therein, or the recent overview by Kuo and Prasad (2000). In this section, we consider the issue of enhancing survivability by strengthening. The problem boils down to a determination of which components to strengthen and by how much. This class of problems is labeled 'reliability allocation' or 'reliability apportionment' about which there has been some, but not much, of a discourse. As of this writing the most recent addition to this literature is a paper by Falk, Singpurwalla and Vladimirsky (2006) upon which the material that follows is based. Our purpose here is not to dwell on the optimization details contained in the above paper. Rather, our focus is on the decision theoretic set-up that the work entails, and a roadmap to the key results obtained.

Because enhanced reliability is a proxy for enhanced survivability, our discussion will be centered on reliability and its apportionment. Thus in what follows, we use the material of section 10.2.1 as an anchor, and follow the notation therein. Specifically, p_i will denote the reliability of the i-th component, $i = 1, \ldots, n$, and p the reliability of the system, for some specified mission time $\tau > 0$. Under the assumption of independence, p will be a function, say h, of only the p_i's; i.e. $p = h(\mathbf{p})$, where $\mathbf{p} = (p_1, \ldots, p_n)$. With dependence, p will be a function of $\mathbf{p}$ as well as some dependency parameters; we denote this as $h(\mathbf{p}, \bullet)$, where $\bullet$ stands for the dependency parameters.

10.5.1 The Decision Theoretic Formulation

Our aim here is to determine an optimum $\mathbf{p}$ for any specified system architecture Φ and a mission time τ. Associated with each p_i, $p_i \in [0, 1]$, is a cost function $c_i(p_i)$, $i = 1, \ldots, n$. This is the cost of producing a component i, whose reliability for mission time τ is p_i. It is reasonable to suppose that $c_i(p_i)$ is non-decreasing in p_i, with $c_i(0) = 0$ and $c_i(1) = \infty$. Let $c(\mathbf{p}) = \sum_{i=1}^{n} c_i(p_i)$ be the cost associated with building a system whose component reliabilities are p_i, $i = 1, \ldots, n$; we have assumed that the costs are additive, though it need not be so, and most likely it is not so.

Suppose that at time τ, $\Phi(\mathbf{X})=1$, where $\mathbf{X}=(X_1,\ldots,X_n)$ with $X_i=1(0)$, if the i-th component is surviving (failed) at τ. With $\Phi(\mathbf{X})=1(0)$, the system survives (fails) by time τ. Suppose that when $\Phi(\mathbf{X})=1(0)$, the builder/operator of the system receives a monetary reward (penalty) of $\pi^+(\pi^-)$, where π^+ and π^- are both non-negative constants. Then with $\Phi(\mathbf{X})=1$, the utility to the builder/operator is $\pi^+ - c(\mathbf{p})$; similarly, with $\Phi(\mathbf{X})=0$, the utility is $-[\pi^- + c(\mathbf{p})]$. It is of course possible that $\pi^+ < c(\mathbf{p})$ so that $[\pi^+ - c(\mathbf{p})] < 0$; that is, the system is not profitable to operate even when it is perfectly functional.

With the above in place, the normative approach to deciding upon what $\mathbf{p}$ should be, boils down to finding that $\mathbf{p}$ with maximizes the total expected utility $\widetilde{U}(\mathbf{p})$, where

$$\widetilde{U}(\mathbf{p}) = h(\mathbf{p},\bullet)[\pi^+ - c(\mathbf{p})] - [1 - h(\mathbf{p},\bullet)][\pi^- + c(\mathbf{p})], \tag{10.19}$$

and $0 \le p_i < 1$ (section 2.8.2). With $\widetilde{U}(\mathbf{p})$ being a function of π^+, π^-, and $c(\mathbf{p})$, we are considering the utility of money and assuming that it is linear. Several plausible forms for the cost function $c(p)$ have been proposed in the literature. Some examples are: $c(p)=a\log(p/b+1)$, where $a, b>0$ are constants – considered by Lloyd and Lipow (1962); $c(p)=ap^w$, for $0<w<1$ and $a>0$, a constant – considered by Tillman *et al.* (1970); $c(p)=a\exp(b/(1-p))$ – considered by Misra and Ljubojevic (1973); $c(p)=b[p/(1-p)]^a$ – considered by Majumdar, Pal and Chaudhuri (1976), and $c(p)=-a\log(1-p)$ with $p\in[0,1)$ considered by Fratta and Montanari (1976). Dale and Winterbottom (1986) have articulated several desirable features that cost functions should possess, and of the cost functions mentioned above, the one by Fratta and Montanari (1976) possesses these features. Thus, in what follows, we take $c_i(p_i)=-a_i\log(1-p_i)$, for $p_i\in[0,1)$, $i=1,\ldots,n$. With this choice, the optimization problem of (10.19) simplifies as maximizing

$$\widetilde{U}(\mathbf{p}) = h(\mathbf{p},\bullet)\lambda - [\pi^- + c(\mathbf{p})], \tag{10.20}$$

with respect to $\mathbf{p}$, under the constraint that $0 \le p_i < 1, i=1,\ldots,n$, with $\lambda \stackrel{\text{def}}{=} [\pi^+ + \pi^-]$, and $c(\mathbf{p}) = \sum_{i=1}^n -a_i\log(1-p_i)$; $a_i>0$ is specified for all i. Since π^- is a specified constant, (10.20) written out explicitly boils down to finding $\mathbf{p}$ with $0 \le p_i < 1, i=1,\ldots,n$, such that

$$U(\mathbf{p}) = \sum_{i=1}^n a_i \log(1-p_i) + \lambda h(\mathbf{p},\bullet), \tag{10.21}$$

is maximized; recall that $c(\mathbf{p}) = -\sum_{i=1}^n a_i\log(1-p_i)$.

General Solution to the Decision Problem

Falk, Singpurwalla and Vladimirsky (2006) in their Theorem 4.1 show that if $h(\mathbf{p},\bullet)$ is continuous on $\{\mathbf{p}: 0 \le p_i < 1, i=1,\ldots,n\}$, and if there exists an $M>0$ such that $h(\mathbf{p},\bullet) \ge M$ for all $\mathbf{p}$ with $\mathbf{0} \le \mathbf{p} < \mathbf{1}$, then a solution to the optimization problem of (10.21) does exist. Furthermore, they also show, via their Theorem 4.2, that when any of the inequalities

$$p_i \le \max\left\{0, 1 - \frac{a_i}{\lambda}\right\}, \quad i=1,\ldots,n, \tag{10.22}$$

get violated, the expected utility of the system (so constructed) is not maximized. Thus the inequality of (10.22) provides, in the case of any coherent system, an upper bound on the component reliabilities. This upper bound is crude; to obtain sharper results we consider, in the subsection below, specific forms for $h(\mathbf{p},\bullet)$.

10.5.2 Reliability Apportionment for Series Systems

The simplest special case to consider is that of a series system with independent lifelengths. Here $h(\mathbf{p}, \bullet) = h(\mathbf{p}) = \prod_{i=1}^{n} p_i$, so that given $\lambda > 0$ and the constants $a_i > 0, i = 1, \ldots, n$, we need to find a $\mathbf{p}$ such that

$$U(\mathbf{p}) = \sum_{i=1}^{n} a_i \log(1 - p_i) + \lambda \prod_{i=1}^{n} p_i, \tag{10.23}$$

gets maximized subject to the constraints $0 \leq p_i < 1, i = 1, \ldots, n$.

Falk, Singpurwalla and Vladimirsky (2006) show, via their Theorem 5.1, that the optimization problem of (10.23) always has a solution, say $\mathbf{p}^*$. Either all the components of $\mathbf{p}^*$ are positive, or all are zero. When $\mathbf{p}^* = (p_1^*, \ldots, p_n^*) > \mathbf{0} = (0, \ldots, 0)$, the necessary conditions for the optimality of (10.23) boil down to finding the $p_i^*, i = 1, \ldots, n$, such that

$$P = F(P), \tag{10.24}$$

where $P = \prod_{i=1}^{n} p_i^*$, and $F(P) = \prod_{i=1}^{n} \lambda P/(a_i + \lambda P)$.

Clearly $F(P) \in [0, 1)$ and since $P \in [0, 1)$, the function F maps [0,1) into [0,1). Thus in solving (10.24) we are seeking a fixed point, say $\widehat{P}$ of F. Once $\widehat{P}$ is found, we can find $p_i^*, i = 1, \ldots, n$, via

$$p_i^* = \frac{\lambda \widehat{P}}{(a_i + \lambda \widehat{P})}.$$

Since $\widehat{P}$ is a solution to (10.24), namely

$$P = \prod_{i=1}^{n} \frac{\lambda P}{(a_i + \lambda P)},$$

$\widehat{P}$ is a root of the following polynomial in P of degree n,

$$\prod_{i=1}^{n} (a_i + \lambda P) - \lambda^n P^{n-1} = 0. \tag{10.25}$$

This polynomial can have at most two roots or no roots. In the latter case, we set each $p_i^* = 0, i = 1, \ldots, n$. When (10.25) has two roots, it is the larger root that leads to the maximization of (10.23). The roots, when they exist, will be positive; the detailed arguments that justify the above claims are given in Falk, Singpurwalla and Vladimirsky (2006).

Thus to summarize, in the case of a series system with independent component lifelengths and for which $c(p_i) = -\sum_{i=1}^{n} a_i \log(1 - p_i)$ with $a_i > 0$ specified for all i, p_i^*, the optimally allocated reliability of component i, $i = 1, \ldots, n$, is given by

$$p_i^* = \frac{\lambda \widehat{P}}{a_i + \lambda \widehat{P}}, \tag{10.26}$$

where $\widehat{P}$ is a solution to the polynomial in P

$$\prod_{i=1}^{n} (a_i + \lambda P) - \lambda^n P^{n-1} = 0.$$

If the polynomial above has no roots then we set each $p_i^* = 0, i = 1, \ldots, n$. If the polynomial has two roots, we choose the larger of the two roots. If the polynomial has positive roots, but if the roots yield p_i^*'s which when plugged into the expected utility $U(\mathbf{p})$ result in negative values of $U(\mathbf{p})$, we again set $\mathbf{p}^* = \mathbf{0}$.

When n is small, say 2 or 3, the polynomial of (10.25) can be analytically solved without much difficulty. When n is large we may choose to numerically solve for the fixed point of $P = F(P)$ using an iterative approach. A proof of convergence of this approach, and details about its implementation are given in section 5.1 and Appendix B of Falk, Singpurwalla and Vladimirsky (2006). The examples below illustrate the workings of the procedure summarized above.

Example 10.1. Suppose that $n = 2$ with $a_1 = 1, a_2 = 2$, and $\lambda = 3$. The polynomial equation is $9P^2 + 2 = 0$; it has no real solution. Thus we set $p_1^* = p_2^* = 0$. That is, we do not build and operate the series system.

Example 10.2. Suppose that in Example 10.1 above, $\lambda = 6$. Then the solution to the ensuing polynomial $36P^2 - 18P + 2 = 0$ is $P_1 = 1/3$ and $P_2 = 1/6$. These roots yield, via (10.26), $(p_1^* = 2/3, p_2^* = 1/2)$ and $(p_1^* = 1/2, p_2^* = 1/3)$ as choices for p_1 and p_2. However, these choices when plugged into (10.23) for $U(\mathbf{p})$ yield -0.4849 and -0.5041 as expected utilities. Since the expected utilities are negative, we again must set $p_1^* = p_2^* = 0$, and not build the system.

Example 10.3. Suppose now that in Examples 10.1 and 10.2 above, $\lambda = 10$. Then the ensuing polynomial equation $100P^2 - 70P + 2 = 0$ has $P_1 = 2.9844 \times 10^{-2}$ and $P_2 = 0.67106$ as its roots. These roots yield $(p_1^* = 0.22985, p_2^* = 0.12984)$ and $(p_1^* = 0.87016, p_2^* = 0.77016)$ as a pair of choices for p_1 and p_2. Plugging this pair into the expression for $U(\mathbf{p})$ yields -0.24089 and 1.7194 as values for the expected utility, respectively. We therefore choose the pair $(p_1^* = 0.87016, p_2^* = 0.77016)$ as an optimal reliability allocation.

10.5.3 Reliability Apportionment for Parallel Redundant Systems

The case of a parallel system with independent lifelengths is most easily described by considering the dual of the normative formulation given in section 10.5.1. Specifically, let $q_i = 1 - p_i, i = 1, \ldots, n$, and let $h(\mathbf{q}) = 1 - h(\mathbf{p})$, where $\mathbf{q} = (q_1, \ldots, q_n)$. Note that $h(\mathbf{q})$ is the unreliability of the system, which in the case of a parallel-redundant system is $h(\mathbf{q}) = \prod_{i=1}^n q_i$. Following arguments that parallel those leading us to (10.19) we can see that the total expected utility, written in terms of $\mathbf{q}$, is

$$\widetilde{U}(\mathbf{q}) = [1 - h(\mathbf{q})][\pi^+ - c(\mathbf{q})] - h(\mathbf{q})[\pi^- + c(\mathbf{q})],$$

where $c_i(q_i) = c_i(1 - p_i)$ and $c(\mathbf{q}) = \sum_{i=1}^n c_i(q_i)$. Setting $c_i(q_i) = -a_i \log(q_i), i = 1, \ldots, n$, with $a_i > 0$, $h(\mathbf{q}) = \prod_{i=1}^n q_i$ and as before $\lambda = [\pi^+ + \pi^-]$, the ensuing optimization problem boils down to finding $q_1, \ldots, q_n$ that maximize

$$U(\mathbf{q}) = -\lambda \prod_{i=1}^{n} q_i + \sum_{i=1}^{n} a_i \log q_i, \tag{10.27}$$

subject to the constraint that $0 < q_i \leq 1$.

The solution to this problem is encapsulated in Theorem 5.4 of Falk, Singpurwalla and Vladimirsky (2006), which says that if $a_j = \min(a_i; i = 1, \ldots, n)$ is unique, then:

(*i*) When $\lambda \leq a_j$, the optimal allocation would be to set $\mathbf{p}^* = \mathbf{0}$, i.e. the parallel redundant system should not be built and operated.

(*ii*) When $\lambda > a_j$, $p_j^* = 1 - a_j/\lambda$ and $p_i^* = 0$ for all $i \neq j$; in this case the expected utility is $a_j \log(a_j/\lambda) - a_j + \pi^+$.

The essential import of (*ii*) above is that the parallel redundant system of n components (with independent lifelengths) should be collapsed to a single component system, namely, the component that is the least expensive to build, and the allocated reliability of this component should be $1 - a_j/\lambda$. That is, it is only the cheapest to build components that should be strengthened – under the cost structure assumed by us here.

10.5.4 Apportioning Node Reliabilities in Networks

In the case of a network, such as say that of Figure 10.2, $h(\mathbf{p})$ takes a cumbersome form, even when the node lifetimes are assumed independent (equation (10.6)). As a consequence, an analytical solution to the optimization problem of (10.21) becomes difficult. One way out of this difficulty would be to consider optimization algorithms, such as the 'branch and bound' of Falk and Soland (1969). The use of this algorithm for the optimum allocation of node reliability for the network of Figure 10.2 is described in Falk, Singpurwalla and Vladimirsky (2006). The essential idea here is to set each p_i as $p_i = 1 - \exp(-w_i)$ for some $w_i \geq 0$, $i = 1, \ldots, n$, and to consider the resulting expression for $U(\mathbf{p})$ in terms of the w_i's. This expression turns out to be the sum of exponential terms upon which the branch and bound algorithm with linear constraints is invoked; the specifics are in section 5.3 of Falk, Singpurwalla and Vladimirsky (2006). With $a_1, \ldots, a_5$ chosen as 3,2,2,3 and 2, respectively, and $\lambda = 30.15$, the optimum allocation of node reliabilities turns out to be $p_1^* = 0.892$, $p_2^* = 0.928$, and $p_3^* = p_4^* = p_5^* = 0$. This tantamounts to allocating all the resources to only one min. path set of the network, namely the set (1,2). Alternatively put, the five-node network gets replaced by a two-node series system. This example illustrates how the reliability allocation problem facilitates system design.

10.5.5 Apportioning Reliability Under Interdependence

For systems with interdependent lifetimes an optimal allocation of component reliability requires that one be willing to specify the dependency parameter. Thus, for example, in the case of a two component series system with dependent lifetimes whose stochastic behavior is described by the bivariate exponential of Marshall and Olkin (1967) (section 4.7.4) – it can be seen that the reliability of the system (the $h(\mathbf{p}, \bullet)$ of section 10.5.1) can be written as

$$R_S(\tau) = R_1(\tau) R_2(\tau) \exp(\theta_{12} \tau),$$

where $R_i(\tau)$ is the marginal reliability of the i-th component, $i = 1, 2$, τ is the mission time, and θ_{12} is the dependency parameter. With θ_{12} specified, we may replace the $\prod_{i=1}^{n} p_i$ of (10.23) with $\exp(\theta_{12}\tau) \prod_{i=1}^{2} R_i(\tau)$, and then invoke the material of section 10.5.2 to obtain optimum values for $R_i(\tau), i = 1, 2$. With $n > 2$ components in series, we need to specify several dependency parameters and follow the above line of reasoning. However, now we may have to use the fixed point iterative approach of section 10.5.2 to solve the ensuing optimization problem.

Because the model for interdependency prescribed by the BVE of Marshall and Olkin (1967) is not parsimonious in the number of dependency parameters, one may prefer to work with alternate, more parsimonious, models for dependency. One such possibility is the bivariate Pareto of Lindley and Singpurwalla (1986) (section 4.7.5) for which it can be seen that for a series system of n components, the reliability for a mission time τ can be written as

$$R_S(\tau) = \left(\sum_{i=1}^{n} (R_i(\tau))^{-1/(a+1)} + n - 1 \right)^{-(a+1)}, \tag{10.28}$$

where $a > -1$ is a dependency parameter that characterizes the nature of the environment in which the system operates, and

$$R_i(\tau) = \left(\frac{b}{b + \lambda_i \tau}\right)^{a+1}, \tag{10.29}$$

is the marginal reliability of the i-th component, $i = 1, \ldots, n$.

Replacing the $\prod_{i=1}^{n} p_i$ term of (10.23) by $R_5(\tau)$ given above, and the p_i's in the terms $\sum_{i=1}^{n} a_i \ln(1 - p_i)$ by the $R_i(\tau)$s given above, it can be shown (Falk, Singpurwalla and Vladimirsky, 2006) that solving the ensuing expected utility maximization problem boils down to finding the fixed points of $P = F^*(P)$, where

$$F^*(P) = \left(\sum_{i=1}^{n} \left(G_i^{-1}(P)\right)^{-1/w} - n + 1\right)^{-w}; \tag{10.30}$$

here $w = a + 1$ and P is a function of $R_i(\tau)$, say $P = G_i(R_i(\tau))$, $i = 1, \ldots, n$.

With $w = a + 1$ specified, I may, via the fixed points of $P = F^*(P)$, find the optimum values of λ_i and b, $i = 1, \ldots n$. These tantamount to finding the optimal values $R_i(\tau)$ – the marginal component reliabilities – $i = 1, \ldots, n$, under an operating environment characterized by w. Denote these optimal values by $R_i^{(w)}(\tau)$, $i = 1, \ldots, n$. The $R_i^{(w)}(\tau)$s when plugged into the expression

$$\sum_{i=1}^{n} a_i \log\left(1 - R_i^{(w)}(\tau)\right) + \lambda \left(\sum_{i=1}^{n} \left(R_i^{(w)}(\tau)\right)^{-1/w} + n - 1\right)^{-w}, \tag{10.31}$$

give us the expected utility yielded by $R_i^{(w)}(\tau)$s for the specified w. Repeating the above exercise for different values of w enables us to chose that w, say w^*, for which the expected utility of (10.31) is a maximum.

Thus the schemata described above enables us to not only allocate component reliabilities $R_i^{w^*}(\tau)$, $i = 1, \ldots, n$, in an optimal manner, but also determine an optimal operating environment for the system. In Falk, Singpurwalla and Vladimirsky (2006), this schemata is illustrated via a numerical example that entails a range of values of w. The general observation therein, is that the assumption of independence does not result in a cost-effective allocation of component reliabilities for series systems operating in a common environment. Accounting for positive dependence leads us to build systems that are cost effective.

Thus to conclude this section, one may claim that the principle of maximization of expected utility when invoked in the context of designing systems for maximum survivability, compels one to incorporate, when appropriate, interdependence. It also enables one to simplify design by eliminating components whose allocated reliability turns out to be zero. In the context of networks this latter feature is known as 'network collapsibility'.

10.6 THE UTILITY OF RELIABILITY: OPTIMUM SYSTEM SELECTION

This section can be viewed as a companion to section 10.5 because it to pertains to decision making in system reliability considerations. However, its scope is more general. In section 10.5 we were interested in optimal system design, an issue that is relevant to designers and makers of systems. Here we consider the matter of optimally choosing one among several competing designs, a matter that is of concern to system procurers or those who select a particular system to perform a specified task. In these latter circumstances, what matters is the utility of reliability, a notion that is a topic of this section. But first some background and perspective.

Recall, that in section 10.5, utilities were simply costs and that the decision variables were the component reliabilities. In actuality, such decisions cannot be executed with precision, because reliabilities are not tangible quantities; i.e. they cannot be directly observed. The material of section 10.5 therefore suffers from a drawback, because it can merely provide guidance on the proportion in which resources can be allocated. In real-life decision making, actions have to be executable, and utilities must also incorporate behavioristic considerations. In the context of reliability, behavioristic considerations generally manifest themselves in the form of risk aversion. For example, many individuals prefer driving long distances to flying, and many societies choose to avoid the generation of nuclear power. In such scenarios cost alone is not the driving criterion. What also matters here is the subjectivistic notion of utility as seen by behavioral scientists, economists, and preference theorists. In the context of system procurement, a question may also arise as to how much more should be paid for a unit improvement in reliability.

To address issues of the kind mentioned above, we need to do two things:

(*a*) Formally introduce the notion of the utility of reliability and specify its functional form; and
(*b*) Embed this notion within the broader framework of decision under uncertainty as is described in section 1.4 and Figure 1.1.

We are able to come to terms with (*b*) because of our claim (section 4.4) that 'reliability is a chance' and that 'probability is our uncertainty about this chance'. Without this distinction between chance and probability it is not possible to prescribe a coherent decision-making procedure that entails the utility of reliability. The scenario described below, in section 10.6.1, clarifies the schemata for doing the above.

10.6.1 Decision-making for System Selection

The following archetypal problem arises in the context of system procurement and selection. A decision maker $\mathcal{D}$ needs to acquire a system that is able to perform a certain task. The task takes τ time units to be performed; i.e. τ is the mission time, assumed known and fixed. $\mathcal{D}$ has a choice between (say) two systems S_1 and S_2. Let $d_1\,(d_2)$ denote $\mathcal{D}$'s action when system S_1 (S_2) is chosen (Figure 10.7). Let $T_1\,(T_2)$ denote the time to failure of $S_1\,(S_2)$, and $R_1(\tau)$ and $R_2(\tau)$ their respective reliabilities (or chances) for mission time τ. Finally, let $\pi(R_i(\tau))$, $i=1,2$, denote a probability density at $R_i(\tau)$, that encapsulates $\mathcal{D}$'s uncertainty about $R_i(\tau)$, $0 < R_i(\tau) < 1$.

As an example of the above, suppose that T_i has an exponential distribution with scale parameter $\lambda_i > 0$, $i=1,2$, where λ_i is unknown. Then $R_i(\tau)=\exp(-\lambda_i\tau)$, and if $\mathcal{D}$'s uncertainty about λ_i is described by a gamma distribution with scale parameter ψ_i and shape parameter α_i,

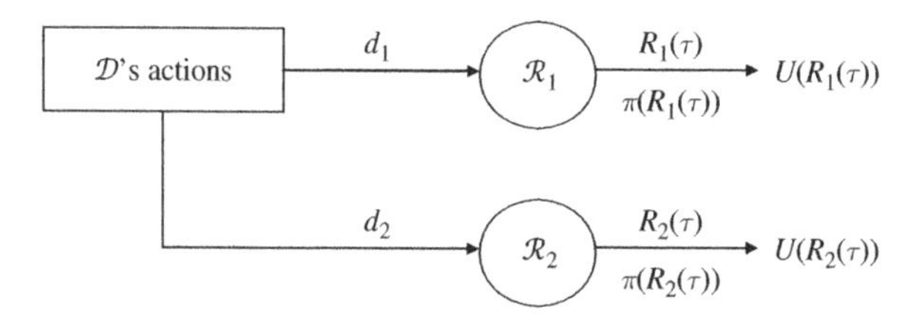

Figure 10.7 Decision tree for system selection.

then $\mathcal{D}$'s uncertainty about $R_i(\tau)$ will be described by a distribution whose probability density at $R_i(\tau)$ will be of the form (cf. equation (5.12))

$$\frac{(R_i(\tau))^{\psi_i/\tau-1}}{\Gamma(\alpha_i)}\left(\frac{\psi_i}{\tau}\right)^{\alpha_i}\left(\log\left(\frac{1}{R_i(\tau)}\right)\right)^{\alpha_i-1};$$

this then is the essence of $\mathcal{D}$'s $\pi(R_i(\tau))$, $i=1,2$. $\mathcal{D}$'s decision tree for choosing between S_1 and S_2 is given in Figure 10.7. For $\mathcal{D}$ to invoke the principle of maximization of expected utility, one additional ingredient is required, namely, $U(R_i(\tau))$, the utility of $R_i(\tau)$, $i=1,2$. Some plausible forms that $U(R_i(\tau))$ can have will be discussed in section 10.6.2. Ignoring (for now) the costs of building S_1 and S_2, $\mathcal{D}$ will take action d_1, i.e. choose system S_1, if

$$U(R_1(\tau))\cdot\pi_1 \geq U(R_2(\tau))\cdot\pi_2; \tag{10.32}$$

otherwise $\mathcal{D}$ will take action d_2. The crux of the problem therefore is a meaningful specification of $U(R_i(\tau))$, $i=1,2$. In writing out the above inequalities we have taken the point of view that when decision d_i is chosen, $R_i(\tau)$ is the unknown state of nature, and that $\pi(R_i(\tau))$ is $\mathcal{D}$'s uncertainty about $R_i(\tau)$, $i=1,2$. This then is the basis of our claim that without the distinction between reliability as a chance, and that probability as our uncertainty about chance, it is not possible to formally specify a paradigm for decision-making that involves the utility of reliability.

10.6.2 The Utility of Reliability

In principle the utilities of any decision maker have to be assessed via gambles of the kind described in utility theory (section 2.8.1). Furthermore, their coherence has to be ensured using methods such as those prescribed by Novick and Lindley (1979). This could be laborious, time consuming, and a one of a kind exercise from which general strategies may be hard to deduce. However, in some cases it may be meaningful to propose some prototype forms for the utility function, and the utility of reliability is one such instance. Another scenario is the utility of money which is often assumed to be concave, such as a logarithmic function with an upper bound. We need only limit our discussion here to bounded utilities.

With the above as a preamble, consider the case of a fixed mission time τ, for which the reliability of a unit is $R(\tau)$, with $0\leq R(\tau)\leq 1$. Let $U(R(\tau))$ denote the utility of $R(\tau)$; clearly, $U(R(\tau)=0)=0$ and set $U(R(\tau)=1)=1$. Also, $U(R(\tau))$ must be non-decreasing in $R(\tau)$; the bigger the reliability, the more valuable the unit. Our conjecture is that for large mission times, $U(R(\tau))$ will be a concave function of $R(\tau)$; $U(R(\tau))$ will be convex for small mission times. With that in mind we assume that for some tuning parameter $\beta\in[0,\infty)$,

$$U(R(\tau))=(R(\tau))^{\beta/\tau}. \tag{10.33}$$

When $\beta>(<)\,\tau$, the utility function is convex (concave) in $R(\tau)$; it is linear when $\beta=\tau$. Thus once a mission time is specified, β encapsulates a decision maker's attitude towards risk. If the decision maker is risk neutral then $\beta=\tau$; if the decision maker is risk averse (prone) then $\beta>(<)\,\tau$. In general, when it comes to matters of reliability, most individuals tend to be risk averse and so their utilities for reliability are most likely to be convex (Figure 10.8).

Incorporating Disutilities Due to Costs

When making decisions about choosing between competing systems, as was described in section 10.6.1, one also needs to incorporate into one's analyses, the cost of building each system.

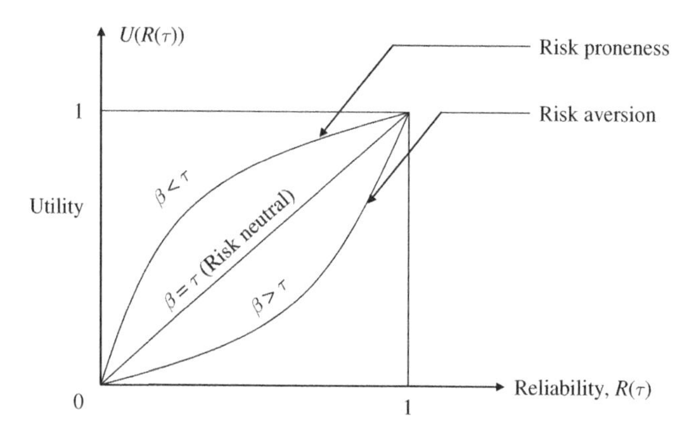

Figure 10.8 Concave and convex utilities of reliability.

Such costs tantamount to a **disutuility** to a decision maker; their effect is to lower the utility. Generally, the more reliable a system, the higher its costs and thus the greater its disutility. In making decisions about system choice, the decision maker trades off between the rewards of higher value versus the penalty of higher costs. To do so, we first need to propose a disutility function. The disutility function should map on the same scale as the utility function so that the two can be combined by addition.

Suppose that $C(R(\tau))$ denotes the cost of producing a unit whose reliability for mission time τ is $R(\tau)$. It is reasonable to suppose that $C(R(\tau))$ increases in $R(\tau)$, taking the value 0 when $R(\tau)=0$ and infinity when $R(\tau)$ goes to 1; it certainty is prohibitively expensive to have! The above features are encapsulated by the function

$$C(R(\tau)) = \frac{\alpha R(\tau)}{1 - R(\tau)}, \tag{10.34}$$

where the scaling constant $\alpha \in [0, \infty]$ is chosen so that the cost of manufacturing to $R(\tau)$ is meaningfully reflected. Figure 10.9 illustrates the behavior of $C(R(\tau))$ for $\alpha = 1$.

Now, it is well accepted that for most individuals the utility of money is concave; indeed the logarithmic function is often assumed. With that in mind we may suppose that the disutility inflicted on the decision maker due to $C(R(\tau))$, the cost of producing to $R(\tau)$, is of the logarithmic form

$$-U(C(R(\tau))) = \log(1 + C(R(\tau))).$$

However this form is unbounded for $R(\tau) = 1$. An alternative functional form that retains the concavity of the disutility of $R(\tau)$, and is bounded above by one, would be

$$-U(C(R(\tau))) = 1 - \exp\left(-\frac{\alpha R(\tau)}{1 - R(\tau)}\right). \tag{10.35}$$

Combining the utility and the disutility of $R(\tau)$, (10.33) and (10.35), we have the net utility provided by a unit that has been produced to deliver $R(\tau)$ as

$$\widetilde{U}(R(\tau)) = (R(\tau))^{\beta/\tau} - \left(1 - \exp\left(-\frac{\alpha R(\tau)}{1 - R(\tau)}\right)\right). \tag{10.36}$$

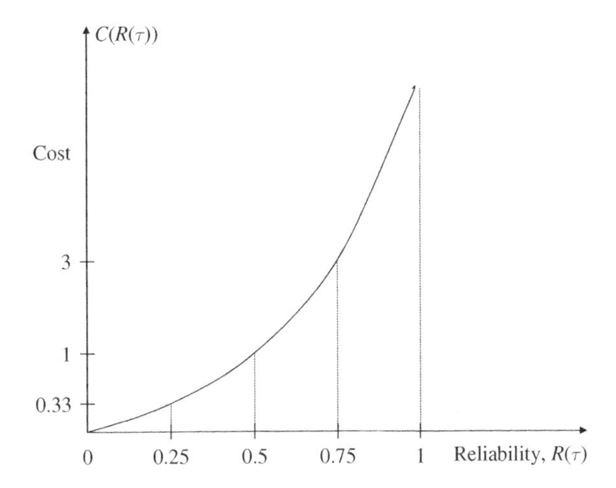

Figure 10.9 Cost of producing to $R(\tau)$.

In the context of scenario of section 10.6.1, $\mathcal{D}$ will replace $U(R_i(\tau))$ with $\widetilde{U}(R_i(\tau))$ in (10.32). That is, system S_1 will be chosen if

$$\widetilde{U}(R_1(\tau)) \cdot \pi_1 \geq \widetilde{U}(R_2(\tau)) \cdot \pi_2; \tag{10.37}$$

otherwise system S_2 will be chosen. $\mathcal{D}$ will chose neither system if both $\widetilde{U}(R_1(\tau))$ and $\widetilde{U}(R_2(\tau))$ turn out to be zero or negative. In such cases the cost of acquisition does not justify the value received.

10.7 MULTI-STATE AND VAGUE STOCHASTIC SYSTEMS

The elements of the theory of coherent systems have been overviewed by us in section 10.2. This theory has been driven by two assertions, both limiting. The first is that a unit (i.e. the system or its components) can exist at any point in time t, in one of only two possible states: failed or functioning, good or bad, etc. With only two states to consider, the calculus of the theory can be based on binary logic. In actuality, units can exist in more than two states, such as functioning, degraded, or failed. More generally, a system with n components can exist in $n+1$ states if the state of the system at t is defined as the number of its surviving components. Systems that can exist in more than two states are called **multi-state systems** (cf. Natvig, 1982). They have been studied by several authors, who in one way or other have continued to lean on binary logic for their study. El-Neweihi and Proschan (1984) give a survey. Thus, in effect, attempts to alleviate the first limitation have been made, albeit in a manner that is not radically novel.

The second limitation of coherent structure theory is that at any time t, each item can exist in only one of the two states. This assertion is not problematic when the states can be sharply delineated, such as functioning or failed, or the number of surviving components at t. When the states cannot be precisely delineated, one is unable to classify the state of an item. When such is the case, it is possible for one to declare that an item simultaneously exists in more than one state. Thus, for example, since it is difficult to delineate the boundary between good and bad, or the boundary between degraded and bad, it is possible to declare that an item is good and bad, or degraded and bad, at the same time. This matter calls for more discussion, but in general, whenever the subsets of a set of states are difficult to define, classifying the state of an item becomes an issue. Sets whose subsets cannot be sharply defined are called vague, and **vague systems** are those whose state space is vague.

The aim of this section is to propose a calculus for the study of multi-state and vague systems. Our view is that many-valued logic provides a common platform for doing the above (Sellers and Singpurwalla, 2006) for the multi-state system scenario. But before dwelling into many-valued logic, it is desirable to articulate more on the matter of vagueness, especially as it relates to the topic of this chapter.

10.7.1 Vagueness or Imprecision

Consider a system with n components. The system and its components exist in a set of states $\mathcal{S} \subseteq [0, 1]$. In the case of binary state systems, $\mathcal{S} = \{0, 1\}$. Consider a generic element, say x, of $\mathcal{S}$. Suppose that at some point in time, we inspect the system and declare that its state is x. If we are able to place this x in a well-defined subset of $\mathcal{S}$, then we claim that the states of the system can be classified with precision. However, there can be scenarios wherein the identification of a state can be done unambiguously, but the classification of this state cannot; this is the case of classification with vagueness. Classification with vagueness typically arises when one uses natural language (Zadeh, 1965) to describe the system, and to make decisions about it based on verbal descriptions. Here is an illustration.

Suppose that $\mathcal{S} = \{0, 1, \ldots, 10\}$, with each element representing the state of the system at any time. Suppose that 10 denotes the ideal state and 0 the most undesirable state. Now suppose that for the purpose of natural language processing I need to partition $\mathcal{S}$ into the subsets of 'good states' and 'bad states'. What then is the subset of good states? For example, is 7 a good state? What about 5; is it a good state or a bad state? Clearly, the subset of good and bad states cannot be sharply defined. As a consequence 5 is simultaneously a good state and a bad state. Thus, if at any point in time the state of the system is 5, then the system can simultaneously exist in multiple states. As another example, should an automobile with, say, 1250 miles on it be classified as a 'new car' or as a 'used car'? Classifications of this kind become germane when setting insurance rates, honoring warranties, or levying taxes. In actuality many decisions are often made on the basis of vague classifications. This is especially true in the health sciences where treatment options are based on classifications that are imprecise, such as 'high blood pressure', or 'bad cholesterol levels'. In the context of reliability, decisions based on imprecise classifications often occur in the arena of maintenance management wherein classifying the state of a system, such as a vehicle, as 'good', 'bad' or 'acceptable' is common. The decision tree of Figure 10.10 illustrates this point. Similarly with the decisions pertaining to quality-of-life considerations (Cox *et al.*, 1992) wherein the responses always tend to be imprecise.

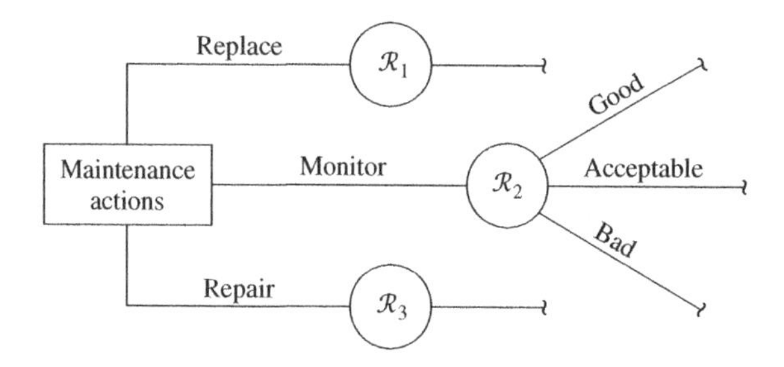

Figure 10.10 Maintenance decisions in a vague environment.

Motivated by the above scenarios, the need to consider a theory of coherent systems based on vague classifications seems appropriate. The existing theory for both binary and multi-state systems, with precise classification as an underlying premise, is unable to deal with the types of scenarios mentioned above. Related concerns have also been voiced by Marshall (1994).

10.7.2 Many-valued Logic: A Synopsis

Binary logic, upon whose foundation the theory of coherent structures has been developed, pertains to propositions that adhere to the Law of Bivalence (or the Law of the Excluded Middle): i.e. all propositions are true or false. Lukasiewicz (1930) recognized the existence of propositions that can be both true and false simultaneously, and thus modified the calculus of binary propositions to develop a calculus of three-valued propositions. His proposal considers two propositions Y and Z, each taking the values 0, $\frac{1}{2}$ and 1. The negation of Y is $Y' = 1 - Y$. When the proposition Y takes the value 1(0) in a truth table, it signals the fact that the proposition is true (false) with certainty. Values of Y intermediate to 1 and 0 signal an uncertainty about the truth or the falsity of Y. The value $\frac{1}{2}$ is chosen arbitrarily, for convenience; any value between 0 and 1 could have been chosen. The other logical connectives in the three-valued logic of Lukasiewicz are conjunction, disjunction, implication and equivalence, denoted $(Y \wedge Z)$, $(Y \vee Z)$, $(Y \rightarrow Z)$ and $(Y \equiv Z)$, respectively. The truth tables for the first two are given in Tables 10.1 and 10.2, respectively.

10.7.3 Consistency Profiles and Probabilities of Vague Sets

The philosopher Black (1939) recognized the inability of binary logic to represent propositions that are neither perfectly true nor false. However, unlike Lukasiewicz (1930), Black did not introduce three-valued propositions. Rather, he defined a vague proposition as one where the possible states of the proposition are not clearly defined with respect to inclusion, and introduced the mechanism of **consistency profiles** as a way of treating vagueness. Black's consistency profile is a graphical portrayal of the degree of membership of some proposition in a set of imprecisely defined states, with 1 representing absolute membership in a state and 0 an absolute lack of membership. The consistency profile of precise propositions is a step function, and that

Table 10.1 Truth table for Lukasiewicz's $Y \wedge Z$

$Y \wedge Z$		Values of proposition Z		
		0	1/2	1
Values of proposition Y	0	0	0	0
	1/2	0	1/2	1/2
	1	0	1/2	1

Table 10.2 Truth table for Lukasiewicz's $Y \vee Z$

$Y \vee Z$		Values of proposition Z		
		0	1/2	1
Values of proposition Y	0	0	1/2	1
	1/2	1/2	1/2	1
	1	1	1	1

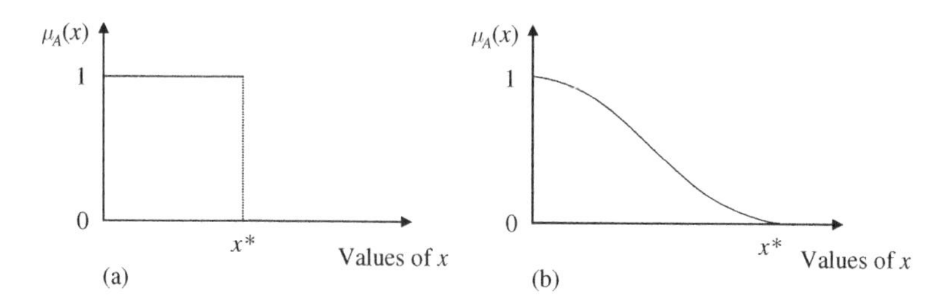

Figure 10.11 Illustration of consistency profiles of (a) a precise set; (b) a vague set.

of vague propositions is a gracefully decreasing function of the kind shown in Figure 10.11(b). The scaling between 0 and 1 is arbitrary; consistency profiles are personal to an individual, or a group, and therefore need not be unique.

For each x, a normalized consistency profile is a function $\mu_A(x)$, where $\mu_A(x)$ describes the degree of containment of x in a set A; $0 \leq \mu_A(x) \leq 1$. When $\mu_A(x) = 1$ or 0, A is a precise set; when $0 \leq \mu_A(x) \leq 1$, A is a vague set. Black's (1939) consistency profile is Zadeh's (1965) membership function, and his vague set is Zadeh's (1965) fuzzy set. For two vague sets A and B with consistency profiles $\mu_A(x)$ and $\mu_B(x)$, respectively, Zadeh (1965) defined the operations:

(a) $\mu_{A\cup B}(x) = \max(\mu_A(x), \mu_B(x))$,
(b) $\mu_{A\cap B}(x) = \min(\mu_A(x), \mu_B(x))$, and
(c) $\mu_{A'}(x) = 1 - \mu_A(x)$.

Thus the union of A and B is a vague set $A \cup B$ whose consistency profile is $\max(\mu_A(x), \mu_B(x))$; similarly the intersection. The important point here is that there is a parallel between the above vague set operations and the conjunction and disjunctive connectives of Lukasiewicz (1930). Later on, in section 10.7.5, I shall use the above operations to define structure functions of vague binary state systems. Thus Lukasiewicz's logic provides a framework for studying vague coherent systems.

Probability Measures of Vague Sets

A strategy for endowing probability measures to vague sets using consistency profiles has been outlined by Singpurwalla and Booker (2004). Their strategy is best explained by conceptualizing an assessor of probabilities (or a decision maker) $\mathcal{D}$, who wants to quantify the uncertainty about any outcome x (of an unknown quantity X) belonging to a vague set A. $\mathcal{D}$'s prior probability that x belongs to A is $\pi_{\mathcal{D}}(x \in A)$; the prior probability here reflects $\mathcal{D}$'s perception as to how nature would classify x. To update this prior probability, $\mathcal{D}$ consults an expert $\mathcal{Z}$, and elicits from $\mathcal{Z}$ a consistency profile $\mu_A(x)$ for A; that is, the degree to which x can belong to A. Assuming X to be discrete, $\mathcal{D}$'s probability measure on the vague set A is

$$P_{\mathcal{D}}(X \in A; \mu_A(x)) = \sum_x \left[1 - \left(1 - \frac{1}{\mu_A(x)}\right) \frac{\pi_{\mathcal{D}}(x \notin A)}{\pi_{\mathcal{D}}(x \in A)}\right]^{-1} P_{\mathcal{D}}(X), \tag{10.38}$$

where $P_{\mathcal{D}}(X)$ is $\mathcal{D}$'s probability that outcome x will occur. The detailed arguments leading upto (10.38) are in Singpurwalla and Booker (2004).

10.7.4 Reliability of Components in Vague Binary States

To invoke the material of the previous section in the context of component reliability, suppose that X denotes the state of a component taking values in $\mathcal{S}=\{x:0\leq x\leq 1\}$, with $x=1$ representing the perfectly functioning state. Let $\mathcal{G}\subset\mathcal{S}$, where $\mathcal{G}=\{x:x \text{ is a 'desirable' state}\}$; interest centers around $\mathcal{G}$. If we are unable to specify an x^* such that $x\geq x^*$ implies that $x\in\mathcal{G}$, and otherwise when $x<x^*$, then $\mathcal{G}$ is a vague set. Let $\mu_{\mathcal{G}}(x)$ be the consistency profile of $\mathcal{G}$. Interest may center around $\mathcal{G}$ for several reasons, one being the need to use 'natural language' for communication with others on matters of repair and replacement, the other being that it may not be possible to observe the actual value of x, but it may be possible to make a statement about the general state of the component. The complement of $\mathcal{G}$, say $\mathcal{G}^c$, is the vague set whose consistency profile is $1-\mu_{\mathcal{G}}(x)$.

With the above in place, we define the ***reliability of a component with a vague state space*** as

$$P_{\mathcal{D}}(X\in\mathcal{G};\mu_{\mathcal{G}}(x)),$$

and use (10.38) to evaluate it.

For the *unreliability* of a component with a vague state space we may use $P_{\mathcal{D}}(X\in\mathcal{G}^c;\mu_{\mathcal{G}}(x))$. Alternatively, we could define a vague set $\mathcal{B}$, where $\mathcal{B}=\{x:x \text{ is an 'undesired' state}\}$, and specify $\mu_{\mathcal{B}}(x)$, the consistency profile of $\mathcal{B}$. Then the unreliability of the component could be defined as $P_{\mathcal{D}}(X\in\mathcal{B};\mu_{\mathcal{B}}(x))$. With either choice as a definition of unreliability, unreliability is not the complement of reliability! This result is in contrast to that for binary state systems and components.

Note that $\mu_{\mathcal{B}}(x)$ need not bear any relationship to $\mu_{\mathcal{G}}(x)$. For example, in Figure 10.12(a), $\mu_{\mathcal{B}}(x)$ and $\mu_{\mathcal{G}}(x)$ are symmetric, whereas in Figure 10.12(b), they are not.

10.7.5 Reliability of Systems in Vague Binary States

In this section, I extend the development of the previous section on binary state components with imprecise classification, to the case of binary state n-component systems with imprecise classification. However, to do so, I first need to define structure functions that relate the vague component states to the vague system states. Motivated by the feature that the structure functions of binary state systems with precise classification (i.e. those overviewed in section 10.2) are the consistency profiles of certain precise sets, I propose the logical connectives of Lukasiewicz (1930) as providing the basis for the required structure functions. The specifics, abstracted from Sellers and Singpurwalla (2006), are outlined below.

Suppose that X_i, the state of the component i, takes values x_i, where $x_i\in\mathcal{S}=\{x:0\leq x\leq 1\}$, $i=1,\ldots,n$. Consider a subset $\mathcal{G}_i$ of $\mathcal{S}$, where $\mathcal{G}_i$ is the vague set $\mathcal{G}_i=\{x_i:x_i \text{ is an 'desirable'}$

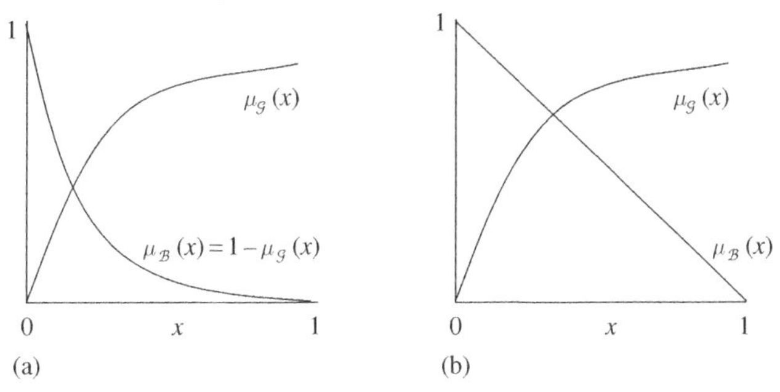

Figure 10.12 Symmetric and asymmetric consistency profiles of vague sets $\mathcal{G}$ and $\mathcal{B}$.

state}. Let $\mu_i(x_i)$ be the consistency profile of $\mathcal{G}_i$. If $\mathbf{X} = (X_1, \ldots, X_n)$, and $\Phi(\mathbf{X})$ is a structure function, then for a series system we define

$$\Phi_S(\mathbf{X}) = \min_i [\mu_i(X_i)], \tag{10.39}$$

and for a parallel system

$$\Phi_P(\mathbf{X}) = \max_i [\mu_i(X_i)]. \tag{10.40}$$

Assuming that the X_is are independent, the notion of independence when X_is takes values in a vague set being articulated in Sellers and Singpurwalla (2006), the reliability of a series system whose component states are vague will be

$$\prod_{i=1}^{n} P_{\mathcal{D}}(X_i \in \mathcal{G}_i; \mu_i(x_i)), \tag{10.41}$$

and that of a corresponding parallel redundant system will be

$$P_{\mathcal{D}}\left[\bigcup_{i=1}^{n}\{X_i \in \mathcal{G}_i; \mu_i(x_i)\}, i = 1, \ldots, n\right], \tag{10.42}$$

an expression that can be evaluated by the inclusion-exclusion formula (Feller, 1968). The quantities $P_{\mathcal{D}}(X_i \in G_i; \mu_i(x_i))$ can be evaluated via (10.38).

10.7.6 Concluding Comments on Vague Stochastic Systems

As a closing comment, I emphasize that the notion of vague component and system states is new and the need for it remains to be more convincingly argued. From a philosophical point of view the matter of making a case for vagueness may be moot, because as stated by Russell (1923), 'all language is more or less vague' so that the law of the excluded middle 'is true when precise symbols are employed but it is not true when symbols are vague, as, in fact, all symbols are'. Furthermore, even if vagueness about the states of components and systems is taken as a given, our approach for assessing the reliability of such components and systems needs to be carefully scrutinized. Are the relationships of (10.39) and (10.40) meaningful? What about (10.41) and (10.42), which are based on the premise that the X_is are independent? Is independence meaningful when the X_is take vales in vague sets? These and several other related issues need to be vetted out. Perhaps what has been described here merely scratches the surface and hopefully does not do a poor job of it. I leave this for the reader to decide. The purpose of including section 10.7 in this chapter is to whet the reader's appetite as to what else is possible when it comes to the survivability of stochastic systems and to give it a more modern twist in the sense that natural language processing could be a wave of the future.

Chapter 11

Reliability and Survival in Econometrics and Finance

Here lies Laplace: His remains are transformed.

11.1 INTRODUCTION AND OVERVIEW

The purpose of this chapter is to describe a platform that develops an interface between the notions and techniques of reliability theory and survival analysis, and the mathematics of economics and finance. This has been made possible because of three features. The first is that there exist relationships between some metrics of reliability and some measures of income inequality such as the Gini index, the Lorenz curve and the Bonferroni index of concentration. The second is that the survival function of reliability bears an isomorphic relationship with the asset pricing formula of a fixed income investment, like a riskless bond. The third feature is motivated by the possibility that the non-parametric methods of survival data analysis can, with some caveats and modifications, be used for the actual pricing of a bond in a competitive environment. Given that risk-free bonds do not fail, whereas the survival function comes into play in the context of failure, the isomorphism mentioned above raises the possibility that the exponentiation formula of (4.9) be re-interpreted from a broader perspective. This we are able to do, so now we may see the exponentiation formula as the law of a diminishing resource. In the context of reliability, the resource is an item's hazard potential (section 4.6); in the context of finance, the resource is a bond's present value. The three features mentioned above enable us to expand the scope of mathematical economics and finance by importing ideas and techniques from reliability and survival analysis to econometrics and financial risk, and vice versa. In what follows we make our case by elaborating on the above claim by specific scenarios and examples. The overall message of this chapter is that some of the material described in the previous chapters may have relevance to the economic and social sciences as well. Its relevance to the actuarial sciences is historical, whereas its relevance to the assurance sciences in the legal context is relatively new; see Singpurwalla (2000b), (2004) for an overview.

Reliability and Risk: A Bayesian Perspective N.D. Singpurwalla
© 2006 John Wiley & Sons, Ltd

11.2 RELATING METRICS OF RELIABILITY TO THOSE OF INCOME INEQUALITY

We start with some preliminaries. Let T denote a non-negative random variable having distribution function $F(t) = P(T \leq t)$; for convenience we have suppressed $\mathcal{H}$, the background information. Let the mean of $F(t)$ be μ, where $\mu < \infty$. Assume that the survival function $\overline{F}(t) = P(T \geq t) = 1 - F(t)$ has a well-defined inverse $\overline{F}^{-1}(t) = \inf_x\{x : F(x) \geq t\}$. Let $T_{(0)} \equiv 0 \leq T_{(1)} \leq \cdots \leq T_{(n)} < \infty$ be the order statistics generated by a sample of size n from $F(\bullet)$ (section 5.4.4). Then, the total time on test to the i-th order statistic is

$$T_n(i) = \sum_{j=1}^{i}(n - j + 1)(T_{(j)} - T_{(j-1)}).$$

We have seen before, in section 5.4.1, that the total time on test statistic plays a key role in life-testing and failure analysis. There are two other quantities of interest that have been spawned by $T_n(i)$; these quantities have played a role in testing for exponentiality – within the frequentist framework. These are: the **scaled total time on test statistic**

$$W_n\left(\frac{i}{n}\right) = \frac{T_n(i)}{\sum_{i=1}^{n} T_{(j)}},$$

and the **cumulative total time on test statistic**

$$V_n = \frac{1}{n-1}\sum_{i=1}^{n-1} W_n\left(\frac{i}{n}\right);$$

note that $W_n\ (i/n)$ is always between 0 and 1. For further discussion on the properties of V_n, see Bergman (1977).

11.2.1 Some Metrics of Reliability and Survival

In reliability theory and survival analysis, the theoretical analogues of $T_n(\bullet)$, $W_n(\bullet)$ and V_n are of interest. These are: the **total time on test transform**

$$H_F^{-1}(t) \stackrel{\text{def}}{=} \int_0^{F^{-1}(t)} \overline{F}(u)\mathrm{d}u;$$

the **scaled total time** on test transform

$$W_F(t) \stackrel{\text{def}}{=} \frac{H_F^{-1}(t)}{H_F^{-1}(1)}; \quad 0 \leq t \leq 1,$$

and the **cumulative total time on test transform**:

$$V_F \stackrel{\text{def}}{=} \int_0^1 W_F(t)\,\mathrm{d}u = \frac{1}{\mu}\int_0^1 H_F^{-1}(u)\,\mathrm{d}u.$$

Also of relevance to us here is the mean residual life, introduced in section 4.2, namely,

$$\epsilon_F(t) = \frac{\int_t^\infty \overline{F}(u)\,\mathrm{d}u}{\overline{F}(t)}.$$

Verify (cf. Chandra and Singpurwalla, 1981) that

$$\epsilon_F(t) = \frac{\mu[1 - W_F(F(t))]}{\overline{F}(t)}.$$

11.2.2 Metrics of Income Inequality

Analogous to the metrics introduced above are some metrics of income inequality used by econometricians. These have been motivated by the sample Lorenz curve, which for $p \in (0, 1)$, is

$$L_n(p) = \frac{\sum_{i=1}^{[np]} T_{(i)}}{\sum_{i=1}^{n} T_{(i)}},$$

and the Gini statistic

$$G_n = \frac{\sum_{i=1}^{n-1} i(n-i)(T_{(i+1)} - T_{(i)})}{(n-1)\sum_{i=1}^{n} T_{(i)}};$$

here $[np]$ is the greatest integer in np.

The theoretical analogs of $L_n(p)$ and G_n are: the *Lorenz curve* of T

$$L_F(p) \stackrel{\text{def}}{=} \frac{1}{\mu}\int_0^p F^{-1}(u)\,\mathrm{d}u, \quad 0 < p < 1,$$

and the Gini index

$$G_F \stackrel{\text{def}}{=} 1 - 2\int_0^1 L_F(p)\,\mathrm{d}p.$$

The Lorenz curve, introduced by Lorenz (1905), has been used by social scientists to characterize income size distributions. Specifically, $L_F(p)$ encapsulates the fraction of total income that the poorest p proportion of the population possesses. Note that since $L_F(p)$ increases in p, with $L_F(0) = 0$ and $L_F(1) = 1$, the Lorenz curve behaves like a distribution function with support on [0,1]. The case $L_F(p) = p$, $0 < p < 1$, is called the 'egalitarian line'. When $L_F(p) = p$, the Gini index G_F is zero; this situation describes an absence of income inequality. Large positive values of G_F connote an inequality of income, the larger the G_F the greater the disparity. Thus the Gini index has come to be a useful device for assessing the concentration of wealth in a population. As an aside, the Gini index and the Lorenz curve have also been used by Gail and Gastwirth (1978a,b) for the testing of exponentiality.

Another quantity that is also of interest to us here is the **cumulative Lorenz curve**

$$(CL)_F \stackrel{\text{def}}{=} \int_0^1 L_F(p)\,\mathrm{d}p = \frac{1}{\mu}\int_0^1\int_0^p F^{-1}(u)\,\mathrm{d}u\mathrm{d}p;$$

this is the area under the Lorenz curve and can be seen as a measure of the extent of poverty in a population.

More recently, the Bonferroni curve and the Bonferroni index have begun to gain some popularity as measures of income inequality (cf. Giorgi, 1998). Their proponents claim that the Bonferroni index is more sensitive that the Gini index to low levels of income distribution (cf. Giorgi and Mondani, 1995). The Bonferroni index B, motivated by a comparison of the partial means with the general mean, was proposed by Bonferroni in 1930. Assuming that $F(t)$ is absolutely continuous with a probability density $f(t)$ at t, Giorgi and Crescenzi (2001) define the 'first incomplete moment', and the 'partial mean' of $F(t)$ as

$$F_1(t) = \frac{1}{\mu}\int_0^t uf(u)\,\mathrm{d}u$$

and

$$\mu_t = \mu\frac{F_1(x)}{F(x)},$$

respectively, and set $B(F(t)) = \mu_t/\mu$. The Bonferroni curve is a plot of $F(t)$ versus $B(F(t))$; it lies within the unit square (Figure 11.1). If we let $F(t) = p$, then the parametric form of the Bonferroni Curve is, for $p \in (0, 1]$,

$$B(p) = \frac{1}{p\mu}\int_0^p F^{-1}(u)\,\mathrm{d}u.$$

Note that since $B(p)$ is undefined when $p \downarrow 0$, one may not claim that the Bonferroni curve necessarily starts from the origin of the $[F(t), B(F(t))]$ plane. Also, since $\mathrm{d}B(p)/\mathrm{d}p > 0$, $B(p)$ is strictly increasing in p; it can be convex in some parts and concave in others (cf. Giorgi and Crescenzi, 2001). The egalitarian line is the line joining the points (0,1) and (1,1), and B, the Bonferroni index of the income concentration is the area between the Bonferroni curve, the egalitarian line, and the ordinate. That is

$$B = \int_0^1 (1 - B(p))\,\mathrm{d}p;$$

clearly, $B \in [0, 1]$ grows as the inequality of income increases. When the concentration of income is a maximum, the Bonferroni curve $B(p)$ becomes 'J-shaped'; i.e. the lines joining the points (0,0) and (0,1), and (0,1) and (1,1). Finally, as is shown in Proposition 3 of Giorgi and Crescenzi (2001), the Bonferroni curve always lies above the Lorenz curve.

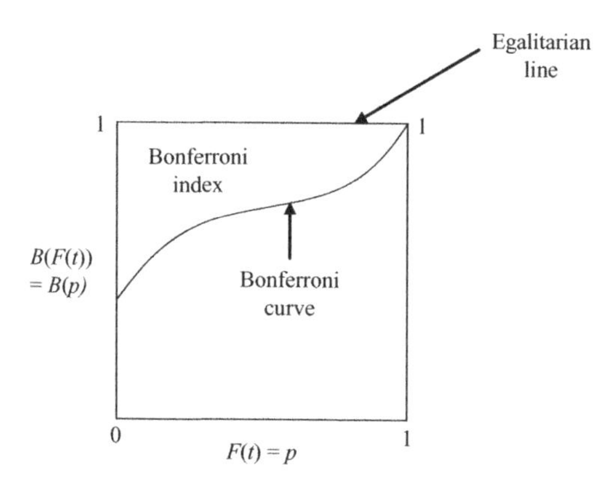

Figure 11.1 The Bonferroni curve and the Bonferroni index.

11.2.3 Relating the Metrics

In the theorems that follow, I show how the metrics of sections 11.2.1 and 11.2.2 relate to each other in a manner that is made precise by the statements of the theorems given below. This enables me to import ideas from reliability theory and survival analysis to the income disparity aspects of economic studies. To keep our exposition short, I refrain from giving proofs of all the results given below, pointing out instead to their original sources.

Theorem 11.1 (Chandra and Singpurwalla (1981), Theorem 2.1).

$$W_n\left(\frac{i}{n}\right) = L_n\left(\frac{i}{n}\right) + \frac{(n-1)T_{(i)}}{\sum_{i=1}^{n} T_{(j)}}, \qquad (a)$$

$$V_n = 1 - G_n, \qquad (b)$$

$$W_F(t) = L_F(t) + \frac{1}{\mu}(1-t)F^{-1}(t), \qquad (c)$$

$$V_F = 2(CL)_F, \qquad (d)$$

$$G_F = 1 - V_F, \qquad (e)$$

$$L_F(F(t)) = 1 - \frac{1}{\mu}\overline{F}(t)[\epsilon_F(t) + t]. \qquad (f)$$

With Theorem 11.1 in place, it is easy to obtain some relationships between some of the metrics of reliability and the Bonferroni metrics. These are summarized in Theorem 11.2 below.

Theorem 11.2

$$B(F(t)) = \frac{1}{F(t)} - \frac{1}{\mu}\frac{\overline{F}(t)}{F(t)}(\epsilon_F(t) + t), \qquad (a)$$

which relates the Bonferroni curve and the mean residual life, and

$$B \leq 1 - \frac{V_F}{2}, \qquad (b)$$

which gives an upper bound on the Bonferroni index in terms of the cumulative total time on test.

Proof. The proof of (a) above follows from part (f) of Theorem 1 and by setting $F(t) = p$ in the parametric form of the Bonferroni curve.

To prove part (b), we start by noting that since the Lorenz curve always lies below the Bonferroni curve (Proposition 3 of Giorgi and Crescenzi (2001)), we have

$$L_F(p) \leq B(p).$$

Thus

$$1 - \int_0^1 L_F(p)\,\mathrm{d}p \geq 1 - \int_0^1 B(p)\,\mathrm{d}p,$$

from which it follows, using the definition of the Gini index, that

$$B \leq \frac{1+G_F}{2}.$$

The required upper bound is now a consequence of part (e) of Theorem 1.

Besides facilitating a proof of Theorem 2, an important role served by Theorem 1 is that is suggests the use of some ideas in reliability for obtaining some interesting inequalities in economics. To see how, we first need to introduce some notions of partial ordering of distribution functions that have proved to be useful in reliability theory. These notions will also be useful in the context of the material of section 11.3.3 on asset pricing. To introduce these notions we start with

Definition 11.1. *Let $\mathcal{F}$ be the class of all continuous distribution functions on $[0, \infty)$; let $F_1, F_2 \in \mathcal{F}$. Then F_1 is said to be* ***convex ordered*** *with respect to F_2, denoted $F_1 \underset{c}{<} F_2$, if $F_2^{-1}(F_1(t))$ is convex on the support of F_1, assumed to be an interval of $[0, \infty)$. Similarly, F_1 is said to be* ***star ordered*** *with respect to F_2, denoted $F_1 \underset{*}{<} F_2$, if $F_2^{-1}(F_1(t))/t$ is non-decreasing on the support of F_1. A generalization of star ordering, denoted $F_1 \underset{*T}{<} F_2$ occurs if $F_2^{-1}(t)/F_1^{-1}(t)$ is increasing for $0 < t < 1$ (Klefsjo, 1984, Definition 2.2).*

The following are some of the consequences of Definition 1 (cf. Barlow and Proschan 1975, p. 107 and Klefsjo, 1984).

(i) $F_1 \underset{c}{<} F_2 \Rightarrow F_1 \underset{*}{<} F_2 \Rightarrow F_1 \underset{*T}{<} F_2$.
(ii) $F_1 \underset{c}{<} F_2 (F_1 \underset{*}{<} F_2)$ and F_2 is an exponential distribution $\Rightarrow F_1$ has a failure rate function that is increasing (increasing on the average).

It is conventional to denote distributions whose failure rate is increasing (decreasing) by IFR (DFR), and those whose failure rate in increasing (decreasing) on the average by IFRA (DFRA). Also, if $F(t)$ is such that if for each $\tau \geq 0$, $F(T+\tau) \geq (\leq) F(T)F(\tau)$, then F is said to be NBU (NWU); NBU (NWU) stands for 'new better (worse) than used'. We may now state the third consequence of Definition 11.1 as

(iii) F IFR (DFR) $\Rightarrow F$ IFRA (DFRA) $\Rightarrow F$ NBU (NWU).

When F_1 and F_2 represent distribution functions of lifetimes, the notions of IFR, IFRA and NBU describe the ageing and wear-out features of the items in question; see, for example, Durham, Lynch and Padgett (1989). Similarly with the notions of DFR, DFRA and NWU. The exponential distribution enjoys the special feature that is sits at the boundary, in the sense that the exponential distribution is both IFR and DFR, IFRA and DFRA, and NBU and NWU. This feature facilitates a comparison between any distribution that belongs to one of these classes (i.e. IFR, DFR, etc.) and the exponential distribution, and also between and two distributions within a class. Such comparisons can be translated to a comparison between two Lorenz curves and inequalities between Gini indices as Theorem 3 below shows. They also enable us to establish a type of preservation property inherited by the Lorenz curve, as Theorem 4 will show.

Theorem 11.3 (Chandra and Singpurwalla (1981), Theorem 4.2). *If $F_1 \overset{<}{*} F_2$ and if F_1 and F_2 have a common mean, then for all $p \in (0, 1)$:*

(i) $L_{F_1}(p) \geq L_{F_2}(p)$, *and*
(ii) $G_{F_1} \leq G_{F_2}$.

The inequality between the Lorenz curves given above suggests that F_1 represents a more equitable distribution of income than does F_2. Economists call this feature 'Lorenz domination'. Similarly with the Gini indices.

The preservation properties mentioned above are summarized in Theorem 4, (*i*) and (*ii*). Part (*i*) of the theorem is due to Chandra and Singpurwalla (1981), Theorem 4.2; part (*ii*) is due to Klefsjo (1984), Theorem 2.1. Before stating Theorem 11.4, it may be useful to remark that L_F^{-1}, the inverse of a Lorenz curve is a concave function that behaves like a distribution function with support on [0, 1]. We now state

Theorem 11.4.

(*i*) $F_1 \underset{c}{<} F_2 \Rightarrow L_{F_1}^{-1} \underset{c}{<} L_{F_2}^{-1}$, *and*
(*ii*) $F_1 \underset{*}{<} TF_2 \Rightarrow L_{F_1}^{-1} \underset{*}{<} TL_{F_2}^{-1}$.

Thus, it is not the Lorenz curves that are preserved under convex and star-T orderings, but it is their inverses that inherit the preservation proprties.

11.2.4 The Entropy of Income Shares

In addition to the Gini and Bonferroni indices there is another measure of income inequality that is motivated by the notion of entropy. Specifically, if Y_i denotes the fraction of the total income in a population earned by the i-th individual, $i = 1, \ldots, N$, then $H(y)$, the entropy of income shares is defined (section 4.6) as

$$H(y) = \sum_{i=1}^{N} Y_i \log \frac{1}{Y_i}.$$

The maximum value that $H(y)$ can take is $\log N$. Accordingly, a measure of income inequality, called the redundancy, is $(\log N - H(y))$. Thiel (1967, p. 96) has proposed a theoretical analogue of the redundancy as

$$R_F = \frac{1}{\mu} \int_0^\infty x \log \frac{x}{\mu} f(x) \mathrm{d}x,$$

where F, the distribution function (of income), has density f and mean μ.

Verify that for a Weibull distribution with shape parameter $\alpha > 1$, R_F decreases as α increases. Similarly for the Pareto distribution $F(x) = 1 - x^{-\alpha}$, with $\alpha > 1$ and $x \geq 1$. In both cases an increase in α signals a decrease in the DFR-ness of the distributions. Motivated by this feature, we have the following inequality property of the redundancy.

Theorem 11.5 Chandra and Singpurwalla (1981), Theorem 4.5). *If F_1 and F_2 have a common mean μ, and if $F_1 \underset{*}{<} F_2$, then $R_{F_1} \leq R_{F_2}$.*

11.2.5 Lorenz Curve Analysis of Failure Data

The role of the total time on test in life-testing and failure data analysis has been mentioned by before at the beginning of this section. In part (*d*) of Theorem 1, we have seen a relationship between the total time on test and the cumulative Lorenz Curve. This relationship motivates us to explore the potential usefulness of using the Lorenz curve for analyzing and interpreting lifetime data. With that in mind we seen, in Figures 11.2 and 11.3, the Lorenz curves of some data reported by Bryson and Siddiqui (1969) and by Doksum (1974), respectively. The former

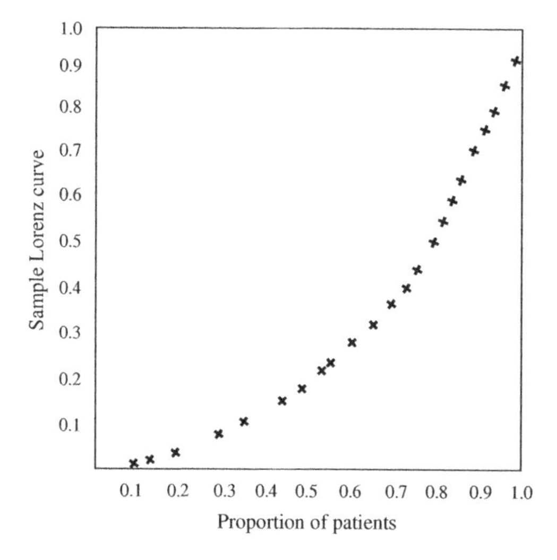

Figure 11.2 Sample Lorenz curve versus proportion of leukemia patients.

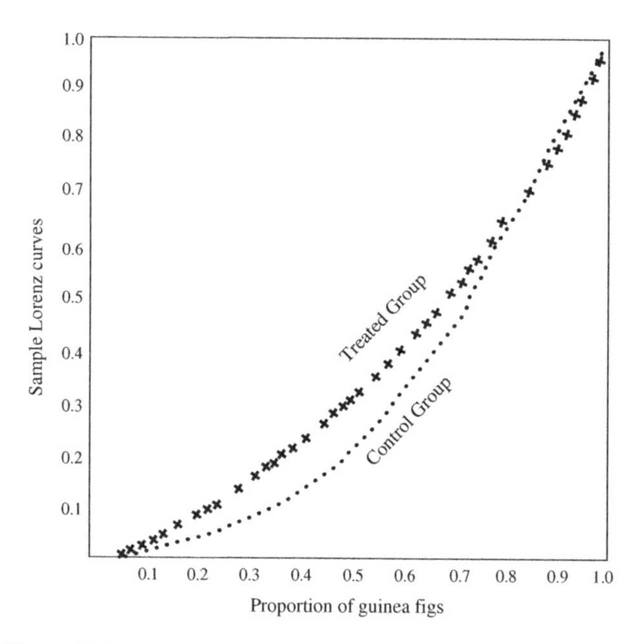

Figure 11.3 Sample Lorenz curves versus proportion of guniea pigs.

consists of survival times of patients suffering from leukemia, and the latter the survival times of a treated and a control group of guinea pigs.

The sample Lorenz curve $L_n(p)$ of Figure 11.2 represents the proportion of the total lifetime contributed by the least fortunate of the $p(100)$ percent of the patients. It is instructive to note, from this figure, that 50% of the patients contribute only 20% of the total lifetime.

The sample Lorenz curves of Figure 11.3 serve an additional purpose from that of Figure 11.2. They enable us to compare the heterogeneity of the survival pattern of the teated and the control group of guinea pigs. For $p < 0.8$, the Lorenz curve for the control group lies below the Lorenz curve for the treated group. This behavior suggests that the treated group is initially at least less heterogenous than the control group. Qualitative results of this type may be insightful with respect to assessing the efficacy of treatments.

11.3 INVOKING RELIABILITY THEORY IN FINANCIAL RISK ASSESSMENT

In the financial risk community, the word 'risk' generally equates with the volatility of an asset's price. By volatility we mean deviation from a benchmark. Is volatility a satisfactory measure of risk? I think not, because the true concern of an investor is loss due to downside volatility. The dictionary definition of risk is the possibility of suffering a harm or a loss; thus the concern of investor loss mentioned above is more in keeping with the dictionary definition of risk. Volatility could encapsulate a temporary divergence of opinions about an asset's price, a state of nervousness, but not necessarily risk itself. By contrast, a survival function does capture the essence of risk in the literal sense of the word, because it describes the degradation (or loss) of survival probability over time. The survival function could therefore be a more meaningful indicator of potential loss. But survival functions are not entities that are natural in finance, save for the case of defaults (i.e. an inability to pay). With this in mind we need to look for ways by which a survival function like behavior can be invoked in the context of finance. This we are able to do via the scenario of a risk-free fixed income asset like a bond. To see how, we start with a review of the pricing of risk-free bonds and conclude with the observation that the present value function of such bonds behaves like the survival function of an item. Once this is recognized we are able to import results from reliability and survival into the arena of finance. The material which follows is exclusively based on Singpurwalla (2006c).

11.3.1 Asset Pricing of Risk-free Bonds: An Overview

A risk-free zero coupon bond pays, with certainty, the bondholder (i.e. the buyer of the bond) \$1 at time T after the time of the purchase of the bond; T is known as the holding period of the bond. The bond is said to mature when the holding period ends. Suppose that the bond holder purchases the bond at some calendar time t, $t \geq 0$, with the intent of holding the bond until time $t+T$; let $P(t, T)$ be the purchase price of this bond. Then, $P(t, T)$ is known as the present value of this bond at time t. Clearly, $P(t, 0) = 1$ and $P(t, T)$ decreases in T. Also, $P(t, T)$ will depend on what the bond holder and the bond issuer think (or speculate) the interest rate will be during the time period $(t, t+T)$.

We assume that the interest rate $r(s)$, $s \geq 0$, changes with time continuously. Consequently, an amount x invested at time s will become (approximately) $x + xr(s)h = x(1 + r(s)h)$ at time $s+h$, assuming that h is small. Let $D(T)$ denote the amount that one has at time T, assuming \$1 is invested at time 0. Then, for small h and $r(s)$, $0 \leq s \leq T$,

$$D(s+h) \approx D(s)(1 + r(s)h),$$

or that the rate of change of the amount at time s is

$$\frac{D(s+h) - D(s)}{h} \approx D(s)\, r(s).$$

In the limit, as $h \downarrow 0$, we have

$$r(s) = \frac{D'(s)}{D(s)},$$

where $D'(s)$ is the derivative of $D(s)$ at time s.

Integrating over s from $[0, T]$, the above becomes

$$\int_0^T r(s)\,\mathrm{d}s = \log D(T) - \log D(0).$$

Since $D(0) = 1$, we have

$$(D(T))^{-1} = \exp\left[-\int_0^T r(s)\,\mathrm{d}s\right].$$

But $(D(T))^{-1}$ is $P(0, T)$, the present value at time 0 of a bond that pays \$1 at time T. Thus, in general, we have the following relationship

$$P(t, T) = \exp\left[-\int_t^{t+T} r(s)\,\mathrm{d}s\right], \tag{11.1}$$

where $P(t, T)$ is the present value at time t of a zero-coupon risk-free bond yielding \$1 at time $t + T$.

Isomorphism with the Survival Function

Equation (11.1) parallels (4.9), the exponentiation formula of reliability and survival, once we look at $r(s)$ as a failure rate function, and $P(t, T)$ as a survival function. Observe that $P(t, 0) = 1$ and $P(t, T)$ asymptotes to 0 as T increases to infinity. In the same vein, if the exponent of the right-hand side of (11.1) is labeled as $R(t, T)$, that is, if

$$R(t, T) \stackrel{\text{def}}{=} \int_t^{t+T} r(s)\,\mathrm{d}s,$$

then $R(t, T)$ can be identified with the cumulative hazard function. Also, the *yield curve* of finance,

$$\widetilde{R}(t, T) \stackrel{\text{def}}{=} \frac{1}{T}\int_t^{t+T} r(s)\,\mathrm{d}s,$$

can be identified with the failure rate average. Recall that the failure rate average was alluded to in section 11.2.3 in the context of introducing the IFRA (DFRA) class of distribution functions.

We now seem to have an isomorphic relationship between the survival function of an item and the present value function of a zero-coupon risk-free bond. But to claim an isomorphism I need to show that the two entities in question have a common genesis. This we do next, in the section which follows. A consequence of the common genesis is that the exponentiation formula of (4.9) represents a phenomenon that is broader in scope than that of ageing and failure.

11.3.2 Re-interpreting the Exponentiation Formula

Suppose that in (11.1), we set $t = 0$, with $P(0, T) = P(T)$, $R(0, T) = R(T)$, and $\widetilde{R}(0, T) = \widetilde{R}(T)$. With the above in place (11.1), when seen in the context of reliability and survival analysis, would imply that $P(T)$ is the survival function and $r(s)$ is the failure rate function. This interpretation of $P(T)$ and $r(s)$ would have an intuitive import that is grounded in the notion of ageing and failure. Since risk-free bonds cannot (by definition) default, one is hard pressed to look at the present value function in the same vein as a survival function. This dilemma motivates us to seek an alternative, more encompassing, way to look at the survival function. This we are able to do by examining the derivation of the exponentiation formula from the first principles.

Suppose then, that X denotes the lifetime of an item, and suppose that $F(x) = P(X \leq x)$ is absolutely continuous with derivative $F'(x) = f(x)$ for (almost) all $x \geq 0$. Consider the quantity

$$P(x < X \leq x + \mathrm{d}x | X > x) = \frac{F(x + \mathrm{d}x) - F(x)}{\overline{F}(x)},$$

where $\overline{F}(x) = 1 - F(x)$. Dividing the above by $\mathrm{d}x$ gives us the rate at which $F(x)$ increases at x, multiplied by $\left(\overline{F}(x)\right)^{-1}$. Taking the limit as $\mathrm{d}x \downarrow 0$, gives us

$$\lim_{\mathrm{d}x \downarrow 0} \frac{F(x + \mathrm{d}x) - F(x)}{\overline{F}(x)\mathrm{d}x} = \frac{f(x)}{\overline{F}(x)} \stackrel{\text{def}}{=} h(x),$$

where $h(x)$ is the failure rate. The qualifier 'failure' is added because the function $F(x)$ whose rate of failure is being discussed represents the probability of failure by x. Thus $h(x)$ reflects the rate at which $F(x)$ increases in x. The exponentiation formula of (11.1) is a consequence of the relationship $h(x) = f(x)/\overline{F}(x)$.

It is important to note that the development above is not contingent on the fact that $F(x)$ be a distribution function. All that has mattered is that $F(x)$ be absolutely continuous and $F(x)$ be non-decreasing. These features enable us to claim that the exponentiation formula is ubiqutious in any scenario involving an absolutely continuous monotonically decreasing function, the interpretation of the function being context dependent. In reliability, the said function is a survival function; in finance it is the present value function.

The Exponentiation Formula as the Law of a Diminishing Resource

The discussion above enables us to remark that when $P(T)$ denotes the present value at time T, and $r(s)$ is the interest rate, then $r(s)\mathrm{d}s$ would represent the proportion of loss in present value at time s, during the time interval $(s, s + \mathrm{d}s)$. Thus one may liken the interest rate as a form of a hazard or risk posed to the present value function vis-à-vis its failure to maintain a particular value at any time. We now have a point of view that connects the interest rate and the failure rate from a common perspective.

Our theme of interpreting the interest rate as the agent for causing a proportion loss in present value has a systematic effect in reliability. Specifically, since $\overline{F}(x)$ decreases in x from $\overline{F}(0) = 1$, the exponentiation formula of (11.1) can be seen as a law which prescribes a lifetime as the consequence of some diminishing resource, with $\overline{F}(0) = 1$ reflecting the item's initial resource. The resource gets depleted over time, with the proportion depleted at x being given as $h(x)\mathrm{d}x$. The notion of a hazard potential, discussed in Section 4.6, provides support for this point of view.

Thus to summarize, the well-known exponentiation formula which arises in the contexts of reliability, survival analysis and the asset pricing of risk-free instruments, can be seen as a law that governs the depletion of a resource, with the proportion loss at x governed by the failure (interest) rate at x. This interpretation is the basis of our claimed isomorphism between the present value function and the survival function. We have now established a platform for assessing financial risk using the more traditional tools of risk analysis, namely, reliability theory and survival analysis. In what follows, I describe how this common platform enables us to import some ideas and notions from reliability to finance and vice versa.

11.3.3 A Characterization of Present Value Functions

In this section, I endeavor to describe the qualitative behavior of $P(T)$, the present value function, when the underlying interest rate $r(s)$, $0 < s \leq T$, or the yield curve $\widetilde{R}(T)$, are monotone (increasing or decreasing) in their respective arguments. The idea here is that when a bond is issued, the precise nature of the interest rate that will prevail during the life of the bond will not be known, but one may speculate upon its general nature as being moving upwards or downwards. A characterization of the present value function enables us to compare present value functions under different forms of interest rate functions, assumed monotonic, and also enable us to obtain bounds and inequalities for different investment horizons. The exercise here parallels that which is done in reliability theory wherein comparison with the exponential survival function has proved to be valuable.

Non-parametric Classes of Present Value Functions

By a non-parametric class of present value functions, we mean a class of functions whose precise form is unknown (i.e. they are not parametrically defined) but about which some general features can be specified. The material which follows parallels that of section 11.2.3 wherein non-parametric classes of life distributions such as IFR (DFR), IFRA (DFRA) and NBU (NWU) were introduced and their consequences explored. We start with

Definition 11.2. *The present value function $P(T)$ is defined to be IIR (DIR), for increasing (decreasing) interest rate – if for each $\tau \geq 0$, $P(T+\tau)/P(T)$ is decreasing (increasing) in $T \geq 0$.*

A consequence of Definition 11.2 is that when $P(T)$ is absolutely continuous, the interest rate function $r(T)$ is increasing (decreasing) in T. Conversely, when $r(T)$ is increasing (decreasing) in T, $P(T)$ is IIR (DIR) (cf. Barlow and Proschan, 1975, p. 54). When $r(t) = \lambda$, a constant greater than 0, $P(T) = \exp(-\lambda T)$, which is both IIR and DIR. All present value functions that display the IIR (DIR) property constitute a class that we label 'IIR (DIR) class'.

Interest rate functions are generally not monotonic even though they may reflect a tendency to edge upwards. They often contain aberrations (or kinks) that are not too severe, in the sense that their average is monotone. In other words, whereas $r(T)$ is not monotone, the yield curve $\widetilde{R}(T)$ is. To bring this feature into play we introduce

Definition 11.3. *The present value function $P(T)$ is defined to be IAIR (DAIR), for increasing (decreasing) average interest rate, if $-[\log P(T)]/T$ is increasing (decreasing) in $T \geq 0$.*

A consequence of Definition 11.3 is that $P(T)$ IAIR (DAIR) is tantamount to $\widetilde{R}(T)$ increasing (decreasing) in $T \geq 0$ (cf. Barlow and Proschan, 1975, p. 84). Analogous to IIR (DIR) class, we define the IAIR (DAIR) class as a collection of functions $P(T)$ that display the IAIR (DAIR) property. Verify that the IAIR class, denoted $\{IAIR\}$, encompass the IIR class, denoted $\{IIR\}$, so that $\{IIR\} \subseteq \{IAIR\}$. Similarly $\{DIR\} \subseteq \{DAIR\}$.

A further generalization of Definitions 11.2 and 11.3, a generalization whose merits will be pointed out later, is given by Definition 11.4 below.

Definition 11.4. *The present value function $P(T)$ is said to display a NWO (NBO), for new worse (better) than old, property if for each τ, $T \geq 0$, $P(T+\tau) \leq (\geq)\, P(T)P(\tau)$.*

It can be shown, details omitted (cf. Barlow and Proschan, 1975, p. 159), that

$$\{IIR\} \subseteq \{IAIR\} \subseteq \{NWO\},$$

and

$$\{DIR\} \subseteq \{DAIR\} \subseteq \{NBO\},$$

where the $\{NWO\}$ and the $\{NBO\}$ classes contain all present value functions that display the NWO and NBO property, respectively.

Financial Interpretation of the NBO (NWO) Feature

Consider the case of equality in Definition11.4. Now

$$P(T+\tau) = P(T)P(\tau), \tag{11.2}$$

and the above relationship holds if and only if $P(T) = \exp(-\lambda T)$, for some $\lambda \geq 0$ and $T \geq 0$. The interest rate function underlying this form of the present value function is $r(s) = \lambda$. Equation (11.2) also implies that

$$\frac{P(T) - P(T+\tau)}{P(T)} = 1 - P(\tau),$$

and since $P(0) = 1$, the above relationship can also be written as

$$\frac{P(T) - P(T+\tau)}{P(T)} = \frac{P(0) - P(\tau)}{P(0)}. \tag{11.3}$$

Because $P(T)$ is a decreasing function of T, the left-hand side of (11.3) describes the proportion loss in present value during a time interval $[0, \tau]$ at the time T, whereas the right-hand side describes the proportion loss in the same time interval, but at time 0. This is an analogue of the memoryless property of the exponential distribution in the context of finance. Its practical consequence is that under a constant interest rate function, there is no reason to prefer one investment horizon over another, so long as the holding period is the same.

Now consider the case of strict inequality. Suppose that $P(T)$ is NWO, so that

$$P(T+\tau) < P(T)P(\tau),$$

and as a consequence

$$\frac{P(T) - P(T+\tau)}{P(T)} < \frac{P(0) - P(\tau)}{P(0)}. \tag{11.4}$$

This means that under (11.4) the proportion loss in present value at some time $T > 0$ is always less than the proportion loss at time 0. Vice-versa when $P(T)$ is NBO and the inequality above is reversed. To a bondholder, the greater the drop in present value, the more attractive is the bond. Consequently, for $P(T)$s that are NWO, an investment for any fixed holding period that is made early on in the life of the bond is more attractive than one (for the same holding period) that is made later on. In the IIR or the IAIR case, the above claim makes intuitive sense because the aforementioned properties are a manifestation of increasing interest rates and increasing yield curves, and $\{IIR\} \subseteq \{IAIR\} \subseteq \{NWO\}$. A similar claim can be made in the case of $P(T)$ that is NBO.

It is of interest to note that our definition of NWO and NBO is a **reverse** of that used in reliability theory, namely, the NBU and NWU classes. This makes sense because a decrease of the present value function is a consequence of an earned resource (namely interest) whereas the decrease of the survival function is a consequence of a depleted resource.

Present Value Functions that are Log Concave and PF$_2$

Suppose that the present value function $P(T)$ belong to one of the several non-parametric classes introduced before, and suppose that the interest rate at time of issue of bond is $\lambda > 0$; i.e. $r(0) = \lambda$. Were the interest rate over the investment horizon T to remain a constant at λ, then the present value function should be of the form $\exp(-\lambda T)$, $T \geq 0$. The purpose of this section is to compare $P(T)$ and $\exp(-\lambda T)$. Such a comparison could provide new insights about desirable asset pricing investment horizons. To do so, we need to introduce the notions of log concavity and Polya Frequency Functions of Order 2 (PF$_2$). These notions have turned out to be useful in reliability theory.

Definition 11.5. *A function $h(x)$, $-\infty < x < \infty$ is said to be PF_2 if: $h(x) \geq 0$ for $-\infty < x < \infty$, and*

$$\begin{vmatrix} h(x_1 - y_1) & h(x_1 - y_2) \\ h(x_2 - y_1) & h(x_2 - y_2) \end{vmatrix} \geq 0$$

for all $-\infty < x_1 < x_2 < \infty$ and $-\infty < y_1 < y_2 < \infty$, or equivalently $\log h(x)$ is concave on $(-\infty, +\infty)$, or equivalently for fixed $\Delta > 0$, $h(x + \Delta)/h(x)$ is decreasing in x for $a \leq x \leq b$, where

$$a = \inf_{h(y)>0} y \quad \text{and} \quad b = \sup_{h(y)>0} y.$$

The above equivalencies are given in Barlow and Proschan (1975, p. 76). Log concavity and PF$_2$ enable us to establish crossing properties of the present value function.

To start with, suppose that $P(\bullet)$ is IIR (DIR). Then, from Definition 11.2 we have that for each $\tau \geq 0$, $P(T + \tau)/P(T)$ is decreasing (increasing) in $T \geq 0$. As a consequence we have

Claim 11.1. *$P(\bullet)$ IIR is equivalent to $P(\bullet)$ being both log-concave and PF_2.*

Since $P(\bullet)$ IIR is equivalent to an increasing interest rate function $r(\bullet)$, and vice-versa, the essence of Claim 11.1 is that increasing interest rate functions lead to log-concave present value functions. What is the behavior of $P(\bullet)$ if instead of the interest rate function being increasing it is the yield curve that is increasing? More generally, suppose that $P(\bullet)$ is IAIR (DAIR). Then, $P^{1/T} \downarrow (\uparrow) T$, for $T \geq 0$ (Definition 11.3). Consequently we have

Claim 11.2. *$P(\bullet)$ IAIR (DAIR) implies that for all $T \geq 0$ and any α, $0 < \alpha < 1$,*

$$P(\alpha T) \geq (\leq) P^{\alpha}(T). \tag{11.5}$$

To interpret (11.5), let $Q(T) = 1/P(T)$. Then $Q(T)$ is the amount received at time T for every unit of money invested at time $T = 0$. Consequently, taking reciprocals in (11.5), we have

$$Q(T/2) \leq (\geq)(Q(T))^{1/2}.$$

Thus, here again, long investment horizons yield more bang for a buck than short horizons when the yield curve is monotonic increasing, and vice-versa when the yield curve is monotone decreasing. Claim 11.2 prescribes how the investment horizon scales.

To explore the crossing properties of present value functions that are IAIR (DAIR), we introduce

Definition 11.6. *A function $h(x)$, $0 \leq x \leq \infty$ is said to be star-shaped if $h(x)/x$ is increasing in x. Otherwise, it is said to be anti star-shaped. Equivalently, $h(x)$ is star-shaped (anti-star-shaped), if for all α, $0 \leq \alpha \leq 1$,*

$$h(\alpha x) \leq (\geq) \alpha h(x).$$

It is easy to verify that any convex function passing through the origin is star-shaped (cf. Barlow and Proschan, 1975, p. 90).

Since $P(\bullet)$ IAIR (DAIR) implies (Definition 11.3) that $-[\log P(T)]/T$ is increasing (decreasing) in $T \geq 0$, it now follows that

Claim 11.3. *$P(\bullet)$ IAIR (DAIR) implies that $T(\widetilde{R}(T))$ is star-shaped (anti-star-shaped).*

Recall that $\widetilde{R}(T)$ is the yield curve. The star-shapedness property, illustrated above, is useful for establishing Theorem 6 which gives bounds on $P(\bullet)$. The essence of the star-shapedness property is that there exists a point from which a ray of light can be drawn to all points of the star-shaped function $T(\widetilde{R}(T)) = \int_0^T r(u)\,du$, with the origin as the point from which the rays of light can be drawn.

It is clear from an examination of Figure 11.4, that a star-shaped function can cross a straight line from the origin at most once, and that if it does so, it will do it from below. Thus we have

Theorem 11.6 *The present value function $P(\bullet)$ is IAIR (DAIR) iff for $T \geq 0$ and each $\lambda > 0$, $(P(T) - \exp(-\lambda T))$, has at most one change of sign, and if a change of sign actually occurs, it occurs from + to − (from − to +).*

A formal proof of this theorem is in Barlow and Proschan (1975, p. 90). Its import is that the present value function under a monotonically increasing yield curve will cross the present value function under a constant interest rate λ, namely $\exp(-\lambda T)$, at most once, and that if it does cross it will do so from above. The reverse is true when the yield curve decreases monotonically.

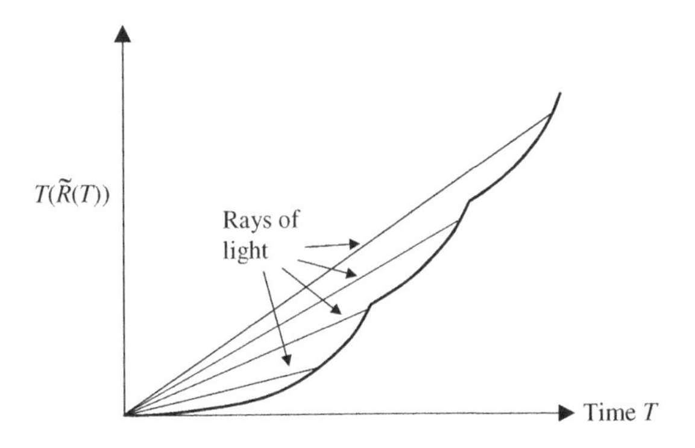

Figure 11.4 Star-shapedness of $T(\widetilde{R}(T))$ when $P(\cdot)$ is IAIR.

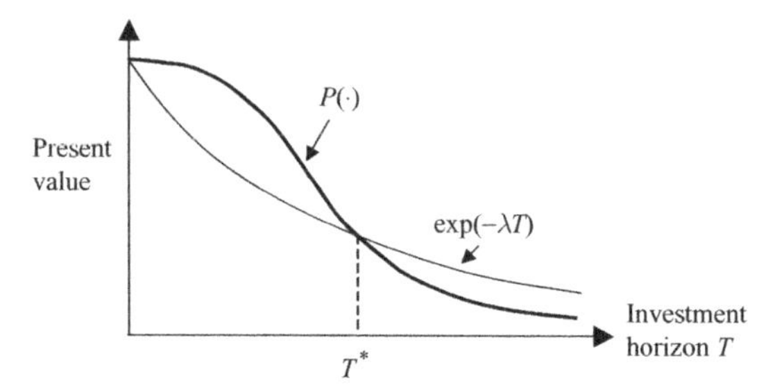

Figure 11.5 Crossing properties of an IAIR present value function.

Figure 11.5 illustrates the aforementioned crossing feature for the case of $P(\bullet)$ IAIR, showing a crossing at some time T^*. In general, T^* is unknown; it will be known only when a specific functional form is assumed for $P(\bullet)$.

The essence of Figure 11.5 is that when the yield curve is predicted to be monotone increasing, and having a spot interest rate $\lambda > 0$ at $T = 0$, then the investment horizon should be at least T^*. Investment horizons smaller than T^* will result in smaller total yields than those greater than T^*. The investment horizon of T^* is an equilibrium point.

The illustration of Figure 11.5 assumes that $P(T)$ and $\exp(-\lambda T)$ cross, whereas Theorem 6 asserts that there is at most one crossing. Thus we need to explore the conditions under which a crossing necessarily occurs and the point at which the crossing occurs. That is, we need to find T^*, assuming that $T^* < \infty$. The result is summarized in Theorem 7. This theorem uses the notion of star ordering which was introduced in Definition 11.1. Theorem 7 is compiled from a collection of results in Barlow and Proschan (1975, p. 107–110).

Theorem 11.7 *Let* $F \underset{*}{<} G$. *Then*

(*i*) $P(\bullet)$ *is IAIR, and*

(*ii*) $P(T)$ *crosses* $\exp(-\lambda T)$ *at most once, and from above, as* $T \uparrow \infty$, *for each* $\lambda > 0$. *Furthermore if* $\int_0^\infty P(u)\mathrm{d}u = 1/\lambda$, *then*

(*iii*) *A single crossing must occur, and* T^*, *the point at which the crossing occurs is greater than* $1/\lambda$. *Finally a crossing will necessarily occur at* $T^* = 1/\lambda$, *if*

(*iv*) $P(u)$ *is DIR and*

$$\int_0^\infty P(u)\,\mathrm{d}u = 1/\lambda.$$

Under (*iv*) above, the interest rate is monotonically decreasing; in this case the investment horizon should be no more than T^*.

In parts (*ii*) and (*iv*) of Theorem 7, we have imposed the requirement that

$$\int_0^\infty P(u)\,\mathrm{d}u = 1/\lambda. \tag{11.6}$$

How must we interpret the condition of (11.6)? To do so, we appeal to the isomorphism of section 11.3.1. Since $P(u)$ behaves like a survival function, with $P(0) = 1$ and $P(T)$ decreasing in T, we may regard T as a random variable with distribution function $(1 - P(\bullet))$. Consequently, the left-hand side of (11.6) is the expected value of T. With this as an interpretation, we may regard the investment horizon as an unknown quantity whose distribution is prescribed by the present value function, and whose mean is $1/\lambda$.

11.3.4 Present Value Functions Under Stochastic Interest Rates

The material of section 11.3.3 was based on the premise that whereas the interest rate over the holding period of a bond is unknown, its general nature, a monotonic increase or decrease, can be speculated. Such speculations may be meaningful for small investment horizons; over the long run interest rates cannot be assumed to be monotonic. In any case, the scenario of section 11.3.3 pertains to the case of deterministic but partially specified interest rates. In this section, we consider the scenario of interest rate functions that are specified up to some unknown constants, or are the realization of a stochastic process. An analogue of these scenarios in reliability theory is a consideration of hazard functions that are stochastic (section 7.1).

Interest Rate Functions with Random Coefficients

Recall (equation (11.1)) the exponentiation formula for the present value function under a specified interest rate function $r(s)$, $s \geq 0$, with $t = 0$, as

$$P(T) = \exp(-R(T)), \tag{11.7}$$

where $R(T)$ is the cumulative interest rate function. Suppose now that $r(s)$, $s \geq 0$ cannot be precisely specified. Then the $R(T)$ of (11.7) becomes a random quantity. Let $\pi[R(T)]$ describe our uncertainty about $R(T)$ for any fixed $T \geq 0$. We require that $\pi(\bullet)$ be assessed and specified. Thus our attention now centers around assessing $P(T;\, \pi)$, the present value function when $\pi[R(T)]$ can be specified for any desired value of T. In other words, $P(T;\, \pi)$ refers to the fact that the present value function depends on π. In what follows I argue that

$$P(T;\, \pi) = E_{\pi}[\exp(-R(T))], \tag{11.8}$$

where E_{π} denotes the expectation with respect to $\pi(\bullet)$. To see why, we may use a strategy used in reliability theory which which begins by noting that

$$\exp(-R(T)) = Pr(X \geq R(T)),$$

where X is a random variable whose distribution function is a unit exponential. Consequently when $R(T)$ is random

$$\begin{aligned} P(T;\, \pi) &= \int_0^\infty Pr(X \geq R(T) | R(T))\pi[R(T)]\, \mathrm{d}R(T) \\ &= \int_0^\infty \exp(-R(T))\pi[R(T)]\, \mathrm{d}R(T) \\ &= E_{\pi}[\exp(-R(T))], \end{aligned}$$

a result that has appeared before as (7.2).

Thus, to obtain the present value function for any investment horizon T, when we are uncertain about interest rate function over the horizon $[0, T]$, all we need do is specify our uncertainty about the cumulative interest rate at T, via $\pi[R(T)]$. What is noteworthy here is that the functional form of $R(t)$, $t \geq 0$ does not matter. All that matters is the value of $R(T)$.

As an illustration of how we may put (11.8) to work, suppose that $r(s) = \lambda$, $s \geq 0$, but that λ is unknown. This means that at time 0^+, the interest rate is to take some value λ, $\lambda \geq 0$ that is unknown and is to remain constant over the life of the bond.

Suppose further that our uncertainty about λ is described by a gamma distribution with scale parameter α and a shape parameter β. Then $U \stackrel{\text{def}}{=} \lambda T$ has a density at u of the form

$$\pi(u; \alpha, \beta) = \frac{\exp(-\alpha u)\alpha^{\beta} u^{\beta-1}}{T^{\beta}\Gamma(\beta)},$$

from which it follows that the present value function is

$$P(T; \alpha, \beta) = \left(\frac{\alpha}{T+\alpha}\right)^{\beta}, \tag{11.9}$$

which is of the same form as the survival function of a Pareto distribution.

It can be verified that the present value function of (11.9) belongs to the DIR class of functions of Definition 11.2. For this class we are able to provide an upper bound on $P(T)$ (Theorem 8). The implication for this theorem is that for scenarios of the type considered here, short investment horizons are to be preferred over long ones.

Theorem 11.8 (Barlow and Proschan (1975), Theorem 5.1). *If $P(T)$ is DIR with mean μ, then*

$$P(T; \mu) \leq \begin{cases} \exp(-T/\mu), & \text{for } T \leq \mu, \\ \frac{\mu}{T} \mathrm{e}^{-1}, & \text{for } T \geq \mu; \end{cases} \tag{11.10}$$

this bound is sharp.

The dark line of Figure 11.6 illustrates the behavior of this bound. It shows that the decay in present value for time horizons smaller than μ is greater than the decay in present value for time horizons greater than μ.

The dotted line of Figure 11.6 shows the behavior of the upper bound had its decay been of the form $\exp(-T/\mu)$ for all values of T. Clearly investment horizons greater than μ would not be of advantage to a holder of the bond.

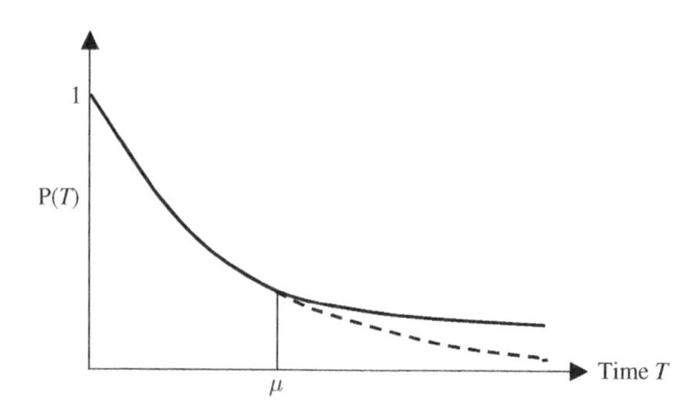

Figure 11.6 Upper bound on $P(T)$ when $P(T)$ is DIR.

For the special case considered here, namely λ unknown with its uncertainty described by $\pi(\lambda;\alpha,\beta)$, $P(T;\alpha,\beta)=(\alpha/(T+\alpha))^{\beta}$. Were $P(T;\alpha,\beta)$ be interpreted as a survival function, then the μ of Theorem 8 would be given as

$$\mu=\int_0^{\infty}\left(\frac{\alpha}{T+\alpha}\right)^{\beta}\mathrm{d}T=\frac{\alpha}{\beta-1};$$

μ exists if $\beta>1$. Consequently, under this $P(T;\alpha,\beta)$ the investment horizon should not exceed $\alpha/(\beta-1)$.

Recall that were λ to be known with certainty, $P(T)$ would be $\exp(-\lambda T)$, $\lambda>0$, $T\geq 0$, and that there would be no restrictions on the investment horizon so that a bond holder could choose any value of T as an investment horizon. With λ unknown, the net effect is to choose shorter investment horizons, namely those that are at most $\alpha/(\beta-1)$. A similar conclusion can also be drawn in the case wherein $\pi(\lambda;\alpha,\beta)$ be a uniform over $[\alpha,\beta]$. It can be verified that in the uniform case

$$P(T;\alpha,\beta)=\frac{\mathrm{e}^{-T\alpha}-\mathrm{e}^{-T\beta}}{T(\beta-\alpha)},$$

and that $P(T;\alpha,\beta)$ is again DIR.

Whereas the above conclusions regarding uncertainty about $r(s)$, $s\geq 0$ causing a lowering of the investment horizon, have been made based on a consideration of a special case, namely $r(s)=\lambda$, $\lambda>0$, $s\geq 0$, the question arises about the validity of this claim, were $r(s)$ to be any other function of s, say $r(s)=\alpha\lambda(\lambda s)^{\alpha-1}$, for some $\lambda>0$, and $\alpha>0$. When α is assumed known, and uncertainty about λ is described by $\pi(\lambda;\bullet)$, then (11.7) would be a scale mixture of exponentials and by Theorem 4.7 of Barlow and Proschan (1975, p. 103), it can be seen that $P(T;\bullet)$ is DIR, so that Theorem 8 comes into play and the inequalities of (11.10) continue to hold. Thus once again, uncertainty about λ causes a lowering of the investment horizon. Indeed, the essence of Theorem 8 will always hold if the cumulative interest rate $R(T)$ is such that any function of T does not entail unknown parameters.

Interest Rates as the Realization of a Stochastic Process

We now consider the case of interest rates that are the realization of a stochastic process. A consideration of stochastic processes for describing interest rate function is not new to the literature in mathematical finance. Indeed much has been written and developed therein. Our focus here, however, is a consideration of a shot-noise process for modeling interest rates and to explore its consequences on the present value function. A use of the shot-noise process for describing the failure rate function has been considered in section 7.2.1. Given below is the adaptation of this process for describing the interest rate function and some justification as to why this could be a meaningful thing to do.

Our rationale for using the shot noise for describing the interest rate function is that interest rates take an upward jump when certain deleterious economic events occur. Subsequent to their upward jump, the interest rates tend to come down, or even remain constant, until the next deleterious event occurs. In Figure 4.16, the deleterious events are shown to occur at times $T_{(1)}, T_{(2)}, T_{(3)}, \ldots$ Such events are assumed to occur at random and are governed by say a Poisson process with rate m, $m>0$. The amount by which the interest rate jumps upward at time $T_{(i)}$ is supposed to be random; let this be denoted by a random variable X_i. Finally, suppose that the rate at which the interest rate decays is governed by the attenuation function, $h(s)$, $s\geq 0$. Then, it is easy to see that at any time $T\geq 0$,

$$r(T)=\sum_{i=1}^{\infty}X_ih(T-T_i),$$

with $h(u)=0$ whenever $u<0$.

In what follows, we suppose that the $T_{(i)}$s and the X_is are serially and contemporaneously independent. We also suppose that the X_is are identically distributed as a random variable X.

If $X=d$, a constant, and if the attenuation function is of the form $h(u)=(1+u)^{-1}$ – i.e. the interest rate decays slowly, then it can be seen (cf. equation (7.7)) that the present value function takes the form

$$P(T; m)=\exp(-mT)(1+T)^{m}. \tag{11.11}$$

If, to the contrary, X has an exponential distribution with scale parameter b, and $h(u)=\exp(-au)$ – that is, the interest rate decays exponentially, then (cf. equation (7.8))

$$P(T; m, a, b)=\exp\left(-\frac{mbT}{1+ab}\right)\left(\frac{1+ab-\exp(-aT)}{ab}\right)^{mb/(1+ab)}. \tag{11.12}$$

The $P(T)$ of (11.11) is the survival function of a Pareto distribution. If in (11.12) we set $a=b=1$, and $m=2$, then a change of time scale from T to $\exp(T)$ would result in the present value function having the form of the survival function of a beta distribution on (0, 1) with parameters 1 and 2.

Summary and Conclusions: Future Work

Equations (11.9) through (11.12) were originally obtained in the context of reliability under dynamic environments. The isomorphism of section 11.3.1 has enabled us to invoke them in the context of finance, and what is given in this section scratches the surface. Much more can be done along these lines. For example, a hierarchical modeling of interest rate is one possibility. Another possibility, and one that is motivated by work of Dykstra and Laud (1981), is to describe the cumulative interest rate by a gamma process or to look at the present value functions as Dirichlet or neutral to the right processes. Another possibility, and one that is motivated by the enormous literature in survival analysis, is to model interest rates as a function of covariates and markers. The purpose of this chapter has been mainly to open the door to other possibilities by creating a suitable platform, which I feel has now been done.

11.4 INFERENTIAL ISSUES IN ASSET PRICING

The statistical analyses of failure and survival time data almost always entail the treatment of observables that are supposed to be generated by some stochastic mechanism that characterizes the lifetimes of a biological or a physical unit. The methods described in Chapters 5 and 9 assume the availability of such data. Can some of these methods be used in the context of inferential issues that arise in the context of asset pricing? If so, what are the inferential issues, and how can the said methods be used? The aim of this section is to address the above and related questions.

My answer to the first question is in the affirmative but there is a caveat to it. The caveat is a consequence of the fact that the available information on asset pricing is generated by behavioristic phenomena that are rooted in the socio-economic principles of uncertainty and utility, not the product of biological or physical mechanisms. Thus, for example, the information at hand could be the declared present value functions over a range of investment horizons, by several institutions like banks and investment houses. When such is the case, the inferential issue boils down to how an individual must pool the several declared present value functions in a coherent manner. Thus the answer to the first part of the second question. An answer to the second part of the question is our claim that the non-parametric Bayes methods of Chapter 9 are a suitable way to achieve the desired pooling. However, in order to do so, we need to transform the socio-economic information of a declared present value function so that it mimics observed data from some probability distribution. An approach for doing so is described in section 11.4.2, but first we need to set up a framework for formulating the inferential problem at hand. Our framework is in the spirit of a decision maker eliciting expert testimonies to update his/her prior beliefs (section 5.3).

11.4.1 Formulating the Inferential Problem

Consider a bond issuer $\mathcal{J}$ who needs to come up with a present value function over a range of investment horizons. The bond issuer has at his/her disposal the following quantities.

(*a*) $P_{\mathcal{J}}(T), T \geq 0$, the bond issuer's present value function, as assessed at the 'now time' $t=0$, based on $\mathcal{J}$s prediction of the future interest rate function $r(s), 0 < s \leq T$.
(*b*) $P_i(T_j)$ the i-th bond issuer's present value function, specified over a collection of investment horizons $T_j, j=1, \ldots, m_i, i=1, \ldots, n$. We have assumed that there are n bond issuers, in addition to $\mathcal{J}$, and that each of the n bond issuers has his/her maximum investment horizon, namely, T_{m_i} for bond issuer i. We also assume that all the n bond issuers use the same time intervals for their investment horizons. That is, the T_js are fixed time points, such as one year two years and so on (Figure 11.7).

The problem faced by $\mathcal{J}$ is how to pool the $P_i(T_j)$s with $\mathcal{J}$'s $P_{\mathcal{J}}(T)$ to arrive upon a $P^*_{\mathcal{J}}(T)$ that can be used by $\mathcal{J}$ as an investment strategy. Since all the $n+1$ bond issuers compete with each other to dominate the investment market for bonds, it is quite possible that once $\mathcal{J}$ declares his/her $P^*_{\mathcal{J}}(T)$ to the public, the n other bond issuers will update their respective $P_i(T_j)$s causing $\mathcal{J}$ to update $P^*_{\mathcal{J}}(T)$, and this process of all the $n+1$ bond issuers constantly revising their present values can go on indefinitely. We do not model here this matter of an 'infinite regress'. Instead, we assume that after a few cycles of updating, each bond issuer stays put at his/her present value function.

11.4.2 A Strategy for Pooling Present Value Functions

A coherent approach for pooling the $n+1$ present value functions mentioned above is via Bayes' law. Since it is $\mathcal{J}$'s whose actions that we need to prescribe, $P_{\mathcal{J}}(T)$ would form the basis of $\mathcal{J}$'s prior. How must $\mathcal{J}$ update $P_{\mathcal{J}}(T)$ using the $P_i(T_j)$s, $j=1, \ldots, m_i, i=1, \ldots, n$? My proposal is to use the non-parametric Bayes method of section 9.6.1 for doing the above. This material

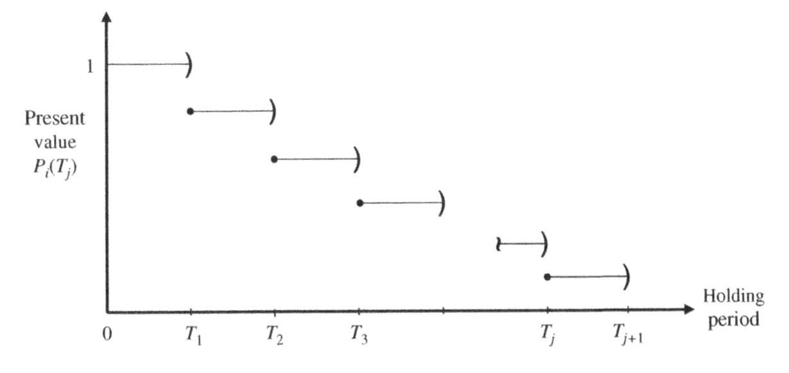

Figure 11.7 Present value function of i-th bond Issuer.

entails the Dirichlet process prior, and data that are assumed to be generated by an underlying distribution which prescribes their stochastic nature. Whereas $P_{\mathcal{J}}(T), T \geq 0$ could form the basis of a prior, we need a way to obtain the required data. Our proposal here is to generate the said data by a simulation from the $P_i(T_j)$s, $i = 1, \ldots, n$, and $j = 1, \ldots, m_i$. Our strategy of obtaining the data via a simulation is not unlike that which one does in a bootstrap (Efron, 1979).

Generating Random Samples from Present Value Functions

A justification for generating the required data via a simulation from the present value functions is provided by the feature that the present value function bears an isomorphic relationship with a survival function (section 11.3.1). Consequently, we can see the likes of Figure 11.7 given below as a proxy for the survival function from which random observations can be generated. Figure 11.7 portrays $P_i(T_j)$, the present value function of the i-th bond issuer for investment horizons $T_1, T_2, \ldots, T_{m_i}$. It is assumed to be a step-function, since investment horizons for bonds are specified for fixed time intervals such as one year, three years, etc.

Suppose that a random variable of size n_i is generated from $P_i(T_j), j = 1, \ldots, m_i$. Denote this random sample by $\mathbf{X}_i = (X_{i1}, \ldots, X_{in_i})$, and repeat this exercise for each of the $nP_i(T_j)$s to obtain $(\mathbf{X}_1, \ldots, \mathbf{X}_n)$ as data. We then update $P_{\mathcal{J}}(T), T \geq 0$, using $(\mathbf{X}_1, \ldots, \mathbf{X}_n)$ via Bayes' law. The updating procedure is straightforward if the $\mathbf{X}_i$s, $i = 1, \ldots, n$, are independent. When such is the case, the updating can be done either sequentially using one $\mathbf{X}_i$ at a time, or simultaneously using the entire set of $\mathbf{X}_1, \ldots, \mathbf{X}_n$. However, the assumption of independence needs to be evaluated with scrutiny. This is because it is reasonable to suppose that $P_1(T_j), \ldots, P_n(T_j)$ will, for each j, bear a relationship to each other. All bond issuers design their investment strategies using common public information. Thus if a common seed (i.e. the set of uniform [0, 1] deviates used to generate a random sample) is used to generate all the n sample vectors, then $\mathbf{X}_1, \ldots, \mathbf{X}_n$ will necessarily be dependent. Independence of the $\mathbf{X}_i$s can be better achieved by using a different seed for each sample.

Prior to Posterior Analysis Using the Dirichlet Process Prior

A pooling by $\mathcal{J}$ of the n present value functions $P_i(T_j)$, $i = 1, \ldots, n$ and $j = 1, \ldots, m_i$, with the aim of updating $\mathcal{J}$'s $P_{\mathcal{J}}(T)$, $T \geq 0$, tantamounts to a prior to posterior analysis using Bayes' law. Here $P_{\mathcal{J}}(T)$ would form the basis of the prior, and the simulated $\mathbf{X}_i$s, $i = 1, \ldots, n$, as the data. Our proposed strategy for doing the above is based on the Dirichlet process prior based methodology of section 9.6.1, with $P_{\mathcal{J}}(T)$ serving as the 'best guess', and Theorem 9.2 as providing a mechanism for the prior to posterior transformation. The strategy can best be described if the updating proceeds sequentially, first updating $P_{\mathcal{J}}(T)$ using $\mathbf{X}_1$, and then updating the updated present value function using $\mathbf{X}_2$, and so on, invoking Theorem 9.2 at each step. The specifics are described below.

To start the process, $\mathcal{J}$ focuses attention on some value of the investment horizon, say $T^* \geq 0$, and per the dictates of Theorem 9.2 supposes that

$$P_{\mathcal{J}}(T^*) \sim Beta\left(\alpha_1\left[0, T^*\right], \alpha_1\left(\mathbb{R}\right) - \alpha_1\left[0, T^*\right]\right).$$

Consequently, the expectation of $P(T^*)$ is $\alpha_1(T^*)/\alpha_1(\mathbb{R})$, and its variance is

$$\frac{\alpha_1(T^*)\left(\alpha_1(\mathbb{R}) - \alpha_1(T^*)\right)}{\left(\alpha_1(T^*) + \alpha_1(\mathbb{R})\right)^2\left(1 + \alpha_1(\mathbb{R})\right)}, \tag{11.13}$$

where $\alpha_1(T^*)$ is shorthand for $\alpha_1[0, T^*]$. Setting $\alpha_1(T^*)/\alpha_1(\mathbb{R}) = P_{\mathcal{J}}(T^*)$ and assigning a number, say σ^2, to the variance (equation (11.13)) $\mathcal{J}$ is able to obtain $\alpha_1(T^*)$ and $\alpha_1(\mathbb{R})$ in terms of $P_{\mathcal{J}}(T^*)$ and σ^2.

With the above in place, $\mathcal{J}$ updates $P_{\mathcal{J}}(T)$ via the likes of (9.27) as

$$P_{\mathcal{J}}^{(1)}(T) = 1 - \frac{\alpha_1(T) + \sum_{j=1}^{n_1} I_{[X_{1j},\infty)}(T)}{\alpha_1(\mathbb{R}) + n_1}, \tag{11.14}$$

where $I_A(t)$ is the indicator of set A, and $T > 0$.

For the second iteration, the $\alpha_1(T)$ of the first iteration becomes $\alpha_2(T) = \alpha_1(T) + \sum_{j=1}^{n_1} I_{[X_{1j},\infty)}(T)$, and $\alpha_1(\mathbb{R})$ becomes $\alpha_1(\mathbb{R}) + n_1$. These, when plugged into (11.14), will give $P_{\mathcal{J}}^{(2)}(T)$. Note that when obtaining $P_{\mathcal{J}}^{(2)}(T)$, the summation term in the numerator of (11.14) will be $\sum_{j=1}^{n_2} I_{[X_{2j},\infty)}(T)$.

Repeating the above n times we obtain $P_{\mathcal{J}}^{(n)}(T)$, $T \geq 0$; $P_{\mathcal{J}}^{(n)}(T)$ is the pooled present value function of $\mathcal{J}$, the pooling being done in a coherent manner. We may now set $P_{\mathcal{J}}^{(n)}(T) = P_{\mathcal{J}}^{*}(T)$, $T \geq 0$; $\mathcal{J}$ will use $P_{\mathcal{J}}^{*}(T)$, $T \geq 0$, to prescribe an investment strategy. An illustrative example showing the workings of this approach is given in section 11.4.3.

Some General Comments on the Pooling Procedure

The pooling procedure described above need not be restricted to the pooling (or fusion) of present value functions. In general, it can be used to pool distribution (survival) functions as well. The pooling of distribution functions is a common exercise that is often performed in the context of decision analysis wherein the fusing of expert testimonies is germane. Some more modern applications of pooling distribution functions occur in the arenas of 'target tracking' and 'sensor fusion'. These activities occur in civilian, military and medical contexts.

As was mentioned before (section 9.6.1) a possible disadvantage of pooling distributions using the Dirichlet process prior is that the pooled distribution is discrete with probability one. A consequence is that the pooled present value function $P_{\mathcal{J}}^{*}(T)$, $T \geq 0$, is not a smooth continuous function of the investment horizon. This may not be a too serious a limitation in the context of asset pricing. It could, in the contexts of target tracking and sensor fusion, be a limitation. To overcome this limitation, one may consider pooling using the neutral to the right (NTR) processes discussed in section 9.6.2. Their disadvantage, however, is an absence of closed form results of the type that the Dirichlet process priors provide.

11.4.3 Illustrative Example: Pooling Present Value Functions

As an illustration of the workings of the material of section 11.4.2, suppose that $P_{\mathcal{J}}(T)$ is of the form $P_{\mathcal{J}}(T) = \exp(-0.2T)$, with $T \geq 0$; this is an exponential survival function. We focus attention on $T^* = 1$, so that $P_{\mathcal{J}}(T^*) = \exp(-0.2) = 0.818$, and suppose that σ^2, $\mathcal{J}$'s measure of uncertainty in specifying $P_{\mathcal{J}}(T^*)$ is $\sigma^2 = 0.01$. These specifications would result in $\alpha_1(\mathbb{R}) = 13.84$ and $\alpha_1(T^*) = 11.332$. With $\alpha_1(\mathbb{R})$ known, we obtain $\alpha_1(T)$ as $\alpha_1(T) = \exp(-0.2T)\alpha_1(\mathbb{R})$.

Now suppose that $P_1(T_j)$, the present value function of another bond issuer, is the piecewise constant function of Figure 11.8. Here $T_j = 1, 3, 5$ and 6.

We generate a sample of size 5 from the $P_1(T)$ of Figure 11.8 regarding the same as a survival function. Denote these as $X_{11} = 1$, $X_{12} = 1$, $X_{13} = 3$, $X_{14} = 5$, and $X_{15} = 5$. We then plug these values into (11.14) to obtain

$$P_{\mathcal{J}}^{(1)}(T) = 1 - \frac{\alpha_1(T) + \sum_{j=1}^{5} I_{[X_{1j},\infty)}(T)}{\alpha_1(\mathbb{R}) + 5},$$

for $T = 0.5, 1, 1.5, 2, \ldots, 9, 9.5$ and 10. In Figure 11.9, I illustrate the behavior of $P_{\mathcal{J}}(T)$ and $P_{\mathcal{J}}^{(1)}(T)$ to show the effect of the other bond issuer's present value function on $\mathcal{J}$'s updated

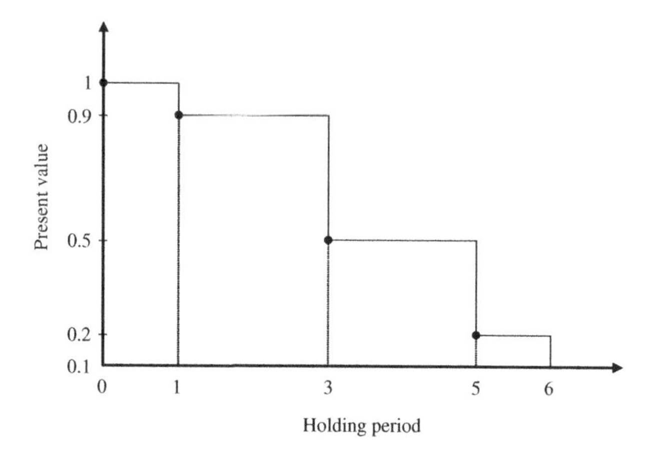

Figure 11.8 $P_1(T)$, a Bond issuer's present value function.

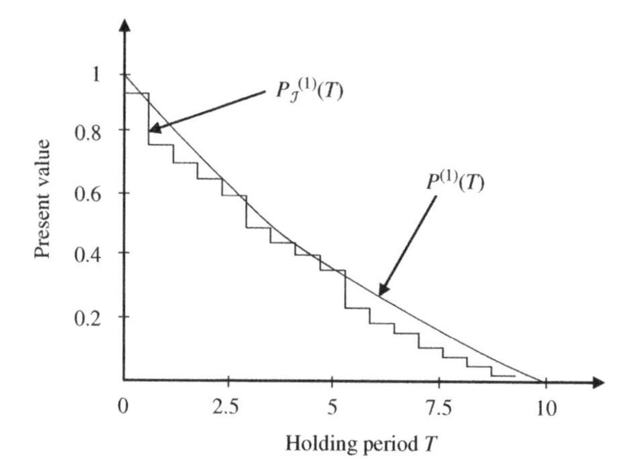

Figure 11.9 $\mathcal{J}$'s Initial and updated present value functions

present value function. The smoothness of $P_{\mathcal{J}}^{(1)}(T)$ will depend on the granularity of T that has been chosen by us. We have chosen the granularity to be 0.5 and n_1, the number of observations to be simulated from $P_1(T)$, to be 5.

11.5 CONCLUDING COMMENTS

Investigators working in reliability and survival analysis are unlikely to view the material given in this chapter as being mainstream vis-à-vis the above topics. Indeed one is hard pressed to name other books in reliability and survival analysis that venture into topics such as economics, finance and other social sciences such as gerontology and longevity. However, if a subject is to thrive then its connections with other subjects should be attractive features. It is with this in mind that I have chosen to include this chapter among the other more subject-related chapters. All the same, what I have presented here is limited to my experience and exposures; i.e. I may have only scratched the surface. Similarly, there may be other topics in physical, social and the mathematical sciences wherein ideas and notions that are unique to reliability and survival analysis may have a role to play. But there is also another side to this coin, namely, that ideas and notions from other disciplines could enhance the state-of-the-art in reliability and survival analysis as well. We have seen a few examples of these, such as the notion of a hazard potential from physics, a use of the Lorenz curve for survival analysis from econometrics, and the hitting times of stochastic hazard rates from finance. Connections with other disciplines also enable us to re-think the conventional ideas from our own discipline. For example, the isomorphism between the survival function and the present value function has enabled us to shed new light on the exponentiation formula. We may now look at this stalwart formula of reliability and survival analysis in the broader framework, as the law of a diminishing resource.

Appendix A

Markov Chain Monté Carlo Simulation

Simulation techniques whose credibility is asserted by the theory of Markov Chains are termed Markov Chain Monté Carlo methods, abbreviated MCMC. Such techniques have proved to be very valuable for implementing Bayesian solutions to many standard statistical problems, including those encountered in reliability and life-testing. MCMC techniques are easily implemented so that Bayesian analyses can be made routinely available for addressing many commonly encountered problems. During the past several years, so much has been written about them and from so many perspectives and points of view that what is given here can be aptly labeled elementary. We shall focus on one algorithm, the 'Gibbs sampling algorithm'.

A.1 THE GIBBS SAMPLING ALGORITHM

This is a Markovian updating scheme which can be used for estimating marginal distributions from a joint distribution and for drawing inferences about non-linear functions of parameters of chance distributions.

To illustrate the workings of this algorithm, suppose that $p(\boldsymbol{\theta}; \mathbf{x})$ denotes the posterior density of $\boldsymbol{\theta} = (\theta_1, \ldots, \theta_k)$ given data $\mathbf{x}$. Suppose that $p(\boldsymbol{\theta}; \mathbf{x})$ can be written out up to a constant of proportionality, as is the case in a Bayesian analysis; i.e. $p(\boldsymbol{\theta}; \mathbf{x}) \propto \mathcal{L}(\boldsymbol{\theta}; \mathbf{x}) p(\boldsymbol{\theta})$, where $\mathcal{L}$ is the likelihood and $p(\boldsymbol{\theta})$ the prior. Let $p(\theta_i | \boldsymbol{\theta}_{-i}; \mathbf{x})$ denote the conditional posterior density of θ_i, conditioned on the remaining elements of $\boldsymbol{\theta}$. Implicit to what follows is the assumption that the above conditionals are able to specify the joint posterior $p(\boldsymbol{\theta}; \mathbf{x})$. This may not always be true; Gelman and Speed (1993) give conditions under which the above is true. The Gibbs sampling algorithm proceeds as follows:

Suppose that $\boldsymbol{\theta}^{(0)} = (\theta_1^0, \ldots, \theta_k^0)$ is some arbitrary (starting) value of $\boldsymbol{\theta}$. Then, θ_1^1 is a random drawing from $p(\theta_1 | \theta_2^0, \ldots, \theta_k^0; \mathbf{x})$ – assuming that this can be done (and here lies the catch), θ_2^1 a drawing from $p(\theta_2 | \theta_1^1, \theta_3^0, \ldots, \theta_k^0; \mathbf{x})$, θ_3^1 a drawing from $p(\theta_3 | \theta_1^1, \theta_2^1, \theta_4^0, \ldots, \theta_k^0; \mathbf{x})$, and so on, so that θ_k is a draw from $p(\theta_k | \theta_{-k}^1; \mathbf{x})$. This process of substituting and drawing completes a transition of $\boldsymbol{\theta}$ from $\boldsymbol{\theta}^0$ to $\boldsymbol{\theta}^{(1)} = (\theta_1^1, \ldots, \theta_k^1)$. After t such transitions we arrive upon $\boldsymbol{\theta}^t = (\theta_1^t, \ldots, \theta_k^t)$. The claim (cf. Smith and Roberts, 1993) is that the transitions $\boldsymbol{\theta}^{(0)}, \boldsymbol{\theta}^{(1)} \ldots, \boldsymbol{\theta}^{(t)}$

Reliability and Risk: A Bayesian Perspective N.D. Singpurwalla
© 2006 John Wiley & Sons, Ltd

constitute a Markov Chain (for example, Çinlar, 1975), which for large values of t has a stationary distribution that is precisely $p(\boldsymbol{\theta}; \mathbf{x})$. Consequently, for t large, $\boldsymbol{\theta}^{(t)}$ converges in distribution to a random vector whose distribution is $p(\boldsymbol{\theta}; \mathbf{x})$, and the elements of $\boldsymbol{\theta}^{(t)}$ converge in distribution to random variables whose distributions are the marginal distributions of $p(\boldsymbol{\theta}; \mathbf{x})$. Some regularity conditions are involved, and these are discussed by Roberts and Smith (1993) (also see Athreya, Doss and Sethuraman, 1996). Repeating the above procedure m times, making sure that the starting values, like our $\boldsymbol{\theta}^{(0)}$, are independently chosen, yields m k-tuples $(\boldsymbol{\theta}_1^t, \ldots, \boldsymbol{\theta}_m^t)$. These m k-tuples can be regarded as a sample of size m from the k-variate posterior distribution $p(\boldsymbol{\theta}; \mathbf{x})$. Furthermore, with $\boldsymbol{\theta}_i^t = (\theta_{1i}^t, \theta_{2i}^t, \ldots, \theta_{ki}^t)$, $i = 1, 2, \ldots, m$, the vector $(\theta_{j1}^t, \theta_{j2}^t, \ldots, \theta_{jm}^t)$ is a sample of size m from the marginal posterior distribution θ_j, $j = 1, \ldots, k$. For m large this latter sample can be used to empirically construct the marginal posterior distribution of θ_j. Furthermore, if ψ is any function of θ_j, then $\frac{1}{m}\sum_{r=1}^{m} \psi(\theta_{jr}^t)$ is a consistent estimator of $E[\psi(\theta_j); \mathbf{x}]$. This completes our discussion of the Gibbs sampling algorithm; some more details can be found in the excellent tutorial article by Casella and George (1992). However, there is some irony to the above scheme. It appears that in empirically estimating posterior distributions, and in appealing to the consistency property of estimates of functions of unknown parameters, a Bayesian uses perfectly honorable frequentist procedures for easing the computational burdens that his/her paradigm imposes. But to conclude this topic, we need discuss two issues, one a matter of concern and the other a virtue.

The first is that it may not be always possible to generate random samples from the conditional distributions $p(\theta_i|\theta_{-i}; x)$. The nature of the likelihood and the prior could be such that their product does not result in a form that is standard. A consequence is that the choice of priors could be influenced by ones ability to generate the required samples. The second matter pertains to the matter of censoring. This could result in a likelihood function that contains integrals. When such is the case, one is unable to generate samples from $p(\theta_i|\theta_{-i}; \mathbf{x})$. To overcome this obstacle, Gelfand, Smith and Lee (1992) propose an ingenious scheme, namely, treating the random variables that are censored as unknowns, just like the unknown $\boldsymbol{\theta}$, and then invoking the Gibbs sampling algorithm. This idea is best illustrated via a specific example.

Suppose that n items are subjected to a life-test and r of these are observed to fail at $\mathbf{x} = (x_1, \ldots, x_r)$, and the remaining $(n - r)$ are censored so that $C_j \leq X_j \leq D_j$, for $j = r + 1, \ldots, n$. The C_j and the D_j are assumed known and the lifetimes of the items have a distribution whose density at x is $f(x|\theta)$; the prior on θ is $\pi(\theta)$. Then, the posterior of θ, in the light of the available information is

$$p(\theta; C, D, \mathbf{x}) \propto \pi(\theta) \prod_{i=1}^{r} f(x_i|\theta) \prod_{j=r+1}^{r} \int_{C_i}^{D_i} f(s_j|\theta) \mathrm{d}s_j,$$

where $C = (C_{r+1}, \ldots, C_n)$ and $D = (D_{r+1}, \ldots, D_n)$.

If the second product term of the likelihood cannot be obtained in closed form, we will not be able to sample from the conditionals, and the Gibbs sampling algorithm cannot be invoked. However, since $\mathbf{X}' = (X_{r+1}, \ldots, X_n)$ have not been observed they are unknown, just like θ, and can therefore be augmented with θ in the Gibbs sampling algorithm. This process is called **data augmentation**, an example of which also appears in section 5.4.7. Thus we need to consider the conditionals of $p(\theta, \mathbf{X}'; C, D, \mathbf{x})$, namely $p(\theta|\mathbf{X}'; C, D, \mathbf{x})$ and $p(\mathbf{X}'|\theta; C, D, \mathbf{x})$. But

$$p(\theta|\mathbf{X}'; C, D, \mathbf{x}) = p(\theta|\mathbf{X}'; \mathbf{x}) \quad \text{and}$$

$$p(\mathbf{X}'|\theta; C, D, \mathbf{x}) = p\left(\mathbf{X}'|\theta; C, D = \prod_{j=r+1}^{n} \int_{C_i}^{D_i} f(s_j|\theta) \mathrm{d}s_j\right),$$

since θ need not depend on C and D, and $\mathbf{X}'$ need not depend on $\mathbf{x}$. A striking feature of the above is that $p(\theta|\mathbf{X}'; \mathbf{x})$ is just the posterior distribution of θ had there been no censoring, so that random variate generation from $p(\theta|\mathbf{X}'; \mathbf{x})$ proceeds along the same lines as in the case of a full sample. Random variate generation from $p(\mathbf{X}'|\theta; C, D, \mathbf{x})$ is simple, since it entails independent draws from the truncated chance distribution $f(x|\theta)$. This would boil down to generating random variates from the untruncated distribution, and retaining only those variates that fall within the C_j and the D_j, $j = r+1, \ldots, n$. Thus we see that in treating the censored observations as unknowns, we have uncoupled the complications in the likelihood into constituents that are more manageable.

Appendix B

Fourier Series Models and the Power Spectrum

B.1 PRELIMINARIES: TRIGONOMETRIC FUNCTIONS

The functions $\sin t$ and $\cos t$ are periodic with period 2π, i.e. $\sin(t+2\pi)=\sin t$, and $\cos(t+2\pi)=\cos t$, for $t \geq 0$. Thus for any k, $k=0, \pm 1, \pm 2, \ldots$, $\sin(t+2k\pi)=\sin t$, and $\cos(t+2k\pi)=\cos t$. It is easy to verify that for any constant $\lambda > 0$, $\sin \lambda t$ and $\cos \lambda t$ are also periodic with period $2\pi/\lambda$. The effect of the scalar λ is to expand or contract the scale. The reciprocal of the period is called the **frequency**; it represents the number of periods per unit of time. The largest (smallest) value that $\sin t$ can take is $+1(-1)$; these occur at $\pi/2$ and $3\pi/2$, respectively. Similarly, the largest (smallest) value that $\cos t$ can take is $+1(-1)$, but they occur at 0 (and 2π) and π, respectively.

A translation of the entire sine or cosine curve is achieved by introducing a shift parameter θ, called the **phase**. Specifically, $\sin(\lambda t+2\pi-\theta)=\sin(\lambda t-\theta)$, and $\cos(\lambda t+2\pi-\theta)=\cos(\lambda t-\theta)$. Since $\cos \lambda t$ has maxima at $\frac{2\pi}{\lambda}k$, for $k=0, \pm 1, \pm 2, \ldots$, $\cos(\lambda t-\theta)$ will have maxima at $(2k\pi+\theta)/\lambda$. Because $\cos(\lambda t-\theta)=\cos \lambda t \cdot \cos \theta + \sin \lambda t \cdot \sin \theta$, $\rho\cos(\lambda t-\theta)$ can be written as $\alpha \cos \lambda t + \beta \sin \lambda t$, where $\alpha=\rho\cos\theta$ and $\beta=\rho\sin\theta$. Thus a translated cosine (or sine) curve is a linear combination of a sine and a cosine function. Furthermore, since $\cos^2\theta+\sin^2\theta=1$, $\alpha^2+\beta^2=\rho^2(\cos^2\theta+\sin^2\theta)=\rho^2$. Finally, $\theta=\tan^{-1}(\beta/\alpha)$, and the maximum value that $\rho\cos(\lambda t-\theta)$ can take is ρ; ρ is called the **amplitude** of $\rho\cos(\lambda t-\theta)$.

The foregoing material forms the basis of much that is to follow. It makes the key point that by a judicious choice of ρ, λ and θ, I can obtain any desired shape of the sine and cosine curves. Such curves enables me to approximate each number in a sequence of numbers by a linear combination of a finite number of sines and cosines with varying amplitudes and frequencies. How this can be done is described later in section B.3, but first an illustration of the $\cos \lambda t$ curve with $\lambda=1/2$, 1 and 2, and the $\cos(2t-\theta)$ curve with $\theta=\pi/2$ could be helpful. This is done in Figure B.1 with the dotted curve showing the effect of the phase θ on $\cos 2t$.

Reliability and Risk: A Bayesian Perspective N.D. Singpurwalla
© 2006 John Wiley & Sons, Ltd

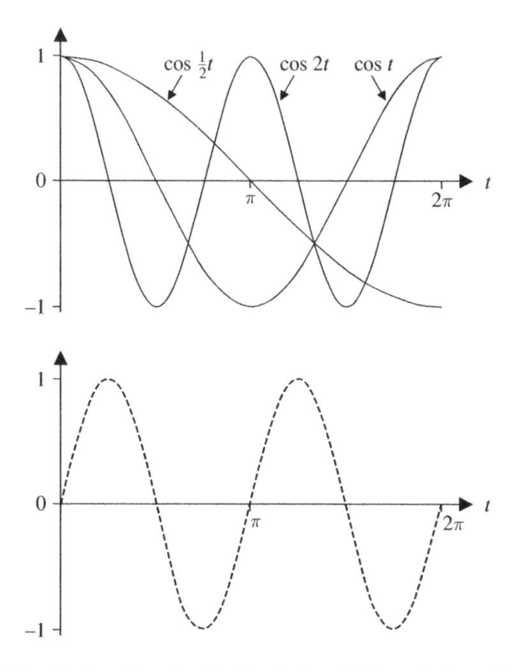

Figure B.1 A plot of $\cos \lambda t$, for $\lambda = 1, 2$ and 1/2, and of $\cos(2t - \pi/2)$.

B.2 ORTHOGONALITY OF TRIGONOMETRIC FUNCTIONS

Trigonometric functions possess an attractive property called orthogonality. This property is stated in (a), (b) and (c) below; a proof can be found in Anderson (1971, p. 94).

Consider an integer, say T, where $T > 0$. Let $[\frac{T}{2}] = T/2$ if T is even; otherwise $[\frac{T}{2}] = (T-1)/2$. Then:

$$\text{(a)}\quad \sum_{t=1}^{T} \cos\frac{2\pi j}{T}t \cdot \cos\frac{2\pi k}{T}t = \begin{cases} 0, & 0 \le k \ne j \le \left[\dfrac{T}{2}\right], \\ \dfrac{T}{2}, & 0 < k = j < \dfrac{T}{2}, \\ T, & k = j = 0 \text{ or } \dfrac{T}{2}; \end{cases}$$

$$\text{(b)}\quad \sum_{t=1}^{T} \cos\frac{2\pi j}{T}t \cdot \sin\frac{2\pi k}{T}t = 0,\ k, j = 0, 1, \ldots, \left[\frac{T}{2}\right];$$

$$\text{(c)}\quad \sum_{t=1}^{T} \sin\frac{2\pi j}{T}t \cdot \sin\frac{2\pi k}{T}t = \begin{cases} 0, & 0 \le k \ne j \le \left[\dfrac{T}{2}\right], \\ \dfrac{T}{2}, & 0 < k = j < \dfrac{T}{2}, \\ 0, & k = j = 0 \text{ or } \dfrac{T}{2}. \end{cases}$$

In (a), (b) and (c) above, the cosine and sine functions have periods T/j and frequencies j/T, for $j = 0, 1, \ldots, [\frac{T}{2}]$. The orthogonality property is advantageously exploited for representing each of a set of T numbers as a finite sum of sines and cosines. This is described next.

B.3 THE FOURIER REPRESENTATION OF A FINITE SEQUENCE OF NUMBERS

Consider a sequence of T numbers, $y_1, \ldots, y_T$, and suppose that T is even. Let $\mathbf{M}$ be the following $T \times T$ matrix:

$$\mathbf{M} = \sqrt{\frac{2}{T}} \begin{bmatrix} \frac{1}{\sqrt{2}} & \cos\frac{2\pi}{T} & \sin\frac{2\pi}{T} & \cos\frac{4\pi}{T} & \cdots & \sin\frac{2\pi}{T}\left(\frac{T}{2}-1\right) & -\frac{1}{\sqrt{2}} \\ \frac{1}{\sqrt{2}} & \cos\frac{2\pi}{T}2 & \sin\frac{2\pi}{T}2 & \cos\frac{4\pi}{T}2 & \cdots & \sin\frac{2\pi}{T}\left(\frac{T}{2}-1\right)2 & \frac{1}{\sqrt{2}} \\ \vdots & \vdots & \vdots & \vdots & & \vdots & \vdots \\ \frac{1}{\sqrt{2}} & 1 & 0 & 1 & \cdots & 0 & \frac{1}{\sqrt{2}} \end{bmatrix}.$$

The orthogonality property ensures that $\mathbf{M}'\mathbf{M} = \mathbf{I}$, where $\mathbf{M}'$ is the transpose of $\mathbf{M}$, and $\mathbf{I}$ is the identity matrix; i.e. all the elements of $\mathbf{I}$ are zero, save the diagonal elements, which are one.

Let $\mathbf{y} = (y_1, \ldots, y_T)'$, and $\mathbf{x} = (x_1, \ldots, x_T)'$, where $\mathbf{x} = \mathbf{M}'\mathbf{y}$. Since $\mathbf{M}'\mathbf{M} = \mathbf{I}$, $\mathbf{y} = \mathbf{M}\mathbf{x}$. From $\mathbf{x} = \mathbf{M}'\mathbf{y}$, I have:

$$x_1 = \frac{1}{\sqrt{T}} \sum_{t=1}^{T} y_t;$$

$$x_{2k} = \sqrt{\frac{2}{T}} \sum_{t=1}^{T} y_t \cos\frac{2\pi k}{T}t, \quad k = 1, 2, \ldots, \frac{T}{2} - 1;$$

$$x_{2k+1} = \sqrt{\frac{2}{T}} \sum_{t=1}^{T} y_t \sin\frac{2\pi k}{T}t, \quad k = 1, 2, \ldots, \frac{T}{2} - 1; \text{ and}$$

$$x_T = \frac{1}{\sqrt{T}} \sum_{t=1}^{T} y_t(-1)^t.$$

The $x_1, \ldots, x_T$ given above are known as **Fourier coefficients**. Since $\mathbf{y} = \mathbf{M}\mathbf{x}$, we may write $y_t, t = 1, \ldots, T$, as

$$y_t = \sqrt{\frac{2}{T}} \left(\frac{x_1}{\sqrt{2}} + x_2 \cos\frac{2\pi}{T}t + x_3 \sin\frac{2\pi}{T}t + \cdots + x_{T-1} \sin\frac{2\pi}{T}\left(\frac{T}{2} - 1\right) + x_T \frac{(-1)^t}{\sqrt{2}} \right). \quad \text{(B.1)}$$

The above representation of y_t is known as the **Fourier representation** with Fourier coefficients $x_1, \ldots, x_T$. The essence of (B.1) is that each y_t can be written as a linear combination of trigonometric functions, the trigonometric functions having the periods $T, T/2, T/3, \ldots,$ and $T/(T/2 - 1)$. Since $T/(T/2 - 1) > 2$, the smallest period is greater than 2; the largest period is T. The Fourier coefficients can be interpreted as weights assigned to the trigonometric functions.

When T is odd, the development proceeds along similar lines, save for the fact that the last column of $\mathbf{M}$ consisting of the $1/\sqrt{2}$ terms is deleted. In this case

$$y_t = \sqrt{\frac{2}{T}}\left(\frac{x_1}{\sqrt{2}} + x_2 \cos\frac{2\pi}{T}t + x_3 \sin\frac{2\pi}{T}t + \cdots + x_{T-1}\sin\frac{2\pi}{T}\left(\frac{T}{2}-1\right)\right). \quad \text{(B.2)}$$

In (B.1) and (B.2) above, let

$$H_{kt} = X_{2k}\cos\frac{2\pi}{T}kt + X_{2k+1}\sin\frac{2\pi}{T}kt,$$

for $k = 1, 2, \ldots, \frac{T}{2} - 1$, in the case of equation (B.1), and $k = 1, 2, \ldots, \frac{T-1}{2}$, in the case of equation (B.2). Observe that H_{kt} is a linear combination of a cosine and a sine curve with periods T/k and frequency k/T. Let

$$\rho(k) = (X_{2k}^2 + X_{2k+1}^2)^{1/2} \text{ and } \theta(k) = \tan^{-1}(X_{2k+1}/X_{2k});$$

then

$$H_{kt} = \rho(k)\cos\left(\frac{2\pi}{T}kt + \theta(k)\right).$$

The quantity H_{kt} is called the k-th harmonic and $\rho(k)$ is called the amplitude of the k-th harmonic. The first harmonic, namely H_{1t}, is called the fundamental harmonic and its frequency $1/T$ is called the fundamental frequency; it is the smallest frequency. Multiples of $1/T$ – the fundamental frequency – are known as harmonics of the fundamental frequency. Observe that H_{kt} has frequency k/T.

B.4 FOURIER SERIES MODELS FOR TIME SERIES DATA

Suppose that $y_1, \ldots, y_T$ represent the observed values of a time series, and our aim is to propose a model that generates these values. More importantly, we need to know if there are (hidden) periodicities in these data, because, in the context of reliability and survival analysis, such periodicities are indicators of possible defeats. To do so, we propose the following model for describing the observations:

$$y_t = f(t) + u_t, \quad t = 1, \ldots, T,$$

where $f(t)$ is an unknown function of t and $\{u_t\}$ is a sequence of independent normally distributed random variables with mean 0 and variance σ^2. If $f(t)$ is periodic with some known period that divides T, then $f(t)$ can be expressed as a linear combination of sines and cosines via either (B.1) or (B.2) depending on whether T is even or odd. However, to do so we need to know the Fourier coefficients $x_1, \ldots, x_{T-1}$; these we do not know, because $f(t)$ is itself unknown. However, the data $y_1, \ldots, y_T$ can be used to assess the unknowns, and to do so we proceed as follows.

Suppose that T is odd, and consider the set of integers $\mathcal{J} = \{1, 2, \ldots, (T-1)/2\}$. Let $\{k_1, \ldots, k_q\}$ be a subset of $\mathcal{J}$, i.e. $q \leq (T-1)/2$. For estimation and inference using a non-Bayesian approach, like least squares, it is necessary that $\{k_1, \ldots, k_q\}$ be a proper subset of $\mathcal{J}$, i.e. $q < (T-1)/2$. With $k_1, \ldots, k_q$ chosen, we consider functions of the form $\sin\frac{2\pi}{T}k_j t$

and $\cos \frac{2\pi}{T} k_j t$, for $j = 1, \ldots, q$. These functions have period T/k_j and frequency k_j/T; for $j = 1, \ldots, q$. Once this is done, we may – inspired by (B.2) – write $f(t)$ as

$$f(t) = \alpha_0 + \sum_{j=1}^{q} \left(\alpha(k_j) \cos \frac{2\pi}{T} k_j t + \beta(k_j) \sin \frac{2\pi}{T} k_j t \right), \tag{B.3}$$

where α_0, $\alpha(k_j)$ and $\beta(k_j)$, $j = 1, \ldots, q$ are unknown constants, the latter pair being reminiscent of Fourier coefficients.

If we let $\rho(k_j) = (\alpha^2(k_j) + \beta^2(k_j))^{1/2}$, and $\theta(k_j) = \tan^{-1}(\beta(k_j)/\alpha(k_j))$, then

$$\alpha(k_j) = \rho(k_j) \cos \theta(k_j) \text{ and } \beta(k_j) = \rho(k_j) \sin \theta(k_j).$$

Consequently, (B.3) can also be written as:

$$f(t) = \alpha_0 + \sum_{j=1}^{q} \rho(k_j) \cos \left[\frac{2\pi}{T} k_j t - \theta(k_j) \right]. \tag{B.4}$$

With (B.3) and (B.4) in place, assessing $f(t)$ boils down to assessing α_0 and the $\alpha(k_j)$s and $\beta(k_j)$s, $j = 1, \ldots, q$.

Note that in the above representations of $f(t)$, the trigonometric terms having period 2 do not appear. To include such terms T needs to be even. When such is the case, I consider proper subsets of $\{1, 2, \ldots, T/2\}$, and now $f(t)$ takes the form

$$f(t) = \alpha_0 + \sum_{j=1}^{q} \rho(k_j) \cos \left[\frac{2\pi}{T} k_j t - \theta(k_j) \right] + \alpha_{T/2}(-1)^t; \tag{B.5}$$

the details are given in Anderson (1971), upon which this Appendix is almost exclusively based. The representation of $f(t)$ via (B.3)–(B.5) is known as a **Fourier Series model**.

As was stated before, the main reason for considering such models is to discover 'hidden periodicities' that may be present in $y_1, \ldots, y_T$. This is because the coefficient $\rho(k_j)$ is a measure of how closely the trigonometric function having frequency k_j/T describes $f(t)$. In particular, if a series of length T has a period, say ϕ, then the value $\rho(k_j)$ corresponding to $k_j = T/\phi$ will tend to be the largest among all the other $\rho(k_j)$s. Thus, for example, the choice $q = 1$ with $k_1 = 1$, reflects the belief that there are no hidden periodicities. In this case

$$f(t) = \alpha_0 + \alpha(1) \cos \frac{2\pi}{T} t + \beta(1) \sin \frac{2\pi}{T} t;$$

this is known as the **simple one-harmonic model**.

B.4.1 The Spectrum and the Periodgram of f(t)

A plot of $\rho^2(k_j) = \alpha^2(k_j) + \beta^2(k_j)$ versus the frequency k_j/T, $j = 1, \ldots, q$ is called the **spectrum** (or the **power spectrum**) of $f(t)$. A plot of $\rho^2(k_j)$ versus T/k_j, the period, is called a **periodgram**. When T is odd, the largest value that q can take is $(T - 1)/2$. Thus the smallest value of the period T/k_q is $2T/(T - 1) > 2$. When T is even, the largest value that q can take is $T/2$. Thus the smallest value of the period is 2. Thus in either case, the periodgram is defined as for periods ≥ 2. Similarly, the spectrum is defined for frequencies ≤ 2. The spectrum has an advantage over the periodgram in the sense that it is evenly spaced over the frequencies. Thus it is the power spectrum, rather than the periodgram, that is often used by engineers as a working tool. It is often the case that in engineering applications, Hertz (Hz) is used as the unit for expressing frequencies. A Hz is the smallest frequency of the spectrum, and it represents the number of cycles (revolutions) per second of rotating machinery.

Appendix C

Network Survivability and Borel's Paradox

C.1 PREAMBLE

The material of this appendix supplements that of section 10.3.4 on prior distributions on the unit hypercube and becomes relevant under certain circumstances. These circumstances will be outlined here. Borel's paradox is a paradox in probability theory which has implications in modeling and inference. The paradox arises when probabilities from a high dimensional non-discrete space get induced to its lower dimensional subspace using conditional arguments that are ill-defined. There are circumstances in system survivability assessment wherein we may be required to induce probabilities from high dimensional spaces to their lower dimensional subspaces. When such is the case, one should be cognizant of encountering Borel's paradox, because in so doing we will be lead to results that are counter to common belief.

To set the stage for the material that follows, we look at (10.10) of Section 10.3.3 for the survivability of the system. Writing $\mathbf{p}$ for $(p_1(\tau), \ldots, p_n(\tau))$ and $\mathrm{d}\mathbf{p}$ for $(\mathrm{d}p_1(\tau), \ldots, \mathrm{d}p_n(\tau))$, we have

$$P(\tau) = \int_{\mathbf{p}} h(\mathbf{p})\pi(\mathbf{p})\,\mathrm{d}\mathbf{p}, \qquad \text{(C.1)}$$

as our expression for system survivability. Recall that $\pi(\mathbf{p})$ encapsulates our uncertainty about $\mathbf{p}$, and is a probability distribution on the unit hypercube. The matter I wish to focus on here pertains to an assessment of $\pi(\mathbf{p})$ via expert testimonies and some nuances that such testimonies can create.

C.2 RE-ASSESSING TESTIMONIES OF EXPERTS WHO HAVE VANISHED

To set the stage for motivating a scenario wherein Borel's paradox can come into play, let us consider the situation wherein an expert $\mathcal{E}$ (or a collection of experts acting as one) is consulted, and $\pi_{\mathcal{E}}(\mathbf{p})$, a prior for $\mathbf{p}$, elicited as the expert's testimony about $\pi(\mathbf{p})$. Suppose that once $\pi_{\mathcal{E}}(\mathbf{p})$ has been elicited $\mathcal{E}$ is no longer available; i.e. $\mathcal{E}$ has 'vanished'. This type of situation is not

© 2006 John Wiley & Sons, Ltd

fictitious and occurs in the context of thermonuclear weapons systems. The original physicists who witnessed actual detonations (which can now not be conducted due to test-ban treaties) have played key roles in specifying the likes of $\pi_{\mathcal{E}}(\mathbf{p})$, but are no longer alive. Suppose further that much after $\pi_{\mathcal{E}}(\mathbf{p})$ gets specified, we receive now unanticipated information which suggests that all the nodes of the networked system are identical so that $p_1(\tau) = p_2(\tau) = \ldots = p_n(\tau) = p$, say. How must we update $\pi_{\mathcal{E}}(\mathbf{p})$ in light of this new additional information? We assume that $\pi_{\mathcal{E}}(\mathbf{p})$ is credible, and so is the new information. Since $\mathcal{E}$ has vanished we are unable to elicit a re-assessed $\pi_{\mathcal{E}}(\mathbf{p})$ from $\mathcal{E}$.

The strategy we propose here is called **retrospective conditioning** (cf. Diaconis and Zabell, 1982); it goes as follows. Since $\pi_{\mathcal{E}}(\mathbf{p})$ is defined on a unit hypercube of dimension $n \geq 2$, and since the new information says that all the $p_i(\tau)$'s be equal to $\mathbf{p}$, we need to induce a one dimensional prior from $\pi_{\mathcal{E}}(\mathbf{p})$. Since $\pi_{\mathcal{E}}(\mathbf{p})$ is continuous, the diagonal joining the points $(0, 0, \ldots, 0)$ and $(1, 1, \ldots, 1)$ has an n-dimensional Lebesgue measure zero. Consequently, to induce a one-dimensional prior from $\pi_{\mathcal{E}}(\mathbf{p})$ we need to condition on a set of measure zero. But conditioning on sets of measure zero entails a limiting argument, which in this case is not unique. Consequently, we obtain different answers for the induced one dimensional prior, and thus different assessments of system survivability. This is known as the Borel–Kolmogorov Paradox (DeGroot 1989, pp. 171–174). An example in two dimensions illustrates this point.

C.3 THE PARADOX IN TWO DIMENSIONS

Let P be a point described by its Cartesian coordinates X and Y, $0 \leq X \leq 1$ and $0 \leq Y \leq 1$; X and Y are random. Suppose that P has a uniform distribution on the unit square. What is the distribution of X were we to be told that P lies on the diagonal; i.e. we are to condition on $X = Y$?

It turns out that our answer depends on how we parameterize the diagonal; by $Z = X/Y = 1$ or by $W = X - Y = 0$. More specifically, if $f_{X|Z=1}(x)$ denotes the probability density of X at x, conditional on $Z = 1$, and $f_{X|W=0}(x)$ the density conditional on $W = 0$, then $f_{X|Z=1}(x) = 2x$ whereas $f_{X|W=0}(x) = 1$. There are two answers to the same question and both are correct. The details are given in Appendix A of Singpurwalla and Swift (2001). An intuitive explanation as to why we obtain two answers to the same question is based on geometrical considerations. It can be appreciated by an examination of Figures C.1 and C.2. These figures illustrate the feature that the conditioning set $X = Y$ can involve two different limiting operations, each leading to a different probability. The shaded regions of these figures indicate the limiting operations. The paradox will not arise if the limiting operation were to be unique or be specified as a protocol.

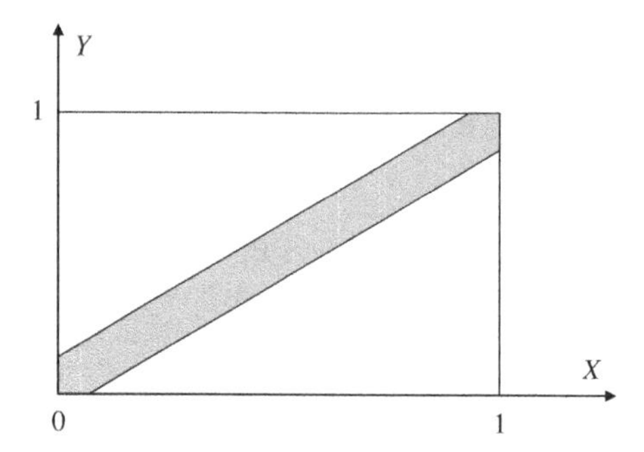

Figure C.1 Diagonal parameterized as $W = X - Y = 0$.

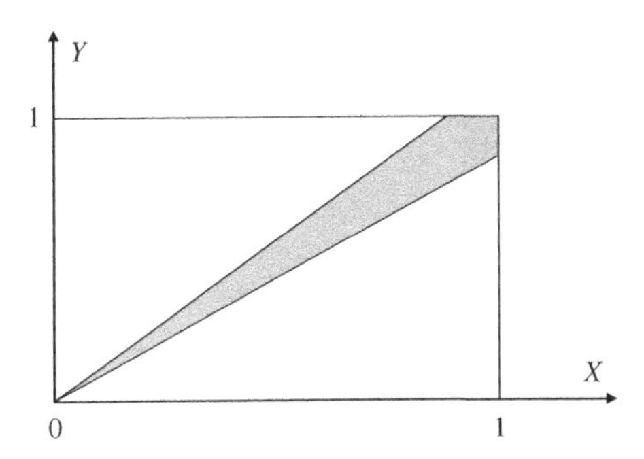

Figure C.2 Diagonal parameterized as $Z = X/Y = 1$

In Figure C.1, values of W in the vicinity of 0 can be described by $|W| \le \delta$, where δ is small. This means that X and Y must satisfy the relationship $X - \delta \le X \le X + \delta$, and since $X, Y \in (0, 1)$, we must have $\max(0, Y - \delta) \le X \le \min(1, Y + \delta)$. The parallel lines of Figure C.1 enveloping the diagonal imply that all values of X (save those at the ends of the diagonal) can be judged equally likely, and so the distribution of X, given that $X = Y$, is uniform; consequently $f_{X|W=0}(x) = 1$. By contrast, suppose that in Figure C.2 values of Z in the vicinity of 1 are described by $1 - \delta \le Z \le 1 + \delta$, where δ is small. Then X and Y must satisfy the inequalities $\max(0, Y - Y\delta) \le X \le \min(0, Y + Y\delta)$; see the shaded region of Figure C.2. The geometry of the cone in the shaded region suggests that values of X in the vicinity of 1 are more likely than those in the vicinity of zero; as a consequence $f_{X|Z=1}(x) = 2x$.

The moral of the story here is that conditioning arguments involving limiting operations that are not uniquely specified lead to a paradox. Thus the cause of the paradox is the freedom (or flexibility) given to us in defining the diagonal and characterizing its vicinity. The paradox will not arise if the manner in which we need to partition the sample space is specified in advance as a protocol. The importance of protocols in conditional probability assessment has been emphasized by Shafer (1985) (also Lindley, 1982a).

C.4 THE PARADOX IN NETWORK SURVIVABILITY ASSESSMENT

To appreciate the effect of Borel's paradox in the context of system survivability assessment, consider a two component series system with component i having reliability p_i, $i = 1, 2$, and $\pi_{\mathcal{E}}(p_1, p_2) = 1$, $0 \le p_i \le 1$. Here (C.1) takes the form

$$P(\tau) = \int_{(p_1, p_2)} p_1 p_2 \,\mathrm{d}p_1\,\mathrm{d}p_2.$$

Suppose now that subsequent to the specification of $\pi_{\mathcal{E}}(p_1, p_2)$ we are told that the two components are identical. This statement can be mathematically cast in one of two possible ways, namely, that

$$P(p_1 \le p) = P(p_2 \le p), \quad \text{for all } p \in (0, 1), \tag{C.2}$$

or that for some $p \in [0, 1)$,

$$P(p_1 \in (p, p + \mathrm{d}p) | p_2 \in (p, p + \mathrm{d}p)) \to 1, \quad \text{as } \mathrm{d}p \downarrow 0,$$

and also that

$$P(p_2 \in (p, p+\mathrm{d}p) | p_1 \in (p, p+\mathrm{d}p)) \rightarrow 1, \quad \text{as } \mathrm{d}p \downarrow 0; \tag{C.3}$$

i.e. $p_1 = p_2 = p$, say.

When (C.2) is invoked, the paradox will not arise. The paradox will arise when (C.3) is invoked. With $p_1 = p_2 = p$ used as a conditioning argument, the induced probability $\pi(p|p_1 = p_2)$ has its mass concentrated on the diagonal so that now

$$P(\tau) = \int_p p^2 \pi(p|p_1 = p_2)\,\mathrm{d}p = \begin{cases} \frac{1}{3}, & \text{if } p_1 = p_2 \text{ is represented as } p_1 - p_2 = 0, \\ \frac{1}{2}, & \text{if } p_1 = p_2 \text{ is represented as } p_1/p_2 = 1. \end{cases}$$

Similarly, in the case of a parallel-redundant system,

$$P(\tau) = \int_p (2p - p^2)\pi(p|p_1 = p_2)\,\mathrm{d}p = \begin{cases} \frac{2}{3}, & \text{if } p_1 = p_2 \text{ is represented as } p_1 - p_2 = 0, \\ \frac{5}{6}, & \text{if } p_1 = p_2 \text{ is represented as } p_1/p_2 = 1. \end{cases}$$

Generalizing to the case of n components in series, the survivability of the system can be shown to be (see Appendix B of Singpurwalla and Swift (2001)) $1/(n+1)$ or $\frac{1}{2}$ depending on how we represent $p_1 = p_2 = \cdots = p_n$. In the parallel redundant case the corresponding answers are $n/(n+1)$ and $1 - (n!)^2/(2n)!$, respectively. As $n \uparrow \infty$, the survivability of the parallel system goes to one, as is expected, though at different rates. However, in the case of a series system, the survivability of the system can stay put at $\frac{1}{2}$, irrespective of how large n becomes, were we to represent $p_1 = p_2 = \cdots = p_n$ via $p_i/p_j = 1, \forall i \neq j$. This result is counter to common belief, namely that the survivability of a series system should decrease to zero as the number of components increases.

Bibliography

Aalen, O.O. (1987) Dynamic modelling and causality. *Scandinavian Actuarial Journal*, 177–90.

Aalen, O.O. (1989) A linear regression model for analysis of life times. *Statistics in Medicine*, **8**, 907–25.

Aalen, O.O. and Gjessing H.K. (2001) Understanding the shape of the hazard rate: A process point of view. *Statistical Science*, **16** (1), 1–13.

Aalen, O.O. and Huseby, E. (1991) statistical analysis of repeated events forming renewal processes. *Statistics in Medicine*, **10**, 1227–240.

Abel, P.S. and N.D. Singpurwalla (1993) To survive or to fail: That is the question. *American Statistician*, **48** (1), 18–21.

Al-Mutairi, D., Y. Chen and N.D. Singpurwalla (1998) An adaptive concatenated failure rate model for software reliability. *Journal of the American Statistical Association*, **93** (443), 1150–63.

Andersen, P.K., O. Borgan, R.D. Gill and N. Keiding (1993) *Statistical Models Based on Counting Processes*. Springer-Verlag, New York, NY.

Anderson, T.W. (1958) *An Introduction to Multivariate Statistical Analysis*, John Wiley, New York, NY.

Anderson, T.W. (1971) *The Statistical Analysis of Time Series*, John Wiley, New York, NY.

Antelman, G. and Savage, I.R. (1965) Characteristic functions of stochastic integrals and reliability problems. *Naval Res. Logist. Quart.*, **12**, 199–22.

Arjas, E. (1981) The failure and hazard processes in multivariate reliability systems. *Math. Oper. Res.*, **6**, 551–62.

Arjas, E. and M. Bhattacharjee (2004) Modeling heterogeneity in repeated failure time data: A hierarchical Bayesian approach, In *Mathematical Reliability* (eds R. Soyer, T. Mazzuchi and N.D. Singpurwalla), Kluwer Academic Publishers, Norwell, MA.

Arjas, E. and Gasbarra, D. (1994) Nonparametric Bayesian inference from right censored survival data, using the Gibbs sampler. *Statistica Sinica*, **4**, 505–24.

Arnauld, A. and P. Nicole (1662) l'Art de Penser, Paris.

Arnold, B., E. Castillo, and J.M. Sarabia (1992) Conditionally specified distributions, *Lecture Notes in Statistics*, Springer-Verlag.

Ascher, H. and H. Fiengold (1984) *Repairable systems reliability*, Dekker, New York, NY.

Athreya, K.B., H. Doss and J. Sethuraman (1996) On the convergence of the Markov chain simulation method. *The Annals of Statistics*, **24** (1), 69–100.

Atkinson, A.C. (1979) A family of switching algorithms for the computer generation of Beta Random Variables. *Biometrika*, **66**, 141–45.

Aven, T. and U. Jensen (1999) *Stochastic Models in Reliability*, Springer-Verlag, New York, NY.

Bagdonavicius, V. and M. Nikulin (2001) *Accelerated Life Models: Modeling and Statistical Analysis*, Chapman and Hall/CRC, London.

Barlow, R.E. (1985) A Bayes explanation of an apparent failure rate paradox. *IEEE Transactions on Reliability*, **R-34**, 107–8.

Barlow, R.E. (1988) Using influence diagrams. *In Accelerated Life Testing and Experts' Opinions in Reliability* (eds C.A. Clarotti and D.V. Lindley), Elsevier, New York, NY, pp. 145–57.

© 2006 John Wiley & Sons, Ltd

Barlow, R.E. and R. Campo (1975) Total time on a test process and applications to failure data analysis. *In Reliability and Fault Tree Analysis: Theoretical and Applied Aspects of System Reliability and Safety Assessment* (eds R.E. Barlow, H.B. Fussell and N.D. Singpurwalla), SIAM, Philadelphia, PA.

Barlow, R.E., C.A. Clarotti and F. Spizzichino (eds) (1993) *Reliability and Decision Making*, Elsevier, New York, NY.

Barlow, R.E., H.B. Fussell and N.D. Singpurwalla (eds) (1975) *Reliability and Fault Tree Analysis: Theoretical and Applied Aspects of System Reliability and Safety Assessment*, SIAM, Philadelphia, PA.

Barlow, R.E. and J.H. Hsiung (1983) Expected information from a life test experiment. *Statistician*, **32**, 35–45.

Barlow, R.E. and M.B. Mendel (1992) de Finetti-type representations for life distributions. *Journal of the American Statistical Association*, **87**, 1116–22.

Barlow, R.E. and C.A.B. Pereira (1990) Conditional independence and probabilistic influence diagrams. In *Topics in Statistical Dependence, IMS Lecture Notes-Monograph Series 16* (eds. H.W. Block, A.R. Sampson and T.H. Savits), 19–33.

Barlow, R.E. and F. Proschan (1965) *Mathematical Theory of Reliability*, John Wiley, New York, NY.

Barlow, R.E. and F. Proschan (1975) *Statistical Theory of Reliability and Life Testing*. Holt, Rinehart, and Winston, Inc., New York.

Barlow, R.E. and F. Proschan (1981) *Statistical Theory of Reliability and Life Testing: Probability Models*, TO BEGIN WITH, Silver Spring, MD.

Barndorff-Nielsen, O., Blaesid, P. and Halgreen, C. (1978) First hitting time models for the generalized inverse Gaussian distribution. *Stochastic Processes aand their Applications*, **7**, 49–54.

Basu, A.P. (1971) Bivariate Failure Rate. *Journal of the American Statistical Association*, **66**, 103–4.

Basu, D. (1975) Statistical information and likelihood (with discussion). **Sankhyä, A** (37), 1–71.

Bayes, T. (1763) An essay towards solving a problem in the doctrine of chances. *The Philosophical Transactions of the Royal Society*, **53**, 370–418. (Reprinted by Biometrika, **45**, 293–315).

Bement, T.A., J.A. Booker, and N.D. Singpurwalla (2002) *Testing the Untestable: Reliability in the 21st Century*. Technical Report GWU/IRRA Serial TR-99/6. The George Washington University.

Bemis, B.M., L.J. Bain, and J.J. Higgins (1972) Estimation and hypotheses testing for the parameters of a bivariate exponential distribution. *Journal of the Royal Statistical Association*, **67**, 927–29.

Bement, T.R., J.M. Booker, S. Keller-McNulty, and N.D. Singapurwalla (2003) Testing the untestable: Reliability in the 21st century. *IEEE Trans. On Reliability*, **52** (1): 118–24.

Berger, J.O. (1985) *Statistical Decision Theory and Bayesian Analysis*, 2nd edn, Springer-Verlag, New York, NY.

Berger, J.O. and J.M. Bernardo (1992) On the Development of Reference Priors. In Bayesian Statistics 4 (eds, Bernardo, J.M., Berger, J.O., Dawid, A.P., Smith, A.F.M.) Oxford University Press, Oxford, pp. 35–60 (with discussion).

Berger, J.O. and D.A. Berry (1988) Statistical analysis and the illusion of objectivity. *American Scientist*, March–April, 159–65.

Berger, J.O. and M. Delampady (1987) Testing of precise hypotheses. *Statistical Science*, **2**, 317–52.

Berger, J.O. and L.R. Pericchi (1996) The intrinsic Bayes factor for model selection and prediction. *Journal of the American Statistical Association*, **91**, 109–22.

Berger, J.O. and R. Wolpert (1988) *The Likelihood Principle, IMS Lecture Notes-Monograph Series 6*, 2nd edn. Institute of Mathematical Statistics, Hayward, CA.

Bergman, B. (1977) Crossings in the total time on test plot. *Scandinavian Journal of Statistics*, **4**, 171–77.

Bernardo, J.M. (1979a) Reference posterior distributions for Bayesian inference. *Journal of the Royal Statistical Society, Ser. B* (41), 113–47.

Bernardo, J.M. (1979b) Expected information as expected utility. *Annals of Statistics*, **7** (3), 686–90.

Bernardo, J.M. (1997) Non-informative priors do not exist: A dialogue with José Bernardo. *Journal Statistical Planning and Inference*, **65**, 158–89.

Bernardo, J.M. and A.F.M. Smith (1994) *Bayesian Theory*, Wiley, Chichester.

Betró, B. and R. Rotondi (1991) On Bayesian inference for the inverse Gaussian distribution, *Statistics and Probability Letters*, **11** (3), 219–24.

Bhattacharyya, G.K. and R.A. Johnson (1973) On a test of independence in a bivariate exponential distribution. *Journal of the American Statistical Association*, **68**, 704–06.

Birnbaum, Z.W. and Saunders, S.C. (1969) A new family of life distributions. *Journal of Applied Probability*, **6**, 319–27.

Black, M. (1939) Vagueness: An exercise in logical analysis. *Philosophy of Science*, 427–55.

Blackwell, D. (1973) Discreteness of ferguson selections. *Annals of Statistics*, **1**, 356–58.

Blackwell, L.M. and N.D. Singpurwalla (1988) Inference from accelerated life tests using filtering in colored noise. *Journal of the Royal Statistical Society, Series B*, **50** (2), 281–92.

Block, H.W. and A.P. Basu (1974) A continuous bivariate exponential extension. *Journal of the American Statistical Association*, **69** (348), 1031–37.

Block, H.W. and T.H. Savits (1997) Burn-In. *Statistical Science*, **12** (1), 1–13.

Block, H.W., T. Savits and H. Singh (1998) The reversed hazard rate function. *Probability in the Engineering and Information Sciences*, **12**, 69–90.

Blumenthal, R.M., Getoor, R.K. and McKean. H.P. (1962) Markov process with identical hitting distributions. *Illinois Journal of Mathematics*, **6**, 402–20.

Bogdanoff, J.L. and Kozin, F. (1985) *Probabilistic Models of Cumulative Damage*. John Wiley & Sons, New York, NY.

Boland, P.J. and F.J. Samaniego (2004) The signature of a coherent system. *In Mathematical Reliability: An Expository Perspective*, (eds Soyer, Mazzuchi and Singpurwalla), Kluwer Academic Press, Norwell, MA, pp. 3–30.

Borel, E. (1914) *Introduction geometrique a quelques theories physiques*, Gauthier-Villars, Paris.

Box, G.E. and G.C. Tiao (1992) *Bayesian Inference in Statistical Analysis*, John Wiley, New York, NY.

Brieman, L. (2001) Statistical modeling: The two cultures. *Statistical Science*, **16** (3), 199–231.

Brooks R.J. (1982) On the loss of information through censoring. *Biometrika*, **69**, 137–44.

Brown, M. and Proschan, F. (1982). Imperfect Repair. *Journal of Applied Probability*, **20**, 851–59.

Bryson, M.C. and Siddiqui, M.M. (1969) Some criteria for ageing. *Journal of American Statistical Association*, **64**, 1472–85.

Cameron, R. and Martin, W. (1944) The Wiener measure of Hilbert neighborhoods in the space of real continuous functions. *Journal of Mathematical physics*, **23**, 195–209.

Campodónico, S. (1993) Bayes analysis of survival times and reliability data: A software package. *In Proceedings of the Annual Reilability and Maintainability Symposium*, Atlanta, GA, pp. 163–166.

Campodónico, S. and N.D. Singpurwalla (1994a) A Bayesian analysis of the logarithmic-Poisson execution time model based on expert opinion and failure data. *IEEE Transactions on Software Engineering*, **20**, 677–83.

Campodónico, S. and N.D. Singpurwalla (1994b) The signature as a covariate in reliability and biometry. *In Aspects of Uncertainty*, (eds. P.R. Freedman and A.F.M. Smith), New York, NY, John Wiley, 119–47.

Campodónico, S. and N.D. Singpurwalla (1995) Inference and predictions from Poisson point processes incorporating expert knowledge, *Journal of the American Statistical Association*, **90** (429), 220–6.

Carnap, R. (1950) *Logical Foundations of Probability*, University of Chicago Press, Chicago, IL.

Casella, G. (1996) Statistical inference and Monte Carlo algorithms. *Test*, **5**, 249–340 (with discussion).

Casella, G. and E.I. George (1992) Explaining the Gibbs sampler. *The American Statistician*, **46** (3), 167–74.

Chaloner, K.M. and G.T. Duncan (1983) Assessment of a beta distribution: PM elicitation. *The Statistician* **32**, 174–80.

Chandra, N.K. and D. Roy (2001) Some results on reversed hazard rate. *Probability in the Engineering and Informational Sciences*, **15**, 95–102.

Chandra, M. and N.D. Singpurwalla (1981) Relationships between notions which are common to reliability theory and economics. *Mathematics of Operations Research*, **6** (1), 113–21.

Chandra, M., N.D. Singpurwalla and M.A. Stephens (1981) Kolmogorov statistics for tests of fit for the extreme-value and Weibull distributions. *Journal of the American Statistical Association*, **76** (375), 729–31.

Chari, M.K. and Colbourn, C.J. (1998) Reliability polynomials: A survey. *Journal of Combinatorics, Information and System Sciences*.

Chaudhuri, G. and J. Sarkar (1998). "On the Crossing Property of the Reliability Functions of a Coherent System in a Random Environment". Technical Report. Department of Mathematical Sciences, Indiana University – Purdue University, Indianapolis.

Chen, J. and Singpurwalla, N.D. (1994a) The notion of 'composite reliability' and its hierarchical Bayes estimation. *JASA*, **91**, 1474–484.

Chen, Y. and N.D. Singpurwalla (1994b) A non-Gaussian Kalman filter model for tracking software reliability. *Statistica Sinica*, **4** (2), 535–48.

Chen, J. and N.D. Singpurwalla (1996) Composite reliability and its hierarchical Bayes estimation. *Journal of the American Statistical Association*, **91** (436), 1474–484.

Chen, Y. and N.D. Singpurwalla (1997) Unification of software reliability models via self-exciting point processes. *Advances in Applied Probabilitiy*, **29** (2), 337–52.

Chernoff, H. (1954) Rational selection of decision functions. *Econometrica*, **22**, 422–43.

Chhikara, R.S. and Folks, J.L. (1977) The inverse Gaussian distribution as a lifetime model. *Technometrics*, **19**, 461–68.

Chipman, H., E.I. George and R.E. McCulloch (2001) The practical implementation of Bayesian model selection. In *Model Selection, IMS Lecture Notes – Monograph Series, Vol. 38.*

Çinlar, E. (1972) Markov additive processes II. *Z. Wahrsch. Verw. Gebiete*, **24**, 94–121.

Çinlar, E. (1975) *Introduction to Stochastic Processes*, Prentice-Hall, Inc, Englewood Cliffs, NJ.

Çinlar, E. (1977) Shock and wear models and Markov additive processes. *The Theory and Applications of Reliability*, Academic, New York, NY.

Çinlar, E. (1979) On increasing continuous processes. *Applied Stochastic Processes*, **9**, 147–54.

Çinlar, E. and S. Ozekici (1987) Reliability of complex devices in random environments. *Probability in the Engineering and Informational Sciences*, **1**, 97–115.

Clarotti, C.A. and F. Spizzichino (1990) Bayes burn-in decision procedures. *Probability in the Engineering and Informational Sciences*, **4**, 437–45.

Clemen, R.T. (1991) *Making Hard Decisions*, PWS-Kent, Boston, MA.

Connor, R.J. and Mosimann, J.E. (1969) Concepts of independence for proportions with a generalization of the Dirichlet distribution. *American Statistical Association Journal*, 194–206.

Cornfield, J. (1957) The estimation of probability of developing a disease in the presence of competing risks. *American Journal of Public Health*, **47**, 601–07.

Cornfield, J. (1969) The Bayesian outlook and its application. *Biometrics*, **25**, 617–41.

Cowell, R.G., A.P. Dawid, S.L. Lauritzen and D.J. Spiegelhalter (1999) *Probabilistic Networks and Expert Systems*, Springer-Verlag, New York, NY.

Cox, D.R. (1972) Regression models and life tables (with discussion). *Journal of the Royal Statistical Society*, B **34**: 187–220.

Cox, D.R., Fitzpatrick, R., Fletcher, A.E., Gore, S.M., Spiegelhalter, D.J. and Jones, D.R. (1992) Quality-of-life assessment: Can we keep it simple?. *Journal of the Royal Statistical Society, Series A*, **155**, 353–93.

Cox, D.R. and V. Isham (1980) *Point Processes*, Chapman & Hall, London.

Crowder, M.J., A.C. Kimber, R.L. Smith, and T.J. Sweeting (1991) *Statistical Analysis of Reliabilitiy Data*, Chapman & Hall, New York, NY.

Cui, Y., C. W. Kong and N.D. Singpurwalla (2002) Information fusion for damage prediction. *In Case Studies in Reliability and Maintenance*, (eds, W. Blischke, and D.N.P. Murthy), John Wiley & Sons, Inc, New York, NY, 251–65.

Currit, A. and N.D. Singpurwalla (1988). "On the Reliability Function of a System of Components Sharing a Common Environment". *Journal of Applied Probability*, 25, 4, 763–771.

Dale, C.J. and Winterbottom, A. (1986) Optimal allocations of effort to improve system reliability. *IEEE Transactions on Reliability*, **R-35** (2), 188–91.

Damien, P., Laud, P.W. and Smith, A.F.M. (1995) Approximate random variate generation from infinitely divisible distributions with applications to Bayesian inference. *JRSS B*, **57** (3), 547–63.

Danaher, P.J. and Hardie, B.G.S. (2005). "Bacon With Your Eggs? Applications of a New Bivariate Beta-Binomial Distribution". *The American Statistician*, 59, 4, 287–291.

Dawid, A.P. (1997) Comments on 'non-informative priors do not exist'. *Journal of Statistical Planning and Inference*, **65**, 159–89.

de Finetti, B. (1937) La prevision: ses lois logiques, ses sources subjectives. *Annales de l'Institut Henri Poincare*, **7**, 1–68.

de Finetti, B. (1938) Sur la condition d'equivalence partielle. *Actualites Scientifiques et Industrielles, 739*, pp. 5–18. Herman, Paris, (Translated in Studies in Inductive Logic and Probability II (ed. R. Jeffrey) University of California, Berkeley, CA.)

de Finetti, B. (1972) *Probability, Induction and Statistics*, Wiley, New York.

de Finetti, B. (1974) *Theory of Probability*, John Wiley, New York, NY.

de Finetti, B. (1976) Probability: beware of falsification!. *Scientia*, **111**, 283–303.

DeGroot, M.H. (1970) *Optimal Statistical Decisions*, McGraw-Hill, New York, NY.

DeGroot, M.H. (1984) Changes in utility as information. *Theory Decision*, **17**, 287–303.

DeGroot, M.H. (1989). *Probability and Statistics*. Addison-Wesley.

De Moivre, A. (1718) *The Doctrine of Chances*, London.

Dempster, A.P. (1968) A generalization of Bayesian inference. *Journal of the Royal Statistical Society, B* **30**, 205– 47.

Dey, J., Erickson, R.V. and R.V. Ramamoorthi (2003) Some aspects of neutral to right priors. *International Statistical Review*, **71** (2), 383–401.

Diaconis, P. (1988) Recent progress on de Finetti's notions of exchangeability. *In Bayesian Statistics* 3, (eds J.M. Bernardo, M.H. DeGroot, D.V. Lindley and A.F.M. Smith), Clarendon, Oxford, 111–25.

Diaconis, P. and D. Freedman (1979) de Finetti's generalization of exchangeability. *In Foundations of Subjective Probabilities*, (ed. R. Jeffrey), University of California Press, Los Angeles, CA, 233–49.

Diaconis, P. and D. Freedman (1980) Finite exchangeable sequences. *The Annals of Probability*, **8**, 745–64.

Diaconis, P. and D. Freedman (1981) Partial exchangeability and sufficiency. *In Statistics: Applications and New Directions*, (eds. J.K. Ghosh and J. Roy), Indian Statistical Institute, Calcutta, India, 205–36.

Diaconis, P. and D. Freedman (1987) A dozen de Finetti-style results in search of a theory. *Annales de l'Institut Henri Poincare Probabilities et Statistiques*, **23**, 397–423.

Diaconis, P. and S.L. Zabell (1982) Updating subjective probability. *Journal of the American Statistical Association*, **77**, 822–30.

Doksum, K. (1974) Tailfree and neutral random probabilities and their posterior distributions. *The Annals of Probability*, **2** (2), 183–201.

Doksum, K.A. (1991) Degradation models for failure time and survival data. *CWI Quarterly, Amsterdam*, **4**, 195–203.

Downton, F. (1970) Bivariate exponential distributions in reliability theory. *Journal of the Royal Statstical Society, Series B*, **33** (3), 408–17.

Durham, S.D. and W.J. Padgett (1997) A cumulative damage model for system failure with application to carbon fibers and composites. *Technometrics*, **39**, 34–44.

Durham, S.D., Lynch J. and W.J. Padgett (1989) A theoretical justification for an increasing failure rate average distribution in fibrous composites. *Naval Research Logistics*, **36**, 655–61.

Dykstra, R.L. and Laud, P. (1981) A Bayesian Approach to Reliability. *The Annals of Statistics*, **9** (2), 356–67.

Ebrahimi, N. and E.S. Soofi (1990) Relative information loss under Type II censored exponential data. *Biometrika*, **77** (2), 429–35.

Efron, B. (1978) Controversies in the foundation of statistics. *American Mathematical Monthly*, **85**, 231–46.

Efron, B. (1979) Bootstrap methods: another look at the Jackknife. *The Annals of Statistics*, **7** (1): 1–26.

Efron, B. (1986) Why isn't everyone a Bayesian? *The American Statistician*, **40**, 1–5.

El-Neweihi, E. and Proschan, F. (1984) Degradable systems: A survey of multistate system theory. *Communications in Statistics: Theory and Methods*, **13**, 405–32.

El-Sayyad, G.M. (1969) Information and sampling from the exponential distribution. *Technometrics*, **11** (1): 41–5.

Esary, J.D., Marshall, A.W. and Proschan, F. (1973) Shock models and wear processes. *Annals of Probability*, 1, 627–49.

Esary, J.D., F. Proschan, and D.W. Walkup (1967) Association of random variables, with applications *The Annals of Mathematical Statistics*, **38** (5): 1466–1474.

Falk, J.E., Singpurwalla, N.D. and Vladimirsky, Y.Y. (2006) Reliability allocations for networks and systems. *Siam Review*, **48**, 43–65.

Falk, J.E. and Soland, R.M. (1969) An algorithm for separable nonconvex programming problems. *Management Science*, **15** (9), 550–69.

Fratta, L. and Montanari, U.G. (1976) Synthesis of available networks. *IEEE Transactions on Reliability*, **R-25**, 81–7.

Feller, W. (1968). *An Introduction to Probability Theory and Its Applications I*. John Wiley & Sons, Inc, New York.

Feller, W. (1966) *An Introduction to Probability Theory and Its Applications II*, John Wiley & Sons, Inc, New York.

Feller, W. (1968) *An Introduction to Probability Theory and Its Applications*, 3rd edn. Vol. 1. John Wiley & Sons, Inc, New York.

Ferguson, T.S. (1973) A Bayesian analysis of some nonparametric problems. *The Annals of Statistics*, **1** (2), 209–30.

Ferguson, T.S. (1974) Prior distributions on spaces of probability measures. *The Annals of Statistics*, **2**, 615–29.

Ferguson, T.S. and Phadia, E.G. (1979) Bayesian nonparametric estimation based on censored data. *The Annals of Statistics*, **7** (1), 163–76.

Ferguson, T.S., Phadia, E.G. and Tiwari, R.C. (1992) Bayesian nonparametric inference. *Current Issues in Statistical Inference: Essays in Honor of D. Basu*, 127–50.

Forman, E.H. and R.F. Dyer (1991) *An Analytical Approach to Marketing Decisions*, Prentice Hall, Englewood Cliffs, NJ.

Fréchet, M. (1951) Sur une application de la statistique mathematique a la biology *Biometrics*, **7** (2), 180–84.

Fréchet, M. (1974) Sur les tableaux de correlation dont les marges sont donnés. *Annales de l'Université de Lyon, Series 3*, **14**, 53–77.

Freedman, D. (1962) Invariants under mixing which generalize de Finetti's theorem. *The Annals of Mathematical Statistics*, **33**, 916–23.

Freund, J.E. (1961) A bivariate extension of the exponential distribution. *Journal of the American Statistical Association*, **56**, 971–77.

Funahashi, K. (1989). "On the Approximate Realization of Continuous Mappings by Neural Networks". *Neural Networks*, 2, 183–192.

Gail, M. and J.L. Gastwirth (1978a) A scale-free goodness-of-fit test for the exponential distribution based on the Lorenz curve. *Journal of the American Statistical Association*, **73**, 787–93.

Gail, M. and J.L. Gastwirth (1978b) A scale-free goodness-of-fit test for the exponential distribution based on the Gini statistic. *Journal of the Royal Statistical Society B*, **40**, 350–57.

Gallager, R.G. (1968) *Information Theory and Reliable Communication*, John Wiley, New York, NY.

Gaver, D.P. (1963) Random hazards in reliability problems. *Technometrics*, 5, 211–26.

Gavrilov L.A. and Gavrilova N.S. (2001) The reliability theory of aging and longevity. *Journal of Theoretical Biology*, **213** (4), 527–45.

Geisser, S. (1984) On prior distributions for binary trials. *The American Statistician*, **38** (4), 244–51.

Geisser, S. (1993) *Predictive Inference: An Introduction*, Chapman & Hall, New York, NY.

Gelfland, A.E. and Kuo, L. (1991) Nonparametric Bayesian bioassay including ordered polytomous responses. Biometrika, **71**, 657–66.

Gelfand, A.E., A.F.M. Smith and T.M. Lee (1992) Bayesian analysis of constrained parameter and truncated data problems using gibbs sampling. *Journal of the American Statistical Association*, **87**, 523–32.

Gelman, A., J.B. Carlin, H.S. Stern and D.B. Rubin (1995). *Bayesian Data Analysis*. Chapman & Hall.

Gelman, A. and T.P. Speed (1993) Characterizing a joint probability distribution by conditionals. *Journal of the Royal Statistical Society* B, **55** (1), 185–88.

Genest, C. and J. MacKay (1986) The joy of copulas: Bivariate distributions with uniform marginals. *American Statistician*, **40**, 280–85.

Gertsbakh, I.B. and Kh.B. Kordonsky (1969) *Models of Failure*, Springer-Verlag, New York, NY.

Ghosh, J.K. and Samanta, T. (2002) Towards a nonsubjective Bayesian paradigm. *Golden Jubilee Volume for Mathematics*, I.I.T Kharagpur (editor J. Misra).

Gilks, W.R. and P. Wild (1992) Adaptive rejection sampling for the Gibbs model. *Applied Statistics*, 41, 337–48.

Gill, R.D. (1984) Understanding Cox's regression model: A martingale approach. *Journal of the American Statistical Association*, **79** (386), 441–47.

Giorgi, G.M. (1998) Concentration index, Bonferroni. *Encyclopedia of Statistical Sciences, Update 2*, Wiley, N.Y. 141–46.

Giorgi, G.M. and Crescenzi, M. (2001) A look at the Bonferroni inequality measure in the reliability framework. *Statistica*, anno LXI, **4**, 571–81.

Giorgi, G.M. and Mondani, R. (1995) Sampling distribution of the Bonferroni inequality index from exponential population. *Sankhya Series B*, **57** (1), 10–8.

Gjessing, H.K., Aalen, O.O. and Hjort, N.L. (2003) Frailty models based on Levy processes. *Advanced Applied Probability*, **35**, 532–50.

Gnedenko, B.V. (1993) *Probability Theory and Mathematical Statistics from Medieval to Modern Time*, Technical Report of SOTAS. SOTAS, Inc., Rockville, MD.

Gnedenko, B.V., Yu.K. Belyaev and A.D. Soloyev (1969) *Mathematical Methods of Reliability Theory*, Academic Press, New York, NY.

Good, I.J. (1950) *Probability and the Weighing of Evidence*, Griffin, London.

Good, I.J. (1961) A casual calculus I. *British Journal of Philosophical Science*, **11**, 305–18; and A casual calculus II. *British Journal of Philosophical Science*, **12**, 43–51.

Good, I.J. (1965) *The Estimation of Probabilities*, The MIT Press, Cambridge, MA.

Good I.J. (1966) A derivation of the probabilistic explication of information. *Journal of the Royal Statistical Society, Series B*, **28**, 578–81.

Good, I.J. (1969) What is the use of a distribution?. *In Multivariate Analysis*, (ed. Krishnaiah), Academic Press, New York, pp. 183–203.

Green, E.J., F.A. Roesch, A.F.M. Smith and W.E. Strawderman (1994) Bayesian estimation for a three-parameter Weibull distribution with tree diameter data. *Biometrics*, **50**, 254–69.

Gumbel, E.J. (1958) *Statistics of Extremes*, Columbia University Press.

Gumbel, E.J. (1960) Bivariate exponential distributions. *Journal of the American Statistical Association*, **50**, 698–707.

Gupta, R.D. and D. Kundu (1998) Hybrid censoring schemes with exponential failure distribution. *Communications in Statistics: Theory and Methods*, **27** (12), 3065–083.

Gurland, J. and J. Sethuraman (1995) How pooling data may reverse increasing failure rates. *Journal of the American Statistical Association*, **90**, 1416–423.

Hacking, I. (1975) *The Emergency of Probability*, Cambridge University Press, London.

Harris, C.M. and Singpurwalla, N.D. (1967) Life distribution derived from stochastic hazard functions. *IEEE Transactions on Reliability, R-17* **2**, 70–9.

Hartigan, J.A. (1983) *Bayes Theory*, Springer-Verlag, New York, NY.

Hawkes, A.G. (1972) A bivariate exponential distribution with applications to reliability theory. *Journal of the Royal Statistical Society, Series B*, **34** (1), 129–31.

Haykin, S. (1999) *Neural Networks: A Comprehensive Foundation*, Prentice-Hall, New Jersey.

Heath, D. and W. Sudderth (1976) de Finetti's theorem on exchangeable variables. *The American Statistician*, **30**, 188–89.

Hesslow, G. (1976). "Two Notes on the Probabilistic Approach to Causality". *Philosophy of Science*, 43, 290–292.

Hesslow, G. (1981). "Causality and Determinism". *Philosophy of Science*, 48, 591–605.

Hill, B.M. (1978) Decision theory. *In Studies in Statistics*, **19**, (ed. R.V. Hogg), 168–209. The Mathematical Association of America.

Hill, B.M. (1982) Comment on Lindley's paradox by G. Shafer. *Journal of the American Statistical Association*, **77**, 344–47.

Hill, B.M. (1993) Dutch books, the Jeffreys-Savage theory of hypothesis testing and Bayesian reliability. *In Reliability and Decision Making*, (eds. R.E. Barlow, C.A. Clarotti and F. Spizzichino), Chapman & Hall, New York, NY, 31–85.

Hjort, N.L. (1990) Nonparametric Bayes estimators based on beta processes in models for life history data. *The Annals of Statistics*, **18** (3), 1259–294.

Hogarth, R.M. (1975) Cognitive processes and the assessment of subjective probability distributions. *Journal of the American Statistician*, **83** (401), 43–51.

Hollander, M. and D.A. Wolfe (1972) *Nonparametric Statistical Methods*, John Wiley & Sons, New York, NY.

Howard, R.A. and J.E. Matheson (1984) Influence diagrams. *In Readings in the Principles and Applications of Decisions Analysis*, 2, (eds. R.A. Howard and J.E. Matheson). Strategic Decision Group, Menlo Park, CA.

Howson, C. and P. Urbach (1989) *Scientific Reasoning: The Bayesian Approach*, Open Co, La Salle, IL.

Hoyland, A. and Rausand, M. (2004) *System Reliability Theory: Models and Statistical Methods*, Wiley-Interscience.

Iglesias, P.L.Z., C.A.B. Pereira and N.I. Tanka (1994) Finite forms of de Finetti's type theorem for univariate uniform distributions. *Unpublished paper of Institute de Matematica e Estatistica*, Universidade de São Paulo, São Paulo, Brazil.

IMS Lecture Notes on Topics in Statistical Dependence (1991) Edited by Block, Sampson, and Savits.

Itô, K. (1969) *Stochastic Processes*, Matematisk Institut, Aarhus University.

James, I.R. and Mosimann, J.E. (1980) A new characterization of the Dirichlet distribution through Neutrality. *Annals of Statistics*, **8**, 183–89.

Jeffreys, H. (1946) An invariant form for the prior probability in estimation problems. *Proceedings of the Royal Statistical Society of London, Ser. A*, **186**, 453–61.

Jeffreys, H. (1961) *Theory of Probability*, 3rd edn. Clarendon, Oxford.

Jeffreys, H. (1980) Some general points in probability theory. *In Bayesian Analysis in Econometrics and Statistics: Essays in Honor of Harold Jeffreys*, (ed. A. Zellner), Amsterdam: North-Holland.

Jeffrey, R. (1965) *The Logic of Decision*, McGraw-Hill, New York, NY.

Jewell, N.P. and Kalbfleisch, J.D. (1996) Marker processes in survival analysis. *Lifetime Data Analysis*, **2**, 15–29.

Johnson, A.L. and S. Kotz (1970) *Distributions in statistics: continuous multivariate distributions*, John Wiley & Sons, Inc, New York, NY.

Johnson, A.L. and S. Kotz (1975) A vector multivariate hazard rate. *Journal of Multivariate Analysis*, 5, 53–66.

Kadane, J.B. and N. Sedransk. (1979) Towards a more ethical trial. *In Bayesian Statistics*, (eds. J.M. Bernardo, M.H. DeGroot, D.V. Lindley and A.F.M. Smith), Valentia, Spain, pp. 329–38.

Kalbfleisch, J.D. (1978) Non-parametric Bayesian analysis of survival time data. *JRSS B*, **40** (2), 214–21.

Kalbfleisch, J.D. and R.L. Prentice (1980) *The Statistical Analysis of Failure Time Data*, John Wiley, New York, NY.

Karlin, S. and Taylor, H.M. (1981) *A second course in stochastic processes*, Academic, New York. NY.

Kashyap, R.L. (1971) Prior probability and uncertainty. *IEEE Transactions on Information Theory*, IT-**14**, 641–50.

Kass, R.E. and A.E. Raftery (1995) Bayes factors. *Journal of the American Statistical Association*, **90** (430), 773–95.

Kass, R.E. and L. Wasserman (1996) The selection of prior distributions by formal rules. *Journal of the American Statistical Association*, **91** (435), 1343–370.

Kebir, Y. (1991) On hazard rate processes. *Naval Res. Logist.*; **38**, 865–76.

Keeney, R.L. and H. Raiffa (1976) *Decisions with Multiple Objectives : Preferences and Value Tradeoffs*, John Wiley & Sons, New York.

Keynes, J.M. (1921) *A Treatise on Probability*, Macmillan, London.

Kingman, J.F.C. (1972) On random sequences with spherical symmetry. *Biometrika*, **59**, 492–94.

Kingman, J.F.C. (1975) Random discrete distributions. *JRSS B*, **37** (1), 1–22.

Kingman, J.F.C. (1978) Use of exchangeability. *The Annals of Probability*, **6**, 183–97.

Klefsjo, B. (1984) Reliability interpretations of some concepts from economics. *Naval Research Logistics Quarterly*, **31**, 301–08.

Klotz, J. (1973) Statistical inference in Bernoulli trials with dependence. *Annals of Statistics*, **1** (2), 373–79.

Kraft, C.H. and van Eeden, C. (1964). "Bayesian Bioassay". *Annals of Mathematical Statistics*, 35, 886–890.

Kolmogorov, A.N. (1963). "On Tables of Random Numbers". *Sankhya*, Series A, 25, 369–376.

Kolmogorov, A.N. and S.V. Fomin (1970) *Introductory real analysis* (Revised English edition translated and edited by R.A. Silverman), Dover, New York, NY.

Kordonsky, K.B. and I. Gertsbakh (1997) Multiple time scales and the lifetime coefficient of variation: Engineering applications. *Lifetime Data Analysis*, **2**, 139–56.

Kotz, S. and N.D. Singpurwalla (1999) On a bivariate distribution with exponential marginals. *Scandinavian Journal of Statistics*, **26**, 451–64.

Kotz, S., N. Balakrishnan and N.L. Johnson (2000). *Continuous Multivariate Distributions. Vol. 1: Models and Applications*. Wiley, New York.

Kumar Joag-Dev (1983) Independence via uncorrelatedness under certain dependence structures *The Annals of Probability*, **11** (4), 1037–1041.

Kumar, S. and Tiwari, R.C. (1989) Bayes estimation of reliability under a random environment governed by a Dirichlet prior. *IEEE Transactions in Reliability*, **38** (2), 218–23.

Kuo, W. and R. Prasad (2000) An annotated overview of system reliability optimization. *IEEE Transactions in Reliability*, **R 49** (2), 176–87.

Kyburg, Jr, H.E. and H.E. Smokler (1964) *Studies in Subjective Probability*, John Wiley, New York, NY.

Lane, D.A. (1987) An epistemic justification of the law of total probability. *In Probability Theory and Applications, Proceedings of the 1st World Congress of the Bernoulli Society*, Eds. Y.A. Prokhorov and V.V. Sazonov, **1**, VNU Science Press, Utrecht. pp. 155–67.

Laplace, P.S. (1774) Mémoire sur la probabilité des causes par les évenemens. Mémoires de mathématique et de physique presentés à l'Académie royale des sciences, par divers savans, & lûs dans ses assemblées, **6**, 621–56.

Laplace, P.S. (1795) *Essai de Philosophique sur les Probabilités*. Paris.

Laplace, P.S. (1812) *Théorie Analytique des Probabilités*. Courcier, Paris.

Laud, P.W., Smith, A.F.M. and Damien, P. (1996) Monte Carlo methods for approximating a posterior hazard rate process. *Statistics and Computing*, **6**, 77–83.

Lauritzen, S.L. and D.J. Spiegelhalter (1988) Local computations with probabilities on graphical structures and their application to expert systems (with discussion). *Journal of the Royal Statistical Society, B* **50**, 157–224.

Lavine, Michael and M.J. Schervish (1999) Bayes factors: what they are and what they are not. *The American Statistician*. **53** (2), 119–22.

Lawless, J.F. (1998) Statistical analysis of product warranty data. *International Statistical Review*, **66** (1), 41–60.

Lee, L. and S.K. Lee (1978) Some results on inference for the Weibull process. *Technometrics*, **20** (1), 41–5.

Lee, L. (1979). "Multivariate Distributions Having Weibull Properties". *Journal of Mulitvariate Analysis*, 9, 267–277.

Lee, P.M. (1989) *Bayesian Statistics: An Introduction*, Oxford, New York, NY.

Lee, M._L.T. and G.A. Whitmore (2006). "Threshold Regression for Survival Analysis". *Statistical Science*. To appear.

Lefebvre, C. and Principe, J. (1999) *Neurosolution*, Wiley, New York.

Lehmann, E.L. (1950) Some principles of the theory of testing hypotheses. *Annals of Mathematical Statistics*, **21**, 1–26.

Lemoine, A.J. and Wenocur, M.L. (1985) On failure modeling. *Naval Res. Logist. Quart.*, **32**, 497–508.

Lemoine, A.J. and M.L. Wenocur (1986) A note on shot-noise and reliability modeling. *Operations Research*, **34**, 320–23.

Lindley, D.V. (1956) On a measure of the information provided by an experiment. *Annals of Mathematical Statistics*, **27**, 986–1005.

Lindley, D.V. (1957a) A statistical paradox. *Biometrika*, **44**, 187–92.

Lindley, D.V. (1957b) Binomial sampling schemes and the concept of information. *Biometrika*, **44**, 179–86.

Lindley, D.V. (1978) *Making Decisions*, Wiley-Interscience, London and New York.

Lindley, D.V. (1982a) The Bayesian approach to statistics. *In Some Recent Advances in Statistics*, (eds. J.T. de Oliveria and B. Epstein), Academic Press, London. 65–87.

Lindley, D.V. (1982b) Scoring rules and the inevitability of probability. *International Statistical Review*, **50**, 1–26

Lindley, D.V. (1983) Reconciliation of probability distributions. *Operatoins Research*, **31**, 866–80.

Lindley, D.V. (1985) *Making Decisions*, 2nd edn, John Wiley, New York, NY.

Lindley, D.V. (1997a) Some comments on Bayes factors. *Journal of Statistical Planning and Inference*, **61**, 181–89.

Lindley, D.V. (1997b) The choice of sample size. *The Statistician*, **46** (2), 129–38.

Lindley, D.V. (2002) Seeing and doing: The concept of causation *International Statistical Review*, **70** (2), 191–214.

Lindley, D.V. and M.R. Novick (1981) The role of exchangeability in inference. *The Annals of Statistics*, **9**, 45–58.

Lindley, D.V. and L.D. Phillips (1976) Inference for a Bernoulli process (a Bayesian view). *The American Statistician*, **30**, 112–19.

Lindley, D.V. and N.D. Singpurwalla (1986) Reliability (and fault tree) analysis using expert opinions. *Journal of the American Statistical Association*, **81**, 87–90.

Lindley, D.V. and N.D. Singpurwalla (1991) On the evidence needed to reach agreed action between adversaries, with application to acceptance sampling. *Journal of the Royal Statistical Association*, **86** (416), 933–37.

Lindley, D.V. and N.D. Singpurwalla (1993) Adversarial life testing. *Journal of the Royal Statistical Society, Series B*, **55** (4), 837–47.

Lindley, D.V. and N.D. Singpurwalla (2002) On exchangeable, causal and cascading failures. *Statistical Science*, vol. 17, No. 2. pp. 209–219.

Lloyd, D.K. and M. Lipow (1962). Reliability: Management, Methods, and Mathematics, Prentice Hall, Englewood Cliffs, New Jersey.

Lochner, R.H. (1975) A generalized Dirichlet distribution in Bayesian life testing. *JRSS B*, **37**, 103–13.

Lorenz, M.O. (1905) Methods of measuring the concentration of wealth. *Publication of the American Statistical Association*, **9**, 209–19.

Lu, C.J., Meeker, W.Q., Escobar, L.A. (1996) Using degradation measurements to estimate a Time-to-Failure distribution. *Statistica Sinica*, **6**, 531–46.

Lukasiewicz, J. (1930) Philosophische bemerkungen zu mehrwertigen systemen des aus-sageenkalkuls. english tr. Philosophical remarks on many valued systems of propositional logic. Polish Logic 1920–1939 (ed. S McCall), Clarendon Press, Oxford. pp. 40–1965, 1967.

Lynn, N.J. and N.D. Singpurwalla (1997) 'Burn-In' makes us feel good. *Statistical Science*, **12** (1), 13–9.

Lynn, N., Singpurwalla, N.D. and Smith, A. (1998) Bayesian assessment of network reliability. *SIAM Review*, **40** (2), 202–27.

Majumdar, D.D., Pal, S.K. and Chaudhuri, B.B. (1976) Fast algorithm for reliability and cost of a complex network. *IEEE Transactions on Reliability*, **R-25** (4), 258.

Makeham, W.M. (1873) On an application of the theory of composition of decremental forces. *Institue of Actuaries Journal*, **18**, 317–22.

Mann, N.R., R.E. Schafer and N.D. Sinpurwalla (1974) *Methods for Statistical Analysis of Reliability and Life Data*, John Wiley, New York, NY.

Marshall, A.W. (1975) Some comments on the hazard gradient. *Stochastic Processes and their Applications*, **3**, 293–300.

Marshall, A.W. (1994) A systems model for reliability studies. *Statistica Sinica*, **4**, 549–65.

Marshall, A.W. and I. Olkin (1967) A multivariate exponential distribution. *Journal of the American Statistical Association*, **62**, 30–44.

Martz, H.F. and R.A. Waller (1982) *Bayesian Reliability Analysis*, John Wiley, New York, NY.

Mastran, D.V. and Singpurwalla, N.D. (1978) A Bayesian estimation of reliability of coherent structures, *Operations Research*, **26**, 633–72.

Mazzuchi, T.A. and Singpurwalla, N.D. (1985) A Bayesian approach to inference for monotone failure rates. *Statistics and Probability Letters*, **3** (3), 135–42.

Mazzuchi, T.A. and Soyer, R. (1993) A Bayes method for assessing product-reliability during development testing. *IEEE Transactions on Reliability*, **42**, 503–10.

McCulloch, W.S. and Pitts, W. (1943) A logical calculus of ideas imminent in nervous activity. *Bulletin of Mathematical Biophysics*, **5**, 115–33.

Meeker, W.Q. and L.A. Escobar (1998) *Statistical methods for reliability data*. John Wiley, New York, NY.

Meinhold, R.J. and N.D. Singpurwalla (1983) Understanding the Kalman filter. The American Statistician, **37** (2), 123–27.

Meinhold, R.J. and N.D. Singpurwalla (1987) A Kalman-filter approach for extrapolations in certain dose-response, damage assessment, and accelerated-life-testing studies. The American Statistician, **41** (2), 101–06.

Meinhold, R.J. and N.D. Singpurwalla (1989) Robustification of Kalman filter models. *Journal of the American Statistical Association*, **84** (406), 479–86.

MIL-STD-105D (1963) *Sampling Procedures and Tables for Inspection by Attributes*, U.S. Government Printing Office, Washington, DC.

MIL-STD-781C (1977) *Reliability Design Qualification and Production Acceptance Tests: Exponential Distribution*, U.S. Government Printing Office, Washington, DC.

Misra, K.B. and M.D. Ljubojevic (1973) Optimal reliability design of a system – A new look. *IEEE Transactions on Reliability*, **R-22**, 256–58.

Montagne, E.R. and N.D. Singpurwalla (1985) Robustness of sequential exponential life testing procedures. *Journal of the American Statistical Association*, **80** (391), 715–19.

Morgenstern, D. (1956) Einfache Beispiele zweidimensionaler Verteilungen. *Mittelungsblatt für Mathematische Statistik*, **8**, 234–35.

Myers, L.E. (1981) Survival functions induced by stochastic covariate processes. *Journal of Applied Probability*, **18**, 523–29.

Nair, V.N. (1998) *Discussion of Estimation of reliability in field performance studies by J.D. Kalbfleisch and J.F. Lawless. Technometrics*, **30**, 379–83.

National Research Council (1998) *Improving the Continued Airworthiness of Civil Aircraft*, National Academy Press, Washington, DC.

Natvig, B. (1982) Two suggestions of how to define a multistate coherent system. *Advances in Applied Probability*, **14**, 434–55.

Nelson, R.B. (1995) Copulas, characterization, correlation and counterexamples. *Mathematics Magazine* **68** (3), 193–98.

Nelson, R.B. (1999) An Introduction to Copulas, Springer-Verlag.

Nelson, W. (1982) Applied Life Data Analysis, John Wiley, New York, NY.

Neyman, J. and Pearson, E.S. (1967) Joint Statistical Papers of J. Neyman and E.S. Pearson, University of California Press, Berkeley, CA.

Nieto-Barajas, L.E. and Walker, S.G. (2002) Markov beta and gamma processes for modelling hazard rates. *Scandinavian Journal of Statistics*, **29**, 413–24.

Novick, M.R. and W.J. Hall (1965) A Bayesian indifference procedure. *Journal of the American Statistical Association*, **60**, 1104–17.

Novick, M.R. and D.V. Lindley (1979) Fixed-state assessment of utility functions. Journal of the American Statistical Association, **74** (366), 306–10.

Oakes, D. (1995). "Multiple Time Scales in Survival Analysis". *Lifetime Data Analysis*, 1, 7–20.

O'Hagan, A. (1995) Fractional Bayes factors for model comparison. *Journal of the Royal Statistical Society. Series B (Methodological)*, **57** (1), 99–138.

Padgett, W.J. and Wei, L.J. (1981) A Bayesian nonparametric estimator of survival probability assuming increasing failure rate. *Communications in Statistics – Theory and Methods*, **A10** (1), 49–63.

Pearl, J. (2000). *Causality: Models, Reasoning, and Inference*. Cambridge University Press, Cambridge.

Peña, E.A. and A.K. Gupta (1990) Bayes estimation for the Marshall-Olkin exponential distribution. *Journal of the Royal Statistical Society, Series B*, **52**, 379–89.

Peña, E.A. and Hollander, M. (2004) Models for recurrent events in reliability and survival analysis. In *Mathematical Reliability: An Expository Perspective* (eds Soyer. *IEEE Transactions on Reliability*. R-34, 69–72.

Peterson, C.R. and L.R. Beach (1967) Man as an intuitive statistician. *Psychological Bulletin*, **68**, 29–64.

Pierce, D.L., Easley, A.R., Windle, J.R. and T.R. Engel (1989) Fast fourier transformation of the entire low amplitude late QRS potential to predict ventricular tachycardia. *Journal of the American College of Cardiology*, **14** (7), 1731–740.

Pitman, J.W. and Speed, T.P. (1973) A note on random times. *Stochastic Processes and their applications*, **1**, 369–74.

Polson, N.G. (1992). "On the Expected Amount of Information from a Non-Linear Model". *Journal of the Royal Statistical Society, Series B*, 54, 3, 889–895.

Press, S.J. (1996) The di Finetti transform. *In Proceedings of the Fifteenth International Workshop on Maximum Entropy and Bayesian Methods*, Kluwer Academic Press, Boston.

Proschan, F. and P. Sullo (1974) Estimating the parameters of a bivariate exponential distribution in several sampling situations. In *Reliability and Biometry* (eds F. Proschan and R.J. Serfling), SIAM.

Puri, P.S. and H. Rubin (1974) On a characterization of the family of distributions with constant multivariate failure rates. *Annals of Probability*, **2**, 738–40.

Raiffa, H. and R. Schaifer (1961) Applied Statistical Decision Theory. Harvard University, Graduate School of Business Administration, Division of Research, Boston.

Ramgopal, P., Laud, P.W. and Smith, A.F.M. (1993) Nonparametric Bayesian bioassay with prior constraints on the shape of the potency curve. *Biometrika*, **80**, 82–6.

Ramsey, F.L. (1972) A Bayesian approach to bioassay. *Biometrics*, **28**, 841–58.

Ramsey, F.P. (1926) Truth and probability. Reprinted in *Studies in Subjective Probability*, (eds. H.E. Kyburg, Jr. and H.E. Smokler), John Wiley, 1964, New York, NY, 61–92.

Rissanen, J. (1983) A universal prior for integers and estimation by minimum description length. *The Annals of Statistics*, **ii**, 416–31.

Roberts, G.O. and A.F.M. Smith (1993) Simple conditions for the convergence of the Gibbs sampler and the metropolis – Hastings algorithms. *Stochastic Processes and their Applications*, **49**, 207–16.

Roberts, H.V. (1967) Informative stopping rules and inferences about population size. *Journal of the American Statistical Association*, **62** (319), 763–75.

Ross, S.M. (1996). *Stochastic Processes*. John Wiley & Sons, Inc., New York.

Rosen, D.A. (1978). "In Defense of a Probabilistic Theory of Causality". *Philosophy of Science*, 45, 604–613.

Royal Society Study Group (1992) *Risk: Analysis, Perception and Management*, The Royal Society, London.

Royden H.L. (1968) *Real Analysis*, MacMillan Publishing Co., Inc., NewYork.

Rubin, H. (1987) A weak system of axioms for 'rational' behavior and the nonseparability of utility from prior. *Statistics and Decisions*, **5**, 47–58.

Russell, B. (1923) Vagueness. *Australasian Journal of Philosophy*, **1**, 88.

Ryu, K. (1993) An extension of Marshall and Olkin's bivariate exponential distribution. *Journal of the American Statistical Association*, **88** (424), 1458–465.

Salmon, W. (1980). "Probabilistic Causality". *Pacific Philosophical Quarterly*, 61, 50–74.

Samaniego, F.J. (1985). On closure of IFR class under formation of coherent systems. *IEEE Transactions on Reliablity*, R-34, 69–72.

San Martini, A. and F. Spezzaferri (1984) A predictive model selection criterion. *Journal of the Royal Statistical Society, Series B*, **46**, 296–303.

Sarkar, S.K. (1987) A continuous bivariate exponential distribution. *Journal of the American Statistical Association*, **82**, 667–75.

Savage, L.J. (1954) *The Foundations of Statistics*, 1st edn. John Wiley, New York, NY.

Savage, L.J. (1971) Elicitation of personal probabilities and expectations. *Journal of the American Statistical Association*, **66**, 783–801.

Savage, L.J. (1972) *The Foundations of Statistics*, 2nd edn. Dover, New York, NY.

Schoenberg, I.J. (1938) Metric spaces and positive definite functions. *Transactions of the American Mathematical Society*, **44**, 522–36.

Schrodinger, E. (1947) The foundation of the theory of probability-I. *Proceedings of the Royal Irish Academy*, **51 A** (5), 51–66.

Sellers, K.F. and Singpurwalla, N.D. (2006) Many-valued Logic in Multistate and Vague Stochastic Systems. Technical Report, The George Washington University. Under review.

Sethuraman, J. (1994) A constructive definition of Dirichlet priors. *Statistica Sinica*, **4** (2), 639–50.

Sethuraman, J. and Tiwari, R.C. (1982) Convergence of Dirichlet measures and the interpretations of their parameters. *Statistical Decision Theory and Related Topics, III*. Academic Press.

Shafer, G. (1976) *A Mathematical Theory of Evidence*, Princeton University Press, Princeton, NJ.

Shafer, G. (1981) Jeffrey's rule of conditioning. *Philosophy of Science*, **48**, 337–62.

Shafer, G. (1982a) Lindley's paradox. *Journal of the American Statistical Association*, **77**, 325–51.

Shafer, G. (1982b) Bayes's two arguments for the rule of conditioning. *The Annals of Statistics*, **10** (4), 1075–089.

Shafer, G. (1985) Conditional probability. *International Statistical Review*, **53** (3), 261–77.

Shafer, G. (1986) The combination of evidence. *International Journal of Intelligent Systems*, **1**, 155–79.

Shafer, G. (1990) The unity and diversity of probability. *Statistical Science*, **5**, 435–62.

Shafer, G. (1991) What Is Probability? Working Paper No. 229. School of Business, The University of Kansas, Lawrence, KS.

Shafer, G. (1996) The significance of Jacob Bernoulli's Ars Conjectandi for the philosophy of probability today. *Journal of Econometrics*, **75**, 15–32.

Shaked, M. (1977) A concept of positive dependence for exchangeable random variables. *The Annals of Statistics*, **5**, 505–15.

Shamseldin, A.A. and S.J. Press (1984) Bayesian parameter and reliability estimation for a bivariate exponential distribution. *Journal of Econometrics*, **24**, 363–78.

Shannon, C.E. (1948) A mathematical theory of communication. *Bell Systems Tech Journal*, **27**, 379–423 and 623–56.

Singpurwalla, N.D. (1983) A unifying perspective on statistical modeling. *SIAM Review*, **31** (4), 560–64.

Singpurwalla, N.D. (1988a) Foundational issues in reliability and risk analysis. SIAM Review, **30**, 264–82.

Singpurwalla, N.D. (1988b) An interactive PC-based procedure for reliability assessment incorporating expert opinion and survival data. *Journal of the American Statisticial Association*, **83** (401), 43–51.

Singpurwalla, N.D. (1989) A unifying perspective on statistical modeling. *SIAM Review*, **31** (4), 560–64.

Singpurwalla, N.D. (1993) Comments on statistical analysis of reliability data by Crowder, M.J., Kimber, A.C., Smith, R.L. and Sweeting T.J. (1991), Chapman Hall, in *SIAM Review*, **35**, 535–38.

Singpurwalla, N.D. (1995a) The failure rate of software: does it exist?. *The IEEE Transitions in Reliability*, **44** (3), 463–69.

Singpurwalla, N.D. (1995b) Survival in dynamic environments. *Statistical Science*, **10** (1), 86–113.

Singpurwalla, N.D. (1996) Entropy and information in reliability. *In Bayesian Analysis in Statistics and Econometrics*, (eds. D.A. Berry, K.M. Chaloner, and J.K. Geweke). John Wiley, New York, NY, pp. 459–69.

Singpurwalla N.D. (1997). "Gamma Processes and their Generalizations: An Overview". In *Engineering Probabilistic Design and Maintenance for Flood Protection* (R. Cooke., M. Medel and H. Vrijling, eds.), 67–75. Kluwer Academic Publishers, Dordrecht.

Singpurwalla, N.D. (1998) A paradigm for modeling and checking reliability growth. *In reliability growth modeling* (eds, K.J. Farquar and A. Mosleh), 121–25.

Singpurwalla, N.D. (2000a) The point process paradox: Where should we extend the conversation?. *The American Statistician*, **54** (2), 119–20.

Singpurwalla, N.D. (2000b). "Contract Warranties and Equilibrium Probabilities". In *Statistical Sciences in the Courtroom* (J.L. Gastwirth, ed.), 363–377. Springer-Verlag, New York.

Singpurwalla, N.D. (2001) Cracks in the empire of chance: flaws in the foundations of reliability. *International Statistical Review*, **10** (1), 53–78.

Singpurwalla, N.D. (2002) On causality and causal mechanisms. *International Statistical Review*, **10** (2), 198–206.

Singpurwalla, N.D. (2004) Warranty: A surrogate of reliability. *Mathematical Reliability – An Expository Perspective* (eds, Soyer, Mazzuchi, and Singpurwalla), Kluwer Academic Publishers.

Singpurwalla, N.D. (2005) Decelerated testing: A hierarchical Bayesian approach. *Technometrics*, **47** (4), 468–77.

Singpurwalla, N.D. (2006a) The hazard potential: Introduction and overview. *Journal of the American Statistical Associtation*. To appear (Dec. 2006).

Singpurwalla, N.D. (2006b) On competing risk and degradation processes. The second Erich Lehmann Symposium – Optimality. Institute of Mathematical Statistics – Monograph Series, Vol. 49 (ed. J. Rojo) pp. 289–304.

Singpurwalla, N.D. (2006c) Reliability and survival in financial risk. In Advances in statistical Modeling and Inference – Essays in Honor of Kjell Doksum (ed. V. Nair), World Scientific Publications. To appear.

Singpurwalla, N.D. and Booker J. (2004) Membership functions and probability measures of fuzzy sets. *Journal of the American Statistical Association*, **99**, 867–77.

Singpurwalla, N.D., J. Eliashberg and S. Wilson (1996) Calculating the reserve for a time and usage indexed warranty. *Management Science*. **43** (7), 966–75.

Singpurwalla, N.D. and C. Kong (2004) Specifying interdependence in networked systems. *IEEE Trans. On Reliability*, **53** (3), 401–05.

Singpurwalla, N.D. and Shaked, M. (1990) A Bayesian approach for quantile and response probability estimation with applications to reliability. *Annals of the Institute of Statistical Mathematics*, **42** (1), 1–19.

Singpurwalla, N.D. and Song, M.S. (1986) An analysis of Weibull lifetime data incorporating expert opinion. *In Probability and Bayesian Statistics* (ed. R. Viertl), New York: Plenum: 431–42.

Singpurwalla, N.D. and R. Soyer (1992) Nonhomogeneous autoregressive processes for tracking (software) reliability growth, and their Bayesian analysis. *Journal of the Royal Statistical Society*, **B 54**, 145–56.

Singpurwalla, N.D. and A. Swift (2001) Network reliability and Borel's paradox. *The American Statistician*, **53** (3), 213–18.

Singpurwalla, N.D. and S. Wilson (1993). "Warranties: Their Statistical and Game Theoretic Aspects". *SIAM Review*, 35, 1, 17–42.

Singpurwalla, N.D. and S.P. Wilson (1995) The exponentiation formula of reliability and survival: does it always hold? *Life Data Analysis*, Kluwer Academic Publishers, **1**, 187–94.

Singpurwalla, N.D. and S.P. Wilson (1998) Failure models indexed by two scales. *Advanced Applied Probability*, **30**, 1058–72.

Singpurwalla, N.D. and S.P. Wilson (1999) Statistical methods in software engineering. Springer, New York, NY.

Singpurwalla, N.D. and P. Wilson (2004) When can finite testing ensure infinite trustworthiness?. *Journal of the Iranian Statistical Society*. **3** (1), 1–37.

Singpurwalla, N.D. and P. Wilson (2006) Item response models for coherent utility assessments. Technical Report, The George Washington University.

Singpurwalla, N.D. and M.A. Youngren (1993) Multivariate distributions induced by dynamic environments. *Scandinavian Journal of Statistics*, **20**, 251–61.

Sinha, S.K. and J.A. Sloan (1988) Bayes estimation of the parameters and reliability function of the three parameter Weibull distribution. *IEEE Transactions on Reliability*, **37**, 364–69.

Sklar, A. (1959) Fonctions de répartition à n dimensions et leurs marges. *Publications de l'Institut Statistique de l'Université de Paris*, **8**, 229–31.

Smith, A.F.M. (1981) On random sequences with centered spherical symmetry. *Journal of the Royal Statistical Society*, **B 43**, 208–09.

Smith, A.F.M. and G.O. Roberts (1993) Bayesian computation via the Gibbs sampler and related Markov Chain Monté Carlo methods. *Journal of the Royal Statistical Society, Series B*, **55**, 2–23.

Smith, C.A.B. (1961) Consistency in statistical inference and decision. *Journal of the Royal Statistical Society*, **B 23**, 213–20.

Smythe, R.T. (2004) Beta distribution in bioassay. *Handbook of Beta Distribution and its Applications*, Marcell Dekker.

Sobczyk, K. (1987) Stochastic models for fatigue damage of materials. *Advanced Applied Probability*, 19, 652–73.

Soland, R.M. (1969). "Bayesian Analysis of the Weibull Process with Unknown Scale and Shape Parameters". *IEEE Transactions in Reliability*, 18, 181–184.

Soofi, E.S. (2000) Principal information theoretic approaches. *Journal of the American Statistical Association*, **95**, (452), 1349–353.

Spizzichino, F. (1988) Symmetry conditions on opinion assessment leading to time-transformed exponential models. *In Accelerated Life Testing and Expert's Opinions in Reliability*, (eds. C.A. Clarotti and D.V. Lindley), Elsevier, New York, NY. 83–97.

Statistical Software Engineering (1996) National Academy Press, Washington, DC 20055.

Stigler, S.M. (1982) Thomas Bayes's Bayesian inference. *Journal of Royal Statistics*, **A 145** (2), 250–58.

Stigler, S.M. (1983) Who discovered Bayes's theorem? *The American Statistician*, **37**, 290–96.

Stigler, S.M. (1986) Laplace's 1774 memoir on inverse probability. *Statistical Science*, **1** (3), 359–78.

Sun, D. (1997) A note on noninformative priors for Weibull distributions. *Journal of Statistical Planning and Inference*, **61**, 319–38.

Suppes, P. (1970) *A Probabilistic Theory of Causality*, North-Holland, Amsterdam.

Susarla, V. and Van Ryzin, J. (1976) Nonparametric Bayesian estimation of survival curves from incomplete observations. *Journal of American Statistical Society*, **71**, 740–54.

Swift, A. (2001) Stochastic models of cascading failures. PhD Thesis, The George Washington University.

Teugels, J.L. (1990) Some representations of multivariate Bernoulli and binomial distributions. *Journal of Multivariate Analysis*, **32**, 256–68.

Tillman, F.A., Hwang, C.L., Fan, L.T. and Lai, K.C. (1970) Optimal reliability of a complex system. *IEEE Transactions on Reliability*, **R-19** (3), 95–100.

Thiel, H. (1967). *Economics and Information Theory*. Nort-Holland, Amsterdam.

Tversky, A. and D. Kahneman (1986) Rational choice and the framing of decisions. *Journal of Business*, **59**, S251–278.

Upadhyay, S.K. and A.F.M. Smith (1994) Modeling complexities in reliability, and the role of simulation in Bayesian computation. *International Journal of Continuing Education*, **4**, 93–104.

Upadhyay, S.K., N. Vasishta and A.F.M. Smith (2001) Bayes inference in life testing and reliability via Markov Chain Monté Carlo simulation. *Sankhya: The Indian Journal of Statistics*, 63, A, Part 1: 15–40.

van der Weide H. (1997). "Gamma Processes". In *Engineering Probabilistic Design and Maintenance for Flood Protection* (R. Cooke., M. Medel and H. Vrijling, eds.), 77-83. Kluwer Academic Publishers, Dordrecht.

Venn, J. (1866) *The Logic of Chance*, Macmillan, London, UK.

Verdinelli, I., N. Polson and N.D. Singpurwalla (1993) Shannon information, Bayesian design in accelerated life-testing. *In Reliability and Decision Making*, (eds. R.E. Barlow, C.A. Clarotti, and F. Spizzichino), Chapman and Hall, London: pp. 247–256.

von Mises, R. (1939) *Probability Statistics and Truth*, Hodge, London, UK. German original 1928.

von Neumann, J. and O. Morgenstern (1944) *Theory of Games and Economic Behavior*, Princeton University Press, Princeton, NJ.

Wald, A. (1950) *Statistical Decision Functions*, John Wiley & Sons, New York, NY.

Walker, S. and Damien, P. (1998) A full Bayesian non-parametric analysis involving a neutral to the right process. *Scandinavian Journal of Statistics*, **25**, 669–80.

Walker, S. and Muliere, P. (1997) Beta-Stacy processes and a generalization of the Polya-Urn Scheme. *The Annals of Statistics*, **25** (4), 1762–780.

Walley, P. (1991) *Statistical Reasoning with Imprecise Probabilities*, Chapman and Hall.

Wenocur, M.L. (1989) A reliability model based on gamma process and its analytic theory. *Advances in Applied Probability*, **21**, 899–918.

West, M. and J. Harrison (1989). *Bayesian Forecasting and Dynamic Models*. Springer-Verlag, New York.

Whitmore, G.A. (1995) Estimating degradation by a Wiener diffusion process subject to measurement error. *Lifetime Data Analysis*, **I**, 307–19.

Whitmore, G.A., Crowder, M.J. and Lawless, J.F. (1998) Failure inference from a marker process based on a bivariate Wiener process. *Lifetime Data Analysis*, **4**, 229–51.

Wilks, S.S. (1962) *Mathematical Statistics*, John Wiley & Sons, Inc., New York, NY.

Woodbury, M. and Manton, K. (1977) A random walk model for human mortality and aging. *Theoretical Population Biology*, **11**, 37–48.

Yang, Y. and Klutke, G.A. (2000) Lifetime-characteristics and inspection-schemes for Levy degradation processes. *IEEE Transactions on Reliability*, **49** (4) 377–282.

Yashin, A. and E. Arjas (1998) A note on random intensities and conditional survival functions. *Journal of Applied Probability*, **25**, 630–35.

Yashin, A.I. and Manton, K.G. (1997) Effects of unobserved and partially observed covariate processes on system failure: a review of models and estimation strategies. *Statistical Science*, **12**, 20–34.

Yor, M. (1992) On some exponential functionals of Brownian motion. *Advanced Applied Probability*, **24**, 509–31.

Zacks, S. (1992) *Introduction to Reliability Analysis: Probability Models and Statistical Methods*, Springer-Verlag, New York, NY.

Zadeh, L. (1979) Possibility theory and soft data analysis, memo. *Technical Report UCB/ERL M79/66*. University of California, Berkeley, CA.

Zadeh, L. (1965) Fuzzy sets. *Information and Control*, **8**, 338–53.

Zellner, A. (1977) Maximal data information prior distributions. *In New Developmentsin the Application of Bayesian Methods*. (eds, A. Aykac and C. Brumat), Amsterdam: North-Holland.

Zellner, A. (1991) Bayesian methods and entropy in economics and econometrics. *In Maximum Entropy and Bayesian Methods*. (eds. W.T. Grandy, Jr. and L.H. Schick). Boston: Kluwer, pp. 17–31.

Zellner, A. and C. Min (1993) Bayesian analysis, model selection and prediction. *In Physics and Probability: Essays in Honor of Edwin T. Jaynes*, (eds. W.T. Grandy, Jr. and P.W. Milonni), Cambridge University Press, Cambridge, UK, pp. 585–647.

Index

Reliability and Risk: A Bayesian Perspective N.D. Singpurwalla
© 2006 John Wiley & Sons, Ltd

WILEY SERIES IN PROBABILITY AND STATISTICS

ESTABLISHED BY WALTER A. SHEWHART AND SAMUEL S. WILKS

Editors
David J. Balding, Peter Bloomfield, Noel A.C. Cressie, Nicholas I. Fisher, Iain M. Johnstone, J.B. Kadane, Geert Molenberghs, Louise M. Ryan, David W. Scott, Adrian F.M. Smith
Editors Emeriti
Vic Barnett, J. Stuart Hunter, David G. Kendall, Jozef L. Teugels

The *Wiley Series in Probability and Statistics* is well established and authoritative. It covers many topics of current research interest in both pure and applied statistics and probability theory. Written by leading statisticians and institutions, the titles span both state-of-the-art developments in the field and classical methods.

Reflecting the wide range of current research in statistics, the series encompasses applied, methodological and theoretical statistics, ranging from applications and new techniques made possible by advances in computerized practice to rigorous treatment of theoretical approaches.

This series provides essential and invaluable reading for all statisticians, whether in academia, industry, government, or research.

ABRAHAM and LEDOLTER. Statistical Methods for Forecasting
AGRESTI. Analysis of Ordinal Categorical Data
AGRESTI. An Introduction to Categorical Data Analysis
AGRESTI. Categorical Data Analysis, *Second Edition*
ALTMAN, GILL, and McDONALD. Numerical Issues in Statistical Computing for the Social Scientist
AMARATUNGA and CABRERA. Exploration and Analysis of DNA Microarray and Protein Array Data
ANDĚL. Mathematics of Chance
ANDERSON. An Introduction to Multivariate Statistical Analysis, *Third Edition*
*ANDERSON. The Statistical Analysis of Time Series
ANDERSON, AUQUIER, HAUCK, OAKES, VANDAELE, and WEISBERG. Statistical Methods for Comparative Studies
ANDERSON and LOYNES. The Teaching of Practical Statistics
ARMITAGE and DAVID (editors). Advances in Biometry
ARNOLD, BALAKRISHNAN, and NAGARAJA. Records
*ARTHANARI and DODGE. Mathematical Programming in Statistics
*BAILEY. The Elements of Stochastic Processes with Applications to the Natural Sciences
BALAKRISHNAN and KOUTRAS. Runs and Scans with Applications
BARNETT. Comparative Statistical Inference, *Third Edition*
BARNETT. Environmental Statistics: Methods & Applications
BARNETT and LEWIS. Outliers in Statistical Data, *Third Edition*
BARTOSZYNSKI and NIEWIADOMSKA-BUGAJ. Probability and Statistical Inference
BASILEVSKY. Statistical Factor Analysis and Related Methods: Theory and Applications
BASU and RIGDON. Statistical Methods for the Reliability of Repairable Systems
BATES and WATTS. Nonlinear Regression Analysis and Its Applications
BECHHOFER, SANTNER, and GOLDSMAN. Design and Analysis of Experiments for Statistical Selection, Screening, and Multiple Comparisons
BELSLEY. Conditioning Diagnostics: Collinearity and Weak Data in Regression
BELSLEY, KUH, and WELSCH. Regression Diagnostics: Identifying Influential Data and Sources of Collinearity
BENDAT and PIERSOL. Random Data: Analysis and Measurement Procedures, *Third Edition*
BERNARDO and SMITH. Bayesian Theory
BERRY, CHALONER, and GEWEKE. Bayesian Analysis in Statistics and Econometrics: Essays in Honor of Arnold Zellner
BHAT and MILLER. Elements of Applied Stochastic Processes, *Third Edition*

*Now available in a lower priced paperback edition in the Wiley Classics Library.

BHATTACHARYA and JOHNSON. Statistical Concepts and Methods
BHATTACHARYA and WAYMIRE. Stochastic Processes with Applications
BIEMER, GROVES, LYBERG, MATHIOWETZ and SUDMAN. Measurement Errors in Surveys
BILLINGSLEY. Convergence of Probability Measures, *Second Edition*
BILLINGSLEY. Probability and Measure, *Third Edition*
BIRKES and DODGE. Alternative Methods of Regression
BLISCHKE and MURTHY (editors). Case Studies in Reliability and Maintenance
BLISCHKE and MURTHY. Reliability: Modeling, Prediction, and Optimization
BLOOMFIELD. Fourier Analysis of Time Series: An Introduction, *Second Edition*
BOLLEN. Structural Equations with Latent Variables
BOLLEN and CURRAN. Latent Curve Models: A Structural Equation Perspective
BOROVKOV. Ergodicity and Stability of Stochastic Processes
BOULEAU. Numerical Methods for Stochastic Processes
BOX. Bayesian Inference in Statistical Analysis
BOX. R. A. Fisher, the Life of a Scientist
BOX and DRAPER. Empirical Model-Building and Response Surfaces
*BOX and DRAPER. Evolutionary Operation: A Statistical Method for Process Improvement
BOX, HUNTER, and HUNTER. Statistics for Experimenters: An Introduction to Design, Data Analysis, and Model Building
BOX, HUNTER, and HUNTER. Statistics for Experimenters: Design, Innovation and Discovery, *Second Edition*
BOX and LUCEÑO. Statistical Control by Monitoring and Feedback Adjustment
BRANDIMARTE. Numerical Methods in Finance: A MATLAB-Based Introduction
BROWN and HOLLANDER. Statistics: A Biomedical Introduction
BRUNNER, DOMHOF, and LANGER. Nonparametric Analysis of Longitudinal Data in Factorial Experiments
BUCKLEW. Large Deviation Techniques in Decision, Simulation, and Estimation
CAIROLI and DALANG. Sequential Stochastic Optimization
CASTILLO, HADI, BALAKRISHNAN and SARABIA. Extreme Value and Related Models with Applications in Engineering and Science
CHAN. Time Series: Applications to Finance
CHATTERJEE and HADI. Sensitivity Analysis in Linear Regression
CHATTERJEE and PRICE. Regression Analysis by Example, *Third Edition*
CHERNICK. Bootstrap Methods: A Practitioner's Guide
CHERNICK and FRIIS. Introductory Biostatistics for the Health Sciences
CHILÈS and DELFINER. Geostatistics: Modeling Spatial Uncertainty
CHOW and LIU. Design and Analysis of Clinical Trials: Concepts and Methodologies, *Second Edition*
CLARKE and DISNEY. Probability and Random Processes: A First Course with Applications, *Second Edition*
*COCHRAN and COX. Experimental Designs, *Second Edition*
CONGDON. Applied Bayesian Modelling
CONGDON. Bayesian Statistical Modelling
CONGDON. Bayesian Models for Categorical Data
CONOVER. Practical Nonparametric Statistics, *Second Edition*
COOK. Regression Graphics
COOK and WEISBERG. Applied Regression Including Computing and Graphics
COOK and WEISBERG. An Introduction to Regression Graphics
CORNELL. Experiments with Mixtures, Designs, Models, and the Analysis of Mixture Data, *Third Edition*
COVER and THOMAS. Elements of Information Theory
COX. A Handbook of Introductory Statistical Methods
*COX. Planning of Experiments
CRESSIE. Statistics for Spatial Data, *Revised Edition*

* Now available in a lower priced paperback edition in the Wiley Classics Library.

CSÖRGÖ and HORVÁTH. Limit Theorems in Change Point Analysis
DANIEL. Applications of Statistics to Industrial Experimentation
DANIEL. Biostatistics: A Foundation for Analysis in the Health Sciences, *Sixth Edition*
*DANIEL. Fitting Equations to Data: Computer Analysis of Multifactor Data, *Second Edition*
DASU and JOHNSON. Exploratory Data Mining and Data Cleaning
DAVID and NAGARAJA. Order Statistics, *Third Edition*
*DEGROOT, FIENBERG, and KADANE. Statistics and the Law
DEL CASTILLO. Statistical Process Adjustment for Quality Control
DEMARIS. Regression with Social Data: Modeling Continuous and Limited Response Variables
DEMIDENKO. Mixed Models: Theory and Applications
DENISON, HOLMES, MALLICK, and SMITH. Bayesian Methods for Nonlinear Classification and Regression
DETTE and STUDDEN. The Theory of Canonical Moments with Applications in Statistics, Probability, and Analysis
DEY and MUKERJEE. Fractional Factorial Plans
DILLON and GOLDSTEIN. Multivariate Analysis: Methods and Applications
DODGE. Alternative Methods of Regression
*DODGE and ROMIG. Sampling Inspection Tables, *Second Edition*
*DOOB. Stochastic Processes
DOWDY and WEARDEN, and CHILKO. Statistics for Research, *Third Edition*
DRAPER and SMITH. Applied Regression Analysis, *Third Edition*
DRYDEN and MARDIA. Statistical Shape Analysis
DUDEWICZ and MISHRA. Modern Mathematical Statistics
DUNN and CLARK. Applied Statistics: Analysis of Variance and Regression, *Second Edition*
DUNN and CLARK. Basic Statistics: A Primer for the Biomedical Sciences, *Third Edition*
DUPUIS and ELLIS. A Weak Convergence Approach to the Theory of Large Deviations
EDLER and KITSOS (editors). Recent Advances in Quantitative Methods in Cancer and Human Health Risk Assessment
*ELANDT-JOHNSON and JOHNSON. Survival Models and Data Analysis
ENDERS. Applied Econometric Time Series
ETHIER and KURTZ. Markov Processes: Characterization and Convergence
EVANS, HASTINGS, and PEACOCK. Statistical Distributions, *Third Edition*
FELLER. An Introduction to Probability Theory and Its Applications, Volume I, *Third Edition*, Revised; Volume II, *Second Edition*
FISHER and VAN BELLE. Biostatistics: A Methodology for the Health Sciences
FITZMAURICE, LAIRD, and WARE. Applied Longitudinal Analysis
*FLEISS. The Design and Analysis of Clinical Experiments
FLEISS. Statistical Methods for Rates and Proportions, *Second Edition*
FLEMING and HARRINGTON. Counting Processes and Survival Analysis
FULLER. Introduction to Statistical Time Series, *Second Edition*
FULLER. Measurement Error Models
GALLANT. Nonlinear Statistical Models.
GELMAN and MENG (editors): Applied Bayesian Modeling and Casual Inference from Incomplete-data Perspectives
GEWEKE. Contemporary Bayesian Econometrics and Statistics
GHOSH, MUKHOPADHYAY, and SEN. Sequential Estimation
GIESBRECHT and GUMPERTZ. Planning, Construction, and Statistical Analysis of Comparative Experiments
GIFI. Nonlinear Multivariate Analysis
GIVENS and HOETING. Computational Statistics
GLASSERMAN and YAO. Monotone Structure in Discrete-Event Systems
GNANADESIKAN. Methods for Statistical Data Analysis of Multivariate Observations,*Second Edition*

*Now available in a lower priced paperback edition in the Wiley Classics Library.

GOLDSTEIN and LEWIS. Assessment: Problems, Development, and Statistical Issues
GREENWOOD and NIKULIN. A Guide to Chi-Squared Testing
GROSS and HARRIS. Fundamentals of Queueing Theory, *Third Edition*
*HAHN and SHAPIRO. Statistical Models in Engineering
HAHN and MEEKER. Statistical Intervals: A Guide for Practitioners
HALD. A History of Probability and Statistics and their Applications Before 1750
HALD. A History of Mathematical Statistics from 1750 to 1930
HAMPEL. Robust Statistics: The Approach Based on Influence Functions
HANNAN and DEISTLER. The Statistical Theory of Linear Systems
HEIBERGER. Computation for the Analysis of Designed Experiments
HEDAYAT and SINHA. Design and Inference in Finite Population Sampling
HELLER. MACSYMA for Statisticians
HINKELMAN and KEMPTHORNE:. Design and Analysis of Experiments, Volume 1: Introduction to Experimental Design
HINKELMANN and KEMPTHORNE. Design and analysis of experiments, Volume 2: Advanced Experimental Design
HOAGLIN, MOSTELLER, and TUKEY. Exploratory Approach to Analysis of Variance
HOAGLIN, MOSTELLER, and TUKEY. Exploring Data Tables, Trends and Shapes
*HOAGLIN, MOSTELLER, and TUKEY. Understanding Robust and Exploratory Data Analysis
HOCHBERG and TAMHANE. Multiple Comparison Procedures
HOCKING. Methods and Applications of Linear Models: Regression and the Analysis of Variance, *Second Edition*
HOEL. Introduction to Mathematical Statistics, *Fifth Edition*
HOGG and KLUGMAN. Loss Distributions
HOLLANDER and WOLFE. Nonparametric Statistical Methods, *Second Edition*
HOSMER and LEMESHOW. Applied Logistic Regression, *Second Edition*
HOSMER and LEMESHOW. Applied Survival Analysis: Regression Modeling of Time to Event Data
HUBER. Robust Statistics
HUBERTY. Applied Discriminant Analysis
HUNT and KENNEDY. Financial Derivatives in Theory and Practice, *Revised Edition*
HUSKOVA, BERAN, and DUPAC. Collected Works of Jaroslav Hajek—with Commentary
HUZURBAZAR. Flowgraph Models for Multistate Time-to-Event Data
IMAN and CONOVER. A Modern Approach to Statistics
JACKSON. A User's Guide to Principle Components
JOHN. Statistical Methods in Engineering and Quality Assurance
JOHNSON. Multivariate Statistical Simulation
JOHNSON and BALAKRISHNAN. Advances in the Theory and Practice of Statistics: A Volume in Honor of Samuel Kotz
JOHNSON and BHATTACHARYYA. Statistics: Principles and Methods, *Fifth Edition*
JUDGE, GRIFFITHS, HILL, LU TKEPOHL, and LEE. The Theory and Practice of Econometrics, *Second Edition*
JOHNSON and KOTZ. Distributions in Statistics
JOHNSON and KOTZ (editors). Leading Personalities in Statistical Sciences: From the Seventeenth Century to the Present
JOHNSON, KOTZ, and BALAKRISHNAN. Continuous Univariate Distributions, Volume 1, *Second Edition*
JOHNSON, KOTZ, and BALAKRISHNAN. Continuous Univariate Distributions, Volume 2, *Second Edition*
JOHNSON, KOTZ, and BALAKRISHNAN. Discrete Multivariate Distributions
JOHNSON, KOTZ, and KEMP. Univariate Discrete Distributions, *Second Edition*
JUREČKOVÁ and SEN. Robust Statistical Procedures: Asymptotics and Interrelations
JUREK and MASON. Operator-Limit Distributions in Probability Theory
KADANE. Bayesian Methods and Ethics in a Clinical Trial Design
KADANE and SCHUM. A Probabilistic Analysis of the Sacco and Vanzetti Evidence

* Now available in a lower priced paperback edition in the Wiley Classics Library.

KALBFLEISCH and PRENTICE. The Statistical Analysis of Failure Time Data, *Second Edition*
KARIYA and KURATA. Generalized Least Squares
KASS and VOS. Geometrical Foundations of Asymptotic Inference
KAUFMAN and ROUSSEEUW. Finding Groups in Data: An Introduction to Cluster Analysis
KEDEM and FOKIANOS. Regression Models for Time Series Analysis
KENDALL, BARDEN, CARNE, and LE. Shape and Shape Theory
KHURI. Advanced Calculus with Applications in Statistics, *Second Edition*
KHURI, MATHEW, and SINHA. Statistical Tests for Mixed Linear Models
*KISH. Statistical Design for Research
KLEIBER and KOTZ. Statistical Size Distributions in Economics and Actuarial Sciences
KLUGMAN, PANJER, and WILLMOT. Loss Models: From Data to Decisions
KLUGMAN, PANJER, and WILLMOT. Solutions Manual to Accompany Loss Models: From Data to Decisions
KOTZ, BALAKRISHNAN, and JOHNSON. Continuous Multivariate Distributions, Volume 1, *Second Edition*
KOTZ and JOHNSON (editors). Encyclopedia of Statistical Sciences: Volumes 1 to 9 with Index
KOTZ and JOHNSON (editors). Encyclopedia of Statistical Sciences: Supplement Volume
KOTZ, READ, and BANKS (editors). Encyclopedia of Statistical Sciences: Update Volume 1
KOTZ, READ, and BANKS (editors). Encyclopedia of Statistical Sciences: Update Volume 2
KOVALENKO, KUZNETZOV, and PEGG. Mathematical Theory of Reliability of Time-Dependent Systems with Practical Applications
KUROWICKA and COOKE. Uncertainty Analysis with High Dimensional Dependence Modelling
LACHIN. Biostatistical Methods: The Assessment of Relative Risks
LAD. Operational Subjective Statistical Methods: A Mathematical, Philosophical, and Historical Introduction
LAMPERTI. Probability: A Survey of the Mathematical Theory, *Second Edition*
LANGE, RYAN, BILLARD, BRILLINGER, CONQUEST, and GREENHOUSE. Case Studies in Biometry
LARSON. Introduction to Probability Theory and Statistical Inference, *Third Edition*
LAWLESS. Statistical Models and Methods for Lifetime Data, *Second Edition*
LAWSON. Statistical Methods in Spatial Epidemiology, *Second Edition*
LE. Applied Categorical Data Analysis
LE. Applied Survival Analysis
LEE and WANG. Statistical Methods for Survival Data Analysis, *Third Edition*
LEPAGE and BILLARD. Exploring the Limits of Bootstrap
LEYLAND and GOLDSTEIN (editors). Multilevel Modelling of Health Statistics
LIAO. Statistical Group Comparison
LINDVALL. Lectures on the Coupling Method
LINHART and ZUCCHINI. Model Selection
LITTLE and RUBIN. Statistical Analysis with Missing Data, *Second Edition*
LLOYD. The Statistical Analysis of Categorical Data
LOWEN and TEICH. Fractal-Based Point Processes
MAGNUS and NEUDECKER. Matrix Differential Calculus with Applications in Statistics and Econometrics, *Revised Edition*
MALLER and ZHOU. Survival Analysis with Long Term Survivors
MALLOWS. Design, Data, and Analysis by Some Friends of Cuthbert Daniel
MANN, SCHAFER, and SINGPURWALLA. Methods for Statistical Analysis of Reliability and Life Data
MANTON, WOODBURY, and TOLLEY. Statistical Applications Using Fuzzy Sets
MARCHETTE. Random Graphs for Statistical Pattern Recognition
MARDIA and JUPP. Directional Statistics
MARONNA, MARTIN, and YOHAI. Robust Statistics: Theory and Methods
MASON, GUNST, and HESS. Statistical Design and Analysis of Experiments with Applications to Engineering and Science, *Second Edition*
MCCULLOCH and SEARLE. Generalized, Linear, and Mixed Models

* Now available in a lower priced paperback edition in the Wiley Classics Library.

MCFADDEN. Management of Data in Clinical Trials
MCLACHLAN. Discriminant Analysis and Statistical Pattern Recognition
MCLACHLAN, DO, and AMBROISE. Analyzing Microarray Gene Expression Data
MCLACHLAN and KRISHNAN. The EM Algorithm and Extensions
MCLACHLAN and PEEL. Finite Mixture Models
MCNEIL. Epidemiological Research Methods
MEEKER and ESCOBAR. Statistical Methods for Reliability Data
MEERSCHAERT and SCHEFFLER. Limit Distributions for Sums of Independent Random Vectors: Heavy Tails in Theory and Practice
MICKEY, DUNN and CLARK. Applied Statistics: Analysis of Variance and Regression, *Third Edition*
*MILLER. Survival Analysis, *Second Edition*
MONTGOMERY, PECK, and VINING. Introduction to Linear Regression Analysis, *Third Edition*
MORGENTHALER and TUKEY. Configural Polysampling: A Route to Practical Robustness
MUIRHEAD. Aspects of Multivariate Statistical Theory
MURRAY. X-STAT 2.0 Statistical Experimentation, Design Data Analysis, and Nonlinear Optimization
MURTHY, XIE, and JIANG. Weibull Models
MYERS and MONTGOMERY. Response Surface Methodology: Process and Product Optimization Using Designed Experiments, *Second Edition*
MYERS, MONTGOMERY, and VINING. Generalized Linear Models. With Applications in Engineering and the Sciences
†NELSON. Accelerated Testing, Statistical Models, Test Plans, and Data Analyses
†NELSON. Applied Life Data Analysis
NEWMAN. Biostatistical Methods in Epidemiology
OCHI. Applied Probability and Stochastic Processes in Engineering and Physical Sciences
OKABE, BOOTS, SUGIHARA, and CHIU. Spatial Tesselations: Concepts and Applications of Voronoi Diagrams, *Second Edition*
OLIVER and SMITH. Influence Diagrams, Belief Nets and Decision Analysis
PALTA. Quantitative Methods in Population Health: Extensions of Ordinary Regressions
PANKRATZ. Forecasting with Dynamic Regression Models
PANKRATZ. Forecasting with Univariate Box-Jenkins Models: Concepts and Cases
*PARZEN. Modern Probability Theory and It's Applications
PEÑA, TIAO, and TSAY. A Course in Time Series Analysis
PIANTADOSI. Clinical Trials: A Methodologic Perspective
PORT. Theoretical Probability for Applications
POURAHMADI. Foundations of Time Series Analysis and Prediction Theory
PRESS. Bayesian Statistics: Principles, Models, and Applications
PRESS. Subjective and Objective Bayesian Statistics, *Second Edition*
PRESS and TANUR. The Subjectivity of Scientists and the Bayesian Approach
PUKELSHEIM. Optimal Experimental Design
PURI, VILAPLANA, and WERTZ. New Perspectives in Theoretical and Applied Statistics
PUTERMAN. Markov Decision Processes: Discrete Stochastic Dynamic Programming
QIU. Image Processing and Jump Regression Analysis
*RAO. Linear Statistical Inference and Its Applications, *Second Edition*
RAUSAND and HØYLAND. System Reliability Theory: Models, Statistical Methods and Applications, *Second Edition*
RENCHER. Linear Models in Statistics
RENCHER. Methods of Multivariate Analysis, *Second Edition*
RENCHER. Multivariate Statistical Inference with Applications
RIPLEY. Spatial Statistics
RIPLEY. Stochastic Simulation
ROBINSON. Practical Strategies for Experimenting

* Now available in a lower priced paperback edition in the Wiley Classics Library.
† Now available in a lower priced paperback edition in the Wiley - Interscience Paperback Series.

ROHATGI and SALEH. An Introduction to Probability and Statistics, *Second Edition*
ROLSKI, SCHMIDLI, SCHMIDT, and TEUGELS. Stochastic Processes for Insurance and Finance
ROSENBERGER and LACHIN. Randomization in Clinical Trials: Theory and Practice
ROSS. Introduction to Probability and Statistics for Engineers and Scientists
ROSSI, ALLENBY and MCCULLOCH. Bayesian Statistics and Marketing
ROUSSEEUW and LEROY. Robust Regression and Outlier Detection
RUBIN. Multiple Imputation for Nonresponse in Surveys
RUBINSTEIN. Simulation and the Monte Carlo Method
RUBINSTEIN and MELAMED. Modern Simulation and Modeling
RYAN. Modern Regression Methods
RYAN. Statistical Methods for Quality Improvement, *Second Edition*
SALTELLI, CHAN, and SCOTT (editors). Sensitivity Analysis
*SCHEFFE. The Analysis of Variance
SCHIMEK. Smoothing and Regression: Approaches, Computation, and Application
SCHOTT. Matrix Analysis for Statistics
SCHOUTENS. Levy Processes in Finance: Pricing Financial Derivatives
SCHUSS. Theory and Applications of Stochastic Differential Equations
SCOTT. Multivariate Density Estimation: Theory, Practice, and Visualization
*SEARLE. Linear Models
SEARLE. Linear Models for Unbalanced Data
SEARLE. Matrix Algebra Useful for Statistics
SEARLE, CASELLA, and McCULLOCH. Variance Components
SEARLE and WILLETT. Matrix Algebra for Applied Economics
SEBER. Multivariate Observations
SEBER and LEE. Linear Regression Analysis, *Second Edition*
SEBER and WILD. Nonlinear Regression
SENNOTT. Stochastic Dynamic Programming and the Control of Queueing Systems
*SERFLING. Approximation Theorems of Mathematical Statistics
SHAFER and VOVK. Probability and Finance: Its Only a Game!
SILVAPULLE and SEN. Constrained Statistical Inference: Inequality, Order, and Shape Restrictions
SINGPURWALLA. Reliability and Risk: A Bayesian Perspective
SMALL and MCLEISH. Hilbert Space Methods in Probability and Statistical Inference
SRIVASTAVA. Methods of Multivariate Statistics
STAPLETON. Linear Statistical Models
STAUDTE and SHEATHER. Robust Estimation and Testing
STOYAN, KENDALL, and MECKE. Stochastic Geometry and Its Applications, *Second Edition*
STOYAN and STOYAN. Fractals, Random Shapes and Point Fields: Methods of Geometrical Statistics
STYAN. The Collected Papers of T. W. Anderson: 1943–1985
SUTTON, ABRAMS, JONES, SHELDON, and SONG. Methods for Meta-Analysis in Medical Research
TANAKA. Time Series Analysis: Nonstationary and Noninvertible Distribution Theory
THOMPSON. Empirical Model Building
THOMPSON. Sampling, *Second Edition*
THOMPSON. Simulation: A Modeler's Approach
THOMPSON and SEBER. Adaptive Sampling
THOMPSON, WILLIAMS, and FINDLAY. Models for Investors in Real World Markets
TIAO, BISGAARD, HILL, PEÑA, and STIGLER (editors). Box on Quality and Discovery: with Design, Control, and Robustness
TIERNEY. LISP-STAT: An Object-Oriented Environment for Statistical Computing and Dynamic Graphics
TSAY. Analysis of Financial Time Series
UPTON and FINGLETON. Spatial Data Analysis by Example, Volume II: Categorical and Directional Data
VAN BELLE. Statistical Rules of Thumb

* Now available in a lower priced paperback edition in the Wiley Classics Library.

VAN BELLE, FISHER, HEAGERTY, and LUMLEY. Biostatistics: A Methodology for the Health Sciences, *Second Edition*
VESTRUP. The Theory of Measures and Integration
VIDAKOVIC. Statistical Modeling by Wavelets
VINOD and REAGLE. Preparing for the Worst: Incorporating Downside Risk in Stock Market Investments
WALLER and GOTWAY. Applied Spatial Statistics for Public Health Data
WEERAHANDI. Generalized Inference in Repeated Measures: Exact Methods in MANOVA and Mixed Models
WEISBERG. Applied Linear Regression, *Second Edition*
WELSH. Aspects of Statistical Inference
WESTFALL and YOUNG. Resampling-Based Multiple Testing: Examples and Methods for p-Value Adjustment
WHITTAKER. Graphical Models in Applied Multivariate Statistics
WINKER. Optimization Heuristics in Economics: Applications of Threshold Accepting
WONNACOTT and WONNACOTT. Econometrics, *Second Edition*
WOODING. Planning Pharmaceutical Clinical Trials: Basic Statistical Principles
WOOLSON and CLARKE. Statistical Methods for the Analysis of Biomedical Data, *Second Edition*
WU and HAMADA. Experiments: Planning, Analysis, and Parameter Design Optimization
YANG. The Construction Theory of Denumerable Markov Processes
*ZELLNER. An Introduction to Bayesian Inference in Econometrics
ZELTERMAN. Discrete Distributions: Applications in the Health Sciences
ZHOU, OBUCHOWSKI, and McCLISH. Statistical Methods in Diagnostic Medicine

* Now available in a lower priced paperback edition in the Wiley Classics Library.

Printed and bound in the UK by
CPI Antony Rowe, Eastbourne

Printed and bound by CPI Group (UK) Ltd, Croydon, CR0 4YY
16/04/2025
14658504-0001